Advanced Dynamics

Advanced Dynamics is a broad and detailed description of the analytical tools of dynamics as used in mechanical and aerospace engineering. The strengths and weaknesses of various approaches are discussed, and particular emphasis is placed on learning through problem solving.

The book begins with a thorough review of vectorial dynamics and goes on to cover Lagrange's and Hamilton's equations as well as less familiar topics such as impulse response, and differential forms and integrability. Techniques are described that provide a considerable improvement in computational efficiency over the standard classical methods, especially when applied to complex dynamical systems. The treatment of numerical analysis includes discussions of numerical stability and constraint stabilization. Many worked examples and homework problems are provided. The book is intended for use in graduate courses on dynamics, and will also appeal to researchers in mechanical and aerospace engineering.

Donald T. Greenwood received his Ph.D. from the California Institute of Technology, and is a Professor Emeritus of aerospace engineering at the University of Michigan, Ann Arbor. Before joining the faculty at Michigan he worked for the Lockheed Aircraft Corporation, and has also held visiting positions at the University of Arizona, the University of California, San Diego, and ETH Zurich. He is the author of two previous books on dynamics.

Advanced Dynamics

Donald T. Greenwood

University of Michigan

CAMBRIDGE
UNIVERSITY PRESS

CAMBRIDGE UNIVERSITY PRESS
Cambridge, New York, Melbourne, Madrid, Cape Town, Singapore, São Paulo

Cambridge University Press
The Edinburgh Building, Cambridge CB2 2RU, UK

Published in the United States of America by Cambridge University Press, New York

www.cambridge.org
Information on this title: www.cambridge.org/9780521826129

First published 2003
This digitally printed first paperback version (with corrections) 2006

A catalogue record for this publication is available from the British Library

Library of Congress Cataloguing in Publication data

Greenwood, Donald T.
Advanced dynamics / Donald T. Greenwood.
    p.   cm.
Includes bibliographical references and index.
ISBN 0-521-82612-8
1. Dynamics.   I. Title.
QA845.G826 2003
531′.11–dc21   2003046078

ISBN-13 978-0-521-82612-9 hardback
ISBN-10 0-521-82612-8 hardback

ISBN-13 978-0-521-02993-3 paperback
ISBN-10 0-521-02993-7 paperback

Contents

1   Appendix

Preface

This is a dynamics textbook for graduate students, written at a moderately advanced level. Its principal aim is to present the dynamics of particles and rigid bodies in some breadth, with examples illustrating the strengths and weaknesses of the various methods of dynamical analysis. The scope of the dynamical theory includes both vectorial and analytical methods. There is some emphasis on systems of great generality, that is, systems which may have nonholonomic constraints and whose motion may be expressed in terms of quasi-velocities. Geometrical approaches such as the use of surfaces in n-dimensional configuration and velocity spaces are used to illustrate the nature of holonomic and nonholonomic constraints. Impulsive response methods are discussed at some length.

Some of the material presented here was originally included in a graduate course in computational dynamics at the University of Michigan. The ordering of the chapters, with the chapters on dynamical theory presented first followed by the single chapter on numerical methods, is such that the degree of emphasis one chooses to place on the latter is optional. Numerical computation methods may be introduced at any point, or may be omitted entirely.

The first chapter presents in some detail the familiar principles of Newtonian or vectorial dynamics, including discussions of constraints, virtual work, and the use of energy and momentum principles. There is also an introduction to less familiar topics such as differential forms, integrability, and the basic theory of impulsive response.

Chapter 2 introduces methods of analytical dynamics as represented by Lagrange's and Hamilton's equations. The derivation of these equations begins with the Lagrangian form of d'Alembert's principle, a common starting point for obtaining many of the principal forms of dynamical equations of motion. There are discussions of ignorable coordinates, the Routhian method, and the use of integrals of the motion. Frictional and gyroscopic forces are studied, and further material is presented on impulsive systems.

Chapter 3 is concerned with the kinematics and dynamics of rigid body motion. Dyadic and matrix notations are introduced. Euler parameters and axis-and-angle variables are used extensively in representing rigid body orientations in addition to the more familiar Euler angles. This chapter also includes material on constrained impulsive response and input-output methods.

The theoretical development presented in the first three chapters is used as background for the derivations of Chapter 4. Here we present several differential methods which have the advantages of simplicity and computational efficiency over the usual Lagrangian methods in the analysis of general constrained systems or for systems described in terms of quasi-velocities. These methods result in a minimum set of dynamical equations which are computationally efficient. Many examples are included in order to compare and explain the various approaches. This chapter also presents detailed discussions of constraints and energy rates by using velocity space concepts.

Chapter 5 begins with a derivation of Hamilton's principle in its holonomic and nonholonomic forms. Stationarity questions are discussed. Transpositional relations are introduced and there follows a further discussion of integrability including Frobenius' theorem. The central equation and its explicit transpositional form are presented. There is a comparison of integral methods by means of examples.

Chapter 6 presents some basic principles of numerical analysis and explains the use of integration algorithms in the numerical solution of differential equations. For the most part, explicit algorithms such as the Runge–Kutta and predictor–corrector methods are considered. There is an analysis of numerical stability of the integration methods, primarily by solving the appropriate difference equations, but frequency response methods are also used. The last portion of the chapter considers methods of representing kinematic constraints. The one-step method of constraint stabilization is introduced and its advantages over standard methods are explained. There is a discussion of the use of energy and momentum constraints as a means of improving the accuracy of numerical computations.

A principal objective of this book is to improve the problem-solving skills of each student. Problem solving should include not only a proper formulation and choice of variables, but also a directness of approach which avoids unnecessary steps. This requires that the student repeatedly attempt the solution of problems which may be kinematically complex and which involve the application of several dynamical principles. The problems presented here usually have several parts that require more than the derivation of the equations of motion for a given system. Thus, insight is needed concerning other dynamical characteristics. Because of the rather broad array of possible approaches presented here, and due in part to the generally demanding problems, a conscientious student can attain a real perspective of the subject of dynamics and a competence in the application of its principles.

Finally, I would like to acknowledge the helpful discussions with Professor J. G. Papastavridis of Georgia Tech concerning the material of Chapters 4 and 5, and with Professor R. M. Howe of the University of Michigan concerning portions of Chapter 6.

1 Introduction to particle dynamics

In the study of dynamics at an advanced level, it is important to consider many approaches and points of view in order that one may attain a broad theoretical perspective of the subject. As we proceed we shall emphasize those methods which are particularly effective in the analysis of relatively difficult problems in dynamics. At this point, however, it is well to review some of the basic principles in the dynamical analysis of systems of particles. In the process, the kinematics of particle motion will be reviewed, and many of the notational conventions will be established.

1.1 Particle motion

The laws of motion for a particle

Let us consider Newton's three laws of motion which were published in 1687 in his *Principia*. They can be stated as follows:

    I. Every body continues in its state of rest, or of uniform motion in a straight line, unless compelled to change that state by forces acting upon it.

    II. The time rate of change of linear momentum of a body is proportional to the force acting upon it and occurs in the direction in which the force acts.

    III. To every action there is an equal and opposite reaction; that is, the mutual forces of two bodies acting upon each other are equal in magnitude and opposite in direction.

In the dynamical analysis of a system of particles using Newton's laws, we can interpret the word "body" to mean a particle, that is, a certain fixed mass concentrated at a point. The first two of Newton's laws, as applied to a particle, can be summarized by the *law of motion*:

$$\mathbf{F} = m\mathbf{a} \tag{1.1}$$

Here $\mathbf{F}$ is the total force applied to the particle of mass m and it includes both direct contact forces and field forces such as gravity or electromagnetic forces. The acceleration $\mathbf{a}$ of the particle must be measured relative to an *inertial* or *Newtonian* frame of reference. An example of an inertial frame is an xyz set of axes which is not rotating relative to the "fixed" stars and has its origin at the center of mass of the solar system. Any other reference frame which is not rotating but is translating at a constant rate relative to an inertial frame is itself an inertial frame. Thus, there are infinitely many inertial frames, all with constant translational velocities relative to the others. Because the relative velocities are constant, the acceleration of a given particle is the same relative to any inertial frame. The force **F** and mass m are also the same in all inertial frames, so Newton's law of motion is identical relative to all inertial frames.

Newton's third law, the law of action and reaction, has a corollary assumption that the interaction forces between any two particles are directed along the straight line connecting the particles. Thus we have the *law of action and reaction*:

When two particles exert forces on each other, these interaction forces
are equal in magnitude, opposite in sense, and are directed along the
straight line joining the particles.

The *collinearity* of the interaction forces applies to all mechanical and gravitational forces. It does not apply, however, to interactions between moving electrically charged particles for which the interaction forces are equal and opposite but not necessarily collinear. Systems of this sort will not be studied here.

An alternative form of the equation of motion of a particle is

$$\mathbf{F} = \dot{\mathbf{p}} \tag{1.2}$$

where the *linear momentum* of the particle is

$$\mathbf{p} = m\mathbf{v} \tag{1.3}$$

and **v** is the particle velocity relative to an inertial frame.

Kinematics of particle motion

The application of Newton's laws of motion to a particle requires that an expression can be found for the acceleration of the particle relative to an inertial frame. For example, the position vector of a particle relative to a *fixed* Cartesian frame might be expressed as

$$\mathbf{r} = x\mathbf{i} + y\mathbf{j} + z\mathbf{k} \tag{1.4}$$

where **i**, **j**, **k** are *unit vectors*, that is, vectors of unit magnitude which have the directions of the positive x, y, and z axes, respectively. When unit vectors are used to specify a vector in 3-space, the three unit vectors are always linearly independent and are nearly always mutually perpendicular. The velocity of the given particle is

$$\mathbf{v} = \dot{\mathbf{r}} = \dot{x}\mathbf{i} + \dot{y}\mathbf{j} + \dot{z}\mathbf{k} \tag{1.5}$$

and its acceleration is

$$\mathbf{a} = \dot{\mathbf{v}} = \ddot{x}\mathbf{i} + \ddot{y}\mathbf{j} + \ddot{z}\mathbf{k} \tag{1.6}$$

relative to the inertial frame.

A force $\mathbf{F}$ applied to the particle may be described in a similar manner.

$$\mathbf{F} = F_x\mathbf{i} + F_y\mathbf{j} + F_z\mathbf{k} \qquad (1.7)$$

where (F_x, F_y, F_z) are the *scalar components* of $\mathbf{F}$. In general, the force components can be functions of position, velocity, and time, but often they are much simpler.

If one writes Newton's law of motion, (1.1), in terms of the Cartesian unit vectors, and then equates the scalar coefficients of each unit vector on the two sides of the equation, one obtains

$$\begin{aligned} F_x &= m\ddot{x} \\ F_y &= m\ddot{y} \\ F_z &= m\ddot{z} \end{aligned} \qquad (1.8)$$

These three scalar equations are equivalent to the single vector equation. In general, the scalar equations are coupled through the expressions for the force components. Furthermore, the differential equations are often nonlinear and are not susceptible to a complete analytic solution. In this case, one can turn to numerical integration on a digital computer to obtain the complete solution. On the other hand, one can often use energy or momentum methods to obtain important characteristics of the motion without having the complete solution.

The calculation of a particle acceleration relative to an inertial Cartesian frame is straightforward because the unit vectors $(\mathbf{i}, \mathbf{j}, \mathbf{k})$ are fixed in direction. It turns out, however, that because of system geometry it is sometimes more convenient to use unit vectors that are not fixed. For example, the position, velocity, and acceleration of a particle moving along a circular path are conveniently expressed using radial and tangential unit vectors which change direction with position.

As a more general example, suppose that an arbitrary vector $\mathbf{A}$ is given by

$$\mathbf{A} = A_1\mathbf{e}_1 + A_2\mathbf{e}_2 + A_3\mathbf{e}_3 \qquad (1.9)$$

where the unit vectors $\mathbf{e}_1$, $\mathbf{e}_2$, and $\mathbf{e}_3$ form a mutually orthogonal set such that $\mathbf{e}_3 = \mathbf{e}_1 \times \mathbf{e}_2$. This unit vector triad changes its orientation with time. It rotates as a rigid body with an angular velocity ω, where the direction of ω is along the axis of rotation and the positive sense of ω is in accordance with the right-hand rule.

The first time derivative of $\mathbf{A}$ is

$$\dot{\mathbf{A}} = \dot{A}_1\mathbf{e}_1 + \dot{A}_2\mathbf{e}_2 + \dot{A}_3\mathbf{e}_3 + A_1\dot{\mathbf{e}}_1 + A_2\dot{\mathbf{e}}_2 + A_3\dot{\mathbf{e}}_3 \qquad (1.10)$$

where

$$\dot{\mathbf{e}}_i = \omega \times \mathbf{e}_i \qquad (i = 1, 2, 3) \qquad (1.11)$$

Thus we obtain the important equation

$$\dot{\mathbf{A}} = (\dot{\mathbf{A}})_r + \omega \times \mathbf{A} \qquad (1.12)$$

Here $\dot{\mathbf{A}}$ is the time rate of change of $\mathbf{A}$, as measured in a nonrotating frame that is usually considered to also be inertial. $(\dot{\mathbf{A}})_r$ is the derivative of $\mathbf{A}$, as measured in a rotating frame in which the unit vectors are fixed. It is represented by the first three terms on the right-hand side of (1.10). The term $\omega \times \mathbf{A}$ is represented by the final three terms of (1.10). In detail, if the angular velocity of the rotating frame is

$$\omega = \omega_1 \mathbf{e}_1 + \omega_2 \mathbf{e}_2 + \omega_3 \mathbf{e}_3 \tag{1.13}$$

then

$$\dot{\mathbf{A}} = (\dot{A}_1 + \omega_2 A_3 - \omega_3 A_2)\,\mathbf{e}_1 + (\dot{A}_2 + \omega_3 A_1 - \omega_1 A_3)\,\mathbf{e}_2$$
$$+ (\dot{A}_3 + \omega_1 A_2 - \omega_2 A_1)\,\mathbf{e}_3 \tag{1.14}$$

Velocity and acceleration expressions for common coordinate systems

Let us apply the general equation (1.12) to some common coordinate systems associated with particle motion.

Cylindrical coordinates

Suppose that the position of a particle P is specified by the values of its cylindrical coordinates (r, ϕ, z). We see from Fig. 1.1 that the position vector $\mathbf{r}$ is

$$\mathbf{r} = r\mathbf{e}_r + z\mathbf{e}_z \tag{1.15}$$

where we notice that r is not the magnitude of $\mathbf{r}$. The angular velocity of the $\mathbf{e}_r \mathbf{e}_\phi \mathbf{e}_z$ triad is

$$\omega = \dot{\phi}\mathbf{e}_z \tag{1.16}$$

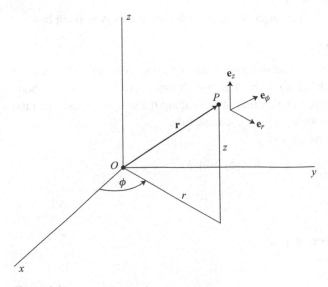

Figure 1.1.

so we find that $\dot{\mathbf{e}}_z$ vanishes and

$$\dot{\mathbf{e}}_r = \boldsymbol{\omega} \times \mathbf{e}_r = \dot{\phi}\mathbf{e}_\phi \qquad (1.17)$$

Thus, the velocity of the particle P is

$$\mathbf{v} = \dot{\mathbf{r}} = \dot{r}\mathbf{e}_r + r\dot{\phi}\mathbf{e}_\phi + \dot{z}\mathbf{e}_z \qquad (1.18)$$

Similarly, noting that

$$\dot{\mathbf{e}}_\phi = \boldsymbol{\omega} \times \mathbf{e}_\phi = -\dot{\phi}\mathbf{e}_r \qquad (1.19)$$

we find that its acceleration is

$$\mathbf{a} = \dot{\mathbf{v}} = (\ddot{r} - r\dot{\phi}^2)\,\mathbf{e}_r + (r\ddot{\phi} + 2\dot{r}\dot{\phi})\,\mathbf{e}_\phi + \ddot{z}\mathbf{e}_z \qquad (1.20)$$

If we restrict the motion such that $\dot{z}$ and $\ddot{z}$ are continuously equal to zero, we obtain the velocity and acceleration equations for plane motion using *polar coordinates*.

Spherical coordinates

From Fig. 1.2 we see that the position of particle P is given by the spherical coordinates (r, θ, ϕ). The position vector of the particle is simply

$$\mathbf{r} = r\mathbf{e}_r \qquad (1.21)$$

The angular velocity of the $\mathbf{e}_r\mathbf{e}_\theta\mathbf{e}_\phi$ triad is due to $\dot{\theta}$ and $\dot{\phi}$ and is equal to

$$\boldsymbol{\omega} = \dot{\phi}\cos\theta\,\mathbf{e}_r - \dot{\phi}\sin\theta\,\mathbf{e}_\theta + \dot{\theta}\,\mathbf{e}_\phi \qquad (1.22)$$

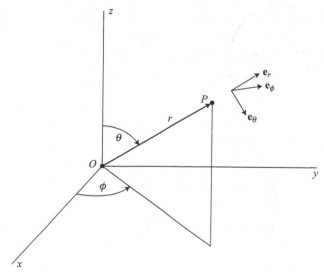

Figure 1.2.

We find that

$$\dot{\mathbf{e}}_r = \boldsymbol{\omega} \times \mathbf{e}_r = \dot{\theta}\mathbf{e}_\theta + \dot{\phi}\sin\theta\ \mathbf{e}_\phi$$
$$\dot{\mathbf{e}}_\theta = \boldsymbol{\omega} \times \mathbf{e}_\theta = -\dot{\theta}\mathbf{e}_r + \dot{\phi}\cos\theta\ \mathbf{e}_\phi \qquad (1.23)$$
$$\dot{\mathbf{e}}_\phi = \boldsymbol{\omega} \times \mathbf{e}_\phi = -\dot{\phi}\sin\theta\ \mathbf{e}_r - \dot{\phi}\cos\theta\ \mathbf{e}_\theta$$

Then, upon differentiation of (1.21), we obtain the velocity

$$\mathbf{v} = \dot{\mathbf{r}} = \dot{r}\mathbf{e}_r + r\dot{\theta}\mathbf{e}_\theta + r\dot{\phi}\sin\theta\ \mathbf{e}_\phi \qquad (1.24)$$

A further differentiation yields the acceleration

$$\mathbf{a} = \dot{\mathbf{v}} = (\ddot{r} - r\dot{\theta}^2 - r\dot{\phi}^2\sin^2\theta)\,\mathbf{e}_r + (r\ddot{\theta} + 2\dot{r}\dot{\theta} - r\dot{\phi}^2\sin\theta\cos\theta)\,\mathbf{e}_\theta$$
$$+ (r\ddot{\phi}\sin\theta + 2\dot{r}\dot{\phi}\sin\theta + 2r\dot{\theta}\dot{\phi}\cos\theta)\,\mathbf{e}_\phi \qquad (1.25)$$

Tangential and normal components

Suppose a particle P moves along a given path in three-dimensional space. The position of the particle is specified by the single coordinate s, measured from some reference point along the path, as shown in Fig. 1.3. It is convenient to use the three unit vectors $(\mathbf{e}_t, \mathbf{e}_n, \mathbf{e}_b)$ where $\mathbf{e}_t$ is tangent to the path at P, $\mathbf{e}_n$ is normal to the path and points in the direction of the center of curvature C, and the binormal unit vector is

$$\mathbf{e}_b = \mathbf{e}_t \times \mathbf{e}_n \qquad (1.26)$$

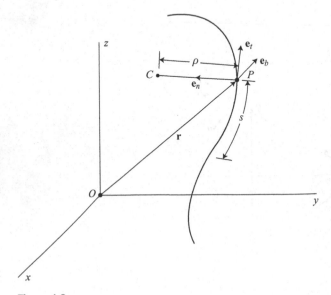

Figure 1.3.

The velocity of the particle is equal to its speed along its path, so

$$\mathbf{v} = \dot{\mathbf{r}} = \dot{s}\mathbf{e}_t \tag{1.27}$$

If we consider motion along an infinitesimal arc of radius ρ surrounding P, we see that

$$\dot{\mathbf{e}}_t = \frac{\dot{s}}{\rho}\mathbf{e}_n \tag{1.28}$$

Thus, we find that the acceleration of the particle is

$$\mathbf{a} = \dot{\mathbf{v}} = \ddot{s}\mathbf{e}_t + \dot{s}\dot{\mathbf{e}}_t = \ddot{s}\mathbf{e}_t + \frac{\dot{s}^2}{\rho}\mathbf{e}_n \tag{1.29}$$

where ρ is the radius of curvature. Here $\ddot{s}$ is the tangential acceleration and $\dot{s}^2/\rho$ is the centripetal acceleration. The angular velocity of the unit vector triad is directly proportional to $\dot{s}$. It is

$$\omega = \omega_t \mathbf{e}_t + \omega_b \mathbf{e}_b \tag{1.30}$$

where ω_t and ω_b are obtained from

$$\begin{aligned}
\dot{\mathbf{e}}_t &= \omega_b \mathbf{e}_n = \frac{\dot{s}}{\rho}\mathbf{e}_n \\
\dot{\mathbf{e}}_b &= -\omega_t \mathbf{e}_n = \dot{s}\frac{d\mathbf{e}_b}{ds}
\end{aligned} \tag{1.31}$$

Note that $\omega_n = 0$ and also that $d\mathbf{e}_b/ds$ represents the *torsion* of the curve.

Relative motion and rotating frames

When one uses Newton's laws to describe the motion of a particle, the acceleration $\mathbf{a}$ must be *absolute*, that is, it must be measured relative to an inertial frame. This acceleration, of course, is the same when measured with respect to any inertial frame. Sometimes the motion of a particle is known relative to a rotating and accelerating frame, and it is desired to find its absolute velocity and acceleration. In general, these calculations can be somewhat complicated, but for the special case in which the moving frame A is not rotating, the results are simple. The absolute velocity of a particle P is

$$\mathbf{v}_P = \mathbf{v}_A + \mathbf{v}_{P/A} \tag{1.32}$$

where $\mathbf{v}_A$ is the absolute velocity of any point on frame A and $\mathbf{v}_{P/A}$ is the velocity of particle P relative to frame A, that is, the velocity recorded by cameras or other instruments fixed in frame A and moving with it. Similarly, the absolute acceleration of P is

$$\mathbf{a}_P = \mathbf{a}_A + \mathbf{a}_{P/A} \tag{1.33}$$

where we note again that the frame A is moving in pure translation.

Now consider the general case in which the moving xyz frame (Fig. 1.4) is translating and rotating arbitrarily. We wish to find the velocity and acceleration of a particle P relative

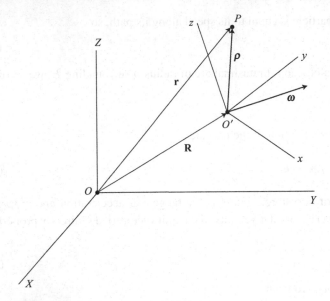

Figure 1.4.

to the inertial XYZ frame in terms of its motion with respect to the noninertial xyz frame. Let the origin O' of the xyz frame have a position vector $\mathbf{R}$ relative to the origin O of the XYZ frame. The position of the particle P relative to O' is ρ, so the position of P relative to XYZ is

$$\mathbf{r} = \mathbf{R} + \rho \tag{1.34}$$

The corresponding velocity is

$$\mathbf{v} = \dot{\mathbf{r}} = \dot{\mathbf{R}} + \dot{\rho} \tag{1.35}$$

Now let us use the basic equation (1.12) to express $\dot{\rho}$ in terms of the motion relative to the moving xyz frame. We obtain

$$\dot{\rho} = (\dot{\rho})_r + \omega \times \rho \tag{1.36}$$

where ω is the angular velocity of the xyz frame and $(\dot{\rho})_r$ is the velocity of P relative to that frame. In detail,

$$\rho = x\mathbf{i} + y\mathbf{j} + z\mathbf{k} \tag{1.37}$$

and

$$(\dot{\rho})_r = \dot{x}\mathbf{i} + \dot{y}\mathbf{j} + \dot{z}\mathbf{k} \tag{1.38}$$

where $\mathbf{i}, \mathbf{j}, \mathbf{k}$ are unit vectors fixed in the xyz frame and rotating with it. From (1.35) and (1.36), the absolute velocity of P is

$$\mathbf{v} = \dot{\mathbf{r}} = \dot{\mathbf{R}} + (\dot{\rho})_r + \omega \times \rho \tag{1.39}$$

The expression for the inertial acceleration **a** of the particle is found by first noting that

$$\frac{d}{dt}(\dot{\rho})_r = (\ddot{\rho})_r + \omega \times (\dot{\rho})_r \tag{1.40}$$

$$\frac{d}{dt}(\omega \times \rho) = \dot{\omega} \times \rho + \omega \times ((\dot{\rho})_r + \omega \times \rho) \tag{1.41}$$

Thus, we obtain the important result:

$$\mathbf{a} = \dot{\mathbf{v}} = \ddot{\mathbf{R}} + \dot{\omega} \times \rho + \omega \times (\omega \times \rho) + (\ddot{\rho})_r + 2\omega \times (\dot{\rho})_r \tag{1.42}$$

where ω is the angular velocity of the xyz frame. The nature of the various terms is as follows. $\ddot{\mathbf{R}}$ is the inertial acceleration of O', the origin of the moving frame. The term $\dot{\omega} \times \rho$ might be considered as a tangential acceleration although, more accurately, it represents a changing tangential velocity $\omega \times \rho$ due to changing ω. The term $\omega \times (\omega \times \rho)$ is a centripetal acceleration directed toward an axis of rotation through O'. These first three terms represent the acceleration of a point coincident with P but fixed in the xyz frame. The final two terms add the effects of motion relative to the moving frame. The term $(\ddot{\rho})_r$ is the acceleration of P relative to the xyz frame, that is, the acceleration of the particle, as recorded by instruments fixed in the xyz frame and rotating with it. The final term $2\omega \times (\dot{\rho})_r$ is the *Coriolis acceleration* due to a velocity relative to the rotating frame. Equation (1.42) is particularly useful if the motion of the particle relative to the moving xyz frame is simple; for example, linear motion or motion along a circular path.

Instantaneous center of rotation

If each point of a rigid body moves in planar motion, it is useful to consider a *lamina*, or slice, of the body which moves in its own plane (Fig. 1.5). If the lamina does not move in pure translation, that is, if $\omega \neq 0$, then a point C exists in the lamina, or in an imaginary

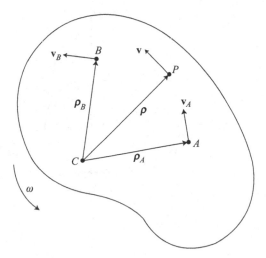

Figure 1.5.

extension thereof, at which the velocity is momentarily zero. This is the *instantaneous center of rotation*.

Suppose that arbitrary points A and B have velocities $\mathbf{v}_A$ and $\mathbf{v}_B$. The instantaneous center C is located at the intersection of the perpendicular lines to $\mathbf{v}_A$ and $\mathbf{v}_B$. The velocity of a point P with a position vector ρ relative to C is

$$\mathbf{v} = \omega \times \rho \qquad (1.43)$$

where ω is the angular velocity vector of the lamina. Thus, if the location of the instantaneous center is known, it is easy to find the velocity of any other point of the lamina at that instant. On the other hand, the acceleration of the instantaneous center is generally not zero. Hence, the calculation of the acceleration of a general point in the lamina is usually not aided by a knowledge of the instantaneous center location.

If there is planar rolling motion of one body on another fixed body without any slipping, the instantaneous center lies at the contact point between the two bodies. As time proceeds, this point moves with respect to both bodies, thereby tracing a path on each body.

Example 1.1 A wheel of radius r rolls in planar motion without slipping on a fixed convex surface of radius R (Fig. 1.6a). We wish to solve for the acceleration of the contact point on the wheel. The contact point C is the instantaneous center, and therefore, the velocity of the wheel's center O' is

$$\mathbf{v} = r\omega\mathbf{e}_\phi \qquad (1.44)$$

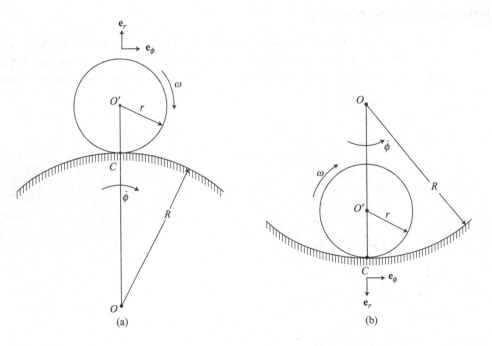

(a)　　　　　　　　(b)

Figure 1.6.

In terms of the angular velocity $\dot{\phi}$ of the radial line OO', the velocity of the wheel is

$$(R + r)\dot{\phi} = r\omega \tag{1.45}$$

so we find that

$$\dot{\phi} = \frac{r\omega}{R + r} \tag{1.46}$$

To show that the acceleration of the contact point C is nonzero, we note that

$$\mathbf{a}_C = \mathbf{a}_{O'} + \mathbf{a}_{C/O'} \tag{1.47}$$

The center O' of the wheel moves in a circular path of radius $(R + r)$, so its acceleration $\mathbf{a}_{O'}$ is the sum of tangential and centripetal accelerations.

$$\mathbf{a}_{O'} = (R + r)\ddot{\phi}\mathbf{e}_\phi - (R + r)\dot{\phi}^2 \mathbf{e}_r$$
$$= r\dot{\omega}\mathbf{e}_\phi - \frac{r^2\omega^2}{R + r}\mathbf{e}_r \tag{1.48}$$

Similarly C, considered as a point on the rim of the wheel, has a circular motion about O', so

$$\mathbf{a}_{C/O'} = -r\dot{\omega}\mathbf{e}_\phi + r\omega^2 \mathbf{e}_r \tag{1.49}$$

Then, adding (1.48) and (1.49), we obtain

$$\mathbf{a}_C = \left(r - \frac{r^2}{R + r} \right)\omega^2 \mathbf{e}_r = \left(\frac{Rr}{R + r} \right)\omega^2 \mathbf{e}_r \tag{1.50}$$

Thus, the instantaneous center has a nonzero acceleration.

Now consider the rolling motion of a wheel of radius r on a concave surface of radius R (Fig. 1.6b). The center of the wheel has a velocity

$$\mathbf{v}_{O'} = r\omega\mathbf{e}_\phi = (R - r)\dot{\phi}\mathbf{e}_\phi \tag{1.51}$$

so

$$\dot{\phi} = \frac{r\omega}{R - r} \tag{1.52}$$

In this case, the acceleration of the contact point is

$$\mathbf{a}_C = \mathbf{a}_{O'} + \mathbf{a}_{C/O'} \tag{1.53}$$

where

$$\mathbf{a}_{O'} = (R - r)\ddot{\phi}\mathbf{e}_\phi - (R - r)\dot{\phi}^2 \mathbf{e}_r$$
$$= r\dot{\omega}\mathbf{e}_\phi - \frac{r^2\omega^2}{R - r}\mathbf{e}_r \tag{1.54}$$
$$\mathbf{a}_{C/O'} = -r\dot{\omega}\mathbf{e}_\phi - r\omega^2 \mathbf{e}_r \tag{1.55}$$

Thus, we obtain

$$\mathbf{a}_C = -\left(r + \frac{r^2}{R - r} \right)\omega^2 \mathbf{e}_r = -\left(\frac{Rr}{R - r} \right)\omega^2 \mathbf{e}_r \tag{1.56}$$

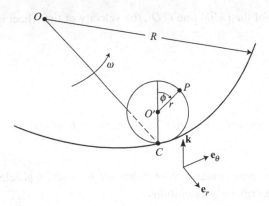

Figure 1.7.

Notice that very large values of $\mathbf{a}_{O'}$ and $\mathbf{a}_C$ can occur, even for moderate values of ω, if R is only slightly larger than r. This could occur, for example, if a shaft rotates in a sticky bearing.

Example 1.2 Let us calculate the acceleration of a point P on the rim of a wheel of radius r which rolls without slipping on a horizontal circular track of radius R (Fig. 1.7). The plane of the wheel remains vertical and the position angle of P relative to a vertical line through the center O' is ϕ.

Let us choose the unit vectors $\mathbf{e}_r$, $\mathbf{e}_\theta$, $\mathbf{k}$, as shown. They rotate about a vertical axis at an angular rate ω which is the rate at which the contact point C moves along the circular path. Since the center O' and C move along parallel paths with the same speed, we can write

$$v_{O'} = r\dot{\phi} = R\omega \tag{1.57}$$

from which we obtain

$$\omega = \frac{r}{R}\dot{\phi}\mathbf{k} \tag{1.58}$$

Choose C as the origin of a moving frame which rotates with the angular velocity $\boldsymbol{\omega}$.

To find the acceleration of P, let us use the general equation (1.42), namely,

$$\mathbf{a} = \ddot{\mathbf{R}} + \dot{\boldsymbol{\omega}} \times \boldsymbol{\rho} + \boldsymbol{\omega} \times (\boldsymbol{\omega} \times \boldsymbol{\rho}) + (\ddot{\boldsymbol{\rho}})_r + 2\boldsymbol{\omega} \times (\dot{\boldsymbol{\rho}})_r \tag{1.59}$$

The acceleration of C is

$$\ddot{\mathbf{R}} = -R\omega^2\mathbf{e}_r + R\dot{\omega}\mathbf{e}_\theta = -\frac{r^2\dot{\phi}^2}{R}\mathbf{e}_r + r\ddot{\phi}\mathbf{e}_\theta \tag{1.60}$$

The relative position of P with respect to C is

$$\boldsymbol{\rho} = r\sin\phi\ \mathbf{e}_\theta + r(1 + \cos\phi)\mathbf{k} \tag{1.61}$$

From (1.58) we obtain

$$\dot{\boldsymbol{\omega}} = \frac{r}{R}\ddot{\phi}\mathbf{k} \tag{1.62}$$

Then

$$\dot{\omega} \times \rho = -\frac{r^2}{R}\ddot{\phi}\sin\phi\, \mathbf{e}_r \tag{1.63}$$

$$\omega \times (\omega \times \rho) = -\frac{r^3}{R^2}\dot{\phi}^2\sin\phi\, \mathbf{e}_\theta \tag{1.64}$$

Upon differentiating (1.61), with $\mathbf{e}_\theta$ and $\mathbf{k}$ held constant, we obtain

$$(\dot{\rho})_r = r\dot{\phi}\cos\phi\, \mathbf{e}_\theta - r\dot{\phi}\sin\phi\, \mathbf{k} \tag{1.65}$$

and

$$2\omega \times (\dot{\rho})_r = -\frac{2r^2}{R}\dot{\phi}^2\cos\phi\, \mathbf{e}_r \tag{1.66}$$

Also,

$$(\ddot{\rho})_r = (r\ddot{\phi}\cos\phi - r\dot{\phi}^2\sin\phi)\mathbf{e}_\theta - (r\ddot{\phi}\sin\phi + r\dot{\phi}^2\cos\phi)\mathbf{k} \tag{1.67}$$

Finally, adding terms, the acceleration of P is

$$\mathbf{a} = -\left[\frac{r^2}{R}\ddot{\phi}\sin\phi + \frac{r^2}{R}\dot{\phi}^2(1 + 2\cos\phi)\right]\mathbf{e}_r + \left[r\ddot{\phi}(1 + \cos\phi) - r\dot{\phi}^2\left(1 + \frac{r^2}{R^2}\right)\sin\phi\right]\mathbf{e}_\theta$$
$$- (r\ddot{\phi}\sin\phi + r\dot{\phi}^2\cos\phi)\mathbf{k} \tag{1.68}$$

Example 1.3 A particle P moves on a plane spiral having the equation

$$r = k\theta \tag{1.69}$$

where k is a constant (Fig. 1.8). Let us find an expression for its acceleration. Also solve for the radius of curvature of the spiral at a point specified by the angle θ.

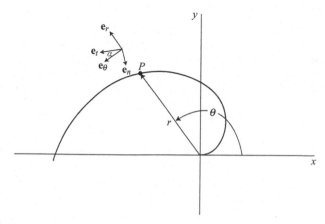

Figure 1.8.

First note that the unit vectors $(\mathbf{e}_r, \mathbf{e}_\theta)$ rotate with an angular velocity

$$\boldsymbol{\omega} = \dot{\theta}\mathbf{k} \tag{1.70}$$

where the unit vector $\mathbf{k}$ points out of the page. We obtain

$$\dot{\mathbf{e}}_r = \boldsymbol{\omega} \times \mathbf{e}_r = \dot{\theta}\mathbf{e}_\theta$$
$$\dot{\mathbf{e}}_\theta = \boldsymbol{\omega} \times \mathbf{e}_\theta = -\dot{\theta}\mathbf{e}_r \tag{1.71}$$

The position vector of P is

$$\mathbf{r} = r\mathbf{e}_r \tag{1.72}$$

and its velocity is

$$\mathbf{v} = \dot{\mathbf{r}} = \dot{r}\mathbf{e}_r + r\dot{\mathbf{e}}_r = \dot{r}\mathbf{e}_r + r\dot{\theta}\mathbf{e}_\theta \tag{1.73}$$

The acceleration of P is

$$\begin{aligned}
\mathbf{a} = \dot{\mathbf{v}} &= \ddot{r}\mathbf{e}_r + \dot{r}\dot{\mathbf{e}}_r + r\ddot{\theta}\mathbf{e}_\theta + \dot{r}\dot{\theta}\mathbf{e}_\theta + r\dot{\theta}\dot{\mathbf{e}}_\theta \\
&= (\ddot{r} - r\dot{\theta}^2)\mathbf{e}_r + (r\ddot{\theta} + 2\dot{r}\dot{\theta})\mathbf{e}_\theta \\
&= (k\ddot{\theta} - k\theta\dot{\theta}^2)\mathbf{e}_r + (k\theta\ddot{\theta} + 2k\dot{\theta}^2)\mathbf{e}_\theta
\end{aligned} \tag{1.74}$$

The radius of curvature at P can be found by first establishing the orthogonal unit vectors $(\mathbf{e}_t, \mathbf{e}_n)$ and then finding the normal component of the acceleration. The angle α between the unit vectors $\mathbf{e}_t$ and $\mathbf{e}_\theta$ is obtained by noting that

$$\tan\alpha = \frac{v_r}{v_\theta} = \frac{\dot{r}}{r\dot{\theta}} = \frac{k\dot{\theta}}{k\theta\dot{\theta}} = \frac{1}{\theta} \tag{1.75}$$

and we see that

$$\sin\alpha = \frac{1}{\sqrt{1 + \theta^2}}$$
$$\cos\alpha = \frac{\theta}{\sqrt{1 + \theta^2}} \tag{1.76}$$

The normal acceleration is

$$a_n = -a_r \cos\alpha + a_\theta \sin\alpha \tag{1.77}$$

where, from (1.74),

$$a_r = k\ddot{\theta} - k\theta\dot{\theta}^2$$
$$a_\theta = k\theta\ddot{\theta} + 2k\dot{\theta}^2 \tag{1.78}$$

Thus, we obtain

$$a_n = \frac{k\dot{\theta}^2}{\sqrt{1 + \theta^2}}(2 + \theta^2) \tag{1.79}$$

From (1.29), using tangential and normal components, we find that the normal acceleration is

$$a_n = \frac{\dot{s}^2}{\rho} = \frac{v^2}{\rho} = \frac{v_r^2 + v_\theta^2}{\rho} = \frac{k^2\dot{\theta}^2(1+\theta^2)}{\rho} \tag{1.80}$$

where ρ is the radius of curvature. Comparing (1.79) and (1.80), the radius of curvature at P is

$$\rho = \frac{k(1+\theta^2)^{3/2}}{2+\theta^2} \tag{1.81}$$

Notice that ρ varies from $\frac{1}{2}k$ at $\theta = 0$ to r for very large r and θ.

1.2   Systems of particles

A system of particles with all its interactions constitutes a dynamical system of great generality. Consequently, it is important to understand thoroughly the principles which govern its motions. Here we shall establish some of the basic principles. Later, these principles will be used in the study of rigid body dynamics.

Equations of motion

Consider a system of N particles whose positions are given relative to an inertial frame (Fig. 1.9). The ith particle is acted upon by an external force $\mathbf{F}_i$ and by $N-1$ internal

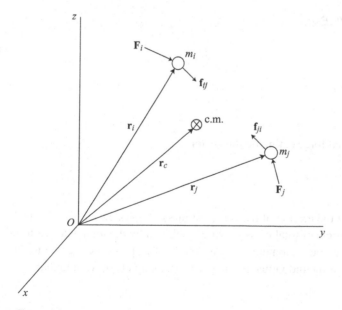

Figure 1.9.

interaction forces $\mathbf{f}_{ij}$ $(j \neq i)$ due to the other particles. The equation of motion for the ith particle is

$$m_i\ddot{\mathbf{r}}_i = \mathbf{F}_i + \sum_{j=1}^{N} \mathbf{f}_{ij} \tag{1.82}$$

The right-hand side of the equation is equal to the total force acting on the ith particle, external plus internal, and we note that $\mathbf{f}_{ii} = 0$; that is, a particle cannot act on itself to influence its motion.

Now sum (1.82) over the N particles.

$$\sum_{i=1}^{N} m_i\ddot{\mathbf{r}}_i = \sum_{i=1}^{N} \mathbf{F}_i + \sum_{i=1}^{N}\sum_{j=1}^{N} \mathbf{f}_{ij} \tag{1.83}$$

Because of Newton's law of action and reaction, we have

$$\mathbf{f}_{ji} = -\mathbf{f}_{ij} \tag{1.84}$$

and therefore

$$\sum_{i=1}^{N}\sum_{j=1}^{N} \mathbf{f}_{ij} = 0 \tag{1.85}$$

The *center of mass* location is given by

$$\mathbf{r}_c = \frac{1}{m}\sum_{i=1}^{N} m_i\mathbf{r}_i \tag{1.86}$$

where the total mass m is

$$m = \sum_{i=1}^{N} m_i \tag{1.87}$$

Then (1.83) reduces to

$$m\ddot{\mathbf{r}}_c = \mathbf{F} \tag{1.88}$$

where the total external force acting on the system is

$$\mathbf{F} = \sum_{i=1}^{N} \mathbf{F}_i \tag{1.89}$$

This result shows that the motion of the center of mass of a system of particles is the same as that of a single particle of total mass m which is driven by the total external force $\mathbf{F}$.

The translational or linear momentum of a system of N particles is equal to the vector sum of the momenta of the individual particles. Thus, using (1.3), we find that

$$\mathbf{p} = \sum_{i=1}^{N} \mathbf{p}_i = \sum_{i=1}^{N} m_i\dot{\mathbf{r}}_i \tag{1.90}$$

where each particle mass m_i is constant. Then, for the system, the rate of change of momentum is

$$\dot{\mathbf{p}} = \sum_{i=1}^{N} \dot{\mathbf{p}}_i = \sum_{i=1}^{N} m_i \ddot{\mathbf{r}}_i = \mathbf{F} \tag{1.91}$$

in agreement with (1.88). Note that if $\mathbf{F}$ remains equal to zero over some time interval, the linear momentum remains constant during the interval. More particularly, if a component of $\mathbf{F}$ in a certain fixed direction remains at zero, then the corresponding component of $\mathbf{p}$ is conserved.

Angular momentum

The angular momentum of a single particle of mass m_i about a fixed reference point O (Fig. 1.10) is

$$\mathbf{H}_i = \mathbf{r}_i \times m_i \dot{\mathbf{r}}_i = \mathbf{r}_i \times \mathbf{p}_i \tag{1.92}$$

which has the form of a moment of momentum. Upon summation over N particles, we find that the angular momentum of the system about O is

$$\mathbf{H}_O = \sum_{i=1}^{N} \mathbf{H}_i - \sum_{i=1}^{N} \mathbf{r}_i \times m_i \dot{\mathbf{r}}_i \tag{1.93}$$

Now consider the angular momentum of the system about an arbitrary reference point P. It is

$$\mathbf{H}_p = \sum_{i=1}^{N} \boldsymbol{\rho}_i \times m_i \dot{\boldsymbol{\rho}}_i \tag{1.94}$$

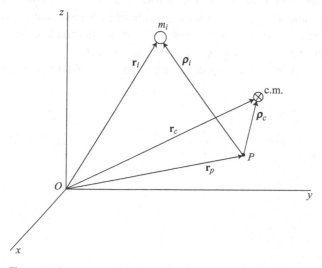

Figure 1.10.

Notice that the velocity $\dot{\rho}_i$ is *measured relative to the reference point P* rather than being an absolute velocity. The use of relative versus absolute velocities in the definition of angular momentum makes no difference if the reference point is either fixed or at the center of mass. There is a difference, however, in the form of the equation of motion for the general case of an accelerating reference point P, which is not at the center of mass. In this case, the choice of relative velocities yields simpler and physically more meaningful equations of motion.

To find the angular momentum relative to the center of mass, we take the reference point P at the center of mass ($\rho_c = 0$) and obtain

$$\mathbf{H}_c = \sum_{i=1}^{N} \rho_i \times m_i \dot{\rho}_i \tag{1.95}$$

where ρ_i is now the position vector of particle m_i relative to the center of mass.

Now let us write an expression for $\mathbf{H}_c$ when P is not at the center of mass. We obtain

$$\mathbf{H}_c = \sum_{i=1}^{N} (\rho_i - \rho_c) \times m_i(\dot{\rho}_i - \dot{\rho}_c)$$

$$= \sum_{i=1}^{N} \rho_i \times m_i \dot{\rho}_i - \rho_c \times m \dot{\rho}_c \tag{1.96}$$

where

$$\sum_{i=1}^{N} m_i \rho_i = m \rho_c \tag{1.97}$$

Then, recalling (1.94), we find that

$$\mathbf{H}_p = \mathbf{H}_c + \rho_c \times m \dot{\rho}_c \tag{1.98}$$

This important result states that the angular momentum about an *arbitrary point P* is equal to the angular momentum about the center of mass plus the angular momentum due to the relative translational velocity $\dot{\rho}_c$ of the center of mass. Of course, this result also applies to the case of a fixed reference point P when $\dot{\rho}_c$ is an absolute velocity.

Now let us differentiate (1.93) with respect to time in order to obtain an equation of motion. We obtain

$$\dot{\mathbf{H}}_O = \sum_{i=1}^{N} \mathbf{r}_i \times m_i \ddot{\mathbf{r}}_i \tag{1.99}$$

where, from Newton's law,

$$m_i \ddot{\mathbf{r}}_i = \mathbf{F}_i + \sum_{j=1}^{N} \mathbf{f}_{ij} \tag{1.100}$$

and we note that

$$\sum_{i=1}^{N} \sum_{j=1}^{N} \mathbf{r}_i \times \mathbf{f}_{ij} = 0 \tag{1.101}$$

since, by Newton's third law, the internal forces $\mathbf{f}_{ij}$ occur in equal, opposite, and collinear pairs. Hence we obtain an equation of motion in the form

$$\dot{\mathbf{H}}_O = \sum_{i=1}^{N} \mathbf{r}_i \times \mathbf{F}_i = \mathbf{M}_O \tag{1.102}$$

where $\mathbf{M}_O$ is the applied moment about the fixed point O due to forces external to the system.

In a similar manner, if we differentiate (1.95) with respect to time, we obtain

$$\dot{\mathbf{H}}_c = \sum_{i=1}^{N} \boldsymbol{\rho}_i \times m_i \dot{\boldsymbol{\rho}}_i \tag{1.103}$$

where $\boldsymbol{\rho}_i$ is the position vector of the ith particle relative to the center of mass. From Newton's law of motion for the ith particle,

$$m_i(\ddot{\mathbf{r}}_c + \ddot{\boldsymbol{\rho}}_i) = \mathbf{F}_i + \sum_{j=1}^{N} \mathbf{f}_{ij} \tag{1.104}$$

Now take the vector product of $\boldsymbol{\rho}_i$ with both sides of this equation and sum over i. We find that

$$\sum_{i=1}^{N} \boldsymbol{\rho}_i \times m_i \ddot{\mathbf{r}}_c = 0 \tag{1.105}$$

since

$$\sum_{i=1}^{N} m_i \boldsymbol{\rho}_i = 0 \tag{1.106}$$

for a reference point at the center of mass. Also,

$$\sum_{i=1}^{N} \sum_{j=1}^{N} \boldsymbol{\rho}_i \times \mathbf{f}_{ij} = 0 \tag{1.107}$$

because the internal forces $\mathbf{f}_{ij}$ occur in equal, opposite, and collinear pairs. Hence we obtain

$$\sum_{i=1}^{N} \boldsymbol{\rho}_i \times m_i \ddot{\boldsymbol{\rho}}_i = \sum_{i=1}^{N} \boldsymbol{\rho}_i \times \mathbf{F}_i = \mathbf{M}_c \tag{1.108}$$

and, from (1.103) and (1.108),

$$\dot{\mathbf{H}}_c = \mathbf{M}_c \tag{1.109}$$

where $\mathbf{M}_c$ is the external applied moment about the center of mass.

At this point we have found that the basic rotational equation

$$\dot{\mathbf{H}} = \mathbf{M} \tag{1.110}$$

applies in each of two cases: (1) the reference point is fixed in an inertial frame; or (2) the reference point is at the center of mass.

Finally, let us consider the most general case of an arbitrary reference point P. Upon differentiating (1.98) with respect to time, we obtain

$$
\begin{aligned}
\dot{\mathbf{H}}_p &= \dot{\mathbf{H}}_c + \boldsymbol{\rho}_c \times m\ddot{\boldsymbol{\rho}}_c \\
&= \mathbf{M}_c + \boldsymbol{\rho}_c \times m\ddot{\boldsymbol{\rho}}_c
\end{aligned}
\tag{1.111}
$$

But, from Newton's law of motion for the system,

$$
m(\ddot{\mathbf{r}}_p + \ddot{\boldsymbol{\rho}}_c) = \mathbf{F}
\tag{1.112}
$$

so we obtain

$$
\dot{\mathbf{H}}_p = \mathbf{M}_c + \boldsymbol{\rho}_c \times (\mathbf{F} - m\ddot{\mathbf{r}}_p)
\tag{1.113}
$$

The applied moment about P is

$$
\mathbf{M}_p = \mathbf{M}_c + \boldsymbol{\rho}_c \times \mathbf{F}
\tag{1.114}
$$

Thus, the rotational equation for this general case is

$$
\dot{\mathbf{H}}_p = \mathbf{M}_p - \boldsymbol{\rho}_c \times m\ddot{\mathbf{r}}_p
\tag{1.115}
$$

We note immediately that this equation reverts to the simpler form of (1.110) if P is a fixed point ($\ddot{\mathbf{r}}_p = 0$) or if P is located at the center of mass ($\boldsymbol{\rho}_c = 0$). The right-hand term also vanishes if $\boldsymbol{\rho}_c$ and $\ddot{\mathbf{r}}_p$ are parallel.

Accelerating frames

Consider a particle of mass m_i and its motion relative to a *noninertial reference frame* that is not rotating but is translating with point P at its origin (Fig. 1.10). The equation of motion is

$$
m_i(\ddot{\mathbf{r}}_p + \ddot{\boldsymbol{\rho}}_i) = \mathbf{F}_i
\tag{1.116}
$$

where $\mathbf{F}_i$ is now the *total force* acting on the particle. Relative to the accelerating frame, the equation of motion has the form

$$
m_i\ddot{\boldsymbol{\rho}}_i = \mathbf{F}_i - m_i\ddot{\mathbf{r}}_p
\tag{1.117}
$$

The term $-m_i\ddot{\mathbf{r}}_p$ can be regarded as an inertia force due to the acceleration of the *frame*. Note that the same equation of motion is obtained if we assume that the frame attached to P is not accelerating, but instead there is a uniform gravitational field with an acceleration of gravity $-\ddot{\mathbf{r}}_p$.

As another example of motion relative to an accelerating reference frame, consider again the rotational equation given in (1.115). We can write it in the form

$$
\dot{\mathbf{H}}_p = \mathbf{M}_p - \sum_{i=1}^{N} \boldsymbol{\rho}_i \times m_i\ddot{\mathbf{r}}_p
\tag{1.118}
$$

since

$$m\rho_c = \sum_{i=1}^{N} m_i \rho_i \tag{1.119}$$

gives the position ρ_c of the center of mass. The last term of (1.118) can be interpreted as the moment about P of individual inertia forces $-m_i \ddot{\mathbf{r}}_p$ that act on each particle m_i, the forces being parallel in the manner of an artificial gravitational field. The total moment of these inertial forces, as given in (1.115), is $-\rho_c \times m\ddot{\mathbf{r}}_p$, which can be considered as a total inertia force $-m\ddot{\mathbf{r}}_p$ acting at the center of mass.

The concept of inertia forces and an artificial gravitational field due to an accelerating reference frame can be expressed as the following principle of relative motion: *All the results and principles derivable from Newton's laws of motion relative to an inertial frame can be extended to apply to an accelerating but nonrotating frame if the inertia forces associated with the acceleration of the frame are considered as additional forces acting on the particles of the system.* This important result is particularly useful if some reference point in the system has an acceleration that is a known function of time. Note that it applies to work and energy principles relative to the accelerating frame without having to solve for the forces causing the acceleration.

Work and energy

The kinetic energy of a particle of mass m_i moving with speed v_i relative to an inertial frame is

$$T_i = \frac{1}{2} m_i v_i^2 \tag{1.120}$$

The total kinetic energy of a system of N particles is found by summing over the particles, resulting in

$$T = \sum_{i=1}^{N} T_i = \frac{1}{2} \sum_{i=1}^{N} m_i v_i^2 \tag{1.121}$$

Let us use the notation that

$$v_i^2 \equiv \dot{\mathbf{r}}_i^2 \equiv \dot{\mathbf{r}}_i \cdot \dot{\mathbf{r}}_i \tag{1.122}$$

and assume a *center of mass* reference point such that

$$\mathbf{r}_i = \mathbf{r}_c + \rho_i \tag{1.123}$$

The total kinetic energy can be written in the form

$$T = \frac{1}{2} \sum_{i=1}^{N} m_i \dot{\mathbf{r}}_i^2 = \frac{1}{2} \sum_{i=1}^{N} m_i (\dot{\mathbf{r}}_c + \dot{\rho}_i) \cdot (\dot{\mathbf{r}}_c + \dot{\rho}_i)$$

$$= \frac{1}{2} m \dot{\mathbf{r}}_c^2 + \frac{1}{2} \sum_{i=1}^{N} m_i \dot{\rho}_i^2 \tag{1.124}$$

where we recall that

$$\sum_{i=1}^{N} m_i \rho_i = 0 \tag{1.125}$$

for this center of mass reference point. Equation (1.124) is an expression of *Koenig's theorem*: *The total kinetic energy of a system of particles is equal to that due to the total mass moving with the velocity of the center of mass plus that due to the motion of individual particles relative to the center of mass.*

As a further generalization, let us consider a system of particles with a general reference point P (Fig. 1.10). Here we have

$$\mathbf{r}_i = \mathbf{r}_p + \rho_i \tag{1.126}$$

and the total kinetic energy is

$$T = \frac{1}{2} \sum_{i=1}^{N} m_i \dot{\mathbf{r}}_i^2 = \frac{1}{2} \sum_{i=1}^{N} m_i (\dot{\mathbf{r}}_p + \dot{\rho}_i) \cdot (\dot{\mathbf{r}}_p + \dot{\rho}_i)$$

$$= \frac{1}{2} m \dot{\mathbf{r}}_p^2 + \frac{1}{2} \sum_{i=1}^{N} m_i \dot{\rho}_i^2 + \dot{\mathbf{r}}_p \cdot m \dot{\rho}_c \tag{1.127}$$

We see that the total kinetic energy is the sum of three parts: (1) the kinetic energy due to the total mass moving at the speed of the reference point; (2) the kinetic energy due to motion relative to the reference point; and (3) the scalar product of the reference point velocity and the linear momentum of the system relative to the reference point. Equation (1.127) is an important and useful result. It is particularly convenient in the analysis of systems having a reference point whose motion is known but which is not at the center of mass.

Now let us look into the relationship between the work done on a system of particles and its kinetic energy. We start with the equation of motion for the ith particle, as in (1.82), namely,

$$m_i \ddot{\mathbf{r}}_i = \mathbf{F}_i + \sum_{j=1}^{N} \mathbf{f}_{ij} \tag{1.128}$$

Assume that the ith particle moves over a path from A_i to B_i. Take the dot product of each side with $d\mathbf{r}_i$ and evaluate the corresponding line integrals. We obtain

$$\int_{A_i}^{B_i} m_i \ddot{\mathbf{r}}_i \cdot d\mathbf{r}_i = \frac{1}{2} m_i \int_{t_A}^{t_B} \frac{d}{dt} \left(\dot{\mathbf{r}}_i^2 \right) dt = \frac{1}{2} m_i \left(v_{B_i}^2 - v_{A_i}^2 \right) \tag{1.129}$$

which is the increase in kinetic energy of the ith particle. The line integral on the right is

$$W_i = \int_{A_i}^{B_i} \left(\mathbf{F}_i + \sum_{j=1}^{N} \mathbf{f}_{ij} \right) \cdot d\mathbf{r}_i \tag{1.130}$$

which is the total work done on the ith particle by the external plus internal forces. Now sum over all the particles. The total work done on the system is

$$W = \sum_{i=1}^{N} W_i = \sum_{i=1}^{N} \int_{A_i}^{B_i} \left(\mathbf{F}_i + \sum_{j=1}^{N} \mathbf{f}_{ij} \right) \cdot d\mathbf{r}_i \tag{1.131}$$

and the increase in the total kinetic energy is

$$T_B - T_A = \frac{1}{2} \sum_{i=1}^{N} m_i \left(v_{B_i}^2 - v_{A_i}^2 \right) \tag{1.132}$$

Thus, equating the line integrals obtained from (1.128), we find that

$$T_B - T_A = W \tag{1.133}$$

This is the *principle of work and kinetic energy: The increase in the kinetic energy of a system of particles over an arbitrary time interval is equal to the work done on the system by external and internal forces during that time.* Since this principle applies continuously to an evolving system, we see that

$$\dot{T} = \dot{W} \tag{1.134}$$

that is, the rate of increase of kinetic energy is equal to the rate of doing work by the forces acting on the system.

If we choose a center of mass reference point, we have

$$\mathbf{r}_i = \mathbf{r}_c + \rho_i \tag{1.135}$$

and the work done on the system can be written in the form

$$W - \int_{A_c}^{B_c} \mathbf{F} \cdot d\mathbf{r}_c + \sum_{i=1}^{N} \int_{A_i}^{B_i} \left(\mathbf{F}_i + \sum_{j=1}^{N} \mathbf{f}_{ij} \right) \cdot d\rho_i \tag{1.136}$$

where A_c and B_c are the end-points of the path followed by the center of mass. The equation of motion for the center of mass is identical in form with that of a single particle; hence, they will have similar work–energy relationships. Therefore, the work done by the total external force $\mathbf{F}$ in moving through the displacement of the center of mass must equal the increase in the kinetic energy associated with the center of mass motion, as given in the first term of (1.124). Then the remaining term in the work expression, representing the work of the external and internal forces in moving through displacements relative to the center of mass, must equal the increase in the kinetic energy of relative motions, that is, in the change in the last term of (1.124).

Conservation of energy

Let us consider a particle whose position (x, y, z) is given relative to an inertial Cartesian frame. Suppose that the work done on the particle in an arbitrary infinitesimal displacement is

$$dW = \mathbf{F} \cdot d\mathbf{r} = F_x dx + F_y dy + F_z dz \tag{1.137}$$

and the right-hand side is equal to the total differential of a function of position. Let us take

$$dW = -dV = -\frac{\partial V}{\partial x}dx - \frac{\partial V}{\partial y}dy - \frac{\partial V}{\partial z}dz \tag{1.138}$$

where the minus sign is chosen for convenience and the *potential energy* function is $V(x, y, z)$. Then, since $d\mathbf{r}$ is arbitrary, we can equate coefficients to obtain

$$F_x = -\frac{\partial V}{\partial x}, \qquad F_y = -\frac{\partial V}{\partial y}, \qquad F_z = -\frac{\partial V}{\partial z} \tag{1.139}$$

or, using vector notation,

$$\mathbf{F} = -\nabla V \tag{1.140}$$

that is, the force is equal to the negative gradient of $V(x, y, z)$.

In accordance with the principle of work and kinetic energy, the increase in kinetic energy is

$$dT = dW = -dV \tag{1.141}$$

so we find that

$$\dot{T} + \dot{V} = 0 \tag{1.142}$$

or, after integration with respect to time,

$$T + V = E \tag{1.143}$$

where the total energy E is a constant. This is the *principle of conservation of energy* applied to a rather simple system.

This principle can easily be extended to apply to a system of N particles whose positions are given by the $3N$ Cartesian coordinates $x_1, x_2, \ldots, x_{3N}$. In this case, the kinetic energy T is the sum of the individual kinetic energies, and the overall potential energy $V(x)$ is a function of the particle positions. The force in the positive x_j direction obtained from V is

$$F_j = -\frac{\partial V}{\partial x_j} \tag{1.144}$$

The system will be *conservative*, that is, the total energy $T + V$ will be constant if it meets the following conditions: (1) the potential energy $V(x)$ is a function of position only and not an explicit function of time; and (2) all forces which do work on the system in the actual motion are obtained from the potential energy in accordance with (1.144); any constraint forces do no work. Later the concept of a conservative system will be extended and generalized.

It sometimes occurs that the forces doing work on a system are all obtained from a potential energy function of the more general form $V(x, t)$ by using (1.144) or (1.140). In this case, the forces are termed *monogenic*, that is, derivable from a potential energy function, whether conservative or not.

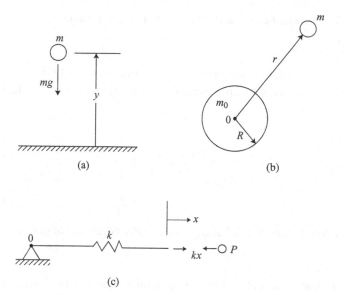

Figure 1.11.

Example 1.4 Let us consider the form of the potential energy function in various common cases.

Uniform gravity Suppose a particle of mass m is located at a distance y above a reference level in the presence of a uniform gravitational field whose gravitational acceleration is g, as shown in Fig. 1.11a. The downward force acting on the particle is its *weight*

$$w = mg \qquad (1.145)$$

From (1.139) we obtain

$$F_y = -\frac{\partial V}{\partial y} = -mg \qquad (1.146)$$

which can be integrated to yield the potential energy

$$V(y) = mgy \qquad (1.147)$$

The constant of integration has been chosen equal to zero in order to give zero potential energy at the reference level.

Inverse-square gravity A uniform spherical body of mass m_0 and radius R exerts a gravitational attraction on a particle of mass m located at a distance r from the center O, as shown in Fig. 1.11b. The radial force on the particle has the form

$$F_r = -\frac{K}{r^2} \qquad (r \geq R) \qquad (1.148)$$

where $K = Gm_0 m$. The *universal gravitational constant* G has the value

$$G = 6.673 \times 10^{-11} \, \text{N} \cdot \text{m}^2 / \text{kg}^2$$

and we have used units of Newtons, meters, and kilograms. Equation (1.148) is a statement of *Newton's law of gravitation* where each attracting body is regarded as a particle. Using (1.144) we obtain

$$-\frac{\partial V}{\partial r} = -\frac{K}{r^2} \tag{1.149}$$

which integrates to

$$V(r) = -\frac{K}{r} \tag{1.150}$$

and we note that the gravitational potential energy is generally negative, but goes to zero as $r \to \infty$.

Linear spring   A commonly encountered form of potential energy is that due to elastic deformation. As an example, consider a particle P which is attached by a linear spring of stiffness k to a fixed point O, as shown in Fig. 1.11c. The force of the spring acting on the particle is

$$F_x = -\frac{\partial V}{\partial x} = -kx \tag{1.151}$$

where x is the elongation of the spring, measured from its unstressed position. Integration of (1.151) yields the potential energy

$$V(x) = \frac{1}{2} k x^2 \tag{1.152}$$

Example 1.5   A particle of mass m is displaced slightly from its equilibrium position at the top of a smooth fixed sphere of radius r, and it slides downward due to gravity (Fig. 1.12). We wish to solve for its velocity as a function of position, and the angle θ at which it loses contact with the sphere.

Rather than writing the tangential equation of motion involving $\ddot{\theta}$ and then integrating, we can solve directly for the velocity of the particle by using conservation of energy. We see that

$$T = \frac{1}{2} m v^2 \tag{1.153}$$

and, using the center O as the reference level,

$$V = mgr \cos \theta \tag{1.154}$$

Conservation of energy results in

$$T + V = \frac{1}{2} m v^2 + mgr \cos \theta = mgr \tag{1.155}$$

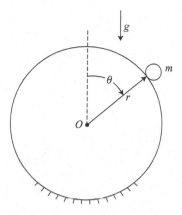

Figure 1.12.

and, solving for the velocity, we obtain

$$v = \sqrt{2gr(1 - \cos\theta)} \tag{1.156}$$

Let N be the radial force of the sphere acting on the particle. The radial equation of motion is

$$ma_r = -\frac{mv^2}{r} = N - mg\cos\theta \tag{1.157}$$

From (1.156) and (1.157) we obtain

$$N = mg\,(3\cos\theta - 2) \tag{1.158}$$

The particle leaves the sphere when the force N decreases to zero, that is, when

$$\cos\theta = \frac{2}{3} \quad \text{or} \quad \theta = 48.19° \tag{1.159}$$

Example 1.6 Particles A and B, each of mass m, are connected by a rigid massless rod of length l, as shown in Fig. 1.13. Particle A is restrained by a linear spring of stiffness k, but can slide without friction on a plane inclined at 45° with the horizontal. We wish to obtain the differential equations for the planar motion of this system under the action of gravity.

It will simplify matters if we can obtain the differential equations of motion without having to solve for either the normal constraint force of the inclined plane acting on particle A, or for the compressive force of the rod acting on both particles. First, let us write Newton's law for the motion of the center of mass in the x-direction. We obtain

$$2m\left[\ddot{x} + \frac{1}{2}l\ddot{\theta}\cos(\theta - 45°) - \frac{1}{2}l\dot{\theta}^2\sin(\theta - 45°)\right] = -kx + 2mg\sin 45°$$

or

$$2m\ddot{x} + \frac{ml}{\sqrt{2}}\ddot{\theta}(\cos\theta + \sin\theta) - \frac{ml}{\sqrt{2}}\dot{\theta}^2(\sin\theta - \cos\theta) = -kx + \sqrt{2}\,mg \tag{1.160}$$

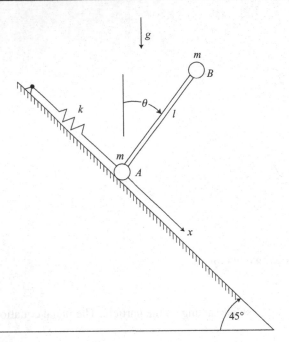

Figure 1.13.

where x is measured from the position of zero force in the spring. This is the x equation of motion.

Next, let us write the rotational equation, using A as a reference point where A has an acceleration $\ddot{x}$ down the plane. Thus, we must use (1.118) for motion relative to an accelerating frame. This means that an inertia force $m\ddot{x}$ is applied at B and is directed upward, parallel to the plane. Then, relative to A, the gravitational and inertial moments are added. Thus, we obtain

$$\dot{H} = ml^2\ddot{\theta} = mgl\,\sin\theta - ml\ddot{x}\cos(\theta - 45°)$$

or

$$ml^2\ddot{\theta} + \frac{ml}{\sqrt{2}}\ddot{x}\,(\cos\theta + \sin\theta) = mgl\,\sin\theta \qquad (1.161)$$

This is the θ equation of motion.

Example 1.7 Three particles, each of mass m, are located at the vertices of an equilateral triangle and are in plane rotational motion about the center of mass (Fig. 1.14). They are subject to mutual pairwise gravitational attractions as the separation $l = l_0$ remains constant. (a) Solve for the required angular velocity $\omega = \omega_0$. (b) Now suppose that with $l = l_0$ and $\dot{l} = 0$, the angular velocity is suddenly changed to $\frac{1}{2}\omega_0$. Find the minimum value of l in the ensuing motion, assuming that an equilateral configuration is maintained continuously.

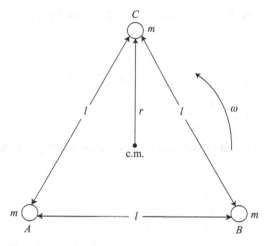

Figure 1.14.

First consider the force acting on particle C due to particles A and B. In accordance with Newton's law of gravitation, the vertical components add to give

$$F_r = -\frac{\sqrt{3}\, Gm^2}{l_0^2} = -\frac{Gm^2}{\sqrt{3}\, r_0^2} \tag{1.162}$$

since $l_0 = \sqrt{3}\, r_0$. From the radial equation of motion for particle C we obtain

$$ma_r = F_r$$

or

$$mr_0\omega_0^2 = \frac{Gm^2}{\sqrt{3}\, r_0^2} \tag{1.163}$$

Thus, we obtain the angular velocity

$$\omega_0 = \left(\frac{Gm}{\sqrt{3}\, r_0^3}\right)^{1/2} = \left(\frac{3Gm}{l_0^3}\right)^{1/2} \tag{1.164}$$

Now let us assume the initial conditions $r(0) = r_0$, $\dot{r}(0) = 0$, $\omega(0) = \frac{1}{2}\omega_0$. We can use conservation of energy and of angular momentum to solve for the minimum value of r, and therefore of l. Let us concentrate on particle C alone since the system values are three times those of a single particle. The potential energy can be obtained by first recalling that the radial force

$$F_r = -\frac{\partial V}{\partial r} = -\frac{Gm^2}{\sqrt{3}\, r^2} \tag{1.165}$$

which can be integrated to yield the general expression

$$V = -\frac{Gm^2}{\sqrt{3}\, r} \tag{1.166}$$

where the integration constant is set equal to zero.

When $r = r_{min}$ we have $\dot{r} = 0$ so the kinetic energy of particle C has the form

$$T = \frac{1}{2}mr^2\omega^2 \tag{1.167}$$

Then, noting the initial conditions, the conservation of energy results in

$$T + V = \frac{1}{2}mr^2\omega^2 - \frac{Gm^2}{\sqrt{3}\,r} = \frac{1}{8}mr_0^2\omega_0^2 - \frac{Gm^2}{\sqrt{3}\,r_0} \tag{1.168}$$

Conservation of angular momentum about the center of mass is expressed by the equation

$$H = mr^2\omega = \frac{1}{2}mr_0^2\,\omega_0 \tag{1.169}$$

Hence

$$\omega = \frac{r_0^2\,\omega_0}{2\,r^2} \tag{1.170}$$

Now substitue for ω from (1.170) into (1.168) and use (1.164). After some algebraic simplification we obtain the quadratic equation

$$7r^2 - 8r_0r + r_0^2 = 0 \tag{1.171}$$

which has the roots

$$r_{1,2} = r_0, \quad \frac{1}{7}r_0 \tag{1.172}$$

one of which is the initial condition. Thus

$$r_{min} = \frac{1}{7}r_0 \quad \text{and} \quad l_{min} = \frac{1}{7}l_0 \tag{1.173}$$

Friction

Systems with friction are characterized by the loss of energy due to relative motion of the particles. Thus, in general, they are not conservative. The two principal types of frictional forces to be considered here are *linear damping* and *Coulomb friction.*

Linear damping

A linear viscous damper with a damping coefficient c is shown connected between two particles in Fig. 1.15. It is assumed to be massless and produces a tensile force proportional to the relative separation rate of the particles in accordance with the equation

$$F_i = c\,(\dot{x}_j - \dot{x}_i) \tag{1.174}$$

In other words, the damper force always opposes any relative motion of the particles and therefore does negative work on the system, dissipating energy whenever a relative velocity exists.

Figure 1.15.

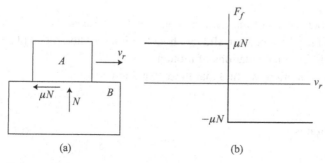

(a)   (b)

Figure 1.16.

Coulomb friction

A Coulomb friction force is dissipative, like other friction forces, but is *nonlinear*; that is, the friction force is a nonlinear function of the relative sliding velocity. As an example, suppose that block A slides with velocity v_r relative to block B, as shown in Fig. 1.16. The force of block B acting on block A has a positive normal component N and a tangential friction component μN, where the *coefficient of sliding friction* μ is independent of the relative sliding velocity v_r. In detail, the Coulomb friction force is

$$F_f = -\mu N \; \text{sgn}(v_r) \tag{1.175}$$

where F_f and v_r are positive in the same direction and where $\text{sgn}(v_r)$ equals ± 1, depending on the sign of v_r. Note that F_f is independent of the contact area.

In the case $v_r = 0$, that is, for no sliding, the force of friction can have any magnitude less than that required to initiate sliding. The actual force in this case is obtained from the equations of statics. Although the force required to begin sliding is actually slightly larger than that required to sustain it, we shall ignore this difference and assume that F_f as a function of v_r is given by Fig. 1.16b or (1.175).

Since the magnitude $|F_f| \leq \mu N$, we see that the *direction* of the total force of block B acting on block A cannot deviate from the normal to the sliding surface by more than an angle ϕ where

$$\tan \phi = \mu \tag{1.176}$$

This leads to the idea of a *cone of friction* having a semivertex angle ϕ. The total force vector must lie on or within the cone of friction. It lies on the cone during sliding.

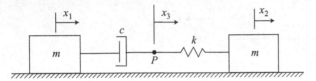

Figure 1.17.

Example 1.8   Two blocks, each of mass m, are connected by a linear damper and spring in series, as shown in Fig. 1.17. They can slide without friction on a horizontal plane. First, we wish to derive the differential equations of motion.

Using Newton's law of motion, we find that the x_1 equation is

$$m\ddot{x}_1 = c(\dot{x}_3 - \dot{x}_1) \tag{1.177}$$

Similarly, the x_2 equation is

$$m\ddot{x}_2 = k(x_3 - x_2) \tag{1.178}$$

We note that the forces in the damper and spring are equal, so we obtain the first-order equation

$$c\,(\dot{x}_3 - \dot{x}_1) = k\,(x_2 - x_3) \tag{1.179}$$

These are the three linear differential equations of motion. They are equivalent to five first-order equations, so the total order of the system is five.

This system is unusual because the coordinate x_3, which is associated with the connecting point P between the damper and spring, does not involve displacement of any mass. This results in an odd total order of the system rather than the usual even number. It is also reflected in the degree of the characteristic equation and in the required number of initial conditions.

Second method   Let us assume the initial conditions

$$x_1(0) = 0, \qquad x_2(0) = 0, \qquad x_3(0) = 0 \tag{1.180}$$

and

$$\dot{x}_1(0) = 0, \qquad \dot{x}_2(0) = v_0 \tag{1.181}$$

We see from (1.179) that $\dot{x}_3(0) = 0$.

The analysis of this system can be simplified if we notice that the center of mass moves at a constant velocity $\frac{1}{2}v_0$ due to conservation of linear momentum. Thus, a frame moving with the center of mass is an inertial frame, and Newton's laws of motion apply relative to this frame. Let us use the coordinates x_{1c}, x_{2c}, x_{3c} for positions relative to the center of mass frame and assume that the origins of the two frames coincide at $t = 0$. Then we have the initial conditions

$$x_{1c}(0) = 0, \qquad x_{2c}(0) = 0, \qquad x_{3c}(0) = 0 \tag{1.182}$$

and

$$\dot{x}_{1c}(0) = \dot{x}_{3c}(0) = -\frac{1}{2}v_0, \qquad \dot{x}_{2c}(0) = \frac{1}{2}v_0 \tag{1.183}$$

In general, x_{1c} and x_{2c} move as mirror images about the center of mass, so we can take

$$x_{1c} = -x_{2c} \tag{1.184}$$

and simplify by writing equations of motion for x_{2c} and x_{3c} only as dependent variables. We obtain, similar to (1.178) and (1.179),

$$m\ddot{x}_{2c} + kx_{2c} - kx_{3c} = 0 \tag{1.185}$$
$$c\dot{x}_{3c} + c\dot{x}_{2c} - kx_{2c} + kx_{3c} = 0 \tag{1.186}$$

To obtain some idea of the nature of the system response, we can assume solutions of the exponential form e^{st}, and obtain the resulting characteristic equation which turns out to be a cubic polynomial in s. One root is zero and the other two have negative real parts. Thus, the solutions for x_{2c} and x_{3c} each consist of two exponentially decaying functions of time plus a constant. As time approaches infinity, the exponential functions will vanish with $x_{2c} = x_{3c}$ and no force in the spring or damper. The *length* of the linear damper, however, will have increased by $2x_{3c}$ compared with its initial value.

Finally, to obtain the solutions in the original inertial frame, we use

$$x_1 = -x_{2c} + \frac{1}{2}v_0t \tag{1.187}$$

$$x_2 = x_{2c} + \frac{1}{2}v_0t \tag{1.188}$$

$$x_3 = x_{3c} + \frac{1}{2}v_0t \tag{1.189}$$

Thus, in their final motion, the blocks each move with a velocity $\frac{1}{2}v_0$ and a separation somewhat larger than the original value.

Example 1.9   Two stacked blocks, each of mass m, slide relative to each other as they move along a horizontal floor, as shown in Fig. 1.18. Suppose that the initial conditions are $x_A(0) = 0$, $\dot{x}_A(0) = v_0$, $x_B(0) = 0$, $\dot{x}_B(0) = 0$. We wish to solve for the time when sliding stops and the final positions of the blocks, assuming a coefficient of friction $\mu = 0.5$ between the two blocks, and $\mu = 0.1$ between block B and the floor.

The Coulomb friction force between blocks A and B is $0.5mg$ whereas the friction force between block B and the floor is $0.2mg$. The friction forces oppose relative motion so the initial accelerations of blocks A and B are

$$\ddot{x}_A = -0.5g \tag{1.190}$$
$$\ddot{x}_B = (0.5 - 0.2)g = 0.3g \tag{1.191}$$

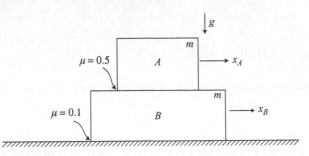

Figure 1.18.

Thus the relative acceleration between blocks A and B is $0.8g$ and the time to stop sliding is

$$t_1 = \frac{v_0}{0.8g} = 1.25 \frac{v_0}{g} \tag{1.192}$$

At this time the common velocity of the two blocks is

$$v_1 = \ddot{x}_B t_1 = 0.375 v_0 \tag{1.193}$$

When $t > t_1$, A and B will slide as a single block of mass $2m$ with an acceleration equal to $-0.1g$. The stopping time for this combination is

$$t_2 = \frac{0.375 v_0}{0.1g} = 3.75 \frac{v_0}{g} \tag{1.194}$$

Thus the time required for sliding to stop completely is

$$t = t_1 + t_2 = 5 \frac{v_0}{g} \tag{1.195}$$

The final displacement of block B is equal to its average velocity multiplied by the total time. We obtain

$$x_B = \frac{1}{2} v_1 t = 0.9375 \frac{v_0^2}{g} \tag{1.196}$$

The final displacement of block A is equal to x_B plus the displacement of A relative to B.

$$x_A = x_B + \frac{1}{2} v_0 t_1 = 1.5625 \frac{v_0^2}{g} \tag{1.197}$$

1.3   Constraints and configuration space

Generalized coordinates and configuration space

Consider a system of N particles. The *configuration* of this system is specified by giving the locations of all the particles. For example, the inertial location of the first particle might be given by the Cartesian coordinates (x_1, x_2, x_3), the location of the second particle by (x_4, x_5, x_6), and so forth. Thus, the configuration of the system would be given by

$(x_1, x_2, \ldots, x_{3N})$. In the usual case, the particles cannot all move freely but are at least somewhat constrained kinematically in their differential motions, if not in their large motions as well. Under these conditions, it is usually possible to give the configuration of the system by specifying the values of fewer than $3N$ parameters. These $n \leq 3N$ parameters are called *generalized coordinates* (qs) and are related to the xs by the *transformation equations*

$$x_k = x_k(q_1, q_2, \ldots, q_n, t) \quad (k = 1, \ldots, 3N) \tag{1.198}$$

The qs are not necessarily uniform in their dimensions. For example, the position of a particle in planar motion may be expressed by the polar coordinates (r, θ) which have differing dimensions. Thus, generalized coordinates may include common coordinate systems. However, a generalized coordinate may also be chosen such that it is not identified with any of the common coordinate systems, but represents a displacement form or shape involving several particles. In this case, the generalized coordinate is defined assuming certain displacement ratios and relative directions among the particles. For example, a generalized coordinate might consist of equal radial displacements of particles at the vertices of an equilateral triangle.

Frequently one attempts to find a set of independent generalized coordinates, but this is not always possible. So, in general, we assume that there are m independent equations of constraint involving the qs and possibly the $\dot{q}$s. If, for the same system, there are l independent equations of constraint involving the $3N$ xs (and possibly the corresponding $\dot{x}$s), then

$$3N - l = n - m \tag{1.199}$$

and this is equal to the number of *degrees of freedom*. The number of degrees of freedom is, in general, a property of the system and not of the choice of coordinates.

Since the configuration of a system is specified by the values of its n generalized coordinates, one can represent any particular configuration by a point in n-dimensional *configuration space* (Fig. 1.19). If the values of all the qs and $\dot{q}$s are known at some initial time t_0, then, as time proceeds, the configuration point C will trace a solution path in configuration space in accordance with the dynamical equations of motion and any constraint equations. For the case of independent qs, the curve will be continuous but otherwise not constrained. If, however, there are holonomic constraints expressed as functions of the qs and possibly time, then the solution point must remain on a hypersurface having fewer than n dimensions, and which may be moving and possibly changing shape. In general, then, one can represent an evolving mechanical system by an n-dimensional vector $\mathbf{q}$, drawn from the origin to the configuration point C, tracing a path in configuration space as time proceeds. This will be discussed further in Chapter 2.

Holonomic constraints

Suppose that the configuration of a system is specified by n generalized coordinates $(q_1, \ldots, q_n)$ and assume that there are m independent equations of constraint of the form

$$\phi_j(q_1, \ldots, q_n, t) = 0 \quad (j = 1, \ldots, m) \tag{1.200}$$

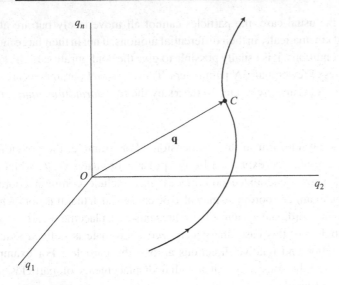

Figure 1.19.

A constraint of this form is called a *holonomic constraint*. A dynamical system whose constraint equations, if any, are all of the holonomic form is called a *holonomic system*.

An example of a holonomic constraint is provided by a particle which is forced to move on a sphere of radius R centered at the origin of a Cartesian frame. In this case the equation of constraint is

$$\phi_j = x^2 + y^2 + z^2 - R^2 = 0 \tag{1.201}$$

where (x, y, z) is the location of the particle. The sphere is a two-dimensional constraint surface which is embedded in a three-dimensional Cartesian space.

The configuration of a holonomic system can always be specified using a minimal set of generalized coordinates equal in number to the degrees of freedom. This is also the number of dimensions of the constraint hypersurface, that is, $n - m$. Hence it is always possible in theory to find a set of *independent* qs describing a holonomic system. For the case of a spherical constraint surface, one could use angles of latitude and longitude to describe the position of a particle. Another possibility might be to use the cylindrical coordinates ϕ and z as qs, where ϕ effectively gives the longitude and z the latitude.

Nonholonomic constraints

Nonholonomic constraints may have the general form

$$f_j(q, \dot{q}, t) = 0 \qquad (j = 1, \ldots, m) \tag{1.202}$$

but usually they have a simpler form which is *linear* in the velocities. Thus, we nearly always assume that nonholonomic constraints have the form

$$f_j = \sum_{i=1}^{n} a_{ji}(q, t)\, \dot{q}_i + a_{jt}(q, t) = 0 \qquad (j = 1, \ldots, m) \tag{1.203}$$

or the alternate differential form

$$\sum_{i=1}^{n} a_{ji}(q,t)\,dq_i + a_{jt}(q,t)\,dt = 0 \qquad (j = 1, \ldots, m) \qquad (1.204)$$

where, in either case, these expressions are *not integrable*. If either expression were integrable, then a function $\phi_j(q,t)$ would exist and (1.200) would apply, indicating that the constraint is actually *holonomic*. In this case, we would have

$$\dot{\phi}_j(q,\dot{q},t) = \sum_{i=1}^{n} \frac{\partial \phi_j}{\partial q_i}\,\dot{q}_i + \frac{\partial \phi_j}{\partial t} = 0 \qquad (1.205)$$

and, comparing (1.203) and (1.205), we find that

$$a_{ji} \equiv \frac{\partial \phi_j}{\partial q_i}, \qquad a_{jt} \equiv \frac{\partial \phi_j}{\partial t} \qquad (i = 1, \ldots, n; \; j = 1, \ldots, m) \qquad (1.206)$$

for this holonomic constraint. We conclude that holonomic and "linear" nonholonomic constraints can be expressed in the forms of (1.203) and (1.204). The holonomic case is distinguished by its integrability. Note that the coefficients $a_{ji}(q,t)$ and $a_{jt}(q,t)$ are generally *nonlinear* in the qs and t.

Other constraint classifications

A constraint is classed as *scleronomic* if the time t does not appear explicitly in the equation of constraint. Otherwise, it is *rheonomic*. Thus, a scleronomic holonomic constraint has the form

$$\phi_j(q) = 0 \qquad (1.207)$$

A scleronomic nonholonomic constraint has the form

$$\sum_{i=1}^{n} a_{ji}(q)\dot{q}_i = 0 \qquad (1.208)$$

where we note that $a_{jt} \equiv 0$.

Constraints having $a_{jt} \neq 0$, or $a_{ji} = a_{ji}(q,t)$, or $\phi_j = \phi_j(q,t)$ are classed as *rheonomic* constraints. Typical examples of rheonomic constraints are a rod of varying length $l(t)$ connecting two particles in the holonomic case, or a knife edge whose orientation angle is an explicit function of time in the nonholonomic case.

A *catastatic* constraint has $\partial \phi_j / \partial t \equiv 0$ if holonomic, or $a_{jt} \equiv 0$ if nonholonomic. Catastatic constraints have an important place in dynamical theory. Note that all scleronomic constraints are also catastatic, but not necessarily *vice versa*. For example, a nonholonomic constraint having the form

$$\sum_{i=1}^{n} a_{ji}(q,t)\,\dot{q}_i = 0 \qquad (1.209)$$

is catastatic but is not scleronomic.

Dynamical *systems* can also be classified as scleronomic or rheonomic. A *scleronomic system* satisfies the conditions that: (1) all constraints, if any, are scleronomic; and (2) the transformation equations, given by (1.198), which relate inertial Cartesian coordinates and generalized coordinates do not contain time explicitly. As an example illustrating the importance of the second condition, consider a particle moving on a spherical surface, centered at the origin, whose radius $R(t)$ is a known function of time. We can use the spherical coordinates θ and ϕ as *independent generalized coordinates*, that is, there are no constraints on the qs. Hence, the first condition is satisfied. The transformation equations, however, are

$$x = R(t)\sin\theta\cos\phi$$
$$y = R(t)\sin\theta\sin\phi \qquad\qquad (1.210)$$
$$z = R(t)\cos\theta$$

These transformation equations do not satisfy the second condition, so the system is rheonomic.

Another classification of dynamical systems involves the categories *catastatic* or *acatastatic*. A catastatic system satisfies the conditions that: (1) all constraints, if any, are catastatic; and (2) the system is capable of being continuously at rest by setting all the $\dot{q}$s equal to zero. This implies that $\partial x_k/\partial t \equiv 0$ for $k = 1, \ldots, 3N$; that is, for all the transformation equations. In other words, $x_k = x_k(q)$ in agreement with the second condition for a scleronomic system. Acatastatic means not catastatic.

Accessibility

Constraint equations, either holonomic or nonholonomic, are *kinematic* in nature; that is, they put restrictions on the possible motions of a system, irrespective of the dynamical equations. There is an important difference in these possible motions that distinguishes holonomic from nonholonomic systems. It lies in the *accessibility* of points in configuration space. For the case of *holonomic constraints*, the configuration point moves in a reduced space of $(n - m)$ dimensions since it must remain on each of m constraint surfaces, that is, on their common intersection. Thus, certain regions of n-dimensional configuration space are no longer accessible.

By contrast, for *nonholonomic constraints*, it is the differential motions which are constrained. Since the differential equations representing these nonholonomic constraints are not integrable, there are no finite constraint surfaces in configuration space and there is no reduction of the accessible region. In other words, by properly choosing the path, it is possible to reach any point in n-dimensional q-space from any other point. As an example, a scleronomic nonholonomic constraint, as given in (1.208), can be represented by an $(n - m)$-dimensional planar differential surface element. Any differential displacement $d\mathbf{q}$ must lie within that plane but is otherwise unconstrained. It is possible, however, to steer the configuration point C and its differential element to any point of configuration space, provided that more than one degree of freedom exists. A scleronomic system with only one degree of freedom is not steerable, and therefore any such system must be integrable and holonomic, and not fully accessible.

Exactness and integrability

Now let us consider the conditions under which a constraint function $f_j(q, \dot{q}, t)$ is integrable. If it is integrable, then a function $\phi_j(q, t)$ exists whose total time derivative is equal to $f_j(q, \dot{q}, t)$, that is,

$$f_j = \dot{\phi}_j(q, \dot{q}, t) = \sum_{i=1}^{n} \frac{\partial \phi_j}{\partial q_i} \dot{q}_i + \frac{\partial \phi_j}{\partial t} \tag{1.211}$$

We see immediately that f_j must be linear in the $\dot{q}$s, so let us consider the linear form

$$f_j = \sum_{i=1}^{n} a_{ji}(q, t) \dot{q}_i + a_{jt}(q, t) \tag{1.212}$$

as in (1.203). By comparing (1.211) and (1.212) we can equate coefficients as follows:

$$u_{ji}(q, t) = \frac{\partial \phi_j}{\partial q_i}, \qquad a_{jt}(q, t) = \frac{\partial \phi_j}{\partial t} \tag{1.213}$$

for all i and j. If $f_j(q, \dot{q}, t)$ is integrable, we know that a function $\phi_j(q, t)$ exists and that

$$\frac{\partial^2 \phi_j}{\partial q_k \partial q_i} = \frac{\partial^2 \phi_j}{\partial q_i \partial q_k}, \qquad \frac{\partial^2 \phi_j}{\partial q_i \partial t} = \frac{\partial^2 \phi_j}{\partial t \, \partial q_i} \tag{1.214}$$

that is, the order of partial differentiation is immaterial. In terms of a_{ji}s we have

$$\begin{aligned}
\frac{\partial a_{ji}}{\partial q_k} &= \frac{\partial a_{jk}}{\partial q_i} \qquad (i, k = 1, \ldots, n) \\
\frac{\partial a_{jt}}{\partial q_i} &= \frac{\partial a_{ji}}{\partial t} \qquad (i = 1, \ldots, n)
\end{aligned} \tag{1.215}$$

These are the *exactness conditions* for integrability. In general, a function $f_j(q, \dot{q}, t)$ is integrable if it has the linear form of (1.212) and if it is either (1) exact as it stands or (2) can be made exact through multiplication by an *integrating factor* of the form $M_j(q, t)$.

There is an alternative form of the exactness conditions for the case in which $f_j(q, \dot{q}, t)$ has the linear form of (1.212). First, we see that

$$\frac{d}{dt}\left(\frac{\partial f_j}{\partial \dot{q}_i}\right) = \dot{a}_{ji} = \sum_{k=1}^{n} \frac{\partial a_{ji}}{\partial q_k} \dot{q}_k + \frac{\partial a_{ji}}{\partial t} \tag{1.216}$$

Also, changing the summing index from i to k in (1.212), we obtain

$$\frac{\partial f_j}{\partial q_i} = \sum_{k=1}^{n} \frac{\partial a_{jk}}{\partial q_i} \dot{q}_k + \frac{\partial a_{jt}}{\partial q_i} \tag{1.217}$$

Hence we find that

$$\frac{d}{dt}\left(\frac{\partial f_j}{\partial \dot{q}_i}\right) - \frac{\partial f_j}{\partial q_i} = \sum_{k=1}^{n}\left(\frac{\partial a_{ji}}{\partial q_k} - \frac{\partial a_{jk}}{\partial q_i}\right)\dot{q}_k + \frac{\partial a_{ji}}{\partial t} - \frac{\partial a_{jt}}{\partial q_i} \qquad (1.218)$$

If the exactness conditions of (1.215) are satisfied, it follows that

$$\frac{d}{dt}\left(\frac{\partial f_j}{\partial \dot{q}_i}\right) - \frac{\partial f_j}{\partial q_i} = 0 \qquad (i = 1, \ldots, n) \qquad (1.219)$$

Conversely, if (1.219) is satisfied for all $\dot{q}$s satisfying the constraints, then, from (1.218), we see that the exactness conditions must apply. Hence, (1.215) and (1.219) are equivalent statements of the exactness conditions.

Another approach to the question of integrability lies in the use of *Pfaffian differential forms*. A Pfaffian differential form Ω in the r variables $x_1, \ldots, x_r$ can be written as

$$\Omega = X_1(x)dx_1 + \cdots + X_r(x)dx_r \qquad (1.220)$$

where the coefficients are functions of the xs, in general. If the differential form is exact it is equal to the total differential $d\Phi$ of a function $\Phi(x)$. The exactness conditions are

$$\frac{\partial X_i}{\partial x_j} = \frac{\partial X_j}{\partial x_i} \qquad (i, j = 1, \ldots, r) \qquad (1.221)$$

that is, for all i and j. The differential form Ω is *integrable* if it is exact, or if it can be made exact through multiplication by an *integrating factor* of the form $M(x)$.

Returning now to a consideration of nonholonomic constraints, we can write the Pfaffian differential form

$$\Omega_j = \sum_{i=1}^{n} a_{ji}(q, t)dq_i + a_{jt}(q, t)dt = 0 \qquad (j = 1, \ldots, m) \qquad (1.222)$$

which we recognize as having been presented previously in (1.204). Of course, the exactness conditions are, as before, those given in (1.215).

Differential forms have wide application in the study of dynamics. A common example is the differential expression

$$dW = \sum_{k=1}^{3N} F_k(x)dx_k \qquad (1.223)$$

for the work done in an inertial Cartesian frame on a system of N particles by the Cartesian force components $F_k(x)$. This differential form may or may not be integrable. The question of integrability is important since it relates to the existence of a potential energy function. If integrable, there exists a potential energy function of the form $V(x)$.

1.4 Work, energy and momentum

With the introduction of generalized coordinates and their use in specifying the kinematic constraints on dynamical systems, we need to consider an expanded, generalized view of work, energy, and momentum.

Virtual displacements

Suppose that the vector $\mathbf{r}_i(q, t)$ gives the location of some point in a mechanical system; for example, it might be the position vector of the ith particle written in terms of the n qs and time. Now consider an *actual* differential displacement

$$d\mathbf{r}_i = \sum_{j=1}^{n} \frac{\partial \mathbf{r}_i}{\partial q_j} dq_j + \frac{\partial \mathbf{r}_i}{\partial t} dt \tag{1.224}$$

which occurs during an infinitesimal time interval dt. If there are m holonomic constraint equations, the dqs must satisfy

$$d\phi_j = \sum_{i=1}^{n} \frac{\partial \phi_j}{\partial q_i} dq_i + \frac{\partial \phi_j}{\partial t} dt = 0 \qquad (j = 1, \ldots, m) \tag{1.225}$$

On the other hand, if there are m nonholonomic constraints, the dqs satisfy

$$\sum_{i=1}^{n} a_{ji}(q, t) \, dq_i + a_{jt}(q, t) \, dt = 0 \qquad (j = 1, \ldots, m) \tag{1.226}$$

as given previously in (1.204).

Now let us hold time fixed by setting $dt = 0$ and imagine a *virtual displacement* $\delta \mathbf{r}_i$ in ordinary three-dimensional space for each of N particles.

$$\delta \mathbf{r}_i = \sum_{j=1}^{n} \frac{\partial \mathbf{r}_i}{\partial q_i} \delta q_i \qquad (i = 1, \ldots, N) \tag{1.227}$$

The virtual displacement of the system can also be described by the n-dimensional vector $\delta \mathbf{q}$ in configuration space. If there are constraints acting on the system, the δqs must satisfy *instantaneous* or *virtual constraint equations* of the form

$$\sum_{i=1}^{n} \frac{\partial \phi_j}{\partial q_i} \delta q_i = 0 \qquad (j = 1, \ldots, m) \tag{1.228}$$

for the holonomic case, or

$$\sum_{i=1}^{n} a_{ji} \, \delta q_i = 0 \qquad (j = 1, \ldots, m) \tag{1.229}$$

for nonholonomic constraints.

A comparison of (1.228) and (1.229) with (1.225) and (1.226) shows that virtual displacements and actual displacements are different, in general, since they satisfy different constraint equations. If the constraints are *catastatic*, however; that is, if any holonomic constraints are of the form $\phi_j(q) = 0$, and if all nonholonomic constraints have $a_{jt} = 0$, then the virtual and actual small displacements satisfy the same set of constraint equations. Of course, the actual motion also satisfies the equations of motion.

The forms of (1.228) and (1.229), which are linear in the δqs, indicate that the virtual displacements lie in an $(n - m)$-dimensional hyperplane at the operating point in n-dimensional configuration space, in accordance with the constraints. For holonomic constraints, the plane is tangent to the constraint surface at the operating point.

Virtual work

The concept of virtual work is fundamental to a proper understanding of dynamical theory. First, it must be emphasized that there is a distinction between work and virtual work. The work done by a force $\mathbf{F}_i$ acting on the ith particle as it moves between points A_i and B_i in an inertial frame is equal to the line integral

$$W_i = \int_{A_i}^{B_i} \mathbf{F}_i \cdot d\mathbf{r}_i \tag{1.230}$$

where $\mathbf{r}_i$ is the position vector of the ith particle. For a system of N particles, the work done in an arbitrary small displacement of the system is

$$dW = \sum_{i=1}^{N} \mathbf{F}_i \cdot d\mathbf{r}_i = \sum_{k=1}^{3N} F_k dx_k \tag{1.231}$$

where $(x_1, \ldots, x_{3N})$ are the Cartesian coordinates of the N particles and the F_k are the corresponding force components applied to the particles.

Now let us transform to generalized coordinates using (1.198) and (1.224). The work done on the system during a small displacement in the time interval dt is

$$dW = \sum_{i=1}^{N} \sum_{j=1}^{n} \mathbf{F}_i \cdot \frac{\partial \mathbf{r}_i}{\partial q_j} dq_j + \sum_{i=1}^{N} \mathbf{F}_i \cdot \frac{\partial \mathbf{r}_i}{\partial t} dt \tag{1.232}$$

or, in terms of Cartesian coordinates,

$$dW = \sum_{k=1}^{3N} \sum_{j=1}^{n} F_k \frac{\partial x_k}{\partial q_j} dq_j + \sum_{k=1}^{3N} F_k \frac{\partial x_k}{\partial t} dt \tag{1.233}$$

At this point it is convenient to introduce the *velocity coefficients* γ_{ij} and γ_{it} defined by

$$\gamma_{ij} = \frac{\partial \mathbf{r}_i}{\partial q_j} = \frac{\partial \dot{\mathbf{r}}_i}{\partial \dot{q}_j}, \qquad \gamma_{it} = \frac{\partial \mathbf{r}_i}{\partial t} \tag{1.234}$$

Note that these coefficients are vector quantities. Now we can write (1.224) in the form

$$d\mathbf{r}_i = \sum_{j=1}^{n} \gamma_{ij} dq_i + \gamma_{it} dt \tag{1.235}$$

The corresponding velocity is

$$\mathbf{v}_i = \sum_{j=1}^{n} \gamma_{ij} \dot{q}_j + \gamma_{it} \tag{1.236}$$

We see that γ_{ij} represents the sensitivity of the velocity $\mathbf{v}_i$ to changes in $\dot{q}_i$, whereas γ_{it} is equal to the velocity $\mathbf{v}_i$ when all qs are held constant.

Returning now to (1.232), we can write

$$dW = \sum_{i=1}^{N} \sum_{j=1}^{n} \mathbf{F}_i \cdot \boldsymbol{\gamma}_{ij} \, dq_j + \sum_{i=1}^{N} \mathbf{F}_i \cdot \boldsymbol{\gamma}_{it} \, dt \tag{1.237}$$

The *virtual work* δW due to the forces $\mathbf{F}_i$ acting on the system is obtained by setting $dt = 0$ and replacing the actual displacements $d\mathbf{r}_i$ by virtual displacements $\delta\mathbf{r}_i$. Thus we obtain the alternate forms

$$\delta W = \sum_{i=1}^{N} \mathbf{F}_i \cdot \delta\mathbf{r}_i = \sum_{k=1}^{3N} F_k \delta x_k \tag{1.238}$$

or

$$\delta W = \sum_{i=1}^{N} \sum_{j=1}^{n} \mathbf{F}_i \cdot \boldsymbol{\gamma}_{ij} \, \delta q_j = \sum_{k=1}^{3N} \sum_{j=1}^{n} F_k \frac{\partial x_k}{\partial q_j} \, \delta q_j \tag{1.239}$$

Let us define the *generalized force* Q_j associated with q_j by

$$Q_j = \sum_{i=1}^{N} \mathbf{F}_i \cdot \boldsymbol{\gamma}_{ij} = \sum_{k=1}^{3N} F_k \frac{\partial x_k}{\partial q_j} \tag{1.240}$$

Then the virtual work can be written in the form

$$\delta W = \sum_{j=1}^{n} Q_j \, \delta q_j \tag{1.241}$$

In general, we assume that the virtual displacements are consistent with any constraints, that is, they satisfy (1.228) or (1.229). But, if the system is holonomic, it is particularly convenient to choose independent δqs.

The question arises concerning why the virtual work δW receives so much attention in dynamical theory rather than the work dW of the actual motion. The reason lies in the nature of constraint forces. An *ideal constraint* is a *workless constraint* which may be either scleronomic or rheonomic. By workless we mean that no work is done by the constraint forces in an arbitrary reversible virtual displacement that satisfies the virtual constraint equations having the form of (1.228) or (1.229). Examples of ideal constraints include frictionless constraint surfaces, or rolling contact without slipping, or a rigid massless rod connecting two particles. Another example is a knife-edge constraint that allows motion in the direction of the knife edge without friction, but does not allow motion perpendicular to the knife edge. Ideal constraint forces, such as the internal forces in a rigid body, may do work on individual particles due to a virtual displacement, but no work is done on the system as a whole because these forces occur in equal, opposite and collinear pairs.

It is convenient to consider the total force acting on the ith particle to be the sum of the *applied force* $\mathbf{F}_i$ and the *constraint force* $\mathbf{R}_i$, by which we mean an ideal constraint force. Thus, all forces that are not ideal constraint forces are classed as applied forces. Frequently the applied forces are known, but the constraint forces either are unknown or are difficult to calculate.

The advantage of using virtual displacements rather than actual displacements in dynamical analyses can be seen by considering the virtual work of all the forces acting on a system of particles. We find that

$$\delta W = \sum_{i=1}^{N} (\mathbf{F}_i + \mathbf{R}_i) \cdot \delta \mathbf{r}_i = \sum_{i=1}^{N} \mathbf{F}_i \cdot \delta \mathbf{r}_i \qquad (1.242)$$

since

$$\sum_{i=1}^{N} \mathbf{R}_i \cdot \delta \mathbf{r}_i = 0 \qquad (1.243)$$

Thus, ideal constraint forces can be ignored in calculating the virtual work of all the forces acting on a system. On the other hand,

$$\sum_{i=1}^{N} \mathbf{R}_i \cdot d\mathbf{r}_i \neq 0 \qquad (1.244)$$

in the general case, indicating that constraint forces can contribute to the work dW resulting from a small actual displacement. To summarize, one can ignore the constraint forces in applying virtual work methods. This advantage will carry over to equations derived using virtual work, an example being Lagrange's equation.

Example 1.10   In this example we will show how a generalized force can be calculated using virtual work. In general, the generalized force Q_i is equal to the virtual work per unit δq_i, assuming that the other δqs are set equal to zero, that is, assuming independent δqs. This is in accordance with (1.241).

Consider a system (Fig. 1.20) consisting of two particles connected by a rigid rod of length L. Let (x, y) be the position of particle 1, and let θ be the angle of the rod relative

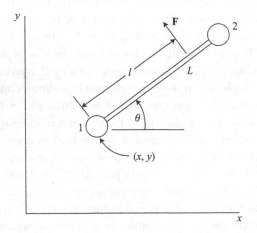

Figure 1.20.

to the x-axis. A force $\mathbf{F}$, perpendicular to the rod, is applied at a distance l from particle 1. We wish to solve for the generalized forces Q_x, Q_y, and Q_θ.

First, we see that

$$\mathbf{F} = -F \sin\theta \, \mathbf{i} + F \cos\theta \, \mathbf{j} \tag{1.245}$$

where $\mathbf{i}$ and $\mathbf{j}$ are Cartesian unit vectors. The virtual displacement $\delta\mathbf{r}$ at the point of application of $\mathbf{F}$ is

$$\delta\mathbf{r} = (\delta x - l \sin\theta \, \delta\theta)\mathbf{i} + (\delta y + l \cos\theta \, \delta\theta)\mathbf{j} \tag{1.246}$$

Thus, the virtual work is

$$\delta W = \mathbf{F} \cdot \delta\mathbf{r} = -F \sin\theta \, \delta x + F \cos\theta \, \delta y + Fl \, \delta\theta \tag{1.247}$$

From (1.241) we have

$$\delta W = Q_x \, \delta x + Q_y \, \delta y + Q_\theta \, \delta\theta \tag{1.248}$$

Then, by comparing coefficients of the δqs, we find that the generalized forces are

$$Q_x = -F \sin\theta, \qquad Q_y = F \cos\theta, \qquad Q_\theta = Fl \tag{1.249}$$

Note that Q_x and Q_y are the x and y components of $\mathbf{F}$, whereas Q_θ is the moment about particle 1.

Principle of virtual work

A system of particles is in static equilibrium if each particle of the system is in static equilibrium. A particle is in static equilibrium if it is motionless at the initial time $t = 0$, and if its acceleration remains zero for all $t \geq 0$.

Now consider a *catastatic system* of particles; that is, all transformation equations from inertial xs to qs do not contain time explicitly. This implies that all particles are at rest if all $\dot{q}$s equal zero. For such a system we can state the *principle of virtual work: The necessary and sufficient condition for the static equilibrium of an initially motionless catastatic system which is subject to ideal bilateral constraints is that zero virtual work is done by the applied forces in moving through an arbitrary virtual displacement satisfying the constraints.*

To explain this principle, first assume that the catastatic system is in static equilibrium. Then

$$\mathbf{F}_i + \mathbf{R}_i = 0 \qquad (i = 1, \ldots, N) \tag{1.250}$$

implying that each particle has zero acceleration. Now take the dot product with $\delta\mathbf{r}_i$ and sum over i. We obtain

$$\sum_{i=1}^{N} (\mathbf{F}_i + \mathbf{R}_i) \cdot \delta\mathbf{r}_i = 0 \tag{1.251}$$

and we assume that the δrs satisfy the constraints. The virtual work of the constraint forces equals zero, that is,

$$\sum_{i=1}^{N} \mathbf{R}_i \cdot \delta \mathbf{r}_i = 0 \tag{1.252}$$

Hence we find that

$$\delta W = \sum_{i=1}^{N} \mathbf{F}_i \cdot \delta \mathbf{r}_i = 0 \tag{1.253}$$

if the system is in static equilibrium. This is the necessary condition.

Now suppose that the system is not in static equilibrium, implying that $\mathbf{F}_i + \mathbf{R}_i \neq 0$ for at least one particle. Then it is always possible to find a virtual displacement such that

$$\sum_{i=1}^{N} (\mathbf{F}_i + \mathbf{R}_i) \cdot \delta \mathbf{r}_i \neq 0 \tag{1.254}$$

since sufficient degrees of freedom remain. Using (1.252) we conclude that a virtual displacement can always be found that results in $\delta W \neq 0$ if the system is not in static equilibrium. Thus, if $\delta W = 0$ for all possible δrs, the system must be in static equilibrium; this is the sufficient condition.

We have assumed a catastatic system. It is possible that a particular system that is not catastatic could, nevertheless, have a position of static equilibrium if a_{jt} and $\partial x_k / \partial t$ are not both identically zero, but are equal to zero at the position of static equilibrium. This is a rare situation, however.

Kinetic energy

Earlier we found that the kinetic energy of a system of N particles can be expressed in the form

$$T = \frac{1}{2} \sum_{i=1}^{N} m_i v_i^2 = \frac{1}{2} \sum_{k=1}^{3N} m_k \dot{x}_k^2 \tag{1.255}$$

where (x_1, x_2, x_3) is the inertial Cartesian position of the first particle whose mass is $m_1 = m_2 = m_3$, and similar notation is used to indicate the position and mass of each of the other particles. We wish to express the same kinetic energy in terms of qs, $\dot{q}$s, and possibly time. To accomplish this, we use the transformation equation (1.198) to obtain

$$\dot{x}_k = \sum_{i=1}^{n} \frac{\partial x_k}{\partial q_i} \dot{q}_i + \frac{\partial x_k}{\partial t} \qquad (k = 1, \ldots, 3N) \tag{1.256}$$

Then we find that the kinetic energy is

$$T(q, \dot{q}, t) = \frac{1}{2} \sum_{k=1}^{3N} m_k \left(\sum_{i=1}^{n} \frac{\partial x_k}{\partial q_i} \dot{q}_i + \frac{\partial x_k}{\partial t} \right)^2 \tag{1.257}$$

Let us express this kinetic energy in terms of homogeneous functions of the $\dot{q}$s, that is, separating the various powers of $\dot{q}_i$. We can write

$$T = T_2 + T_1 + T_0 \tag{1.258}$$

where the quadratic portion is

$$T_2 = \frac{1}{2} \sum_{i=1}^{n} \sum_{j=1}^{n} m_{ij} \, \dot{q}_i \dot{q}_j \tag{1.259}$$

The mass coefficients m_{ij} are

$$m_{ij}(q, t) = m_{ji} = \sum_{k=1}^{3N} m_k \frac{\partial x_k}{\partial q_i} \frac{\partial x_k}{\partial q_j} \qquad (i, j = 1, \ldots, n) \tag{1.260}$$

Note that

$$m_{ij} = \frac{\partial^2 T}{\partial \dot{q}_i \, \partial \dot{q}_j} \tag{1.261}$$

Similarly, the portion that is linear in the $\dot{q}$s is

$$T_1 = \sum_{i=1}^{n} a_i \dot{q}_i \tag{1.262}$$

where

$$a_i(q, t) = \sum_{i=1}^{3N} m_k \frac{\partial x_k}{\partial q_i} \frac{\partial x_k}{\partial t} \qquad (i = 1, \ldots, n) \tag{1.263}$$

Finally, that portion of the kinetic energy which is not a function of the $\dot{q}$s is

$$T_0(q, t) = \frac{1}{2} \sum_{k=1}^{3N} m_k \left(\frac{\partial x_k}{\partial t} \right)^2 \tag{1.264}$$

For a scleronomic or catastatic system, $\partial x_k / \partial t = 0$, so both T_1 and T_0 vanish and $T = T_2$.

Now recall from (1.236) that the velocity of the kth particle, expressd in terms of velocity coefficients is

$$\mathbf{v}_k = \sum_{i=1}^{n} \gamma_{ki}(q, t) \, \dot{q}_i + \gamma_{kt}(q, t) \tag{1.265}$$

where

$$\gamma_{ki} = \frac{\partial \mathbf{v}_k}{\partial \dot{q}_i}, \qquad \gamma_{kt} = \frac{\partial \mathbf{r}_k}{\partial t} \tag{1.266}$$

Thus, the kinetic energy is

$$T = \frac{1}{2} \sum_{k=1}^{N} m_k \left(\sum_{i=1}^{n} \gamma_{ki} \, \dot{q}_i + \gamma_{kt} \right)^2 \tag{1.267}$$

and we find that

$$m_{ij} = m_{ji} = \sum_{k=1}^{N} m_k \gamma_{ki} \cdot \gamma_{kj} \tag{1.268}$$

$$a_i = \sum_{k=1}^{N} m_k \gamma_{ki} \cdot \gamma_{kt} \tag{1.269}$$

$$T_0 = \frac{1}{2} \sum_{k=1}^{N} m_k \gamma_{kt}^2 \tag{1.270}$$

Note that $\gamma_{kt} \equiv 0$ for a scleronomic or catastatic system.

Potential energy

For a system of N particles whose configuration is given in terms of $3N$ Cartesian coordinates, the force F_k obtained from a potential energy function $V(x, t)$ is

$$F_k = -\frac{\partial V}{\partial x_k} \tag{1.271}$$

in accordance with (1.144). The virtual work due to these applied forces is

$$\delta W = \sum_{k=1}^{3N} F_k \, \delta x_k = -\sum_{k=1}^{3N} \frac{\partial V}{\partial x_k} \delta x_k = -\delta V \tag{1.272}$$

Now transform from xs to qs using (1.198). The potential energy has the form $V(q, t)$ and we obtain

$$\delta W = -\delta V = -\sum_{j=1}^{n} \frac{\partial V}{\partial q_j} \delta q_j \tag{1.273}$$

In general, we know from (1.241) that

$$\delta W = \sum_{j=1}^{n} Q_j \, \delta q_j \tag{1.274}$$

Upon comparing coefficients of the δqs, we find that the generalized force Q_j due to $V(q, t)$ is

$$Q_j = -\frac{\partial V}{\partial q_j} \tag{1.275}$$

For the particular case in which $V = V(q)$ and all the applied forces are obtained using (1.275), the total energy $T + V$ is conserved.

Finally, from (1.273) and the principle of virtual work we see that an initially motionless conservative system with bilateral constraints is in static equilibrium if and only if

$$\delta V = 0 \tag{1.276}$$

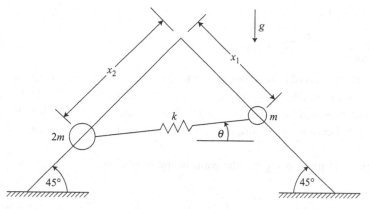

Figure 1.21.

for all virtual displacements consistent with the constraints. Any such static equilibrium configuration is *stable* if it occurs at a local minimum of the potential energy, considered as a function of the qs.

Example 1.11 Particles of mass m and $2m$ can slide freely on two rigid rods inclined at 45° with the horizontal (Fig. 1.21). They are connected by a linear spring of stiffness k. We wish to solve for the inclination θ of the spring and its tensile force F at the position of static equilibrium.

Let us use the principle of virtual work. The applied forces acting on the system are the spring force F and those due to gravity. The virtual work is

$$\delta W = \left[\frac{mg}{\sqrt{2}} - F\cos\left(\frac{\pi}{4} + \theta\right)\right]\delta x_1 + \left[\sqrt{2}\,mg - F\cos\left(\frac{\pi}{4} - \theta\right)\right]\delta x_2 = 0 \qquad (1.277)$$

Since δx_1 and δx_2 are independent virtual displacements, their coefficients must each be zero. Thus we obtain

$$F = \frac{mg}{\sqrt{2}\cos\left(\frac{\pi}{4} + \theta\right)} = \frac{\sqrt{2}\,mg}{\cos\left(\frac{\pi}{4} - \theta\right)} \qquad (1.278)$$

Hence, at equilibrium,

$$\cos\left(\frac{\pi}{4} - \theta\right) = 2\cos\left(\frac{\pi}{4} + \theta\right) \qquad (1.279)$$

or

$$\cos\theta + \sin\theta = 2\left(\cos\theta - \sin\theta\right) \qquad (1.280)$$

Thus,

$$\tan\theta = \frac{1}{3} \quad \text{or} \quad \theta = 18.43° \qquad (1.281)$$

Then, from (1.278), the tensile force F is

$$F = \frac{2mg}{\cos\theta + \sin\theta} = \frac{\sqrt{10}mg}{2} = 1.5811mg \tag{1.282}$$

Notice that, at the equilibrium configuration, the angle θ and force F do not depend on the spring stiffness k. So let us consider the case in which k is infinite; that is, replace the spring by a massless rigid rod of length L. Then δx_1 and δx_2 are not longer independent. The force F becomes a constraint force which does not enter into the calculation of θ.

The principle of virtual work gives the equilibrium condition

$$\delta V = -\frac{mg}{\sqrt{2}}(\delta x_1 + 2\delta x_2) = 0 \tag{1.283}$$

The constraint relation between δx_1 and δx_2 can be found by noting that the length of the rod is unchanged during a virtual displacement.

$$\delta x_1 \cos\left(\frac{\pi}{4} + \theta\right) + \delta x_2 \cos\left(\frac{\pi}{4} - \theta\right) = 0$$

or

$$(\cos\theta - \sin\theta)\,\delta x_1 + (\cos\theta + \sin\theta)\,\delta x_2 = 0 \tag{1.284}$$

We need to express an arbitrary virtual displacement in terms of $\delta\theta$ which is unconstrained. To accomplish this, consider a small rotation of the system about its instantaneous center. This results in

$$\delta\theta = \frac{\sqrt{2}\,\delta x_2}{L\,(\cos\theta - \sin\theta)} = \frac{-\sqrt{2}\,\delta x_1}{L\,(\cos\theta + \sin\theta)} \tag{1.285}$$

Then the equilibrium condition of (1.283) can be written in the form

$$\delta V = \frac{mgL}{2}(3\sin\theta - \cos\theta)\,\delta\theta = 0 \tag{1.286}$$

since $\delta\theta \neq 0$, in general, we conclude that

$$\tan\theta = \frac{1}{3} \tag{1.287}$$

as we obtained previously. The expression for δV can be integrated to yield the potential energy.

$$V = -\frac{mgL}{2}(\sin\theta + 3\cos\theta) \tag{1.288}$$

Then we find that

$$\frac{\partial^2 V}{\partial\theta^2} = \frac{mgL}{2}(\sin\theta + 3\cos\theta) = \sqrt{2.5}mgL \tag{1.289}$$

Since this result is positive, the equilibrium is stable for the system consisting of two particles connected by a rigid rod.

Generalized momentum

The kinetic energy of a system of particles, written as a function of $(q, \dot{q}, t)$, has the form of (1.258) which, in detail, is

$$T = \frac{1}{2} \sum_{i=1}^{n} \sum_{j=1}^{n} m_{ij}(q, t)\, \dot{q}_i \dot{q}_j + \sum_{i=1}^{n} a_i(q, t)\, \dot{q}_i + T_0(q, t) \tag{1.290}$$

The *generalized momentum* associated with the generalized coordinate q_i is

$$p_i = \frac{\partial T}{\partial \dot{q}_i} \tag{1.291}$$

or

$$p_i = \sum_{j=1}^{n} m_{ij}\dot{q}_j + a_i \tag{1.292}$$

For a simple system such as a single particle whose position is given by the Cartesian coordinates (x, y), the x-component of generalized momentum is just the x-component of linear momentum $m\dot{x}$. On the other hand, if the position of the particle is given by the polar coordinates (r, θ), then p_θ is equal to $mr^2\dot{\theta}$, which is the angular momentum about the origin. For more general choices of coordinates, the generalized momentum may not have an easily discerned physical meaning. Because the generalized coordinates associated with a given system do not necessarily have the same units or dimensions, neither will the corresponding generalized momenta. However, the product $p_i \dot{q}_i$ will always have the units of energy.

Example 1.12 Two particles, each of mass m are connected by a rigid massless rod of length l to form a dumbbell that can move in the xy-plane. The position of the first particle is (x, y) and the direction of the second particle relative to the first is given by the angle θ (Fig. 1.22). We wish to find the kinetic energy and the generalized momenta.

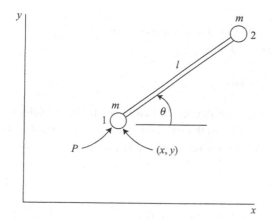

Figure 1.22.

Choose the first particle as the reference point P and use the general kinetic energy expression

$$T = \frac{1}{2}m\dot{\mathbf{r}}_p^2 + \frac{1}{2}\sum_{i=1}^{N} m_i\dot{\boldsymbol{\rho}}_i^2 + \dot{\mathbf{r}}_p \cdot m\dot{\boldsymbol{\rho}}_c \tag{1.293}$$

which is of the form

$$T = T_p + T_r + T_c \tag{1.294}$$

Here T_p, the kinetic energy due to the reference point motion, is

$$T_p = \frac{1}{2}m\dot{\mathbf{r}}_p^2 \tag{1.295}$$

where m is the total mass. The kinetic energy due to motion relative to the reference point is

$$T_r = \frac{1}{2}\sum_{i=1}^{N} m_i\dot{\boldsymbol{\rho}}_i^2 \tag{1.296}$$

Finally, the kinetic energy due to coupling between the two previous motions is

$$T_c = \dot{\mathbf{r}}_p \cdot m\dot{\boldsymbol{\rho}}_c \tag{1.297}$$

If these general equations are applied to the present system, we obtain

$$T = m(\dot{x}^2 + \dot{y}^2) + \frac{1}{2}ml^2\dot{\theta}^2 + ml\dot{\theta}(\dot{y}\cos\theta - \dot{x}\sin\theta) \tag{1.298}$$

Then, using (1.291), we find that

$$p_x = \frac{\partial T}{\partial \dot{x}} = 2m\dot{x} - ml\dot{\theta}\sin\theta \tag{1.299}$$

$$p_y = \frac{\partial T}{\partial \dot{y}} = 2m\dot{y} + ml\dot{\theta}\cos\theta \tag{1.300}$$

These are the x and y components of the total linear momentum.

The generalized momentum associated with θ is

$$p_\theta = \frac{\partial T}{\partial \dot{\theta}} = ml^2\dot{\theta} + ml(\dot{y}\cos\theta - \dot{x}\sin\theta) \tag{1.301}$$

This is equal to absolute angular momentum about P, that is, ml times the velocity component of particle 2 which is perpendicular to the rod. Note that p_θ is not equal to H_p as it is ordinarily defined. To obtain H_p we use the relative kinetic energy T_r.

$$H_p = \frac{\partial T_r}{\partial \dot{\theta}} = \frac{\partial}{\partial \dot{\theta}}\left(\frac{1}{2}ml^2\dot{\theta}^2\right) = ml^2\dot{\theta} \tag{1.302}$$

This is the angular momentum relative to the reference point P.

On the other hand, if we use (x_c, y_c, θ) as generalized coordinates, where (x_c, y_c) are the Cartesian coordinates of the center of mass, the total energy of the system is, in accordance with Koenig's theorem,

$$T = m\left(\dot{x}_c^2 + \dot{y}_c^2\right) + \frac{1}{4}ml^2\dot{\theta}^2 \tag{1.303}$$

The generalized momenta

$$p_x = \frac{\partial T}{\partial \dot{x}_c} = 2m\dot{x}_c, \qquad p_y = \frac{\partial T}{\partial \dot{y}_c} = 2m\dot{y}_c \tag{1.304}$$

are once again the x and y components of the total linear momentum. The third generalized momentum is

$$p_\theta = \frac{\partial T}{\partial \dot{\theta}} = \frac{1}{2}ml^2\dot{\theta} \tag{1.305}$$

This is the angular momentum about the center of mass.

1.5   Impulse response

Linear impulse and momentum

Consider a system of N particles whose position vectors $\mathbf{r}_i$ are measured relative to an inertial frame. Newton's law of motion for the system is

$$\dot{\mathbf{p}} = \mathbf{F} \tag{1.306}$$

where $\mathbf{F}$ is the total external force and $\mathbf{p}$ is the total linear momentum.

$$\mathbf{p} = \sum_{i=1}^{N} m_i \dot{\mathbf{r}}_i = m\dot{\mathbf{r}}_c \tag{1.307}$$

where m is the total mass and $\mathbf{r}_c$ is the position vector of the center of mass.

Now integrate (1.307) with respect to time over an interval t_1 to t_2. We obtain

$$\Delta\mathbf{p} = \mathbf{p}_2 - \mathbf{p}_1 = \hat{\mathbf{F}} \tag{1.308}$$

where the *linear impulse* $\hat{\mathbf{F}}$ is given by

$$\hat{\mathbf{F}} = \int_{t_1}^{t_2} \mathbf{F}\, dt \tag{1.309}$$

Equation (1.308) is a statement of the *principle of linear impulse and momentum: The change in the total linear momentum of a system of particles over a given time interval is equal to the total impulse of the external forces acting on the system.* Notice that the interval is arbitrary and not necessarily small.

Since a vector equation is involved in this principle, a similar scalar equation must apply to each component, provided that the component represents a fixed direction in inertial space.

If any component of the total impulse is zero, the change in the corresponding component of linear momentum is also zero. If **F** or one of its components is continuously equal to zero, the corresponding linear momentum is constant. In this case, there is *conservation of linear momentum*.

Angular impulse and momentum

Again consider a system of N particles and take a reference point P which is either (1) a fixed point, or (2) at the center of mass. The differential equation for the angular momentum has the simple form

$$\dot{\mathbf{H}} = \mathbf{M} \tag{1.310}$$

Integrating this equation with respect to time over the interval t_1 to t_2, we obtain

$$\Delta \mathbf{H} = \mathbf{H}_2 - \mathbf{H}_1 = \hat{\mathbf{M}} \tag{1.311}$$

where the angular impulse $\hat{\mathbf{M}}$ due to the external forces is

$$\hat{\mathbf{M}} = \int_{t_1}^{t_2} \mathbf{M} dt \tag{1.312}$$

Note that the internal forces between particles occur in equal, opposite and collinear pairs. Hence, they do not contribute to **M** or $\hat{\mathbf{M}}$.

Equation (1.311) is a statement of the *principle of angular impulse and momentum*. Note that $\Delta \mathbf{H}$ can be zero even if **M** is not continuously zero. On the other hand, if **M** is continuously zero, then $\hat{\mathbf{M}} = 0$ over any interval, and there is *conservation of angular momentum*. Similarly, if a certain component of **M** remains equal to zero, even though **M** itself is not zero, then there is conservation of that component of **H**. Here we assume that the direction of the component is fixed in inertial space.

Now let us choose an *arbitrary reference point* P for angular momentum. The equation of motion obtained earlier has the form

$$\dot{\mathbf{H}}_p = \mathbf{M}_p - \boldsymbol{\rho}_c \times m\ddot{\mathbf{r}}_p \tag{1.313}$$

Assume that very large forces are applied to the system of particles over an infinitesimal time $\Delta t = t_2 - t_1$ and integrate (1.313) with respect to time over this interval. We obtain

$$\int_{t_1}^{t_1 + \Delta t} \dot{\mathbf{H}}_p dt = \int_{t_1}^{t_1 + \Delta t} (\mathbf{M}_p - \boldsymbol{\rho}_c \times m\ddot{\mathbf{r}}_p) \, dt \tag{1.314}$$

or

$$\Delta \mathbf{H}_p = \hat{\mathbf{M}}_p - \boldsymbol{\rho}_c \times m\Delta\dot{\mathbf{r}}_p \tag{1.315}$$

During the interval Δt the configuration is unchanged but $\mathbf{M}_p$ and $\ddot{\mathbf{r}}_p$ may be very large, resulting in finite $\hat{\mathbf{M}}_p$ and $\Delta\dot{\mathbf{r}}_p$. But if the reference point P is chosen such that $\ddot{\mathbf{r}}_p$ is finite during the impulse, and therefore $\Delta\dot{\mathbf{r}}_p$ is zero, then the simpler form given in (1.311) applies.

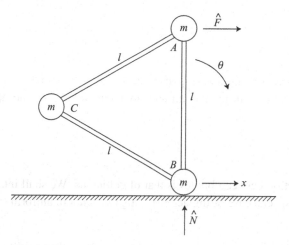

Figure 1.23.

Example 1.13 Three particles, each of mass m, are connected by rigid massless rods in the form of an equilateral triangle (Fig. 1.23). The system is motionless with particle A directly above particle B when a horizontal impulse $\hat{F}$ is suddenly applied to particle A, causing particle B to slide without friction on the horizontal floor. Solve for the values of $\dot{x}$ and $\dot{\theta}$ immediately after the impulse. Evaluate the constraint impulse $\hat{N}$.

First let us use the linear impulse and momentum principle in the x direction to obtain

$$3m\dot{x} + \frac{3}{2}ml\dot{\theta} = \hat{F} \tag{1.316}$$

where we use (x, θ) as generalized coordinates and consider the θ rotation to take place about particle B. If we choose a fixed point that is initially coincident with particle B as the reference point for angular momentum, we can find $\dot{x}$ and $\dot{\theta}$ without having to solve for the constraint impulse $\hat{N}$. Thus, we can write

$$\Delta \mathbf{H} = \hat{\mathbf{M}} \tag{1.317}$$

or

$$\frac{3}{2}ml\dot{x} + 2ml^2\dot{\theta} = \hat{F}l \tag{1.318}$$

Solving (1.316) and (1.318) we obtain

$$\dot{x} = \frac{2\hat{F}}{15m}, \qquad \dot{\theta} = \frac{2\hat{F}}{5ml} \tag{1.319}$$

To find the value of $\hat{N}$ we use linear impulse and momentum in the vertical direction, noting that C is the only particle with vertical motion. We obtain

$$\frac{\sqrt{3}}{2}ml\dot{\theta} = \hat{N} \tag{1.320}$$

giving

$$\hat{N} = \frac{\sqrt{3}}{5}\hat{F} \tag{1.321}$$

Gravitational forces don't affect the impulse response because they are finite and therefore result in a negligible impulse during the infinitesimal interval Δt of the applied impulse $\hat{F}$.

Collisions

An important mode of interaction between bodies is that of collisions. We shall introduce the theory by considering the special case of impact involving two smooth uniform spheres (Fig. 1.24).

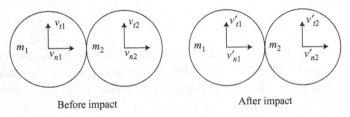

Before impact   After impact

Figure 1.24.

Assume that the spheres are moving in the plane of the figure just before impact. At the moment of impact, the interaction forces are normal to the tangent plane at the point of contact, and therefore are directed along the line of centers. Let us assume that these forces are impulsive in nature, that is, they are very large in magnitude and very short in duration. As a result, there can be sudden changes in translational velocities, but rotational motion is not involved.

Let v_{n1} and v_{n2} represent the velocity components along the line of centers just before impact, and assume that $v_{n1} > v_{n2}$. Let v_{t1} and v_{t2} be the corresponding tangential components; that is, they are perpendicular to the line of centers. Because the spheres are smooth, there are no tangential interaction forces, and therefore no changes in tangential velocities. So we obtain

$$v'_{t1} = v_{t1} \tag{1.322}$$

$$v'_{t2} = v_{t2} \tag{1.323}$$

which implies the conservation of tangential momentum of each particle.

The interaction impulses along the line of centers are equal and opposite, so there is conservation of the total normal momentum. Thus we have

$$m_1 v'_{n1} + m_2 v'_{n2} = m_1 v_{n1} + m_2 v_{n2} \tag{1.324}$$

Now let us introduce the *coefficient of restitution e* in accordance with the equation

$$v'_{n2} - v'_{n1} = e(v_{n1} - v_{n2}) \qquad (0 \le e \le 1) \tag{1.325}$$

In other words, considering normal components, the relative velocity of separation after impact is equal to e times the relative velocity of approach before impact. Solving (1.324) and (1.325) for v'_{n1} and v'_{n2}, we obtain the final normal velocity components.

$$v'_{n1} = \frac{m_1 - em_2}{m_1 + m_2} v_{n1} + \frac{(1+e)m_2}{m_1 + m_2} v_{n2} \qquad (1.326)$$

$$v'_{n2} = \frac{(1+e)m_1}{m_1 + m_2} v_{n1} + \frac{m_2 - em_1}{m_1 + m_2} v_{n2} \qquad (1.327)$$

If $e = 0$, termed *inelastic impact*, we see that $v'_{n1} = v'_{n2}$ and the normal separation velocity is zero. On the other hand, if $e = 1$, we have *perfectly elastic impact*, and the normal relative velocities of approach and separation are equal. There is conservation of energy during perfectly elastic impact, but there is some energy loss for $0 \le e < 1$.

During the impact of two smooth spheres, there are equal and opposite impulses acting on the spheres along the line of centers. These impulses are of magnitude

$$\hat{F} = m_1(v_{n1} - v'_{n1}) = m_2(v'_{n2} - v_{n2}) \qquad (1.328)$$

Now suppose that Coulomb friction with a coefficient μ is introduced at the impacting surfaces. If sliding occurs throughout the impact, there will be equal and opposite friction impulses $\mu\hat{F}$ acting to oppose the relative sliding motion. In other words, the total interaction impulse of each body will lie along a cone of friction. It may turn out, however, that the magnitude $\mu\hat{F}$ of the calculated friction impulse is larger than that required to stop the relative sliding motion. In this case, the relative sliding motion will not reverse, but will stop at zero with the friction impulse reduced accordingly.

We have assumed that the initial and final velocity vectors, as well as the line of centers, all lie in the same plane. If this is not true, one can choose an inertial frame translating with sphere 2 just before impact as the reference frame. Then, relative to this second reference frame, v_{n2} and v_{t2} are both equal to zero. The relative velocity vector of particle 1 and the line of centers determine the plane in which the action takes place and for which the standard equations such as (1.326) and (1.327) apply. After solving for the primed velocity vectors relative to the second reference frame, one can finally transform to the original inertial frame by adding vectorially the velocity of this second frame.

Example 1.14 A dumbbell is formed of two particles, each of mass m, which are connected by a rigid massless rod of length l (Fig. 1.25). The dumbbell, with $\theta = 45°$ and $\dot{\theta} = 0$, is falling vertically downwards with velocity v_0 when it has inelastic impact with a smooth horizontal surface. We wish to find the angular velocity $\dot{\theta}$ and the velocity v_1 of particle 1 immediately after impact.

First method Let us take the reference point P fixed in particle 1 and use (1.315). Any impulse acting on the system must be applied at P, so

$$\hat{\mathbf{M}}_p = 0 \qquad (1.329)$$

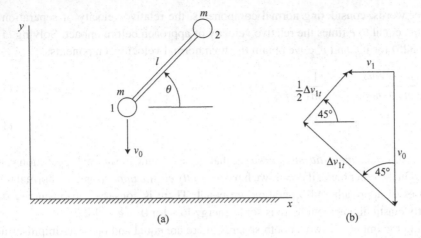

Figure 1.25.

The initial angular momentum about P is zero since $\dot{\theta} = 0$ before impact. The final angular momentum is

$$\Delta \mathbf{H}_p = ml^2 \dot{\theta} \mathbf{k} \tag{1.330}$$

where $\mathbf{k}$ is a unit vector in the z direction.

The impulse applied to particle 1 by the surface must be vertical since the horizontal surface is frictionless. Hence there is no horizontal impulse applied to the system, and its linear momentum in the x direction remains equal to zero. This implies that the x-components of velocity of particles 1 and 2 are equal and opposite. Hence the velocity of particle 1 immediately after impact is

$$\mathbf{v}_1 = \frac{l\dot{\theta}}{2\sqrt{2}} \mathbf{i} \tag{1.331}$$

The vertical velocity of particle 1 after impact is zero because of the assumption of inelastic impact. Thus, we find that change in velocity of particle 1 is

$$\Delta \dot{\mathbf{r}}_p = \frac{l\dot{\theta}}{2\sqrt{2}} \mathbf{i} + v_0 \mathbf{j} \tag{1.332}$$

Note that the horizontal velocity component of the center of mass is continuously zero.

The position of the center of mass relative to the reference point P at particle 1 is

$$\rho_c = \frac{l}{2\sqrt{2}} (\mathbf{i} + \mathbf{j}) \tag{1.333}$$

The total mass of the system is $2m$, so the term $\rho_c \times m \Delta \dot{\mathbf{r}}_p$ in (1.315) is given by

$$\rho_c \times m \Delta \dot{\mathbf{r}}_p = 2m \left(\frac{l v_0}{2\sqrt{2}} - \frac{l^2 \dot{\theta}}{8} \right) \mathbf{k} \tag{1.334}$$

Then substitution into (1.315) yields

$$\frac{3}{4}ml^2\ddot{\theta} = -\frac{mlv_0}{\sqrt{2}} \tag{1.335}$$

resulting in the angular velocity

$$\dot{\theta} = -\frac{2\sqrt{2}\,v_0}{3l} \tag{1.336}$$

The velocity of particle 1, from (1.331), is

$$\mathbf{v}_1 = -\frac{v_0}{3}\,\mathbf{i} \tag{1.337}$$

Second method   Let the reference point P be inertially fixed at the impact point on the surface. Any external impulse on the system acts through P so

$$\hat{\mathbf{M}}_p = 0 \tag{1.338}$$

and we have conservation of momentum about P.

The angular momentum before impact is due entirely to translation of particle 2.

$$\mathbf{H}_1 = -\frac{mlv_0}{\sqrt{2}}\,\mathbf{k} \tag{1.339}$$

After impact, the center of mass moves vertically downward, while particle 1 moves horizontally away from P due to the inelastic impact assumption. Thus, from the geometry, the velocity of particle 2 immediately after impact is

$$\mathbf{v}_2 = \frac{l\dot{\theta}}{\sqrt{2}}\left(-\frac{1}{2}\,\mathbf{i} + \mathbf{j}\right) \tag{1.340}$$

The resulting total angular momentum about P is

$$\mathbf{H}_2 = \rho_2 \times m\mathbf{v}_2 \tag{1.341}$$

where

$$\rho_2 = \frac{l}{\sqrt{2}}\,(\mathbf{i} + \mathbf{j}) \tag{1.342}$$

Thus we obtain

$$\mathbf{H}_2 = \frac{3}{4}ml^2\dot{\theta}\mathbf{k} \tag{1.343}$$

Finally, setting $\mathbf{H}_1 = \mathbf{H}_2$, we find that

$$\dot{\theta} = -\frac{2\sqrt{2}\,v_0}{3l} \tag{1.344}$$

in agreement with (1.336). The center of mass has no horizontal motion, so $\mathbf{v}_1$ is the negative of the x component of $\mathbf{v}_2$.

$$\mathbf{v}_1 = \frac{l\dot{\theta}}{2\sqrt{2}}\,\mathbf{i} = -\frac{v_0}{3}\,\mathbf{i} \tag{1.345}$$

Third method   Consider the effect of the vertical reaction impulse $\hat{R}$ acting on particle 1 and due to its impact with the fixed frictionless surface. Take longitudinal and transverse components of $\hat{\mathbf{R}}$, relative to the dumbbell, where each component is of magnitude $\hat{R}/\sqrt{2}$. The transverse impulse results in a sudden change Δv_{1t} in the transverse velocity of particle 1, as shown in Fig. 1.25b.

$$\Delta v_{1t} = \frac{\hat{R}}{\sqrt{2}m} \tag{1.346}$$

The velocity of particle 2 is not changed by this transverse impulse. On the other hand, the longitudinal component of $\hat{\mathbf{R}}$ affects both particles having a total mass $2m$.

$$\Delta v_{1l} = \Delta v_{2l} = \frac{\hat{R}}{2\sqrt{2}m} = \frac{1}{2}\Delta v_{1t} \tag{1.347}$$

The final velocity of particle 1 is equal to its initial velocity plus the changes due to the reaction impulse $\hat{\mathbf{R}}$, as shown in Fig. 1.25b. From the vertical components we obtain

$$\Delta v_{1t} = \frac{2\sqrt{2}}{3}v_0 \tag{1.348}$$

Then, from the horizontal components we obtain

$$\mathbf{v}_1 = -\frac{\Delta v_{1t}}{2\sqrt{2}}\,\mathbf{i} = -\frac{v_0}{3}\,\mathbf{i} \tag{1.349}$$

The angular velocity $\dot{\theta}$ is zero before impact. During impact, the only impulse acting on particle 2 is longitudinal in direction, so its transverse velocity is unchanged. Therefore, the angular velocity immediately after impact is

$$\dot{\theta} = -\frac{\Delta v_{1t}}{l} = -\frac{2\sqrt{2}\,v_0}{3l} \tag{1.350}$$

The most straightforward method discussed here is the second method. It uses conservation of angular momentum about a fixed point chosen so that the applied angular impulse equals zero.

Generalized impulse and momentum

The configuration of a system of N particles is given by the position vectors $\mathbf{r}_k(q, t)$. The corresponding velocities are

$$\mathbf{v}_k = \frac{d\mathbf{r}_k}{dt} = \sum_{j=1}^{n} \frac{\partial \mathbf{r}_k}{\partial q_j}\,\dot{q}_j + \frac{\partial \mathbf{r}_k}{\partial t} \qquad (k = 1, \ldots, N) \tag{1.351}$$

Recall from (1.234) that the velocity coefficients are

$$\gamma_{kj} = \frac{\partial \mathbf{r}_k}{\partial q_j}, \qquad \gamma_{kt} = \frac{\partial \mathbf{r}_k}{\partial t} \tag{1.352}$$

Thus we obtain

$$\mathbf{v}_k = \sum_{j=1}^{n} \gamma_{kj} \dot{q}_j + \gamma_{kt} \quad (k = 1, \ldots, N) \tag{1.353}$$

The equation of linear momentum for the kth particle is

$$m_k \, \Delta \mathbf{v}_k = \hat{\mathbf{F}}_k \tag{1.354}$$

where $\hat{\mathbf{F}}_k$ is the total impulse acting on the particle. Now take the dot product with γ_{ki} and sum over the particles to obtain

$$\sum_{k=1}^{N} m_k \Delta \mathbf{v}_k \cdot \gamma_{ki} = \sum_{k=1}^{N} \hat{\mathbf{F}}_k \cdot \gamma_{ki} \tag{1.355}$$

Let us assume that all the $\hat{\mathbf{F}}_k$ impulses occur during an infinitesimal interval Δt, implying that the configuration and the values of the velocity coefficients remain constant during this interval. From (1.353) we see that

$$\Delta \mathbf{v}_k = \sum_{j=1}^{n} \gamma_{kj} \, \Delta \dot{q}_j \tag{1.356}$$

Substitute this expression for $\Delta \mathbf{v}_k$ into (1.355) and obtain

$$\sum_{j=1}^{n} \sum_{k=1}^{N} m_k \gamma_{ki} \cdot \gamma_{kj} \, \Delta \dot{q}_j = \sum_{k=1}^{N} \hat{\mathbf{F}}_k \cdot \gamma_{ki} \tag{1.357}$$

Recall from (1.268) that the generalized mass coefficient m_{ij} can be expressed as

$$m_{ij} = m_{ji} = \sum_{k=1}^{N} m_k \gamma_{ki} \cdot \gamma_{kj} \tag{1.358}$$

Furthermore, similar to (1.240), the generalized impulse $\hat{Q}_i$ is given by

$$\hat{Q}_i = \sum_{k=1}^{N} \hat{\mathbf{F}}_k \cdot \gamma_{ki} \quad (i = 1, \ldots, n) \tag{1.359}$$

Finally, we obtain the *equation of generalized impulse and momentum*:

$$\sum_{j=1}^{n} m_{ij} (q, t) \, \Delta \dot{q}_j = \hat{Q}_i \quad (i = 1, \ldots, n) \tag{1.360}$$

In general, $\hat{Q}_i$ includes the contributions of applied and constraint impulses, both internal and external to the system. However, if the qs are *independent*, the $\hat{Q}$s are due to the applied impulses only.

Example 1.15 Two particles with masses m_1 and m_2 are connected by a rigid massless rod of length l. Initially the system is motionless with a vertical orientation, as shown in Fig. 1.26. Then a horizontal impulse $\hat{\mathbf{F}}$ is applied to the upper particle of mass m_1. We wish to solve for the velocity $\dot{x}$ and the angular velocity $\dot{\theta}$ immediately after the impulse.

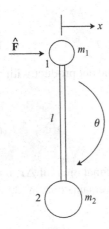

Figure 1.26.

At the outset we see that the center of mass must move horizontally due to $\hat{\mathbf{F}}$, and therefore neither particle can have any vertical velocity initially. Let us choose (x, θ) as generalized coordinates consistent with this motion. The kinetic energy at this time is

$$T = \frac{1}{2}m_1\dot{x}^2 + \frac{1}{2}m_2(\dot{x} - l\dot{\theta})^2 \tag{1.361}$$

From (1.261), the mass coefficients are

$$m_{xx} = \frac{\partial^2 T}{\partial \dot{x}^2} = m_1 + m_2$$

$$m_{x\theta} = m_{\theta x} = \frac{\partial^2 T}{\partial \dot{x}\partial \dot{\theta}} = -m_2 l$$

$$m_{\theta\theta} = \frac{\partial^2 T}{\partial \dot{\theta}^2} = m_2 l^2 \tag{1.362}$$

The velocity coefficients for particle 1 are

$$\gamma_{1x} = \mathbf{i}, \qquad \gamma_{1\theta} = 0 \tag{1.363}$$

where $\mathbf{i}$ is a horizontal unit vector. Then (1.359) gives the generalized impulses

$$\hat{Q}_x = \hat{F}, \qquad \hat{Q}_\theta = 0 \tag{1.364}$$

A substitution into the equation of generalized impulse and momentum yields the following equations:

$$(m_1 + m_2)\Delta\dot{x} - m_2 l \Delta\dot{\theta} = \hat{F} \tag{1.365}$$
$$-m_2 l \Delta\dot{x} + m_2 l^2 \Delta\dot{\theta} = 0 \tag{1.366}$$

The solutions are

$$\Delta\dot{x} = \frac{\hat{F}}{m_1}, \qquad \Delta\dot{\theta} = \frac{\hat{F}}{m_1 l} \tag{1.367}$$

which are the values of $\dot{x}$ and $\dot{\theta}$ immediately after the impulse. Note that these results are independent of the mass m_2 of the lower particle which remains momentarily motionless.

Impulse and energy

Consider a particle of mass m and velocity $\mathbf{v}_0$ which is subject to a total impulse $\hat{\mathbf{F}}$. From the principle of linear impulse and momentum we know that

$$m(\mathbf{v} - \mathbf{v}_0) = \hat{\mathbf{F}} \tag{1.368}$$

where $\mathbf{v}$ is the velocity of the particle immediately after the impulse. Now take the dot product with $\frac{1}{2}(\mathbf{v} + \mathbf{v}_0)$. We find that the increase in kinetic energy is

$$\frac{1}{2}m(v^2 - v_0^2) = \frac{1}{2}\hat{\mathbf{F}} \cdot (\mathbf{v} + \mathbf{v}_0) \tag{1.369}$$

Then, by the principle of work and kinetic energy, the work done on the particle is

$$W = \frac{1}{2}\hat{\mathbf{F}} \cdot (\mathbf{v} + \mathbf{v}_0) \tag{1.370}$$

This result is valid even if the force $\mathbf{F}$ and the time interval are finite and arbitrary because these assumptions apply to the principle of linear impulse and momentum (1.368). For motion in a straight line, the work done on a particle is equal to the impulse times the mean of the initial and final velocities.

Example 1.16 A flexible inextensible rope of length L and linear density ρ initially lies straight and motionless on a smooth horizontal floor (Fig. 1.27). Then, at $t = 0$, end A is suddenly moved to the right with a constant speed v_0. Considering the rope as a system of particles, let us look into the applicable impulse and momentum principles, as well as work and kinetic energy relations.

Assuming that $x(0) = 0$, $y(0) = 0$, we see that

$$y = 2x \tag{1.371}$$

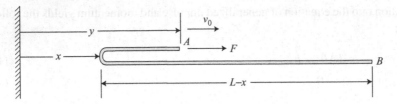

Figure 1.27.

and therefore

$$\dot{y} = 2\dot{x} = v_0 \tag{1.372}$$

Consider a mass element ρdx which suddenly has its velocity changed from zero to v_0. Using the principle of linear impulse and momentum, the required impulse is

$$F\, dt = \rho v_0\, dx \tag{1.373}$$

Thus, the force is

$$F = \rho v_0 \dot{x} = \frac{1}{2}\rho v_0^2 \tag{1.374}$$

The remainder of the moving rope has zero acceleration, so the force F is constant and is due to successive elements ρdx undergoing a sudden change in velocity. This continues over the interval $0 < t < 2L/v_0$, that is, until the rope is again straight with end A leading B. For $t > 2L/v_0$, the applied force equals zero. Over the entire interval, the applied impulse is

$$\hat{F} = Ft = \left(\frac{1}{2}\rho v_0^2\right)\left(\frac{2L}{v_0}\right) = \rho L v_0 \tag{1.375}$$

This is equal to the increase in momentum.

Now consider the work done on the system by the force F.

$$W = 2FL = \rho L v_0^2 \tag{1.376}$$

This is equal to twice the final kinetic energy, where

$$T = \frac{1}{2}\rho L v_0^2 \tag{1.377}$$

The energy lost in the process is $W - T$ and is due to what amounts to inelastic impact as each element suddenly changes its velocity from zero to v_0.

Another viewpoint is obtained by considering an inertial frame in which end A is fixed for $t > 0$ and end B moves to the left with velocity v_0. In this case,

$$F = \frac{1}{2}\rho v_0^2 \tag{1.378}$$

as before, but F does no work since A is fixed. The lost energy is equal to the initial kinetic energy $\frac{1}{2}\rho L v_0^2$ because the final energy is zero. We see that, although the work of F as well as the initial and final kinetic energies are different in the two inertial frames, the lost energy is unchanged. Relative to this second inertial frame, it is more obvious that the lost energy is due to inelastic impact, since the velocity of each element ρdx has its velocity suddenly changed from v_0 to zero.

1.6   Bibliography

Desloge, E. A. *Classical Mechanics*, Vol. 1. New York: John Wiley and Sons, 1982.

Ginsberg, J. H. *Advanced Engineering Dynamics*, 2nd edn. Cambridge, UK: Cambridge University Press, 1995.

Greenwood, D. T. *Principles of Dynamics*, 2nd edn. Englewood Cliffs, NJ: Prentice Hall, 1988.

1.7   Problems

1.1. A particle of mass m can slide without friction on a semicircular depression of radius r in a block of mass m_0. The block slides without friction on a horizontal surface. Consider planar motion starting from rest with the initial conditions $\theta(0) = \pi/2$, $x(0) = 0$. (a) Find the values of $\dot{\theta}$ and $\dot{x}$ when the particle first passes through $\theta = 0$. (b) What is the maximum value of x in the entire motion?

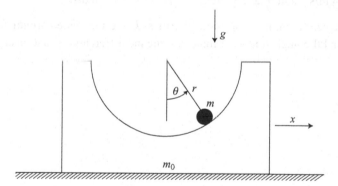

Figure P 1.1.

1.2. A simple pendulum of mass m and length l is initially motionless in the downward equilibrium position. Then its support point is given a constant horizontal acceleration $a = \sqrt{3}g$, where g is the acceleration of gravity. Find the maximum angular deviation from the initial position of the pendulum.

1.3. Two smooth spheres, each of mass m and radius r, are motionless and touching when a third sphere of mass m, radius r, and velocity v_0 strikes them in perfectly elastic impact, as shown. Solve for the velocities of the three spheres after impact.

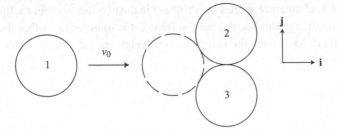

Figure P 1.3.

1.4. Three particles, each of mass m, can slide without friction on a fixed wire in the form of a horizontal circle of radius r. Initially the particles are equally spaced with two particles motionless and the third moving with velocity v_0. Assuming that all collisions occur with a coefficient of restitution $e = 0.9$, solve for the final velocities of the particles as time t approaches infinity.

1.5. A particle is shot with an initial velocity v_0 at an angle $\theta = \theta_0$ above the horizontal. It moves in a uniform gravitational field g. Solve for the radius of curvature ρ of the trajectory as a function of the flight path angle θ.

1.6. A river at 60° north latitude flows due south with a speed $v = 10$ km/h. Determine the difference in water level between the east and west banks if the river is $\frac{1}{2}$ km wide. Assume a spherical earth with radius $R = 6373$ km, acceleration of gravity $g = 9.814$ m/s^2, and rotation rate $\omega_e = 72.92 \times 10^{-6}$ rad/s.

1.7. An elastic pendulum has a spring of stiffness k and unstressed length L. Using the length l and the angle θ as coordinates, write the differential equations of motion.

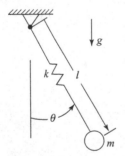

Figure P 1.7.

1.8. Point A moves with a constant velocity v_0. It is connected to a particle of mass m by a linear damper whose tensile force is $F = c(v_0 - \dot{x})$. (a) Assuming the initial conditions $x(0) = 0$, $\dot{x}(0) = 0$, solve for the particle velocity $\dot{x}$ as a function of time. (b) Find the work W done on the system by the force F acting on A. (c) Show that the work W approaches twice the value of the kinetic energy T as time t approaches infinity.

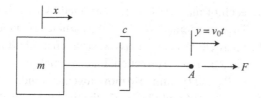

Figure P 1.8.

1.9. A linear damper with a damping coefficient c connects two particles, as shown in Fig. 1.15 on page 31. Show that, for an arbitrary linear motion of the two particles over some time interval Δt, the magnitude of the impulse $\hat{F}$ applied by the damper to either particle is directly proportional to the change in length Δl of the damper.

1.10. Ten identical spheres are arranged in a straight line with a small space between adjacent spheres. The spheres are motionless until sphere 1 is given an impulse causing it to move with velocity v_0 toward sphere 2. The spheres collide in sequence with a coefficient of restitution $e = 0.9$. (a) What is the final velocity of sphere 10? (b) What is the velocity of sphere 9 after its second collision?

1.11. Consider the system of Example 1.14 on page 57 and use (x, y, θ) as generalized coordinates, where the Cartesian coordinates of particle 1 are (x, y). Use the equation of generalized impulse and momentum to solve for: (a) $\dot{x}$ and $\dot{\theta}$ immediately after inelastic impact; (b) the impulse acting on the system at the time of impact due to the horizontal surface.

1.12. Two particles, each of mass m, are connected by a rigid massless rod of length l. The system is moving longitudinally with a velocity v_0, as shown, and strikes a smooth fixed surface with perfectly elastic impact, $e = 1$. (a) Find the velocity components $(\dot{x}, \dot{y})$ of the center of mass and also the angular velocity ω immediately after impact. (b) Solve for the impulse $\hat{F}$ in the rod at impact.

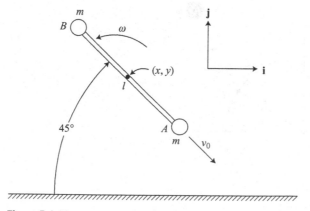

Figure P 1.12.

1.13. Particles A, B, and C are each of mass m and are connected by two rigid massless rods of length l, as shown. There is a joint at B. The three particles are motionless in a straight configuration when particle D, also of mass m, strikes particle B inelastically as it moves with velocity v_0 in a transverse direction. Later, particles A and C collide, also inelastically, as particles B and D continue to move together. Find: (a) the angular velocity of rod AB just before A and C collide and (b) the final velocity of particle A.

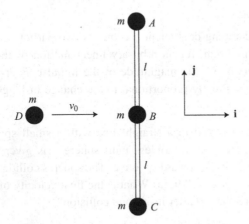

Figure P 1.13.

1.14. A particle of mass m is attached by a string of length l to a point O' which moves in a circular path of radius r at a constant angular rate $\dot{\theta} = \omega_0$. (a) Find the differential equation for ϕ, assuming that the string remains taut and all motion occurs in the xy-plane. (b) Assume the initial conditions $\phi(0) = 0$, $\dot{\phi}(0) = -\omega_0$ and let $l = r$. Find ϕ_{max} in the motion which follows.

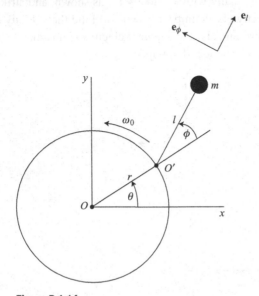

Figure P 1.14.

1.15. A massless disk of radius r has a particle of mass m attached at a distance l from its center. The disk rolls without slipping on a horizontal floor. Then, for some position angle θ, a horizontal impulse $\hat{F}$ is applied at the highest point A on the disk. Solve for the change $\Delta\omega$ in angular velocity, as well as the constraint impulse components $\hat{R}_v$ and $\hat{R}_h$.

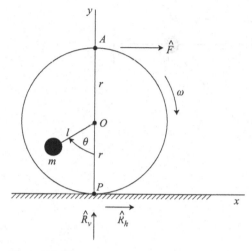

Figure P 1.15.

1.16. Consider the system of Example 1.13 on page 55. Assume the same initial conditions and the same applied impulse $\hat{F}$. Now suppose that Coulomb friction is added for the sliding of particle B relative to the floor. (a) Assuming a friction coefficient μ, solve for the values of $\dot{x}$ and $\dot{\theta}$ immediately after the impulse. (b) What is the minimum value of μ such that there is no sliding?

1.17. Two particles, each of mass m, are connected by a massless cord of length $\frac{1}{2}\pi r$ on a fixed horizontal cylinder of radius r. There is a coefficient of friction μ for sliding of the particles on the cylinder. (a) Obtain the equation of motion, assuming $\dot{\theta} > 0$ and the particles remain in contact with the cylinder. (b) Solve for the tension P in the cord as a function of θ. Assume that $0 \leq \theta \leq \frac{1}{2}\pi$ and $0 < \mu < 1$.

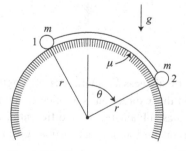

Figure P 1.17.

1.18. A particle P of mass m can slide without friction on a rigid fixed wire. It is connected to a fixed point O through a linear damper with a damping coefficient c. Assuming the initial conditions $x(0) = 0$, $\dot{x}(0) = v_0$, solve for the initial velocity v_0 which will cause the particle to stop at $x = h$.

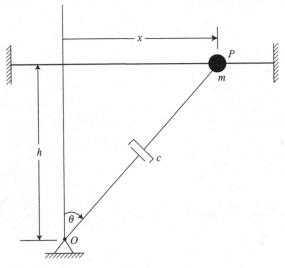

Figure P 1.18.

1.19. Two particles, each of mass m, are connected by a rigid massless rod of length l. Particle 1 can slide without friction on a fixed straight wire. (a) Using (x, θ) as generalized coordinates for motion in the horizontal plane, solve for $\ddot{x}, \ddot{\theta}$, and P as functions of $(\theta, \dot{\theta})$ where P is the tensile force in the rod. (b) Assume the initial conditions $x(0) = 0$, $\dot{x}(0) = 0$, $\theta(0) = 0$, $\dot{\theta}(0) = v_0/l$. Solve for $\dot{x}$ and $\dot{\theta}$ as functions of θ.

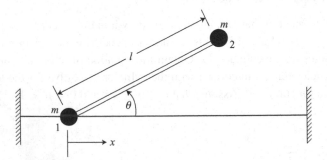

Figure P 1.19.

1.20. Two particles, connected by a spring of stiffness k and unstressed length L, move in the horizontal xy-plane. A force F in the x direction is applied to the first particle. (a) Using (x_1, y_1, x_2, y_2) as coordinates, obtain the four differential equations of motion. (b) Now introduce a constraint on the second particle given by $x_2 = y_2$. Eliminate y_2 and write equations of motion for x_1, y_1, and x_2.

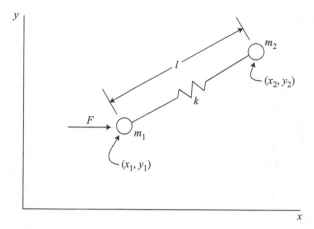

Figure P 1.20.

1.21. Two particles, each of mass m, can slide on smooth parallel fixed wires which are separated by a distance h. The particles are connected by a linear spring of stiffness k and unstressed length h. (a) Assume the initial conditions $x_1(0) = x_2(0) = 0$ and $\dot{x}_1(0) = \dot{x}_2(0) = 0$. At $t = 0$, a constant force $F = \frac{1}{4}kh$ is applied to the first particle. In the motion which follows, solve for the maximum value of θ. (b) Assume the initial conditions $\theta(0) = \theta_0$ and $\dot{\theta}(0) = 0$. Find the value of θ such that it remains constant during the motion.

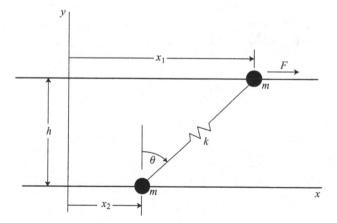

Figure P 1.21.

1.22. A flexible inextensible rope of uniform linear density ρ and length L is initially at rest with $y(0) = \frac{1}{3}L$. (a) Find the differential equation for y in the following two cases: (1) upward motion with $F = \frac{1}{2}\rho Lg$; (2) downward motion with $F = \frac{1}{6}\rho Lg$. (b) Show that in each case the magnitude of the energy loss rate is $\frac{1}{4}\rho|\dot{y}|^3$.

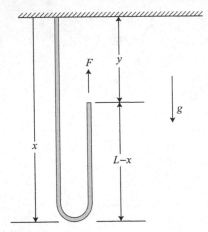

Figure P 1.22.

1.23. A particle P of mass m can slide without friction in a slot having the form of an arc of radius r which is cut in a circular impeller that rotates with a constant angular velocity ω_0. (a) Write the differential equation of motion in terms of the relative position angle θ. (b) Assuming that the particle is released from rest at $\theta = 0$, find the value of $\dot{\theta}$ as the particle leaves the impeller. (c) What is the force on the particle just before it leaves?

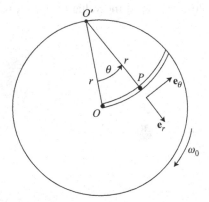

Figure P 1.23.

2 Lagrange's and Hamilton's equations

In our study of the dynamics of a system of particles, we have been concerned primarily with the Newtonian approach which is *vectorial* in nature. In general, we need to know the magnitudes and directions of the forces acting on the system, including the forces of constraint. Frequently the constraint forces are not known directly and must be included as additional unknown variables in the equations of motion. Furthermore, the calculation of particle accelerations can present kinematical difficulties.

An alternate approach is that of *analytical dynamics*, as represented by Lagrange's equations and Hamilton's equations. These methods enable one to obtain a complete set of equations of motion by differentiations of a single scalar function, namely the Lagrangian function or the Hamiltonian function. These functions include kinetic and potential energies, but ideal constraint forces are not involved. Thus, orderly procedures for obtaining the equations of motion are available and are applicable to a wide range of problems.

2.1   D'Alembert's principle and Lagrange's equations

D'Alembert's principle

Let us begin with Newton's law of motion applied to a system of N particles. For the ith particle of mass m_i and inertial position $\mathbf{r}_i$, we have

$$\mathbf{F}_i + \mathbf{R}_i - m_i \ddot{\mathbf{r}}_i = 0 \tag{2.1}$$

where $\mathbf{F}_i$ is the applied force and $\mathbf{R}_i$ is the constraint force. Now take the scalar product with a virtual displacement $\delta \mathbf{r}_i$ and sum over i. We obtain

$$\sum_{i=1}^{N} (\mathbf{F}_i + \mathbf{R}_i - m_i \ddot{\mathbf{r}}_i) \cdot \delta \mathbf{r}_i = 0 \tag{2.2}$$

This result is valid for arbitrary $\delta \mathbf{r}$s; but now assume that the $\delta \mathbf{r}$s satisfy the *instantaneous or virtual constraint equations*, namely,

$$\sum_{i=1}^{3N} a_{ji}(x, t)\delta x_i = 0 \qquad (j = 1, \ldots, m) \tag{2.3}$$

where the δxs are the Cartesian components of the $\delta \mathbf{r}$s. The virtual work of the constraint forces must vanish, that is,

$$\sum_{i=1}^{N} \mathbf{R}_i \cdot \delta \mathbf{r}_i = 0 \tag{2.4}$$

Then (2.2) reduces to

$$\sum_{i=1}^{N} (\mathbf{F}_i - m_i \ddot{\mathbf{r}}_i) \cdot \delta \mathbf{r}_i = 0 \tag{2.5}$$

This important result is the *Lagrangian form of d'Alembert's principle*. It states that the virtual work of the applied forces plus the inertia forces is zero for all virtual displacements satisfying the instantaneous constraints, that is, with time held fixed. The forces due to ideal constraints do not enter into these equations. This important characteristic will be reflected in various dynamical equations derived using d'Alembert's principle, including Lagrange's and Hamilton's equations.

In terms of Cartesian coordinates, d'Alembert's principle has the form

$$\sum_{k=1}^{3N} (F_k - m_k \ddot{x}_k) \delta x_k = 0 \tag{2.6}$$

where the δxs satisfy (2.3). Equation (2.6) is valid for any set of δxs which satisfy the instantaneous constraints. There are $(3N - m)$ independent sets of δxs, each being conveniently expressed in terms of δx_k ratios. This results in $(3N - m)$ equations of motion. An additional m equations are obtained by differentiating the constraint equations of (1.203) or, in this case,

$$\sum_{i=1}^{3N} a_{ji} \dot{x}_i + a_{jt} = 0 \qquad (j = 1, \dots, m) \tag{2.7}$$

with respect to time. Altogether, there are now $3N$ second-order differential equations which can be solved, frequently by numerical integration, for the xs as functions of time. Note that the constraint forces have been eliminated from the equations of motion.

Example 2.1 A particle of mass m can move without friction on the inside surface of a paraboloid of revolution (Fig. 2.1)

$$\phi = x^2 + y^2 - z = 0 \tag{2.8}$$

under the action of a uniform gravitational field in the negative z direction. We wish to find the differential equations of motion using d'Alembert's principle. The applied force components are

$$F_x = 0, \qquad F_y = 0, \qquad F_z = -mg \tag{2.9}$$

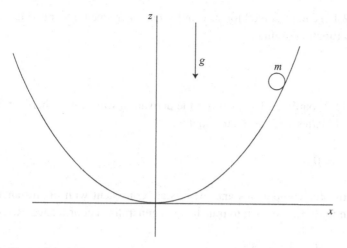

Figure 2.1.

In order to obtain the constraint equation for the virtual displacements, first write (2.8) in the differential form

$$\phi = 2x\dot{x} + 2y\dot{y} - \dot{z} = 0 \tag{2.10}$$

The instantaneous constraint equation, obtained using (2.3), is

$$2x\delta x + 2y\delta y - \delta z = 0 \tag{2.11}$$

We need to find $3N - m = 2$ independent virtual displacements which satisfy (2.11). We can take

$$\delta y = 0, \qquad \delta z = 2x\delta x \tag{2.12}$$

and

$$\delta x = 0, \qquad \delta z = 2y\delta y \tag{2.13}$$

Then d'Alembert's principle, (2.6), results in

$$[-m\ddot{x} - 2mx(\ddot{z} + g)]\,\delta x = 0 \tag{2.14}$$
$$[-m\ddot{y} - 2my(\ddot{z} + g)]\,\delta y = 0 \tag{2.15}$$

Now δx is freely variable in (2.14), as is δy in (2.15), so their coefficients must equal zero. Thus, after dividing by $-m$, we obtain

$$\ddot{x} + 2x(\ddot{z} + g) = 0 \tag{2.16}$$
$$\ddot{y} + 2y(\ddot{z} + g) = 0 \tag{2.17}$$

These are the dynamical equations of motion. A kinematical second-order equation is obtained by differentiating (2.10) with respect to time, resulting in

$$2x\ddot{x} + 2y\ddot{y} - \ddot{z} + 2\dot{x}^2 + 2\dot{y}^2 = 0 \tag{2.18}$$

Equations (2.16)–(2.18) can be solved for $\ddot{x}$, $\ddot{y}$ and $\ddot{z}$, which are then integrated to give the particle motion as a function of time.

Lagrange's equations

Consider a system of N particles. Let us begin the derivation with d'Alembert's principle expressed in terms of Cartesian coordinates, that is,

$$\sum_{k=1}^{3N} (F_k - m_k \ddot{x}_k) \, \delta x_k = 0 \tag{2.19}$$

The F_ks are applied force components and the δxs are consistent with the instantaneous constraints, as given in (2.3). We wish to transform to generalized coordinates. Recall that

$$\delta x_k = \sum_{i=1}^{n} \frac{\partial x_k}{\partial q_i} \delta q_i \qquad (k = 1, \ldots, 3N) \tag{2.20}$$

Hence, we obtain

$$\sum_{i=1}^{n} \sum_{k=1}^{3N} \left(F_k \frac{\partial x_k}{\partial q_i} - m_k \ddot{x}_k \frac{\partial x_k}{\partial q_i} \right) \delta q_i = 0 \tag{2.21}$$

where the δqs conform to any constraints. But since $x_k = x_k(q, t)$,

$$\dot{x}_k = \sum_{i=1}^{n} \frac{\partial x_k}{\partial q_i} \dot{q}_i + \frac{\partial x_k}{\partial t} \tag{2.22}$$

and we see that

$$\frac{\partial \dot{x}_k}{\partial \dot{q}_i} = \frac{\partial x_k}{\partial q_i} \tag{2.23}$$

Furthermore,

$$\frac{d}{dt} \left(\frac{\partial x_k}{\partial q_i} \right) = \sum_{j=1}^{n} \frac{\partial^2 x_k}{\partial q_j \, \partial q_i} \dot{q}_j + \frac{\partial^2 x_k}{\partial t \, \partial q_i} = \frac{\partial \dot{x}_k}{\partial q_i} \tag{2.24}$$

The kinetic energy of the system is

$$T = \frac{1}{2} \sum_{k=1}^{3N} m_k \dot{x}_k^2 \tag{2.25}$$

and the generalized momentum p_i is

$$p_i = \frac{\partial T}{\partial \dot{q}_i} = \sum_{k=1}^{3N} m_k \dot{x}_k \frac{\partial \dot{x}_k}{\partial \dot{q}_i} = \sum_{k=1}^{3N} m_k \dot{x}_k \frac{\partial x_k}{\partial q_i} \tag{2.26}$$

Therefore, we obtain

$$\frac{d}{dt} \left(\frac{\partial T}{\partial \dot{q}_i} \right) = \sum_{k=1}^{3N} m_k \ddot{x}_k \frac{\partial x_k}{\partial q_i} + \sum_{k=1}^{3N} m_k \dot{x}_k \frac{\partial \dot{x}_k}{\partial q_i} \tag{2.27}$$

However, from (2.25),

$$\frac{\partial T}{\partial q_i} = \sum_{k=1}^{3N} m_k \dot{x}_k \frac{\partial \dot{x}_k}{\partial q_i} \tag{2.28}$$

so we find that

$$\sum_{k=1}^{3N} m_k \ddot{x}_k \frac{\partial x_k}{\partial q_i} = \frac{d}{dt}\left(\frac{\partial T}{\partial \dot{q}_i}\right) - \frac{\partial T}{\partial q_i} \tag{2.29}$$

This result is the negative of the generalized inertia force Q_i^*. In other words,

$$Q_i^* = -\frac{d}{dt}\left(\frac{\partial T}{\partial \dot{q}_i}\right) + \frac{\partial T}{\partial q_i} \tag{2.30}$$

From (1.240), we recall that the generalized applied force is

$$Q_i = \sum_{k=1}^{3N} F_k \frac{\partial x_k}{\partial q_i} \tag{2.31}$$

Then d'Alembert's principle, as written in (2.21) results in

$$\sum_{i=1}^{n}(Q_i + Q_i^*)\delta q_i = 0 \tag{2.32}$$

for all δqs satisfying any constraints. Thus, the virtual work of the applied plus the inertial generalized forces is equal to zero.

Using (2.30) and (2.31) with a change of sign, (2.32) takes the form

$$\sum_{i=1}^{n}\left[\frac{d}{dt}\left(\frac{\partial T}{\partial \dot{q}_i}\right) - \frac{\partial T}{\partial q_i} - Q_i\right]\delta q_i = 0 \tag{2.33}$$

where, again, the δqs satisfy the instantaneous constraints. This is d'Alembert's principle expressed in terms of generalized coordinates. We will call it *Lagrange's principle*. It applies to both holonomic and nonholonomic systems, requiring only that T and Q_i be written as functions of $(q, \dot{q}, t)$. Note that constraint forces do not enter the Q_is.

Now let us assume a *holonomic system* with *independent* generalized coordinates. Then the coefficient of each δq_i must be equal to zero. We obtain

$$\frac{d}{dt}\left(\frac{\partial T}{\partial \dot{q}_i}\right) - \frac{\partial T}{\partial q_i} = Q_i \qquad (i = 1, \ldots, n) \tag{2.34}$$

This is the *fundamental holonomic form of Lagrange's equation*. It results in n second-order ordinary differential equations. The Qs are generalized applied forces from any source.

Let us make the further restriction that the Qs are derivable from a potential energy function $V(q, t)$ in accordance with

$$Q_i = -\frac{\partial V}{\partial q_i} \tag{2.35}$$

Now define the *Lagrangian function*

$$L(q, \dot{q}, t) = T(q, \dot{q}, t) - V(q, t) \tag{2.36}$$

Then we can write Lagrange's equation in the *standard holonomic form*:

$$\frac{d}{dt}\left(\frac{\partial L}{\partial \dot{q}_i}\right) - \frac{\partial L}{\partial q_i} = 0 \qquad (i = 1, \dots, n) \tag{2.37}$$

Equivalently, one can use

$$\frac{d}{dt}\left(\frac{\partial T}{\partial \dot{q}_i}\right) - \frac{\partial T}{\partial q_i} + \frac{\partial V}{\partial q_i} = 0 \qquad (i = 1, \dots, n) \tag{2.38}$$

If a portion of the applied generalized force is not obtained from a potential function, we can write

$$Q_i = -\frac{\partial V}{\partial q_i} + Q'_i \tag{2.39}$$

where Q'_i is the nonpotential part of Q_i. Then we obtain Lagrange's equation in the form

$$\frac{d}{dt}\left(\frac{\partial L}{\partial \dot{q}_i}\right) - \frac{\partial L}{\partial q_i} = Q'_i \qquad (i = 1, \dots, n) \tag{2.40}$$

or

$$\frac{d}{dt}\left(\frac{\partial T}{\partial \dot{q}_i}\right) - \frac{\partial T}{\partial q_i} = -\frac{\partial V}{\partial q_i} + Q'_i \qquad (i = 1, \dots, n) \tag{2.41}$$

The basic forms of Lagrange's equations, given in (2.34) and (2.37) apply to holonomic systems described in terms of *independent* generalized coordinates.

When one considers nonholonomic systems, there must be more generalized coordinates than degrees of freedom. Hence, the generalized coordinates are not completely independent and, as a result, there will be generalized constraint forces C_i which are nonzero, in general. In terms of the constraint forces acting on the individual particles, we have, similar to (1.240),

$$C_i = \sum_{j=1}^{N} \mathbf{R}_j \cdot \gamma_{ji} = \sum_{k=1}^{3N} R_k \frac{\partial x_k}{\partial q_i} \tag{2.42}$$

where the R_k are Cartesian constraint force components.

Let us use the *Lagrange multiplier method* to evaluate the generalized constraint forces. First, we note that the virtual work of the constraint forces must equal zero, that is,

$$\sum_{i=1}^{n} C_i \delta q_i = 0 \tag{2.43}$$

provided that the δqs satisfy the instantaneous constraints in the form

$$\sum_{i=1}^{n} a_{ji} \delta q_i = 0 \qquad (j = 1, \dots, m) \tag{2.44}$$

Now let us multiply (2.44) by a *Lagrange multiplier* λ_j and sum over j. We obtain

$$\sum_{j=1}^{m}\sum_{i=1}^{n}\lambda_j a_{ji}\delta q_i = 0 \tag{2.45}$$

Subtract (2.45) from (2.43) with the result

$$\sum_{i=1}^{n}\left(C_i - \sum_{j=1}^{m}\lambda_j a_{ji}\right)\delta q_i = 0 \tag{2.46}$$

Up to this point, the m λs have been considered to be arbitrary, whereas the n δqs satisfy the instantaneous constraints. However, it is possible to choose the λs such that the coefficient of each δq_i vanishes. Thus, the generalized constraint force C_i is

$$C_i = \sum_{j=1}^{m}\lambda_j a_{ji} \qquad (i = 1, \dots, n) \tag{2.47}$$

and the δqs can be considered to be arbitrary.

To understand how m λs can be chosen to specify an arbitrary constraint force in accordance with (2.47), let us first consider a single constraint. The corresponding constraint force is perpendicular to the constraint surface at the operating point; that is, it is in the direction of the vector $\mathbf{a}_j$ whose components in n-space are the coefficients a_{ji}. This is expressed by (2.44) and we note that any virtual displacement $\delta\mathbf{q}$ must lie in the tangent plane at the operating point. The Lagrange multiplier λ_j applies equally to all components a_{ji} and so expresses the magnitude of the constraint force $\mathbf{C}_j = \lambda_j\mathbf{a}_j$.

If there are m constraints, the total constraint force $\mathbf{C}$ is found by summing the individual constraint forces $\mathbf{C}_j$. We can consider the $\mathbf{a}_j$s as m independent basis vectors with the λs representing the scalar components of $\mathbf{C}$ in this m-dimensional subspace. Hence a set of m λs can always be found to represent any possible total constraint force $\mathbf{C}$.

For a system with n generalized coordinates and m nonholonomic constraints, we note first that the generalized constraint force C_i is no longer zero, in general, and must be added to Q_i to obtain

$$\frac{d}{dt}\left(\frac{\partial T}{\partial \dot{q}_i}\right) - \frac{\partial T}{\partial q_i} = Q_i + C_i \qquad (i = 1, \dots, n) \tag{2.48}$$

Then, using (2.47), the result is the *fundamental nonholonomic form of Lagrange's equation.*

$$\frac{d}{dt}\left(\frac{\partial T}{\partial \dot{q}_i}\right) - \frac{\partial T}{\partial q_i} = Q_i + \sum_{j=1}^{m}\lambda_j a_{ji} \qquad (i = 1, \dots, n) \tag{2.49}$$

If we again assume that the Qs are obtained from a potential function $V(q, t)$, we can write

$$\frac{d}{dt}\left(\frac{\partial L}{\partial \dot{q}_i}\right) - \frac{\partial L}{\partial q_i} = \sum_{j=1}^{m}\lambda_j a_{ji} \qquad (i = 1, \dots, n) \tag{2.50}$$

This is the *standard nonholonomic form of Lagrange's equation*. Equivalently, we have

$$\frac{d}{dt}\left(\frac{\partial T}{\partial \dot{q}_i}\right) - \frac{\partial T}{\partial q_i} + \frac{\partial V}{\partial q_i} = \sum_{j=1}^{m} \lambda_j a_{ji} \qquad (i = 1, \ldots, n) \tag{2.51}$$

If nonpotential generalized applied forces Q_i' are present, Lagrange's equation has the form

$$\frac{d}{dt}\left(\frac{\partial L}{\partial \dot{q}_i}\right) - \frac{\partial L}{\partial q_i} = Q_i' + \sum_{j=1}^{m} \lambda_j a_{ji} \qquad (i = 1, \ldots, n) \tag{2.52}$$

The nonholonomic forms of Lagrange's equations presented here are also applicable to holonomic systems in which the qs are not independent. For example, if there are m holonomic constraints of the form

$$\phi_j(q, t) = 0 \qquad (j = 1, \ldots, m) \tag{2.53}$$

we take

$$a_{ji}(q, t) = \frac{\partial \phi_j}{\partial q_i} \tag{2.54}$$

and then use a nonholonomic form of Lagrange's equation such as (2.49) or (2.51).

Example 2.2 Particles A and B (Fig. 2.2), each of mass m, are connected by a rigid massless rod of length l. Particle A can move without friction on the horizontal x-axis, while particle B can move without friction on the vertical y-axis. We desire the differential equations of motion.

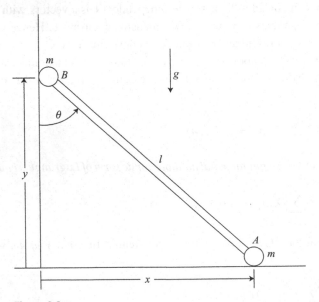

Figure 2.2.

First method   Let us use the nonholonomic form of Lagrange's equation given by (2.51). The coordinates (x, y) have the holonomic constraint

$$\phi = x^2 + y^2 - l^2 = 0 \tag{2.55}$$

Differentiating with respect to time, we obtain

$$\dot{\phi} = 2x\dot{x} + 2y\dot{y} = 0 \tag{2.56}$$

which leads to

$$a_{11} = 2x, \qquad a_{12} = 2y \tag{2.57}$$

The kinetic energy is

$$T = \frac{1}{2}m(\dot{x}^2 + \dot{y}^2) \tag{2.58}$$

and the potential energy is

$$V = mgy \tag{2.59}$$

The use of (2.51) results in the equations of motion

$$m\ddot{x} = 2\lambda x \tag{2.60}$$
$$m\ddot{y} = 2\lambda y - mg \tag{2.61}$$

We need a third differential equation since there are three variables (x, y, λ). Differentiating (2.56) with respect to time, and dividing by two, we obtain

$$x\ddot{x} + y\ddot{y} + \dot{x}^2 + \dot{y}^2 = 0 \tag{2.62}$$

This equation, plus (2.60) and (2.61) are a complete set of second-order differential equations. One can solve for λ and obtain

$$\lambda = \frac{m}{2l^2}[gy - (\dot{x}^2 + \dot{y}^2)] \tag{2.63}$$

Then one can numerically integrate (2.60) and (2.61). We see that the compressive force in the rod is $2\lambda l$.

Second method   A simpler approach is to choose θ as a single generalized coordinate with no constraints. We see that

$$x = l\sin\theta \tag{2.64}$$
$$y = l\cos\theta \tag{2.65}$$
$$\dot{x} = l\dot{\theta}\cos\theta \tag{2.66}$$
$$\dot{y} = -l\dot{\theta}\sin\theta \tag{2.67}$$

The kinetic energy is

$$T = \frac{1}{2}m(\dot{x}^2 + \dot{y}^2) = \frac{1}{2}ml^2\dot{\theta}^2 \tag{2.68}$$

and the potential energy is

$$V = mgy = mgl \cos\theta \qquad (2.69)$$

Then we can use Lagrange's equation in the form

$$\frac{d}{dt}\left(\frac{\partial T}{\partial \dot{q}_i}\right) - \frac{\partial T}{\partial q_i} + \frac{\partial V}{\partial q_i} = 0 \qquad (2.70)$$

to obtain the differential equation for θ which is

$$ml^2\ddot{\theta} - mgl \sin\theta = 0 \qquad (2.71)$$

Third method   Another possibility is to use the holonomic equation of constraint to eliminate one of the variables. For example, we might eliminate y and its derivatives and then consider x to be an independent coordinate. From (2.55), we have

$$y = \pm\sqrt{l^2 - x^2} \qquad (2.72)$$

Assuming the positive sign, we obtain

$$\dot{y} = \frac{-x\dot{x}}{\sqrt{l^2 - x^2}} \qquad (2.73)$$

and the kinetic energy is

$$T = \frac{1}{2}m\dot{x}^2\left(1 + \frac{x^2}{l^2 - x^2}\right) = \frac{1}{2}m\dot{x}^2\left(\frac{l^2}{l^2 - x^2}\right) \qquad (2.74)$$

The potential energy is

$$V = mgy = mg\sqrt{l^2 - x^2} \qquad (2.75)$$

Then (2.70) leads to a differential equation for x, namely,

$$ml^2\left[\frac{\ddot{x}}{l^2 - x^2} + \frac{x\dot{x}^2}{(l^2 - x^2)^2}\right] - mg\frac{x}{\sqrt{l^2 - x^2}} = 0 \qquad (2.76)$$

If y is actually negative, the sign of the last term is changed.

For a system with m *holonomic* constraints, one can use the constraint equations to solve for m *dependent* qs and $\dot{q}$s in terms of the corresponding $(n - m)$ independent quantities. Then one can write T and V in terms of the independent quantities only, and use simple forms of Lagrange's equation such as (2.70) to obtain $(n - m)$ equations of motion. In this example, we let x be independent and y dependent. After integrating (2.76) numerically to obtain x as a function of time, one can use (2.72) and (2.73) to obtain the motion in y.

It should be noted, however, that this procedure does not produce correct results if *nonholonomic* constraints are involved. Various approaches which do not involve Lagrange multipliers, but are applicable to nonholonomic systems, will be discussed in Chapter 4.

This example illustrates that the choice of generalized coordinates has a strong effect on the complexity of the equations of motion. In this instance, the second method is by far the simplest approach if one is interested in solving for the motion. It is not very helpful, however, in solving for various internal forces. The first method involving Lagrange multipliers is the most direct method to solve for forces.

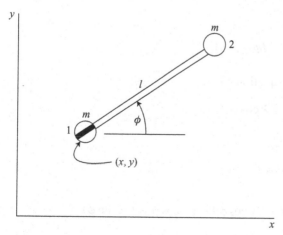

Figure 2.3.

Example 2.3 A dumbbell consists of two particles, each of mass m, connected by a rigid massless rod of length l. There is a knife-edge at particle 1, resulting in the nonholonomic constraint

$$-\dot{x} \sin \phi + \dot{y} \cos \phi = 0 \tag{2.77}$$

which states that the velocity normal to the knife-edge is zero (Fig. 2.3). Assume that the xy-plane is horizontal and therefore the potential energy V is constant. We wish to find the differential equations of motion.

There are no applied forces acting on the system, so we can use Lagrange's equation in the form

$$\frac{d}{dt}\left(\frac{\partial T}{\partial \dot{q}_i}\right) - \frac{\partial T}{\partial q_i} = \sum_{j=1}^{m} \lambda_j a_{ji} \tag{2.78}$$

In this case the qs are (x, y, ϕ). The kinetic energy for the unconstrained system is obtained by using (1.127), with the result

$$T = m\left(\dot{x}^2 + \dot{y}^2 + \frac{1}{2}l^2\dot{\phi}^2 - l\dot{x}\dot{\phi}\sin\phi + l\dot{y}\dot{\phi}\cos\phi\right) \tag{2.79}$$

From the constraint equation we note that

$$a_{11} = -\sin\phi, \qquad a_{12} = \cos\phi, \qquad a_{13} = 0 \tag{2.80}$$

Now

$$\frac{d}{dt}\left(\frac{\partial T}{\partial \dot{x}}\right) = m(2\ddot{x} - l\ddot{\phi}\sin\phi - l\dot{\phi}^2\cos\phi) \tag{2.81}$$

and $\partial T / \partial x = 0$. Using (2.78), the x equation is

$$m(2\ddot{x} - l\ddot{\phi}\sin\phi - l\dot{\phi}^2\cos\phi) = -\lambda\sin\phi \tag{2.82}$$

Similarly,

$$\frac{d}{dt}\left(\frac{\partial T}{\partial \dot{y}}\right) = m(2\ddot{y} + l\ddot{\phi}\cos\phi - l\dot{\phi}^2\sin\phi) \tag{2.83}$$

and $\partial T / \partial y = 0$. Hence, the y equation is

$$m(2\ddot{y} + l\ddot{\phi}\cos\phi - l\dot{\phi}^2\sin\phi) = \lambda\cos\phi \tag{2.84}$$

Finally,

$$\frac{\partial T}{\partial \dot{\phi}} = m(l^2\dot{\phi} - l\dot{x}\sin\phi + l\dot{y}\cos\phi) \tag{2.85}$$

$$\frac{d}{dt}\left(\frac{\partial T}{\partial \dot{\phi}}\right) = m(l^2\ddot{\phi} - l\ddot{x}\sin\phi + l\ddot{y}\cos\phi - l\dot{x}\dot{\phi}\cos\phi - l\dot{y}\dot{\phi}\sin\phi) \tag{2.86}$$

$$\frac{\partial T}{\partial \phi} = m(-l\dot{x}\dot{\phi}\cos\phi - l\dot{y}\dot{\phi}\sin\phi) \tag{2.87}$$

Thus, the ϕ equation is

$$m(l^2\ddot{\phi} - l\ddot{x}\sin\phi + l\ddot{y}\cos\phi) = 0 \tag{2.88}$$

Equations (2.82), (2.84), and (2.88) are the dynamical equations of motion. In addition, we can differentiate the constraint equation with respect to time and obtain

$$-\ddot{x}\sin\phi + \ddot{y}\cos\phi - \dot{x}\dot{\phi}\cos\phi - \dot{y}\dot{\phi}\sin\phi = 0 \tag{2.89}$$

These four equations are linear in $(\ddot{x}, \ddot{y}, \ddot{\phi}, \lambda)$ and can be solved for these variables, which are then integrated to yield $x(t)$, $y(t)$, $\phi(t)$, and $\lambda(t)$.

In general, for a system with n generalized coordinates and m nonholonomic constraint equations, the Lagrangian method results in n second-order differential equations of motion plus m equations of constraint. These $(n + m)$ equations are solved for the n qs and m λs as functions of time. In later chapters, we will discuss methods which are more efficient in the sense of requiring fewer equations to describe a nonholonomic system.

2.2   Hamilton's equations

Canonical equations

The generalized momentum p_i, as given by (1.291), can be made more general by defining

$$p_i = \frac{\partial L}{\partial \dot{q}_i} \tag{2.90}$$

where $L(q, \dot{q}, t) = T(q, \dot{q}, t) - V(q, \dot{q}, t)$, and $V(q, \dot{q}, t)$ is a *velocity-dependent* potential energy function. This might be used, for example, to account for electromagnetic forces acting on moving charged particles. In our development, however, we will continue to assume that the potential energy has the form $V(q, t)$, and therefore that

$$p_i = \sum_{j=1}^{n} m_{ij}(q, t)\dot{q}_j + a_i(q, t) \qquad (i = 1, \ldots, n) \tag{2.91}$$

in agreement with (1.292). We assume that the inertia matrix $[m_{ij}]$ is positive definite and hence has an inverse. Thus, (2.91) can be solved for the $\dot{q}$s with the result

$$\dot{q}_j = \sum_{i=1}^{n} b_{ji}(q, t)(p_i - a_i) \qquad (j = 1, \ldots, n) \tag{2.92}$$

where the matrix $[b_{ji}] = [m_{ij}]^{-1}$.

The *Hamiltonian function* is defined by

$$H(q, p, t) = \sum_{i=1}^{n} p_i \dot{q}_i - L(q, \dot{q}, t) \tag{2.93}$$

The $\dot{q}$s on the right-hand side of (2.93) are expressed in terms of ps by using (2.92). Consider an arbitrary variation of the Hamiltonian function $H(q, p, t)$.

$$\delta H = \sum_{i=1}^{n} \frac{\partial H}{\partial q_i}\delta q_i + \sum_{i=1}^{n} \frac{\partial H}{\partial p_i}\delta p_i + \frac{\partial H}{\partial t}\delta t \tag{2.94}$$

From (2.93), we have

$$\delta H = \sum_{i=1}^{n} p_i \delta \dot{q}_i + \sum_{i=1}^{n} \dot{q}_i \delta p_i - \sum_{i=1}^{n} \frac{\partial L}{\partial q_i}\delta q_i - \sum_{i=1}^{n} \frac{\partial L}{\partial \dot{q}_i}\delta \dot{q}_i - \frac{\partial L}{\partial t}\delta t$$

$$= \sum_{i=1}^{n} \dot{q}_i \delta p_i - \sum_{i=1}^{n} \frac{\partial L}{\partial q_i}\delta q_i - \frac{\partial L}{\partial t}\delta t \tag{2.95}$$

where (2.90) has been used.

Now assume that δqs, δps, and δt are independently variable; that is, there are no kinematic constraints. Then, equating corresponding coefficients in (2.94) and (2.95), we obtain

$$\dot{q}_i = \frac{\partial H}{\partial p_i} \qquad (i = 1, \ldots, n) \tag{2.96}$$

$$\frac{\partial L}{\partial q_i} = -\frac{\partial H}{\partial q_i} \qquad (i = 1, \ldots, n) \tag{2.97}$$

and

$$\frac{\partial L}{\partial t} = -\frac{\partial H}{\partial t} \tag{2.98}$$

Next, introduce the standard holonomic form of Lagrange's equation, (2.37), which can be written as

$$\dot{p}_i = \frac{\partial L}{\partial q_i} \qquad (i = 1, \ldots, n) \tag{2.99}$$

Then we can write

$$\dot{q}_i = \frac{\partial H}{\partial p_i}, \qquad \dot{p}_i = -\frac{\partial H}{\partial q_i} \qquad (i = 1, \ldots, n) \tag{2.100}$$

These $2n$ first-order equations are known as *Hamilton's canonical equations*. The first n equations express the $\dot{q}$s as linear functions of the ps, as in (2.92). The final n equations contain the laws of motion for the system. Because of the symmetry in form of Hamilton's canonical equations, there is a tendency to accord the qs and ps equal status and think of the qs and ps together as a $2n$-dimensional *phase vector*. Thus, the motion of a system can be represented by a path or trajectory in $2n$-dimensional *phase space*.

Comparing the standard holonomic form of Lagrange's equations with Hamilton's canonical equations, we find that they are equivalent in that both require independent qs and apply to the same mechanical systems. Hamilton's equations are $2n$ first-order equations rather than the n second-order equations of Lagrange. However, it should be pointed out that most computer representations require the conversion of higher-order equations to a larger number of first-order equations.

Form of the Hamiltonian function

Let us return to a further consideration of the Hamiltonian function $H(q, p, t)$, as given by (2.93). Using (2.91), we see that

$$\sum_{i=1}^{n} p_i \dot{q}_i = \sum_{i=1}^{n} \sum_{j=1}^{n} m_{ij} \dot{q}_i \dot{q}_j + \sum_{i=1}^{n} a_i \dot{q}_i$$
$$= 2T_2 + T_1 \tag{2.101}$$

where we recall the expressions for T_2 and T_1 given in (1.259) and (1.262). Now

$$L = T - V = T_2 + T_1 + T_0 - V \tag{2.102}$$

Hence, using (2.93), we find that

$$H = T_2 - T_0 + V \tag{2.103}$$

For the particular case of a *scleronomic system*, $T_1 = T_0 = 0$ and $T = T_2$, so

$$H = T + V \tag{2.104}$$

that is, the Hamiltonian function is equal to the total energy.

Now let us write T_2 in the form

$$T_2 = \frac{1}{2}\sum_{i=1}^{n}\sum_{j=1}^{n} m_{ij}\dot{q}_i\dot{q}_j = \frac{1}{2}\sum_{i=1}^{n}\sum_{j=1}^{n} b_{ij}(p_i - a_i)(p_j - a_j)$$

$$= \frac{1}{2}\sum_{i=1}^{n}\sum_{j=1}^{n} b_{ij}p_ip_j - \sum_{i=1}^{n}\sum_{j=1}^{n} b_{ij}a_ip_j + \frac{1}{2}\sum_{i=1}^{n}\sum_{j=1}^{n} b_{ij}a_ia_j \qquad (2.105)$$

where we note that the matrix $[b_{ij}]$ is symmetric and the inverse of $[m_{ij}]$. Then, using (2.103), we can group the terms in the Hamiltonian function according to their degree in p. We can write

$$H = H_2 + H_1 + H_0 \qquad (2.106)$$

where

$$H_2 = \frac{1}{2}\sum_{i=1}^{n}\sum_{j=1}^{n} b_{ij}p_ip_j \qquad (2.107)$$

$$H_1 = -\sum_{i=1}^{n}\sum_{j=1}^{n} b_{ij}a_ip_j \qquad (2.108)$$

$$H_0 = \frac{1}{2}\sum_{i=1}^{n}\sum_{j=1}^{n} b_{ij}a_ia_j - T_0 + V \qquad (2.109)$$

and we note that b_{ij}, a_i, T_0 and V are all functions of (q, t).

Other Hamiltonian equations

We have seen that Hamilton's canonical equations apply to the same systems as the standard holonomic form of Lagrange's equation, (2.37). Other forms of Lagrange's equations have their Hamiltonian counterparts. For example, if nonpotential generalized forces Q_i' are present in a holonomic system, we have the Hamiltonian equations

$$\dot{q}_i = \frac{\partial H}{\partial p_i}, \qquad \dot{p}_i = -\frac{\partial H}{\partial q_i} + Q_i' \qquad (i = 1, \ldots, n) \qquad (2.110)$$

For a *nonholonomic system*, we have, corresponding to (2.50),

$$\dot{q}_i = \frac{\partial H}{\partial p_i}, \qquad \dot{p}_i = -\frac{\partial H}{\partial q_i} + \sum_{j=1}^{m} \lambda_j a_{ji} \qquad (i = 1, \ldots, n) \qquad (2.111)$$

This assumes that all the applied forces arise from a potential energy $V(q, t)$. On the other hand, if there are nonpotential forces in a nonholonomic system, Hamilton's equations have the form

$$\dot{q}_i = \frac{\partial H}{\partial p_i}, \qquad \dot{p}_i = -\frac{\partial H}{\partial q_i} + Q_i' + \sum_{j=1}^{n} \lambda_j a_{ji} \qquad (i = 1, \ldots, n) \qquad (2.112)$$

The nonholonomic forms of Hamilton's equations are solved in conjunction with constraint equations of the form

$$\sum_{i=1}^{n} a_{ji}\dot{q}_i + a_{jt} = 0 \qquad (j = 1, \ldots, m) \tag{2.113}$$

Thus, we have a total of $(2n + m)$ first-order differential equations to solve for n qs, n ps, and m λs as functions of time.

It is well to notice that, as in the Lagrangian case, the nonholonomic forms of Hamilton's equations also apply to holonomic systems in which the motions of the qs are restricted by holonomic constraints.

Example 2.4   A massless disk of radius r has a particle of mass m embedded at a distance $\frac{1}{2}r$ from the center O (Fig. 2.4). The disk rolls without slipping down a plane inclined at an angle α from the horizontal. We wish to obtain the differential equations of motion.

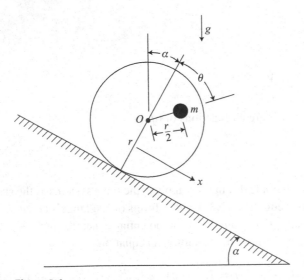

Figure 2.4.

First method   Let us employ Hamilton's canonical equations, as given by (2.100), and choose θ as the single generalized coordinate. The particle velocity v is the vector sum of the velocity of the center O and the velocity of the particle relative to O. Thus, we obtain

$$v^2 = \left(r\dot{\theta}\right)^2 + \left(\frac{1}{2}r\dot{\theta}\right)^2 + r^2\dot{\theta}^2 \cos\theta \tag{2.114}$$

and the kinetic energy is

$$T = \frac{1}{2}mv^2 = \frac{1}{2}mr^2 \left(\frac{5}{4} + \cos\theta\right)\dot{\theta}^2 \tag{2.115}$$

Now the generalized momentum p_θ is

$$p_\theta = \frac{\partial T}{\partial \dot\theta} = mr^2 \left(\frac{5}{4} + \cos\theta \right) \dot\theta \tag{2.116}$$

and we obtain

$$\dot\theta = \frac{p_\theta}{mr^2\left(\frac{5}{4} + \cos\theta\right)} \tag{2.117}$$

The kinetic energy as a function of (q, p) is

$$T = \frac{p_\theta^2}{2mr^2\left(\frac{5}{4} + \cos\theta\right)} \tag{2.118}$$

The potential energy is

$$V = mg\left[-r\theta \sin\alpha + \frac{1}{2}r\cos(\theta + \alpha) \right] \tag{2.119}$$

The Hamiltonian function is, in general,

$$H = T_2 - T_0 + V \tag{2.120}$$

but, in this case, it is equal to the total energy.

$$H = \frac{p_\theta^2}{2mr^2\left(\frac{5}{4} + \cos\theta\right)} + mg\left[-r\theta \sin\alpha + \frac{1}{2}r\cos(\theta + \alpha) \right] \tag{2.121}$$

The first canonical equation results in

$$\dot\theta = \frac{\partial H}{\partial p_\theta} = \frac{p_\theta}{mr^2\left(\frac{5}{4} + \cos\theta\right)} \tag{2.122}$$

which is a restatement of (2.117). The second canonical equation is

$$\dot p_\theta = -\frac{\partial H}{\partial \theta} = \frac{-p_\theta^2 \sin\theta}{2mr^2\left(\frac{5}{4} + \cos\theta\right)^2} + mgr\left[\sin\alpha + \frac{1}{2}\sin(\theta + \alpha) \right] \tag{2.123}$$

These two canonical equations are together equivalent to the single second-order equation in θ which would be obtained by using Lagrange's equation.

Second method   Let us use the same system to illustrate the use of Lagrange multipliers with Hamilton's equations, as in (2.111). We will use (x, θ) as generalized coordinates, where x is the displacement of the center of the disk. There is a holonomic constraint

$$\dot x - r\dot\theta = 0 \tag{2.124}$$

which expresses the nonslipping condition. The kinetic and potential energies, however, must be written for the *unconstrained* system, that is, assuming the possibility of slipping.
   The kinetic energy is

$$T = \frac{1}{2}m\left(\dot x^2 + \frac{1}{4}r^2\dot\theta^2 + r\dot x\dot\theta \cos\theta \right) \tag{2.125}$$

and the potential energy is

$$V = mg \left[-x \sin \alpha + \frac{1}{2} r \cos(\theta + \alpha) \right] \tag{2.126}$$

The generalized momenta are

$$p_x = \frac{\partial T}{\partial \dot{x}} = m\dot{x} + \frac{1}{2} mr\dot{\theta} \cos \theta \tag{2.127}$$

$$p_\theta = \frac{\partial T}{\partial \dot{\theta}} = \frac{1}{2} mr\dot{x} \cos \theta + \frac{1}{4} mr^2 \dot{\theta} \tag{2.128}$$

Equations (2.127) and (2.128) can be solved for $\dot{x}$ and $\dot{\theta}$. We obtain

$$\dot{x} = \frac{rp_x - 2p_\theta \cos \theta}{mr \sin^2 \theta} \tag{2.129}$$

$$\dot{\theta} = \frac{4p_\theta - 2rp_x \cos \theta}{mr^2 \sin^2 \theta} \tag{2.130}$$

Substituting these expressions for $\dot{x}$ and $\dot{\theta}$ into the kinetic energy equation, we obtain, after some algebraic simplification, the Hamiltonian function

$$H = T + V = \frac{1}{2mr^2 \sin^2 \theta} \left(r^2 p_x^2 + 4p_\theta^2 - 4rp_x p_\theta \cos \theta \right) + V \tag{2.131}$$

where V is given by (2.126).

We shall use Hamilton's equations in the form

$$\dot{q}_i = \frac{\partial H}{\partial p_i}, \qquad \dot{p}_i = -\frac{\partial H}{\partial q_i} + \sum_{j=1}^{m} \lambda_j a_{ji} \qquad (i = 1, \ldots, n) \tag{2.132}$$

where, from (2.124), the constraint coefficients are

$$a_{11} = 1, \qquad a_{12} = -r \tag{2.133}$$

The $\dot{x}$ equation is

$$\dot{x} = \frac{\partial H}{\partial p_x} = \frac{rp_x - 2p_\theta \cos \theta}{mr \sin^2 \theta} \tag{2.134}$$

The $\dot{p}_x$ equation is

$$\dot{p}_x = -\frac{\partial H}{\partial x} + \lambda = -\frac{\partial V}{\partial x} + \lambda = mg \sin \alpha + \lambda \tag{2.135}$$

The $\dot{\theta}$ equation is

$$\dot{\theta} = \frac{\partial H}{\partial p_\theta} = \frac{4p_\theta - 2rp_x \cos \theta}{mr^2 \sin^2 \theta} \tag{2.136}$$

Finally, the $\dot{p}_\theta$ equation is

$$
\dot{p}_\theta = -\frac{\partial H}{\partial \theta} - r\lambda
$$

$$
= \frac{1}{mr^2 \sin^3 \theta}\left[(r^2 p_x^2 + 4 p_\theta^2)\cos\theta - 2r p_x p_\theta(1 + \cos^2 \theta)\right]
$$

$$
+ \frac{1}{2}mgr \sin(\theta + \alpha) - r\lambda \tag{2.137}
$$

These four first-order Hamiltonian equations plus the constraint equation (2.124) can be integrated numerically to solve for the 2 qs, 2 ps, and λ as functions of time.

For a *scleronomic system* such as the one we have been considering, relatively simple matrix equations can be used to obtain the Hamiltonian function. The kinetic energy is quadratic in the $\dot{q}$s and of the form

$$
T = \frac{1}{2}\dot{\mathbf{q}}^{\mathrm{T}}\mathbf{m}\dot{\mathbf{q}} \tag{2.138}
$$

where $\mathbf{m}$ is the $n \times n$ matrix of generalized mass coefficients. The generalized momenta are given by the matrix equation

$$
\mathbf{p} = \mathbf{m}\dot{\mathbf{q}} \tag{2.139}
$$

and, conversely,

$$
\dot{\mathbf{q}} = \mathbf{b}\mathbf{p} \tag{2.140}
$$

where $\mathbf{b} = \mathbf{m}^{-1}$. For scleronomic systems, the Hamiltonian function is equal to the total energy, or

$$
H = T + V = \frac{1}{2}\mathbf{p}^{\mathrm{T}}\mathbf{b}\mathbf{p} + V \tag{2.141}
$$

For the system of this example, we have

$$
\mathbf{m} = \begin{bmatrix} m & \frac{1}{2}mr\cos\theta \\ \frac{1}{2}mr\cos\theta & \frac{1}{4}mr^2 \end{bmatrix} \tag{2.142}
$$

and

$$
\mathbf{b} = \frac{4}{mr^2 \sin^2 \theta}\begin{bmatrix} \frac{1}{4}r^2 & -\frac{1}{2}r\cos\theta \\ -\frac{1}{2}r\cos\theta & 1 \end{bmatrix} \tag{2.143}
$$

in agreement with (2.131) and (2.141).

2.3   Integrals of the motion

We have found that for a dynamical system whose configuration is given by n independent generalized coordinates, the Lagrangian method results in n second-order differential equations of motion with time as the independent variable. Any general analytical solution of these equations of motion contains $2n$ constants of integration which are usually evaluated from the $2n$ initial conditions. One method of expressing the general solution is to find $2n$ independent functions of the form

$$g_j(q, \dot{q}, t) = \alpha_j \qquad (j = 1, \ldots, 2n) \tag{2.144}$$

where the αs are constants. The $2n$ functions are called *integrals or constants of the motion*. Each function g_j maintains a constant value α_j during the actual motion of the system. In principle, these $2n$ equations can be solved for the qs and $\dot{q}$s as functions of the αs and t, that is,

$$q_i = q_i(\alpha, t) \qquad (i = 1, \ldots, n) \tag{2.145}$$
$$\dot{q}_i = \dot{q}_i(\alpha, t) \qquad (i = 1, \ldots, n) \tag{2.146}$$

where these solutions satisfy (2.144).

Usually it is not possible to obtain a full set of $2n$ integrals of the motion by any direct process. Nevertheless, the presence of a few integrals such as those representing conservation of energy or momentum are very useful in characterizing the motion of a system.

If one uses the Hamiltonian approach to the equations of motion, integrals of the motion have the form

$$f_j(q, p, t) = \alpha_j \qquad (j = 1, \ldots, 2n) \tag{2.147}$$

Under the proper conditions, the Hamiltonian function itself can be an integral of the motion.

Conservative system

A common example of an integral of the motion is the *energy integral* $E(q, \dot{q}, t)$ which is quadratic in the $\dot{q}$s and is expressed in units of energy. It satisfies the equation

$$E(q, \dot{q}, t) = h \tag{2.148}$$

where h is a constant that is normally evaluated from initial conditions. For scleronomic systems with $T = T_2$, the energy integral, if it exists, is equal to the sum of the kinetic and potential energies. More generally, however, the energy integral is not equal to the total energy. Usually it is not an explicit function of time.

Let us define a *conservative system* as a dynamical system for which an energy integral can be found. To obtain sufficient conditions for the existence of an energy integral, let us consider a system which is described by the standard nonholonomic form of Lagrange's equation.

$$\frac{d}{dt} \left(\frac{\partial L}{\partial \dot{q}_i} \right) - \frac{\partial L}{\partial q_i} = \sum_{j=1}^{m} \lambda_j a_{ji} \qquad (i = 1, \ldots, n) \tag{2.149}$$

This equation is valid for holonomic or nonholonomic systems whose applied forces are derivable from a potential energy function $V(q, t)$.

Multiply (2.149) by $\dot{q}_i$ and sum over i. The result is

$$\sum_{i=1}^{n} \left[\frac{d}{dt} \left(\frac{\partial L}{\partial \dot{q}_i} \right) - \frac{\partial L}{\partial q_i} \right] \dot{q}_i = \sum_{i=1}^{n} \sum_{j=1}^{m} \lambda_j a_{ji} \dot{q}_i \tag{2.150}$$

We note that

$$\sum_{i=1}^{n} \frac{d}{dt} \left(\frac{\partial L}{\partial \dot{q}_i} \right) \dot{q}_i = \frac{d}{dt} \left[\sum_{i=1}^{n} \frac{\partial L}{\partial \dot{q}_i} \dot{q}_i \right] - \sum_{i=1}^{n} \frac{\partial L}{\partial \dot{q}_i} \ddot{q}_i \tag{2.151}$$

where

$$\frac{d}{dt} \left[\sum_{i=1}^{n} \frac{\partial L}{\partial \dot{q}_i} \dot{q}_i \right] = 2\dot{T}_2 + \dot{T}_1 \tag{2.152}$$

Furthermore,

$$\sum_{i=1}^{n} \frac{\partial L}{\partial \dot{q}_i} \ddot{q}_i = \dot{L} - \sum_{i=1}^{n} \frac{\partial L}{\partial q_i} \dot{q}_i - \frac{\partial L}{\partial t} \tag{2.153}$$

and

$$\dot{L} = \dot{T}_2 + \dot{T}_1 + \dot{T}_0 - \dot{V} \tag{2.154}$$

Hence, from (2.150)–(2.154), we obtain

$$\dot{T}_2 - \dot{T}_0 + \dot{V} = \sum_{i=1}^{n} \sum_{j=1}^{m} \lambda_j a_{ji} \dot{q}_i - \frac{\partial L}{\partial t} \tag{2.155}$$

If the right-hand side of (2.155) remains equal to zero as the motion proceeds, then $E = T_2 - T_0 + V$ is an energy integral and the system is conservative. The first term on the right-hand side will be zero if the nonholonomic constraint equations of the form of (2.113) have all $a_{jt} = 0$; that is; if they are all *catastatic*. The second term on the right-hand side will be zero if neither T nor V is an explicit function of time.

To summarize, a system having holonomic or nonholonomic constraints will be *conservative* if it meets the following conditions:

1. The standard form of Lagrange's equation, as given by (2.149) applies.
2. All constraints can be written in the form

$$\sum_{i=1}^{n} a_{ji} \dot{q}_i = 0 \qquad (j = 1, \ldots, m) \tag{2.156}$$

that is, all $a_{jt} = 0$.
3. The Lagrangian function $L = T - V$ is not an explicit function of time.

These are *sufficient conditions* for a conservative system. For systems with *independent* qs, Lagrange's equation has the simpler form

$$\frac{d}{dt}\left(\frac{\partial L}{\partial \dot{q}_i}\right) - \frac{\partial L}{\partial q_i} = 0 \qquad (i = 1, \ldots, n) \tag{2.157}$$

If this equation applies, and if $\partial L/\partial t = 0$, then the system is *conservative*.

Now let us consider the Hamiltonian approach to conservative systems. Suppose that a system can be described by Hamilton's equations of the form

$$\dot{q}_i = \frac{\partial H}{\partial p_i}, \qquad \dot{p}_i = -\frac{\partial H}{\partial q_i} + \sum_{j=1}^{n} \lambda_j a_{ji} \qquad (i = 1, \ldots, n) \tag{2.158}$$

The Hamiltonian function will be a constant of the motion if

$$\dot{H} = \sum_{i=1}^{n} \frac{\partial H}{\partial q_i}\dot{q}_i + \sum_{i=1}^{n} \frac{\partial H}{\partial p_i}\dot{p}_i + \frac{\partial H}{\partial t} = 0 \tag{2.159}$$

Substituting from (2.158) into (2.159), we find that

$$\dot{H} = \sum_{i=1}^{n}\sum_{j=1}^{m} \lambda_j a_{ji}\dot{q}_i + \frac{\partial H}{\partial t} \tag{2.160}$$

We see that the system will be conservative and $H(q, p)$ will be a constant of the motion if

$$\sum_{i=1}^{n} a_{ji}\dot{q}_i = 0 \qquad (j = 1, \ldots, m) \tag{2.161}$$

that is, if $a_{jt} = 0$ for all j, and if the Hamiltonian function is not an explicit function of time, implying that

$$\frac{\partial H}{\partial t} = 0 \tag{2.162}$$

These conditions are equivalent to those found earlier with the Lagrangian approach. Thus, the energy integral is

$$H = T_2 - T_0 + V = E \tag{2.163}$$

Finally, it should be noted from the first equality of (2.159) that if the canonical equations (2.100) apply, then

$$\dot{H} = \frac{\partial H}{\partial t} \tag{2.164}$$

whether the system is conservative or not.

Example 2.5   A particle of mass m can slide without friction on a rigid wire in the form of a circle of radius r, as shown in Fig. 2.5. The circular wire rotates about a vertical axis

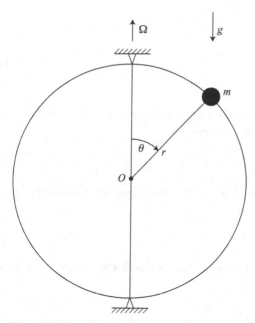

Figure 2.5.

through the center O with a constant angular velocity Ω. We wish to determine if this system is conservative.

First, notice that this is a rheonomic holonomic system. It is possible, however, to choose a single independent generalized coordinate θ. The system is described by Lagrange's equation of the form

$$\frac{d}{dt}\left(\frac{\partial L}{\partial \dot{\theta}}\right) - \frac{\partial L}{\partial \theta} = 0 \tag{2.165}$$

as in (2.157). The Lagrangian function is

$$L = T - V = \frac{1}{2}m\left(r^2\dot{\theta}^2 + r^2\Omega^2 \sin^2\theta\right) - mgr\cos\theta \tag{2.166}$$

which is not an explicit function of time. Hence, the sufficient conditions for a conservative system are met. The energy integral, which is constant during the motion is

$$E = T_2 - T_0 + V = \frac{1}{2}mr^2\dot{\theta}^2 - \frac{1}{2}mr^2\Omega^2 \sin^2\theta + mgr\cos\theta \tag{2.167}$$

This, of course, is different from the total energy $T + V$.

It is interesting to study the same system using the Cartesian coordinates (x, y, z) to specify the position of the particle. Let the Cartesian frame be fixed in space with its origin at the center O and with the positive z-axis pointing upward and lying on the axis of rotation. Choose the time reference such that, at $t = 0$, the circular wire lies in the xz-plane. The transformation equations are

$$x = r \sin \theta \cos \Omega t \tag{2.168}$$
$$y = r \sin \theta \sin \Omega t \tag{2.169}$$
$$z = r \cos \theta \tag{2.170}$$

Notice the explicit functions of time, confirming that the system is rheonomic.

The Lagrangian function has the simple form

$$L = T - V = \frac{1}{2} m (\dot{x}^2 + \dot{y}^2 + \dot{z}^2) - mgz \tag{2.171}$$

which is not an explicit function of time. There are two holonomic constraints, namely,

$$\phi_1 = x^2 + y^2 + z^2 - r^2 = 0 \tag{2.172}$$
$$\phi_2 = x \tan \Omega t - y = 0 \tag{2.173}$$

Upon differentiation with respect to time, we obtain forms that are linear in the $\dot{q}$s, that is,

$$\dot{\phi}_1 = 2 (x\dot{x} + y\dot{y} + z\dot{z}) = 0 \tag{2.174}$$
$$\dot{\phi}_2 = \dot{x} \tan \Omega t - \dot{y} + \Omega x \sec^2 \Omega t = 0 \tag{2.175}$$

The second constraint equation has $a_{jt} \neq 0$, so the sufficient conditions for a conservative system are not met with this Cartesian formulation. Here the energy function would equal the total energy $T + V$, which is not constant because there is a nonzero driving moment about the vertical axis.

Nevertheless, the energy function found earlier in (2.167) remains constant and is a valid energy integral. When expressed in terms of Cartesian coordinates, it is

$$E = \frac{1}{2} m [(\dot{x} \cos \Omega t + \dot{y} \sin \Omega t)^2 + \dot{z}^2] - \frac{1}{2} m \Omega^2 (x^2 + y^2) + mgz \tag{2.176}$$

We conclude that the system should be classed as *conservative* even though it does not meet the sufficient conditions in the Cartesian formulation.

Ignorable coordinates

Consider a *holonomic* system which is described by Hamilton's canonical equations.

$$\dot{q}_i = \frac{\partial H}{\partial p_i}, \qquad \dot{p}_i = -\frac{\partial H}{\partial q_i} \qquad (i = 1, \ldots, n) \tag{2.177}$$

Now suppose that the Hamiltonian function has the form $H(q_{k+1}, \ldots, q_n, p_1, \ldots, p_n, t)$, that is, the first k qs do not appear. Then we find that

$$\dot{p}_i = -\frac{\partial H}{\partial q_i} = 0 \qquad (i = 1, \ldots, k) \tag{2.178}$$

and therefore the first k generalized momenta are

$$p_i = \beta_i \qquad (i = 1, \ldots, k) \tag{2.179}$$

where the βs are constants. The k qs which do not appear in the Hamiltonian function are called *ignorable coordinates*. The constant ps are *integrals of the motion*, in accordance with (2.147).

The Hamiltonian function can now be written in the form $H(q_{k+1}, \ldots, q_n, p_{k+1}, \ldots, p_n, \beta_1, \ldots, \beta_k, t)$. The canonical equations, assuming k ignorable coordinates, are

$$\dot{q}_i = \frac{\partial H}{\partial p_i}, \qquad \dot{p}_i = -\frac{\partial H}{\partial q_i} \qquad (i = k+1, \ldots, n) \tag{2.180}$$

Thus, k degrees of freedom corresponding to the k ignorable coordinates have been removed from the equations of motion. The motion of the ignorable coordinates can be recovered by integrating

$$\dot{q}_i = \frac{\partial H}{\partial \beta_i} \qquad (i = 1, \ldots, k) \tag{2.181}$$

Now let us consider ignorable coordinates from the Lagrangian viewpoint. We assume a holonomic system whose equations of motion have the standard Lagrangian form

$$\frac{d}{dt}\left(\frac{\partial L}{\partial \dot{q}_i}\right) - \frac{\partial L}{\partial q_i} = 0 \qquad (i = 1, \ldots, n) \tag{2.182}$$

First, recall from (2.97) that

$$\frac{\partial L}{\partial q_i} = -\frac{\partial H}{\partial q_i} \qquad (i = 1, \ldots, n) \tag{2.183}$$

Hence, if a certain ignorable coordinate q_i is missing from the Hamiltonian function, it will also be missing from the Lagrangian function. If the first k qs are ignorable, the Lagrangian L will be a function of $(q_{k+1}, \ldots, q_n, \dot{q}_1, \ldots, \dot{q}_n, t)$. We would like to eliminate the ignorable $\dot{q}$s from the Lagrangian formulation in order to reduce the number of degrees of freedom in the equations of motion.

This goal may be accomplished by first defining a *Routhian function* $R(q_{k+1}, \ldots, q_n, \dot{q}_{k+1}, \ldots, \dot{q}_n, \beta_1, \ldots, \beta_k, t)$ as follows:

$$R = L - \sum_{i=1}^{k} \beta_i \dot{q}_i \tag{2.184}$$

where the ignorable $\dot{q}$s have been eliminated by solving the k equations

$$\frac{\partial L}{\partial \dot{q}_i} = \beta_i \qquad (i = 1, \ldots, k) \tag{2.185}$$

for $(\dot{q}_1, \ldots, \dot{q}_k)$ in terms of $(q_{k+1}, \ldots q_n, \dot{q}_{k+1}, \ldots, \dot{q}_n, \beta_1, \ldots, \beta_k, t)$.

Now let us make an arbitrary variation in the Routhian function, including variations in the βs and time. We obtain

$$\delta R = \sum_{i=k+1}^{n} \frac{\partial R}{\partial q_i}\delta q_i + \sum_{i=k+1}^{n} \frac{\partial R}{\partial \dot{q}_i}\delta \dot{q}_i + \sum_{i=1}^{k} \frac{\partial R}{\partial \beta_i}\delta \beta_i + \frac{\partial R}{\partial t}\delta t \tag{2.186}$$

Next, take the variation of the right-hand side of (2.184). We have

$$\delta \left(L - \sum_{i=1}^{k} \beta_i \dot{q}_i \right) = \sum_{i=k+1}^{n} \frac{\partial L}{\partial q_i} \delta q_i + \sum_{i=1}^{k} \frac{\partial L}{\partial \dot{q}_i} \delta \dot{q}_i + \sum_{i=k+1}^{n} \frac{\partial L}{\partial \dot{q}_i} \delta \dot{q}_i + \frac{\partial L}{\partial t} \delta t$$

$$- \sum_{i=1}^{k} \beta_i \delta \dot{q}_i - \sum_{i=1}^{k} \dot{q}_i \delta \beta_i \qquad (2.187)$$

Using (2.185), this simplifies to

$$\delta \left(L - \sum_{i=1}^{k} \beta_i \dot{q}_i \right) = \sum_{i=k+1}^{n} \frac{\partial L}{\partial q_i} \delta q_i + \sum_{i=k+1}^{n} \frac{\partial L}{\partial \dot{q}_i} \delta \dot{q}_i - \sum_{i=1}^{k} \dot{q}_i \delta \beta_i + \frac{\partial L}{\partial t} \delta t \qquad (2.188)$$

We assume that the varied quantities in (2.186) and (2.188) are independent, and therefore the corresponding coefficients must be equal. Thus,

$$\frac{\partial L}{\partial q_i} = \frac{\partial R}{\partial q_i} \qquad (i = k+1, \ldots, n) \qquad (2.189)$$

$$\frac{\partial L}{\partial \dot{q}_i} = \frac{\partial R}{\partial \dot{q}_i} \qquad (i = k+1, \ldots, n) \qquad (2.190)$$

$$\dot{q}_i = -\frac{\partial R}{\partial \beta_i} \qquad (i = 1, \ldots, k) \qquad (2.191)$$

$$\frac{\partial L}{\partial t} = \frac{\partial R}{\partial t} \qquad (2.192)$$

Now let us substitute from (2.189) and (2.190) into Lagrange's equation, (2.182). We obtain

$$\frac{d}{dt} \left(\frac{\partial R}{\partial \dot{q}_i} \right) - \frac{\partial R}{\partial q_i} = 0 \qquad (i = k+1, \ldots, n) \qquad (2.193)$$

These equations are of the form of Lagrange's equation with the Routhian function replacing the Lagrangian function. There are $(n-k)$ second-order equations in the nonignorable variables. Thus, the Routhian procedure has succeeded in eliminating the ignorable coordinates from the equations of motion and, in effect, has reduced the number of degrees of freedom to $(n-k)$. Usually there is no need to solve for the ignorable coordinates, but, if necessary, they can be recovered by integrating (2.191). The k integrals of the motion associated with the ignorable coordinates are given by (2.185) and are equal to the corresponding generalized momenta.

Example 2.6   Consider the same system as in Example 2.5 on page 94 (Fig. 2.5) except that it can rotate *freely* about the fixed vertical axis through the center, the angle of rotation being ϕ. We find that (ϕ, θ) are the generalized coordinates and the Lagrangian function is

$$L = T - V = \frac{1}{2} mr^2 \dot{\phi}^2 \sin^2 \theta + \frac{1}{2} mr^2 \dot{\theta}^2 - mgr \cos \theta \qquad (2.194)$$

We see that ϕ is an ignorable coordinate since it does not appear in the Lagrangian function. The corresponding generalized momentum is

$$p_\phi = \frac{\partial L}{\partial \dot{\phi}} = mr^2 \dot{\phi} \sin^2 \theta = \beta_\phi \tag{2.195}$$

where β_ϕ is a constant that is usually evaluated from initial conditions. Solving for $\dot{\phi}$ from (2.195), we obtain

$$\dot{\phi} = \frac{\beta_\phi}{mr^2 \sin^2 \theta} \tag{2.196}$$

This expression is used in obtaining the Routhian function.

$$R = L - \beta_\phi \dot{\phi} = -\frac{\beta_\phi^2}{2mr^2 \sin^2 \theta} + \frac{1}{2} mr^2 \dot{\theta}^2 - mgr \cos \theta \tag{2.197}$$

The θ equation of motion is obtained from

$$\frac{d}{dt} \left(\frac{\partial R}{\partial \dot{\theta}} \right) - \frac{\partial R}{\partial \theta} = 0 \tag{2.198}$$

It is

$$mr^2 \ddot{\theta} - \frac{\beta_\phi^2 \cos \theta}{mr^2 \sin^3 \theta} - mgr \sin \theta = 0 \tag{2.199}$$

Notice that the ignorable coordinate ϕ and its derivatives are missing from this θ equation. Thus, it can be integrated to give θ as a function of time if the initial conditions are known. On the other hand, the Lagrangian procedure will result in two coupled second-order differential equations in ϕ and θ and their derivatives. Of course, either procedure will yield the same results when the equations of motion are completely integrated.

2.4   Dissipative and gyroscopic forces

Gyroscopic forces and some dissipative forces are represented in the differential equations of motion by terms which are *linear in the $\dot{q}$s*. Suppose, for example, that the differential equations for a system have the form

$$\sum_{j=1}^{n} m_{ij}(q, t) \ddot{q}_j + \sum_{j=1}^{n} f_{ij}(q, t) \dot{q}_j + h_i(q, \dot{q}, t) = 0 \qquad (i = 1, \ldots, n) \tag{2.200}$$

The $n \times n$ coefficient matrix $\mathbf{f}$ can be expressed as the sum of a symmetric matrix $\mathbf{c}$ and a skew-symmetric matrix $\mathbf{g}$ where

$$f_{ij}(q, t) = c_{ij}(q, t) + g_{ij}(q, t) \qquad (i, j = 1, \ldots, n) \tag{2.201}$$

and

$$c_{ij} = c_{ji} \qquad (i, j = 1, \ldots, n) \tag{2.202}$$

$$g_{ij} = -g_{ji} \qquad (i, j = 1, \ldots, n) \tag{2.203}$$

Assuming that the **c** matrix is positive definite or positive semidefinite, the terms in the equations of motion of the form $c_{ij}\dot{q}_j$ are *dissipative* in nature while the terms $g_{ij}\dot{q}_j$ are *gyroscopic* and nondissipative.

Rayleigh's dissipation function

Dissipative terms are not present in equations of motion obtained by using the standard holonomic or nonholonomic forms of Lagrange's equation because these terms are not derivable from a potential energy function $V(q, t)$. Rather, dissipative terms are often introduced through an applied generalized force Q_i' where

$$Q_i' = -\sum_{j=1}^{n} c_{ij}(q, t)\dot{q}_j \qquad (i = 1, \ldots, n) \tag{2.204}$$

The cs are called *damping coefficients* and form a symmetric matrix which is positive definite or positive semidefinite for systems with passive linear dampers.

Now let us define *Rayleigh's dissipation function* $F(q, \dot{q}, t)$ using the equation

$$F = \frac{1}{2} \sum_{i=1}^{n} \sum_{j=1}^{n} c_{ij}\dot{q}_i\dot{q}_j \tag{2.205}$$

Then we see that

$$Q_i' = -\frac{\partial F}{\partial \dot{q}_i} \qquad (i = 1, \ldots, n) \tag{2.206}$$

We can obtain the equations of motion from

$$\frac{d}{dt}\left(\frac{\partial L}{\partial \dot{q}_i}\right) - \frac{\partial L}{\partial q_i} + \frac{\partial F}{\partial \dot{q}_i} = 0 \qquad (i = 1, \ldots, n) \tag{2.207}$$

where we assume that the damping forces are the only applied generalized forces which are not derived from a potential energy function.

The rate at which these linear friction forces dissipate energy is

$$-\sum_{i=1}^{n} Q_i'\dot{q}_i = \sum_{i=1}^{n} \sum_{j=1}^{n} c_{ij}\dot{q}_i\dot{q}_j = 2F \tag{2.208}$$

Thus, Rayleigh's dissipation function is equal to half the instantaneous rate of dissipation of the total mechanical energy. For passive dampers, this dissipation rate must be positive or zero at all times.

Example 2.7 Given the system shown in Fig. 2.6 which, we note, is a linear system with damping. First, let us obtain the differential equations of motion. The kinetic and potential energies are

$$T + \frac{1}{2} m\left(\dot{x}_1^2 + \dot{x}_2^2\right) \tag{2.209}$$

$$V = \frac{1}{2} k\left(x_1^2 + x_2^2\right) \tag{2.210}$$

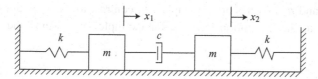

Figure 2.6.

The dissipation function is equal to half the energy dissipation rate, or

$$F = \frac{1}{2}c(\dot{x}_1 - \dot{x}_2)^2, \qquad c > 0 \tag{2.211}$$

Let us use (2.207) in the form

$$\frac{d}{dt}\left(\frac{\partial T}{\partial \dot{q}_i}\right) - \frac{\partial T}{\partial q_i} + \frac{\partial V}{\partial q_i} + \frac{\partial F}{\partial \dot{q}_i} = 0 \tag{2.212}$$

We obtain the following equations of motion:

$$m\ddot{x}_1 + c(\dot{x}_1 - \dot{x}_2) + kx_1 = 0 \tag{2.213}$$
$$m\ddot{x}_2 + c(\dot{x}_2 - \dot{x}_1) + kx_2 = 0 \tag{2.214}$$

Second method   As an alternative approach let us introduce the generalized coordinates

$$q_1 = \frac{1}{2}(x_1 + x_2)$$

$$q_2 = \frac{1}{2}(x_1 - x_2) \tag{2.215}$$

The corresponding transformation equations are

$$x_1 = q_1 + q_2$$
$$x_2 = q_1 - q_2 \tag{2.216}$$

We see that pure q_1 motion with $q_1 = 1$, $q_2 = 0$ implies that $x_1 = x_2 = 1$. On the other hand, pure q_2 motion with $q_2 = 1$, $q_1 = 0$ means that $x_1 = -x_2 = 1$.

The expressions for T and V in terms of generalized coordinates are

$$T = m(\dot{q}_1^2 + \dot{q}_2^2) \tag{2.217}$$
$$V = k(q_1^2 + q_2^2) \tag{2.218}$$

The dissipation function involves relative velocities only and is given by half the dissipation rate, or

$$F = 2c\dot{q}_2^2 \tag{2.219}$$

Then, using (2.212), the equations of motion can be written in the form

$$m\ddot{q}_1 + kq_1 = 0 \tag{2.220}$$
$$m\ddot{q}_2 + 2c\dot{q}_2 + kq_2 = 0 \tag{2.221}$$

We notice that the q_1 and q_2 motions are uncoupled, in contrast to the x motions, and that damping is applied to the q_2 motion only. If, for example, we assume the initial conditions

$$x_1(0) = A, \qquad x_2(0) = A, \qquad \dot{x}_1(0) = \dot{x}_2(0) = 0 \tag{2.222}$$

the solution of (2.220) is

$$q_1 = A \cos \sqrt{\frac{k}{m}} t \tag{2.223}$$

This sinusoidal oscillation continues indefinitely, but any damped q_2 motion disappears as time approaches infinity.

Gyroscopic forces

Gyroscopic terms occur in the differential equations of motion and are of the form $g_{ij}\dot{q}_j$, where the coefficients $g_{ij}(q, t)$ are skew-symmetric, that is, $g_{ij} = -g_{ji}$. A *gyroscopic system* has equations of motion containing gyroscopic terms. These gyroscopic terms arise from T_1 terms in the kinetic energy (or comparable Routhian R_1 terms) when Lagrange's equations are applied.

As an example, recall that T_1 has the form

$$T_1 = \sum_{i=1}^{n} a_i(q, t)\dot{q}_i \tag{2.224}$$

Then

$$\frac{d}{dt}\left(\frac{\partial T_1}{\partial \dot{q}_i}\right) = \dot{a}_i = \sum_{j=1}^{n} \frac{\partial a_i}{\partial q_j}\dot{q}_j + \frac{\partial a_i}{\partial t} \tag{2.225}$$

and

$$\frac{\partial T_1}{\partial q_i} = \sum_{j=1}^{n} \frac{\partial a_j}{\partial q_i}\dot{q}_j \tag{2.226}$$

Hence

$$\frac{d}{dt}\left(\frac{\partial T_1}{\partial \dot{q}_i}\right) - \frac{\partial T_1}{\partial q_i} = \sum_{j=1}^{n}\left(\frac{\partial a_i}{\partial q_j} - \frac{\partial a_j}{\partial q_i}\right)\dot{q}_j + \frac{\partial a_i}{\partial t} \tag{2.227}$$

and we find that the equations of motion contain the gyroscopic terms

$$\sum_{j=1}^{n} g_{ij}\dot{q}_j \equiv \sum_{j=1}^{n}\left(\frac{\partial a_i}{\partial q_j} - \frac{\partial a_j}{\partial q_i}\right)\dot{q}_j \tag{2.228}$$

that is, the *gyroscopic coefficients* are

$$g_{ij} = -g_{ji} = \frac{\partial a_i}{\partial q_j} - \frac{\partial a_j}{\partial q_i} \tag{2.229}$$

Gyroscopic terms, when shifted to the right-hand side of the equation can be considered to be the *gyroscopic forces*

$$G_i = -\sum_{j=1}^{n} g_{ij}\dot{q}_j \qquad (i = 1, \ldots, n) \tag{2.230}$$

Note that G_i is not an applied force, but is *inertial* in nature and always involves coupling between two or more nonignored degrees of freedom. If the gyroscopic terms arise from T_1, the system must be rheonomic. On the other hand, if the Routhian procedure is used, R_1 terms can appear in a scleronomic system.

The rate of doing work by the gyroscopic forces is

$$\sum_{i=1}^{n} G_i\dot{q}_i = -\sum_{i=1}^{n}\sum_{j=1}^{n} g_{ij}\dot{q}_i\dot{q}_j = 0 \tag{2.231}$$

the zero result being due to the skew-symmetry of g_{ij}. In n-dimensional configuration space, the gyroscopic force **G** and the velocity $\dot{\mathbf{q}}$ are orthogonal.

Example 2.8 The Cartesian xy frame rotates at a constant rate Ω relative to the inertial XY frame (Fig. 2.7). A particle of mass m moves in the xy-plane under the action of arbitrary force components $F_x(t)$ and $F_y(t)$. We wish to find the differential equations for the motion of the particle relative to the xy frame.

We can take $V = 0$, so Lagrange's equation has the form

$$\frac{d}{dt}\left(\frac{\partial T}{\partial \dot{q}_i}\right) - \frac{\partial T}{\partial q_i} = Q_i \tag{2.232}$$

The kinetic energy is

$$\begin{aligned} T &= \frac{1}{2}m[(\dot{x} - \Omega y)^2 + (\dot{y} + \Omega x)^2] \\ &= \frac{1}{2}m(\dot{x}^2 + \dot{y}^2) + m\Omega(x\dot{y} - y\dot{x}) + \frac{1}{2}m\Omega^2(x^2 + y^2) \end{aligned} \tag{2.233}$$

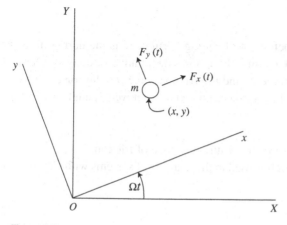

Figure 2.7.

We find that

$$\frac{d}{dt}\left(\frac{\partial T}{\partial \dot{x}}\right) = m\ddot{x} - m\Omega\dot{y} \tag{2.234}$$

$$\frac{\partial T}{\partial x} = m\Omega\dot{y} + m\Omega^2 x \tag{2.235}$$

The x equation is

$$m\ddot{x} - 2m\Omega\dot{y} - m\Omega^2 x = F_x \tag{2.236}$$

Similarly,

$$\frac{d}{dt}\left(\frac{\partial T}{\partial \dot{y}}\right) = m\ddot{y} + m\Omega\dot{x} \tag{2.237}$$

$$\frac{\partial T}{\partial y} = -m\Omega\dot{x} + m\Omega^2 y \tag{2.238}$$

and the y equation is

$$m\ddot{y} + 2m\Omega\dot{x} - m\Omega^2 y = F_y \tag{2.239}$$

The *gyroscopic terms* are $-2m\Omega\dot{y}$ in the x equation and $2m\Omega\dot{x}$ in the y equation. Note that they arise from the T_1 (middle) term of the kinetic energy expression, they represent coupling terms, and the coefficients satisfy skew-symmetry. We frequently associate gyroscopic effects with rotating and precessing axially symmetric bodies. In this example, however, the gyroscopic terms for a single particle are due to *Coriolis* accelerations. Nevertheless, if we should consider the detailed motions of particles in a precessing gyroscope, we would find that it is the Coriolis accelerations of these particles that produce the gyroscopic moment.

Coulomb friction

The nature of Coulomb friction was introduced in Chapter 1. Briefly, the friction force F_f between two blocks in relative sliding motion is

$$F_f = -\mu N \text{sgn}(v_r) \tag{2.240}$$

where the constant μ is the coefficient of sliding friction, N is the normal force between the blocks, and the direction of F_f directly opposes the relative velocity v_r. During sliding the vector sum of the normal force N and the friction force μN lies on a cone of friction whose axis is normal to the sliding surfaces and whose semivertex angle ϕ is given by

$$\tan\phi = \mu \tag{2.241}$$

If there is no sliding, the total force lies within the cone of friction.

When the Lagrangian approach is used in the analysis of systems with Coulomb friction, one can write

$$\frac{d}{dt}\left(\frac{\partial L}{\partial \dot{q}_i}\right) - \frac{\partial L}{\partial q_i} = Q_i' \qquad (i = 1, \dots, n) \tag{2.242}$$

and include the generalized Coulomb friction forces in the nonpotential Q_i'. With this approach, Q_i' might actually be a frictional moment, for example.

Another approach which is particularly applicable to systems with rotating sliding surfaces is to consider the effect of tangential frictional stresses at every point on the surface. The frictional stress σ_f resembles a shear stress and is equal in magnitude to $\mu\sigma_n$ where σ_n is the normal stress or pressure acting at the given point on the sliding surface. The total frictional force or moment is found by integrating the frictional stress over the entire sliding contact area.

Example 2.9   A vertical shaft, with a hemispherical joint of radius R rotates at $\dot{\phi}$ rad/s (Fig. 2.8). The shaft exerts a downward force F, resulting in a normal stress on the surface of the joint given by

$$\sigma_n = \sigma_0 \cos\theta \tag{2.243}$$

where σ_0 is the maximum normal stress. Assuming a coefficient of sliding friction μ, we wish to solve for the frictional moment Q_ϕ'.

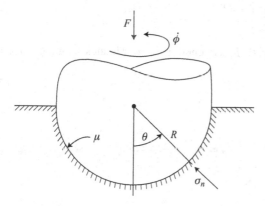

Figure 2.8.

First, let us establish a relation between σ_0 and the applied force F. Consider the vertical component of the force due to the normal stress σ_n acting on a circular strip of circumference $2\pi R \sin\theta$ and width $R d\theta$. This leads to the integral

$$F = \int_0^{\pi/2} 2\pi R^2 \sigma_0 \sin\theta \cos^2\theta \, d\theta = \frac{2}{3}\pi R^2 \sigma_0 \tag{2.244}$$

and we find that the maximum normal stress is

$$\sigma_0 = \frac{3F}{2\pi R^2} \tag{2.245}$$

The frictional stress is

$$\sigma_f = \mu\sigma_n = \frac{3\mu F \cos\theta}{2\pi R^2} \tag{2.246}$$

and acts circumferentially on each circular strip with a moment arm equal to $R \sin \theta$. Thus, we obtain the integral

$$Q'_\phi = -\int_0^{\pi/2} 2\pi \sigma_f R^3 \sin^2 \theta \, d\theta$$

$$= -\int_0^{\pi/2} 3\mu F R \sin^2 \theta \cos \theta \, d\theta = -\mu F R \tag{2.247}$$

The negative sign indicates that the frictional moment Q'_ϕ opposes the rotation ϕ.

Example 2.10 A particle of mass m can slide inside a straight tube which is inclined at an angle θ from the vertical and is rigidly attached to a vertical shaft which rotates at a constant rate Ω (Fig. 2.9). There is a Coulomb friction coefficient μ between the particle and the tube. We wish to find the differential equation of motion.

Let us use Lagrange's equation in the general form

$$\frac{d}{dt}\left(\frac{\partial T}{\partial \dot{q}_i}\right) - \frac{\partial T}{\partial q_i} + \frac{\partial V}{\partial q_i} = \sum_{j=1}^m \lambda_j a_{ji} + Q'_i \tag{2.248}$$

Choose the spherical coordinates (r, θ, ϕ) as generalized coordinates with the constraints

$$\dot{\theta} = 0, \qquad \dot{\phi} - \Omega = 0 \tag{2.249}$$

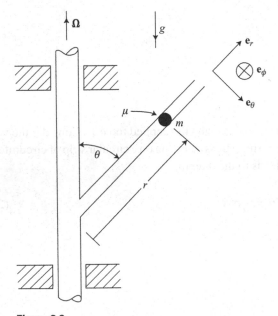

Figure 2.9.

This approach enables one to obtain the normal constraint forces from the λs. The unconstrained kinetic energy is

$$T = \frac{m}{2}(\dot{r}^2 + r^2\dot{\theta}^2 + r^2\dot{\phi}^2\sin^2\theta) \tag{2.250}$$

and the potential energy is

$$V = mgr\cos\theta \tag{2.251}$$

The friction force is entirely in the radial (e_r) direction so we find that

$$Q'_\theta = Q'_\phi = 0, \qquad Q'_r = -\mu N\,\mathrm{sgn}(\dot{r}) \tag{2.252}$$

where N is the normal force of the tube acting on the particle.

Lagrange's equation results in the following r equation:

$$m\ddot{r} - mr\dot{\phi}^2\sin^2\theta + mg\cos\theta = Q'_r$$

or

$$m\ddot{r} - mr\Omega^2\sin^2\theta + mg\cos\theta = -\mu N\,\mathrm{sgn}(\dot{r}) \tag{2.253}$$

The normal tube force is

$$\mathbf{N} = N_\theta\mathbf{e}_\theta + N_\phi\mathbf{e}_\phi \tag{2.254}$$

and has a magnitude

$$N = \sqrt{N_\theta^2 + N_\phi^2} \geq 0 \tag{2.255}$$

In order to find expressions for N_θ and N_ϕ we need to write the differential equations for θ and ϕ. The θ equation, obtained by using (2.248), is

$$mr^2\ddot{\theta} + 2mr\dot{r}\dot{\theta} - mr^2\dot{\phi}^2\sin\theta\cos\theta - mgr\sin\theta = \lambda_1$$

or

$$-mr^2\Omega^2\sin\theta\cos\theta - mgr\sin\theta = \lambda_1 = rN_\theta \tag{2.256}$$

Since θ is an angle, λ_1 must be a moment which is equal to the virtual work per unit $\delta\theta$. Similarly, the ϕ equation is

$$mr^2\ddot{\phi}\sin^2\theta + 2mr\dot{r}\dot{\phi}\sin^2\theta + 2mr^2\dot{\theta}\dot{\phi}\sin\theta\cos\theta = \lambda_2$$

or

$$2mr\dot{r}\Omega\sin^2\theta = \lambda_2 = r\sin\theta N_\phi \tag{2.257}$$

From (2.256) and (2.257) we obtain

$$N_\theta = -mr\Omega^2\sin\theta\cos\theta - mg\sin\theta \tag{2.258}$$

$$N_\phi = 2m\Omega\dot{r}\sin\theta \tag{2.259}$$

Then, from (2.255) we obtain

$$N = m \sin\theta [4\Omega^2 \dot{r}^2 + (r\Omega^2 \cos\theta + g)^2]^{1/2} \tag{2.260}$$

The required differential equation of motion is furnished by the r equation of (2.253) with the expression for N from (2.260) substituted in its right-hand side.

Now suppose that $0 < \theta < \frac{\pi}{2}$ and there are initial conditions $r(0) = r_0$, $\dot{r}(0) = 0$. We desire to find the limits on Ω^2 for incipient motion either outward or inward. The differential equation for incipient outward motion is

$$\ddot{r} = r_0\Omega^2 \sin^2\theta - \mu r_0\Omega^2 \sin\theta \cos\theta - \mu g \sin\theta - g \cos\theta > 0 \tag{2.261}$$

or

$$\Omega^2 > \frac{g}{r_0} \left[\frac{\cos\theta + \mu \sin\theta}{\sin\theta(\sin\theta - \mu \cos\theta)} \right], \qquad 0 \le \mu < \tan\theta \tag{2.262}$$

The differential equation for incipient inward motion is

$$\ddot{r} = r_0\Omega^2 \sin^2\theta + \mu r_0\Omega^2 \sin\theta \cos\theta + \mu g \sin\theta - g \cos\theta < 0 \tag{2.263}$$

or

$$\Omega^2 < \frac{g}{r_0} \left[\frac{\cos\theta - \mu \sin\theta}{\sin\theta(\sin\theta + \mu \cos\theta)} \right], \qquad 0 \le \mu < \cot\theta \tag{2.264}$$

The limits on μ can be visualized in terms of the cone of friction. For incipient motion outward, the total reaction force of the tube acting on the particle cannot have a downward component since it could not be counteracted by either gravity or centrifugal force effects. Similarly, for incipient inward motion, the reaction force of the tube on the particle cannot have an outward component since there is no counteracting force available. In both cases, the symmetry axis of the cone of friction is normal to the tube and passes through the axis of rotation.

Example 2.11   A particle of mass m is embedded at a distance $\frac{1}{2}r$ from the center of a massless disk of radius r (Fig. 2.10). The disk can roll down a plane inclined at an angle $\alpha = 30°$ with the horizontal. First, we wish to find the differential equation for the rotation angle θ, assuming no slipping. Then, for the initial conditions $\theta(0) = 0$, $\dot{\theta}(0) = 0$, and assuming a Coulomb friction coefficient $\mu = \frac{1}{2}$, we wish to find the values of the angle θ at which slipping first begins and then ends.

Let us obtain the differential equation of motion by using Lagrange's equation in the form

$$\frac{d}{dt}\left(\frac{\partial T}{\partial \dot{\theta}}\right) - \frac{\partial T}{\partial \theta} + \frac{\partial V}{\partial \theta} = 0 \tag{2.265}$$

The velocity of the particle as it rotates about the contact point C is

$$v = \left[r^2 + \left(\frac{1}{2}r\right)^2 + 2r\left(\frac{1}{2}r\right)\cos\theta \right]^{\frac{1}{2}} \dot{\theta} \tag{2.266}$$

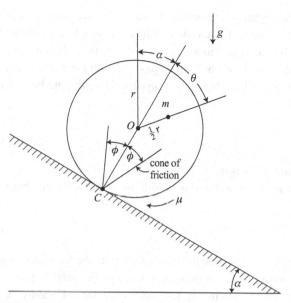

Figure 2.10.

where the coefficient of $\dot{\theta}$ is equal to the distance from C to the particle.
The kinetic energy is

$$T = \frac{1}{2}mv^2 = \frac{1}{2}mr^2\dot{\theta}^2\left(\frac{5}{4} + \cos\theta\right) \qquad (2.267)$$

The potential energy is

$$V = mg\left[-r\theta\sin\alpha + \frac{1}{2}r\cos(\theta + \alpha)\right]$$

$$= mgr\left(-\frac{1}{2}\theta + \frac{\sqrt{3}}{4}\cos\theta - \frac{1}{4}\sin\theta\right) \qquad (2.268)$$

We find that

$$\frac{d}{dt}\left(\frac{\partial T}{\partial\dot{\theta}}\right) = mr^2\ddot{\theta}\left(\frac{5}{4} + \cos\theta\right) - mr^2\dot{\theta}^2\sin\theta \qquad (2.269)$$

$$\frac{\partial T}{\partial\theta} = -\frac{1}{2}mr^2\dot{\theta}^2\sin\theta \qquad (2.270)$$

$$\frac{\partial V}{\partial\theta} = -mgr\left(\frac{1}{2} + \frac{\sqrt{3}}{4}\sin\theta + \frac{1}{4}\cos\theta\right) \qquad (2.271)$$

From (2.265) the differential equation of motion, assuming no slipping, is found to be

$$mr^2\ddot{\theta}\left(\frac{5}{4} + \cos\theta\right) - \frac{1}{2}mr^2\dot{\theta}^2\sin\theta - mgr\left(\frac{1}{2} + \frac{\sqrt{3}}{4}\sin\theta + \frac{1}{4}\cos\theta\right) = 0 \qquad (2.272)$$

The angle θ at which slipping begins can be obtained directly by using the cone of friction. First, note that since the disk is massless there is no resistance to angular acceleration about the particle. Hence, there can be no applied moment about the particle. This means that the line of action of the force at the contact point C must pass through the particle. If this line of action lies within the cone of friction, there is no slipping. But slipping begins when θ reaches 90°, at which time

$$\tan \phi = \frac{\frac{1}{2}r}{r} = \mu \tag{2.273}$$

and the particle lies on the cone of friction.

Slipping will continue as long as the particle is outside the cone of friction but will end when it re-enters the cone at

$$\theta = 90° + 2\phi = 143.1° \tag{2.274}$$

During the period when the particle is outside the cone of friction, the force at C must drop to zero since the line of action of any force must pass through the particle and lie on or within that cone. Thus, the particle moves in free fall under the action of gravity during the slipping period, and continuously approaches the inclined plane. It is reasonable to assume that the disk maintains contact with the plane during the free fall of the particle since $\ddot{\theta}$ is continuously positive and one might consider the effect of an infinitesimal rotational inertia about the particle.

At the instant when the particle re-enters the cone of friction, the sliding suddenly stops and the particle velocity must be perpendicular to the cone of friction. This means that an impulse must be applied to the particle to cause a sudden change in velocity, namely, the velocity component toward C due to sliding must be canceled. Finally, notice that the range of values of θ over which slipping occurs turns out to be independent of the inclination angle α.

2.5   Configuration space and phase space

Configuration space

In Chapter 1 we introduced the idea of configuration space. We defined the *configuration* of a system to be specified by the values of its n generalized coordinates. Thus, at any given time, the configuration is represented by a configuration point C having a position vector $\mathbf{q}$ in an n-dimensional *configuration space* or q-*space*. As time proceeds, the point C traces a solution path or trajectory in configuration space (Fig. 1.19).

If the system has m *holonomic constraints* of the form

$$\phi_j(q, t) = 0 \qquad (j = 1, \ldots, m) \tag{2.275}$$

each constraint is represented by a surface in configuration space on which the configuration point C must move. Thus, for m independent holonomic constraints, the point C is confined to the common intersection of these surfaces, which is itself a subspace of $(n - m)$ dimensions. In general, the constraint surfaces are moving, but they are fixed if the constraints are scleronomic of the form $\phi_j(q)$.

It is always possible to give the location of C within the constraint intersection subspace by defining $(n - m)$ new generalized coordinates which are *independent* and specify the configuration of the system. In other words, it is always possible, in theory, to find a set of $(n - m)$ *independent* generalized coordinates which are consistent with the constraints and define the configuration of a holonomic system. These independent qs can then be used in writing equations of motion by means of Lagrange's equations.

A virtual displacement $\delta \mathbf{q}$ which is consistent with the holonomic constraint

$$\phi_j(q_1, \ldots, q_n, t) = 0 \tag{2.276}$$

must satisfy the instantaneous constraint equation

$$\sum_{i=1}^{n} \frac{\partial \phi_j}{\partial q_i} \delta q_i = 0 \tag{2.277}$$

The coefficients of $\partial \phi_j / \partial q_i$ are components of the gradient vector which is perpendicular to the constraint surface. Equation (2.277) states that the dot product of the gradient vector and $\delta \mathbf{q}$ in configuration space is equal to zero. Hence, the two vectors are orthogonal and $\delta \mathbf{q}$ must lie in the tangent plane at the operating point C (Fig. 2.11).

Now let us suppose there are m *nonholonomic constraints* acting on the system. The constraints have the general form

$$f_j(q, \dot{q}, t) = 0 \qquad (j = 1, \ldots, m) \tag{2.278}$$

or, in the usual linear case,

$$\sum_{i=1}^{n} a_{ji}(q, t)\dot{q}_i + a_{jt}(q, t) = 0 \qquad (j = 1, \ldots, m) \tag{2.279}$$

There are no constraint surfaces in configuration space because the constraint equations are not integrable. Since there are no constraint surfaces, the entire n-dimensional space is accessible to the configuration point C. We see that nonholonomic constraints, for any

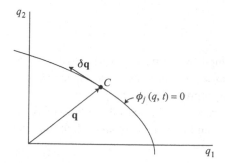

Figure 2.11.

given configuration and time, are essentially restrictions on the possible velocities of the system. For the common case of *catastatic* constraints ($a_{jt} = 0$), the restrictions are on the possible directions of $\dot{\mathbf{q}}$ rather than on its magnitude.

Any virtual displacements of a nonholonomic system must satisfy the *Chetaev* equation

$$\sum_{i=1}^{n} \frac{\partial f_j}{\partial \dot{q}_i} \delta q_i = 0 \qquad (j = 1, \ldots, m) \tag{2.280}$$

or, more commonly,

$$\sum_{i=1}^{n} a_{ji}(q, t) \delta q_i = 0 \qquad (j = 1, \ldots, m) \tag{2.281}$$

which is the instantaneous or virtual constraint equation. The coefficients a_{ji} represent components of a vector associated with the jth constraint and which is normal to the differential surface in which any virtual displacement $\delta \mathbf{q}$ must lie. By properly steering the configuration point C and its associated $(n - m)$-dimensional differential surface, it is always kinematically possible to go between any two points of configuration space if all the constraints are nonholonomic and there are at least two degrees of freedom.

It is sometimes convenient to represent solution paths in an *extended configuration space* or *event space* of $n + 1$ dimensions consisting of the n qs and time. In this space, a solution path will never cross itself. Also, constraint surfaces corresponding to rheonomic holonomic constraints will be fixed rather than moving. When one considers variational methods in dynamics, some procedures are more easily visualized in extended configuration space.

Phase space

We have seen that Lagrange's equations for a dynamical system are n second-order differential equations whose solutions $q_i(t)$ are conveniently expressed as paths in configuration space. For a given dynamical system, there is more than one possible path through a given configuration point because of the variety of possible velocities.

Now consider Hamilton's canonical equations for a holonomic system. They are $2n$ first-order differential equations giving $\dot{q}$s and $\dot{p}$s as functions of (q, p, t). We can consider the qs and ps together to form a $2n$-vector $\mathbf{x} = (q_1, \ldots, q_n, p_1, \ldots, p_n)$ and then the equations of motion have the form

$$\dot{x}_i = X_i(x, t) \qquad (i = 1, \ldots, 2n) \tag{2.282}$$

The motion of the system can be represented by the path of a *phase point P* moving in the $2n$-dimensional *phase space*. Note that a point in phase space specifies not only the configuration but also the state of motion as represented by the ps.

Phase space is particularly convenient in presenting the possible motions of a *conservative holonomic system*. In this case the Hamiltonian function is not an explicit function of time, and the equations of motion have the form

$$\dot{x}_i = X_i(x) \qquad (i = 1, \ldots, 2n) \tag{2.283}$$

or, in detail,

$$\dot{q}_i = \frac{\partial H(q, p)}{\partial p_i}, \qquad \dot{p}_i = -\frac{\partial H(q, p)}{\partial q_i} \qquad (i = 1, \ldots, 2n) \qquad (2.284)$$

Because the system is conservative the Hamiltonian function is constant, that is

$$H(q, p) = h \qquad (2.285)$$

where h is usually evaluated from the initial conditions. This equation represents a surface in phase space, and the corresponding trajectory must lie entirely on this surface. From (2.283) or (2.284) we see that the direction of motion is given at all ordinary points where the velocity is not zero. Hence, there is only one trajectory through each point, and every trajectory is fixed. Thus, the whole of phase space with its trajectories represents the totality of all possible motions of a conservative system.

A point in phase space at which all the $\dot{q}$s and $\dot{p}$s are zero is known as an *equilibrium point* or *singular point*. The corresponding trajectory consists of the equilibrium point only. If we make the usual assumption that the Hamiltonian function has at least two partial derivatives with respect to the qs and ps, it can be shown that an infinite time is required to enter or leave an equilibrium point. Thus, although more than one trajectory may apparently pass through an equilibrium point, it cannot do so in a finite time.

As a simple example of phase space methods, consider the phase plane diagram of Fig. 2.12 which shows the possible solution paths in qp-space for a simple pendulum. Here q is the pendulum angle and p is the corresponding angular momentum. There is a position of stable equilibrium at A, corresponding to $\theta = 0$. Points B and C represent unstable equilibrium positions at $\theta = \pm\pi$.

Now suppose that solution curves are traced in a $2n$-dimensional *state space* consisting of the n qs and n $\dot{q}$s. If one knows the *state* of a system at a certain initial time t_0, then the differential equations will determine its further motion. This is equivalent to knowing the

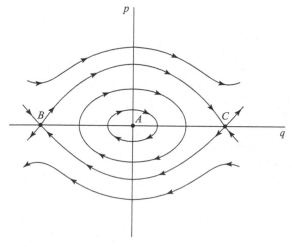

Figure 2.12.

initial values of the qs and ps in phase space. Thus, the trajectories in state space are similar to those in phase space. In particular, the equilibrium points of conservative holonomic systems occur at the same values of qs in both spaces.

Velocity space

The concept of velocity space is important in the analysis of nonholonomic systems. Nonholonomic constraints are essentially constraints on the *velocities* ($\dot{q}$s) of a system for a given configuration and time, whereas holonomic constraints restrict the possible *configurations* (qs) at a given time. Thus, the role of $\dot{q}$s for nonholonomic constraints is similar to that of qs for holonomic constraints.

Velocity space is the n-dimensional space of the $\dot{q}$s. Nonholonomic constraints can be represented as surfaces in velocity space, where the qs and t are regarded as parameters. For example, consider a general nonholonomic constraint of the form

$$f_j(q, \dot{q}, t) = 0 \tag{2.286}$$

which is represented as a curved surface in velocity space (Fig. 2.13).

The constraint force $\mathbf{C}_j$ is perpendicular to the constraint surface; that is, its components are proportional to $\partial f_j / \partial \dot{q}_i$ ($i = 1, \ldots, n$) which are the components of the gradient vector in velocity space. A *virtual velocity* vector $\delta \mathbf{w}$ must lie in the tangent plane at the operating point P. Therefore, $\delta \mathbf{w}$ and $\mathbf{C}_j$ are orthogonal, implying that

$$\sum_{i=1}^{n} C_{ji} \delta w_i = 0 \tag{2.287}$$

similar to a virtual work expression, or

$$\sum_{i=1}^{n} \frac{\partial f_j}{\partial \dot{q}_i} \delta w_i = 0 \tag{2.288}$$

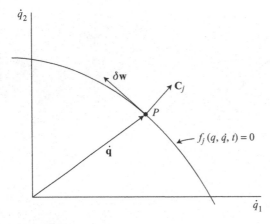

Figure 2.13.

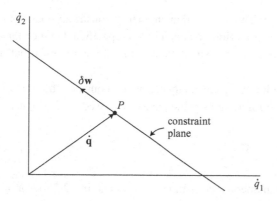

Figure 2.14.

Comparing (2.288) with the Chetaev expression of (2.280) for virtual displacements, we see that the virtual velocity vector $\delta\mathbf{w}$ and the virtual displacement vector $\delta\mathbf{q}$ are constrained to the same allowable directions in space.

Now consider the common case where there are nonholonomic constraints of the linear form

$$\sum_{i=1}^{n} a_{ji}(q,t)\dot{q}_i + a_{jt}(q,t) = 0 \qquad (j = 1, \ldots, m) \tag{2.289}$$

Each constraint is represented by a plane in velocity space (Fig. 2.14). The virtual velocity $\delta\mathbf{w}$ can be considered as a variation $\delta\dot{\mathbf{q}}$ with q and t fixed, and must satisfy

$$\sum_{i=1}^{n} a_{ji}(q,t)\delta w_i = 0 \qquad (j = 1, \ldots, m) \tag{2.290}$$

Therefore, it must lie in the $(n - m)$-dimensional intersection of the constraint planes.

The distance d from the origin of a constraint plane in velocity space is

$$d = \frac{|a_{jt}|}{\left[\sum_i a_{ji}^2\right]^{\frac{1}{2}}} \tag{2.291}$$

We see that if the nonholonomic constraint is catastatic ($a_{jt} = 0$), then the constraint plane goes through the origin. In this case, if a velocity $\dot{\mathbf{q}}$ satisfies the constraints, then $\dot{\mathbf{q}}$ multiplied by an arbitrary scalar constant also satisfies the constraints. Thus, for the rather common case of *catastatic constraints*, the effect of nonholonomic constraints on a system is to restrict the possible directions of $\dot{\mathbf{q}}$ but not its magnitude.

When one considers conservative systems which also have catastatic nonholonomic constraints, there is a temptation to consider the energy integral as an additional nonholonomic constraint. There are, however, qualitative differences between ordinary kinematic constraints and energy constraints. For a given configuration and time, an energy constraint is actually a constraint on the kinetic energy, and is represented in velocity space by a closed surface which surrounds the origin. This means that an energy constraint does not restrict the direction of $\dot{\mathbf{q}}$, but for any given direction, its magnitude is specified. On the other hand, we found that linear catastatic constraints restrict the possible directions of $\dot{\mathbf{q}}$ but not the magnitude.

As a natural application of velocity space methods, let us consider the derivation of Jourdain's principle. We start with Lagrange's principle as expressed in (2.33), namely,

$$\sum_{i=1}^{n} \left[\frac{d}{dt} \left(\frac{\partial T}{\partial \dot{q}_i} \right) - \frac{\partial T}{\partial q_i} - Q_i \right] \delta q_i = 0 \tag{2.292}$$

where the δqs satisfy the instantaneous constraints expressed in (2.280) or (2.281). Lagrange's principle applies to holonomic or nonholonomic systems and leads to $(n - m)$ differential equations of motion if there are m constraints. A comparison of (2.281) which constrains the δqs in configuration space with (2.290) which constrains the δws in velocity space shows that the corresponding δqs and δws are proportional. Hence, we can substitute δw_i for δq_i in (2.292). We obtain

$$\sum_{i=1}^{n} \left[\frac{d}{dt} \left(\frac{\partial T}{\partial \dot{q}_i} \right) - \frac{\partial T}{\partial q_i} - Q_i \right] \delta w_i = 0 \tag{2.293}$$

This is *Jourdain's principle*. The variation $\delta \mathbf{w}$ is considered to take place in velocity space at an operating point P on the common intersection of the constraint planes. By choosing $(n - m)$ independent sets of δws which satisfy the constraints expressed in (2.290), one can use Jourdain's principle to obtain $(n - m)$ second-order differential equations of motion. In addition, there are m constraint equations, making a total of n equations to solve for the n qs.

Example 2.12 Let us apply Jourdain's principle to the nonholonomic system (Fig. 2.3) which was studied previously in Example 2.3 on page 83. The configuration of a dumbbell that moves in the horizontal xy-plane is given by the generalized coordinates (x, y, ϕ). There is a knife-edge constraint at particle 1 which has the Cartesian coordinates (x, y), and the relative position of particle 2 with respect to particle 1 is given by the angle ϕ.

The nonholonomic constraint equation is

$$-\dot{x} \sin \phi + \dot{y} \cos \phi = 0 \tag{2.294}$$

which states that the velocity of particle 1 is limited to the longitudinal direction. Hence, the virtual velocity components must satisfy

$$-\sin\phi \, \delta w_1 + \cos\phi \, \delta w_2 = 0 \tag{2.295}$$

where $\delta w_1 \equiv \delta \dot{x}$ and $\delta w_2 \equiv \delta \dot{y}$. The kinetic energy of the unconstrained system is

$$T = m \left(\dot{x}^2 + \dot{y}^2 + \frac{1}{2} l^2 \dot{\phi}^2 - l\dot{x}\dot{\phi} \sin \phi + l\dot{y}\dot{\phi} \cos \phi \right) \tag{2.296}$$

and we obtain

$$\frac{d}{dt}\left(\frac{\partial T}{\partial \dot{x}}\right) - \frac{\partial T}{\partial x} = m(2\ddot{x} - l\ddot{\phi}\sin\phi - l\dot{\phi}^2\cos\phi) \tag{2.297}$$

$$\frac{d}{dt}\left(\frac{\partial T}{\partial \dot{y}}\right) - \frac{\partial T}{\partial y} = m(2\ddot{y} + l\ddot{\phi}\cos\phi - l\dot{\phi}^2\sin\phi) \tag{2.298}$$

$$\frac{d}{dt}\left(\frac{\partial T}{\partial \dot{\phi}}\right) - \frac{\partial T}{\partial \phi} = ml(-\ddot{x}\sin\phi + \ddot{y}\cos\phi + l\ddot{\phi}) \tag{2.299}$$

In this example, $n = 3$ and $m = 1$ so there are two independent sets of δws which satisfy (2.295). We can choose

$$\delta\mathbf{w} = (\cos\phi, \sin\phi, 0) \tag{2.300}$$

and

$$\delta\mathbf{w} = (0, 0, 1) \tag{2.301}$$

Notice that these components of $\delta\mathbf{w}$ are not necessarily infinitesimal.

Using Jourdain's principle and the first set of δws, we obtain

$$m(2\ddot{x}\cos\phi + 2\ddot{y}\sin\phi - l\dot{\phi}^2) = 0 \tag{2.302}$$

which is the first equation of motion. The second equation of motion is obtained by using the second set of δws, resulting in

$$ml(-\ddot{x}\sin\phi + \ddot{y}\cos\phi + l\ddot{\phi}) = 0 \tag{2.303}$$

These differential equations of motion are the same as those obtained by first using the nonholonomic Lagrange equations with Lagrange multipliers, and then algebraically eliminating the λs. However, the use of Jourdain's principle appears to be more direct.

In addition to the two equations of motion, (2.302) and (2.303), a third equation is obtained by differentiating the constraint equation with respect to time, with the result

$$-\ddot{x}\sin\phi + \ddot{y}\cos\phi - \dot{x}\dot{\phi}\cos\phi \quad \dot{y}\dot{\phi}\sin\phi = 0 \tag{2.304}$$

These three equations are sufficient to solve for $x(t)$, $y(t)$, and $\phi(t)$.

2.6   Impulse response, analytical methods

In the previous discussion of impulse methods in Chapter 1, Newton's law and vectorial methods were used. In this chapter we will use the analytical methods of Hamilton and Lagrange, and will pay particular attention to constrained systems.

Hamiltonian approach

Consider an *unconstrained system* whose dynamical equation of motion has the Hamiltonian form

$$\dot{p}_i = -\frac{\partial H}{\partial q_i} + Q_i' \qquad (i = 1, \ldots, n) \tag{2.305}$$

where Q_i' is the ith generalized applied force which is not derived from the potential energy portion of the Hamiltonian function.

Now let us assume that very large forces are applied to the system over an infinitesimal time interval Δt, beginning at time t_1. Let us integrate (2.305) over the interval Δt. We obtain

$$\int_{t_1}^{t_1+\Delta t} \left(\dot{p}_i + \frac{\partial H}{\partial q_i} \right) dt = \int_{t_1}^{t_1+\Delta t} Q_i' dt \quad (i = 1, \ldots, n) \tag{2.306}$$

The term $\partial H/\partial q_i$ remains finite during the impulse, so its integral can be neglected. Thus we obtain

$$\Delta p_i = \hat{Q}_i \quad (i = 1, \ldots, n) \tag{2.307}$$

for this unconstrained system. The generalized impulse is

$$\hat{Q}_i = \int_{t_1}^{t_1+\Delta t} Q_i' dt \quad (i = 1, \ldots, n) \tag{2.308}$$

and

$$\Delta p_i = \int_{t_1}^{t_1+\Delta t} \dot{p}_i dt = \sum_{j=1}^{n} m_{ij} \Delta \dot{q}_j \quad (i = 1, \ldots, n) \tag{2.309}$$

where we assume that the qs are fixed during the impulse. Thus, the $\Delta \dot{q}$s can be obtained by solving

$$\sum_{j=1}^{n} m_{ij} \Delta \dot{q}_j = \hat{Q}_i \quad (i = 1, \ldots, n) \tag{2.310}$$

as we found earlier in (1.360).

Now consider an unconstrained system of N particles whose positions are given by the Cartesian coordinates $x_1, \ldots x_{3N}$. Suppose that corresponding impulses $\hat{F}_1, \ldots, \hat{F}_{3N}$ are applied to the system during the infinitesimal interval Δt. The generalized impulse associated with q_i is

$$\hat{Q}_i = \sum_{k=1}^{3N} \frac{\partial x_k}{\partial q_i} \hat{F}_k = \sum_{j=1}^{N} \hat{\mathbf{F}}_j \cdot \gamma_{ji} \tag{2.311}$$

in agreement with (1.359).

Lagrangian approach

Let us begin with the fundamental form of Lagrange's equation, namely,

$$\frac{d}{dt} \left(\frac{\partial T}{\partial \dot{q}_i} \right) - \frac{\partial T}{\partial q_i} = Q_i \quad (i = 1, \ldots, n) \tag{2.312}$$

where the Qs are due to the applied forces and the qs are independent. Now integrate with respect to time over the interval Δt. We obtain

$$\int_{t_1}^{t_1+\Delta t} \left[\frac{d}{dt}\left(\frac{\partial T}{\partial \dot{q}_i}\right) - \frac{\partial T}{\partial q_i}\right] dt = \int_{t_1}^{t_1+\Delta t} Q_i dt \tag{2.313}$$

Note that the integral of $\partial T/\partial q_i$ can be neglected and that

$$\frac{d}{dt}\left(\frac{\partial T}{\partial \dot{q}_i}\right) = \dot{p}_i \tag{2.314}$$

Hence we obtain

$$\int_{t_1}^{t_1+\Delta t} \dot{p}_i dt = \int_{t_1}^{t_1+\Delta t} Q_i dt \tag{2.315}$$

or

$$\Delta p_i = \hat{Q}_i \qquad (i = 1, \ldots, n) \tag{2.316}$$

as before.

Constrained impulsive motion

Let us assume that the system has m constraints of the linear form

$$\sum_{i=1}^{n} a_{ji}(q, t)\dot{q}_i + a_{jt}(q, t) = 0 \qquad (j = 1, \ldots, m) \tag{2.317}$$

Jourdain's principle applies in this case, so we can write

$$\sum_{i=1}^{n} \left[\frac{d}{dt}\left(\frac{\partial T}{\partial \dot{q}_i}\right) - \frac{\partial T}{\partial q_i} - Q_i\right] \delta w_i = 0 \tag{2.318}$$

where the virtual velocities δw_i satisfy the instantaneous constraints, that is,

$$\sum_{i=1}^{n} a_{ji}(q, t)\delta w_i = 0 \qquad (j = 1, \ldots, m) \tag{2.319}$$

Now assume that impulsive forces are applied to the system over an infinitesimal interval Δt, beginning at time t_1. Integrate (2.318) with respect to time over the interval Δt for this impulsive case, and again note that the integral of $\partial T/\partial q_i$ can be neglected since it is finite. We obtain

$$\sum_{i=1}^{n} [\Delta p_i - \hat{Q}_i]\delta w_i = 0 \tag{2.320}$$

or

$$\sum_{i=1}^{n} \left[\sum_{j=1}^{n} m_{ij}\Delta \dot{q}_j - \hat{Q}_i\right] \delta w_i = 0 \tag{2.321}$$

where

$$\Delta \dot{q}_j = \dot{q}_j - \dot{q}_{j0} \tag{2.322}$$

and $\dot{q}_j$ is evaluated just after the impulse whereas $\dot{q}_{j0}$ is the initial value just before the impulse. Finally, we can write the general equation for constrained impulsive motion which is

$$\sum_{i=1}^{n} \left[\sum_{j=1}^{n} m_{ij}(\dot{q}_j - \dot{q}_{j0}) - \hat{Q}_i \right] \delta w_i = 0 \tag{2.323}$$

It is assumed that the configuration and the values of the $\dot{q}_{j0}$s are known just before the impulses $\hat{Q}_i$ are applied. There are $(n - m)$ independent sets of δws which satisfy (2.319). In addition, there are m actual constraint equations of the form

$$\sum_{i=1}^{n} a_{ji}(q, t) \dot{q}_i + a_{jt}(q, t) = 0 \qquad (j = 1, \dots, m) \tag{2.324}$$

Thus, there are a total of n equations to solve for the n $\dot{q}$s immediately after the impulses are applied. The qs remain unchanged.

Lagrange multiplier form

Instead of using virtual velocities, we can analyze constrained impulsive motion by starting with the general equation

$$\Delta p_i = \hat{Q}_i + \hat{C}_i \qquad (i = 1, \dots, n) \tag{2.325}$$

where $\hat{Q}_i$ is the generalized applied impulse and $\hat{C}_i$ is the corresponding constraint impulse. Now, by integrating (2.47) with respect to time over the interval Δt of the impulses, we obtain

$$\hat{C}_i = \sum_{k=1}^{m} \hat{\lambda}_k a_{ki} \qquad (i = 1, \dots, n) \tag{2.326}$$

where $\hat{\lambda}_k$ is an *impulsive* Lagrangian multiplier. Thus, using (2.309), (2.325), and (2.326), we obtain

$$\sum_{j=1}^{m} m_{ij}(\dot{q}_j - \dot{q}_{j0}) = \hat{Q}_i + \sum_{k=1}^{m} \hat{\lambda}_k a_{ki} \qquad (i = 1, \dots, n) \tag{2.327}$$

This is the Lagrange multiplier form of the constrained impulsive motion equation. These n equations plus the m constraint equations from (2.324) are a total of $(n + m)$ equations to solve for the n $\dot{q}$s and the m $\hat{\lambda}$s. After finding the $\hat{\lambda}$s, the constraint impulses can be obtained from (2.326).

Impulsive constraints

We have assumed that the coefficients a_{ji} and a_{jt} in the constraint equations are continuous functions of the qs and t. Now consider the case of *impulsive constraints* where one or more of the as may be discontinuous at some time t_1. This allows for the sudden appearance of a constraint or a sudden change in its motion. For example, the sudden appearance of a fixed constraint would be represented by the sudden change of the a_{ji} coefficients for that constraint from zero to nonzero values, whereas a_{jt} remains equal to zero. On the other hand, a sudden change in a_{jt} represents a change in the velocity of a moving constraint.

We shall assume that a sudden change in a constraint is not accompanied by an applied impulse $\hat{Q}_i$ at exactly the same time. Thus, we assume that

$$\hat{Q}_i = 0 \qquad (i = 1, \ldots, n) \tag{2.328}$$

Usually the sudden change in the values of the a_{ji} or a_{jt} coefficients results in $\hat{C}_i$ constraint impulses. An exception occurs when a constraint suddenly disappears, that is, when its as suddenly go to zero. In this case, there are no constraint impulses.

In general, for *impulsive constraints*, (2.323) becomes

$$\sum_{i=1}^{n} \sum_{j=1}^{n} m_{ij}(\dot{q}_j - \dot{q}_{j0})\delta w_i = 0 \tag{2.329}$$

where the virtual velocities δw_i satisfy (2.319). The Lagrange multiplier form, given by (2.327), becomes

$$\sum_{j=1}^{n} m_{ij}(\dot{q}_j - \dot{q}_{j0}) = \sum_{k=1}^{m} \hat{\lambda}_k a_{ki} \qquad (i = 1, \ldots, n) \tag{2.330}$$

When one uses (2.329) and (2.330) a question arises concerning which of the discontinuous values of the as are to be used in the associated constraint equations. The initial velocities $\dot{q}_{j0}$ must satisfy (2.324) where the as are evaluated at t_1^-, that is, just before the discontinuity. On the other hand, the as are evaluated at t_1^+ for constraint equations involving $\dot{q}_j$, δw_i, and for the coefficients of $\hat{\lambda}_k$.

For the case in which a constraint suddenly disappears, we see from (2.330) that, since there are no constraint impulses, the $\dot{q}$s are continuous, that is, $\dot{q}_j = \dot{q}_{j0}$.

Example 2.13   Three particles, each of mass m, are rigidly connected in the form of an equilateral triangle by massless rigid rods of length l (Fig. 2.15). The system moves downward with velocity v_0 in pure translational motion and particle 1 hits a smooth floor at $y = 0$ inelastically. We wish to solve for the velocities of the particles immediately after impact. Assume that particle 1 is directly below particle 3 at the time of impact.

We wish to illustrate the use of the general equations (2.323) and (2.327) for this system which has three ordinary holonomic constraints and one impulsive holonomic constraint. Let us specify the configuration of the system by the Cartesian coordinates $(x_1, y_1, x_2, y_2, x_3, y_3)$. The values of these coordinates at the time of impact are

$$x_1 = x_3, \qquad x_2 - x_1 = \frac{\sqrt{3}}{2}l, \qquad y_1 = 0, \qquad y_2 = \frac{1}{2}l, \qquad y_3 = l \tag{2.331}$$

The corresponding velocities just before impact are

$$\dot{x}_{10} = \dot{x}_{20} = \dot{x}_{30} = 0, \qquad \dot{y}_{10} = \dot{y}_{20} = \dot{y}_{30} = -v_0 \tag{2.332}$$

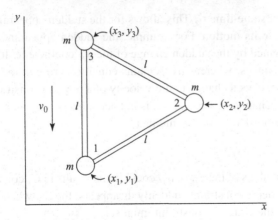

Figure 2.15.

The three holonomic constraints which specify the particle separations are

$$(x_2 - x_1)^2 + (y_2 - y_1)^2 - l^2 = 0 \tag{2.333}$$

$$(x_2 - x_3)^2 + (y_3 - y_2)^2 - l^2 = 0 \tag{2.334}$$

$$(x_3 - x_1)^2 + (y_3 - y_1)^2 - l^2 = 0 \tag{2.335}$$

Upon differentiation with respect to time, and using (2.331), these holonomic constraint equations take the form

$$-\frac{\sqrt{3}}{2}\dot{x}_1 - \frac{1}{2}\dot{y}_1 + \frac{\sqrt{3}}{2}\dot{x}_2 + \frac{1}{2}\dot{y}_2 = 0 \tag{2.336}$$

$$\frac{\sqrt{3}}{2}\dot{x}_2 - \frac{1}{2}\dot{y}_2 - \frac{\sqrt{3}}{2}\dot{x}_3 + \frac{1}{2}\dot{y}_3 = 0 \tag{2.337}$$

$$-\dot{y}_1 + \dot{y}_3 = 0 \tag{2.338}$$

These equations are valid continuously before and after the impact.

The impulsive constraint equation is

$$\dot{y}_1 = 0 \tag{2.339}$$

This equation applies during and after impact, but not before.

First method   Let us use impulsive Lagrange multipliers. Since all the $\hat{Q}$s equal zero, we can use (2.330). The as are the coefficients in (2.336)–(2.339), the constraint equations linear in the $\dot{q}$s. There is no inertial coupling between these Cartesian coordinates, and $m_{ij} = m$ for $i = j$. Thus, noting that $\dot{y}_1 = \dot{y}_3 = 0$, the six dynamical equations are

$$m\dot{x}_1 = -\frac{\sqrt{3}}{2}\hat{\lambda}_1 \tag{2.340}$$

$$mv_0 = -\frac{1}{2}\hat{\lambda}_1 - \hat{\lambda}_3 + \hat{\lambda}_4 \tag{2.341}$$

$$m\dot{x}_2 = \frac{\sqrt{3}}{2}\hat{\lambda}_1 + \frac{\sqrt{3}}{2}\hat{\lambda}_2 \tag{2.342}$$

$$m(\dot{y}_2 + v_0) = \frac{1}{2}\hat{\lambda}_1 - \frac{1}{2}\hat{\lambda}_2 \tag{2.343}$$

$$m\dot{x}_3 = -\frac{\sqrt{3}}{2}\hat{\lambda}_2 \tag{2.344}$$

$$mv_0 = \frac{1}{2}\hat{\lambda}_2 + \hat{\lambda}_3 \tag{2.345}$$

These equations plus (2.336) and (2.337) can be solved for the four unknown $\dot{q}$s and the four $\hat{\lambda}$s. The velocities immediately after the impact are

$$\dot{x}_1 = -\frac{\sqrt{3}}{5}v_0, \qquad \dot{x}_2 = 0, \qquad \dot{x}_3 = \frac{\sqrt{3}}{5}v_0$$

$$\dot{y}_1 = 0, \qquad \dot{y}_2 = -\frac{3}{5}v_0, \qquad \dot{y}_3 = 0 \tag{2.346}$$

The impulsive Lagrange multipliers are

$$\hat{\lambda}_1 = \frac{2}{5}mv_0, \qquad \hat{\lambda}_2 = -\frac{2}{5}mv_0, \qquad \hat{\lambda}_3 = \frac{6}{5}mv_0, \qquad \hat{\lambda}_4 = \frac{12}{5}mv_0 \tag{2.347}$$

The first three $\hat{\lambda}$s are impulses in rods 12, 23, and 31, respectively, positive being compressive. Impulse $\hat{\lambda}_4$ is the floor reaction acting on particle 1.

We see that immediately after impact, the system rotates clockwise about the center of the vertical rod with an angular velocity

$$\omega = \frac{2\sqrt{3}}{5}\frac{v_0}{l} \tag{2.348}$$

Second method   Lagrange multipliers can be avoided by using (2.329). In accordance with (2.319), the virtual velocities (δws) must satisfy

$$-\frac{\sqrt{3}}{2}\delta\dot{x}_1 - \frac{1}{2}\delta\dot{y}_1 + \frac{\sqrt{3}}{2}\delta\dot{x}_2 + \frac{1}{2}\delta\dot{y}_2 = 0 \tag{2.349}$$

$$\frac{\sqrt{3}}{2}\delta\dot{x}_2 - \frac{1}{2}\delta\dot{y}_2 - \frac{\sqrt{3}}{2}\delta\dot{x}_3 + \frac{1}{2}\delta\dot{y}_3 = 0 \tag{2.350}$$

$$-\delta\dot{y}_1 + \delta\dot{y}_3 = 0 \tag{2.351}$$

$$\delta\dot{y}_1 = 0 \tag{2.352}$$

There are two independent sets of δws which satisfy these constraint equations since $n - m = 2$. Let set 1 be

$$\delta w_i = (1, 0, 1, 0, 1, 0) \tag{2.353}$$

corresponding to uniform translation in the x-direction. We will take set 2 to be

$$\delta w_i = (1, 0, 0, \sqrt{3}, -1, 0) \tag{2.354}$$

which represents rotational motion about the midpoint of the vertical rod.

Using (2.329) with set 1, we obtain

$$\dot{x}_1 + \dot{x}_2 + \dot{x}_3 = 0 \tag{2.355}$$

This is equivalent to adding (2.340), (2.342), and (2.344). Similarly, using set 2, we obtain

$$\dot{x}_1 + \sqrt{3}(\dot{y}_2 + v_0) - \dot{x}_3 = 0 \tag{2.356}$$

These two equations plus the four constraint equations, namely, (2.336)–(2.339), can be solved for the six Cartesian components of the particle velocities immediately after impact. The results agree with (2.346).

We have used general equations for the impulse response of constrained systems of particles. For the particular system under consideration, the analysis for its motion could have been simplified considerably if we had considered the rigidly connected particles to be a rigid body. We will study the impulse response of rigid bodies as one of the topics in the next chapter.

Energy relations

Consider a system with n qs and m constraints of the linear form

$$\sum_{i=1}^{n} a_{ki}(q, t)\dot{q}_i + a_{kt}(q, t) = 0 \qquad (k = 1, \ldots, m) \tag{2.357}$$

where the as are continuous functions of (q, t). The general impulse equation can be written in the form

$$\sum_{j=1}^{n} m_{ij}(\dot{q}_j - \dot{q}_{j0}) = \hat{Q}_i + \hat{C}_i = \hat{Q}_i + \sum_{k=1}^{m} \hat{\lambda}_k a_{ki} \tag{2.358}$$

Multiply by $\frac{1}{2}(\dot{q}_i + \dot{q}_{i0})$ and sum over i. We obtain

$$\frac{1}{2} \sum_{i=1}^{n} \sum_{j=1}^{n} m_{ij}(\dot{q}_i\dot{q}_j - \dot{q}_i\dot{q}_{j0} + \dot{q}_j\dot{q}_{i0} - \dot{q}_{i0}\dot{q}_{j0})$$

$$= \frac{1}{2} \sum_{i=1}^{n} \hat{Q}_i(\dot{q}_i + \dot{q}_{i0}) + \frac{1}{2} \sum_{i=1}^{n} \sum_{k=1}^{n} \hat{\lambda}_k a_{ki}(\dot{q}_i + \dot{q}_{i0}) \tag{2.359}$$

If we note that $m_{ij} = m_{ji}$ and require that both the $\dot{q}$s and $\dot{q}_0$s satisfy the constraints of (2.357), we find that (2.359) reduces to

$$\frac{1}{2} \sum_{i=1}^{n} \sum_{j=1}^{n} m_{ij}(\dot{q}_i\dot{q}_j - \dot{q}_{i0}\dot{q}_{j0}) = \frac{1}{2} \sum_{i=1}^{n} \hat{Q}_i(\dot{q}_i + \dot{q}_{i0}) - \sum_{k=1}^{n} \hat{\lambda}_k a_{kt} \tag{2.360}$$

where the last term on the right is equal to the energy input due to constraint impulses.

Now assume that all constraints are *catastatic*, that is, $a_{kt} = 0$. Then

$$\Delta T_2 = \frac{1}{2} \sum_{i=1}^{n} \hat{Q}_i(\dot{q}_i + \dot{q}_{i0}) \tag{2.361}$$

where T_2 is quadratic in the $\dot{q}$s and is given by

$$T_2 = \frac{1}{2} \sum_{i=1}^{n} \sum_{j=1}^{n} m_{ij} \dot{q}_i \dot{q}_j \tag{2.362}$$

Finally, assuming a *scleronomic system*, we have $T_1 = T_0 = 0$ and $T = T_2$. Then, by the principle of work and kinetic energy, the work done on the system by the applied impulses $\hat{Q}_i$ over an infinitesimal interval Δt is

$$W = \Delta T = \frac{1}{2} \sum_{i=1}^{n} \hat{Q}_i(\dot{q}_i + \dot{q}_{i0}) \tag{2.363}$$

Impulsive constraints

Now let us consider energy relations in the case of impulsive constraints for which the a_{ji} or a_{jt} coefficients suddenly change value. First consider the case of *catastatic constraints* having the form

$$\sum_{i=1}^{n} a_{ji}(q, t) \dot{q}_i = 0 \qquad (j = 1, \ldots, m) \tag{2.364}$$

and assume that the as are discontinuous at time t_1. We assume that all $\hat{Q}_i$s are equal to zero and start with the impulsive constraint equation in the form

$$\sum_{i=1}^{n} \sum_{j=1}^{n} m_{ij}(\dot{q}_j - \dot{q}_{j0}) \delta w_i = 0 \tag{2.365}$$

where the δws satisfy

$$\sum_{i=1}^{n} a_{ji} \delta w_i = 0 \qquad (j = 1, \ldots, m) \tag{2.366}$$

Comparing (2.364) and (2.366), we see that any $\dot{q}_i$ which satisfies (2.364) at time t_1^+ will also satisfy (2.366). Then let us take $\delta w_i = \dot{q}_i$ in (2.365) and note that the δws or $\dot{q}$s need not be infinitesimal. We obtain

$$\sum_{i=1}^{n} \sum_{j=1}^{n} m_{ij}(\dot{q}_j - \dot{q}_{j0}) \dot{q}_i = 0 \tag{2.367}$$

The $\dot{q}_{j0}$s satisfy (2.364) with the as evaluated at t_1^-. From (2.367), and recalling that $m_{ij} = m_{ji}$, we see that

$$\sum_{i=1}^{n} \sum_{j=1}^{n} m_{ij} \dot{q}_i \dot{q}_j = \sum_{i=1}^{n} \sum_{j=1}^{n} m_{ij} \dot{q}_i \dot{q}_{j0} = \sum_{i=1}^{n} \sum_{j=1}^{n} m_{ij} \dot{q}_{i0} \dot{q}_j \tag{2.368}$$

Now define the *kinetic energy of relative motion* as follows:

$$K = \frac{1}{2} \sum_{i=1}^{n} \sum_{j=1}^{n} m_{ij}(\dot{q}_i - \dot{q}_{i0})(\dot{q}_j - \dot{q}_{j0}) \tag{2.369}$$

Note that K is positive or zero. A substitution of (2.368) into (2.369) results in

$$K = \frac{1}{2} \sum_{i=1}^{n} \sum_{j=1}^{n} m_{ij}\dot{q}_{i0}\dot{q}_{j0} - \frac{1}{2} \sum_{i=1}^{n} \sum_{j=1}^{n} m_{ij}\dot{q}_i\dot{q}_j = -\Delta T_2 \tag{2.370}$$

Thus, we find that the loss of kinetic energy T_2 due to *impulsive catastatic constraints* is equal to the kinetic energy of relative motion. If $T_1 = T_0 = 0$, this is the total energy loss.

The general catastatic case is illustrated in Fig. 2.16a. A sudden change of $a_{ji}(q, t)$ at time t_1 results in a change in the orientation of the constraint plane which passes through

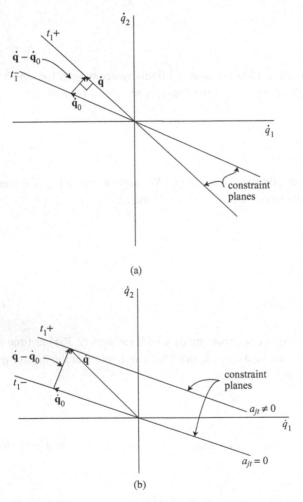

(a)

(b)

Figure 2.16.

the origin in n-dimensional velocity space. The change in generalized momentum is due entirely to the constraint impulse $\hat{\mathbf{C}}$, that is,

$$\sum_{j=1}^{n} m_{ij}(\dot{q}_j - \dot{q}_{j0}) = \hat{C}_i \qquad (i = 1, \ldots, n) \tag{2.371}$$

This constraint impulse is normal to the final constraint plane at t_1^+, so we see that

$$\sum_{i=1}^{n} \hat{C}_i \dot{q}_i = 0 \tag{2.372}$$

for any final velocity $\dot{\mathbf{q}}$ which must lie in that constraint plane. From (2.371) and (2.372) we obtain (2.367), the basic equation which was obtained earlier by other means.

A simple example of an impulsive catastatic constraint is provided by the sudden appearance of a fixed frictionless constraint surface. For example, a particle might suddenly hit a fixed wall with inelastic impact.

Now let us consider the case of impulsive constraints which are *not catastatic*. Specifically, let us assume that the a_{ji} coefficients are continuous at time t_1, but the a_{jt}s change from zero to nonzero values. This could occur, for example, when a constraint suddenly begins to move. In this case, at t_1^- the $\dot{q}_{i0}$s satisfy

$$\sum_{i=1}^{n} a_{ji}\dot{q}_{i0} = 0 \qquad (j = 1, \ldots, m) \tag{2.373}$$

which has the same form as (2.366) that is satisfied by the δws. Hence, substituting $\dot{q}_{i0}$ for δw_i in (2.365), we obtain

$$\sum_{i=1}^{n} \sum_{j=1}^{n} m_{ij}(\dot{q}_j - \dot{q}_{j0})\dot{q}_{i0} = 0 \tag{2.374}$$

and, again noting that $m_{ij} = m_{ji}$, we find that

$$\sum_{i=1}^{n} \sum_{j=1}^{n} m_{ij}\dot{q}_{i0}\dot{q}_j = \sum_{i=1}^{n} \sum_{j=1}^{n} m_{ij}\dot{q}_{j0}\dot{q}_i = \sum_{i=1}^{n} \sum_{j=1}^{n} m_{ij}\dot{q}_{i0}\dot{q}_{j0} \tag{2.375}$$

Then, using (2.369), the kinetic energy of relative motion is equal to

$$K = \frac{1}{2} \sum_{i=1}^{n} \sum_{j=1}^{n} m_{ij}\dot{q}_i\dot{q}_j - \frac{1}{2} \sum_{i=1}^{n} \sum_{j=1}^{n} m_{ij}\dot{q}_{i0}\dot{q}_{j0} = \Delta T_2 \tag{2.376}$$

We conclude that the increase in kinetic energy T_2 due to sudden changes in the a_{jt}s from zero to nonzero values is equal to the kinetic energy of relative motion.

The velocity space representation is shown in Fig. 2.16b. Again the constraint impulse components $\hat{C}_i$ are given by (2.371) so (2.374) becomes

$$\sum_{i=1}^{n} \hat{C}_i \dot{q}_{i0} = 0 \tag{2.377}$$

This equation states that the constraint impulse $\hat{C}$ and the resulting change in generalized momentum are normal to the parallel constraint planes, before and after time t_1 of the impulse. Of course, any possible velocity $\dot{q}_0$ at t_1^- must lie in the constraint plane.

Another possibility would be for the a_{jt} coefficients to jump from nonzero to zero values, corresponding to a sudden stop of moving constraints. This would reverse roles of $\dot{q}_0$ and $\dot{q}$ in Fig. 2.16b and the resulting change in T_2 would be a loss equal to K, the kinetic energy of relative motion.

Example 2.14 Two particles, each of mass m, are connected by a massless rod of length l. Particle 1 can slide without friction on a rod that may rotate about O while particle 2 slides without friction along a vertical wall (Fig. 2.17). Let (r, y) be the generalized coordinates and assume the initial conditions $\theta = \pi/2, \dot{\theta} = 0, r = y = l/\sqrt{2}, \dot{r} = -\dot{y} = v_0$ at time $t = 0^-$. Then, at $t = 0$, the rod $O1$ suddenly starts rotating with an angular velocity ω. For this case of an impulsive constraint we wish to solve for the values of $\dot{r}$ and $\dot{y}$ at $t = 0^+$ as well as the constraint impulses $\hat{R}_1$ and $\hat{R}_2$.

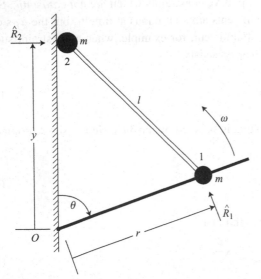

Figure 2.17.

The holonomic constraint equation obtained from the cosine law is

$$r^2 + y^2 - 2ry\cos\theta - l^2 = 0 \tag{2.378}$$

After differentiation with respect to time we obtain

$$r\dot{r} + y\dot{y} - (\dot{r}y + r\dot{y})\cos\theta + ry\dot{\theta}\sin\theta = 0 \tag{2.379}$$

At $t = 0^+$ we have $\dot{\theta} = -\omega$, and (2.379) has the form

$$\dot{r} + \dot{y} - \frac{l\omega}{\sqrt{2}} = 0 \tag{2.380}$$

The last term is the a_{jt} term which becomes nonzero at $t = 0$.

Let us use the basic equation (2.365), namely,

$$\sum_{i=1}^{n}\sum_{j=1}^{n} m_{ij}(\dot{q}_j - \dot{q}_{j0})\delta w_i = 0 \tag{2.381}$$

and thereby avoid having to evaluate Lagrange multipliers. The kinetic energy at $t = 0^+$ is

$$T = \frac{1}{2} m(\dot{r}^2 + \dot{y}^2 + r^2\omega^2) \tag{2.382}$$

which leads to

$$m_{rr} = m_{yy} = m, \qquad m_{ry} = m_{yr} = 0 \tag{2.383}$$

In accordance with (2.366) the virtual velocities must satisfy

$$\delta w_r + \delta w_y = 0 \tag{2.384}$$

Let us take $\delta w_r = 1$ and $\delta w_y = -1$. Then (2.381) results in

$$m\left[(\dot{r} - v_0) - (\dot{y} + v_0)\right] = 0$$

or

$$\dot{r} - \dot{y} - 2v_0 = 0 \tag{2.385}$$

Upon solving (2.380) and (2.385) we obtain

$$\dot{r} = \frac{l\omega}{2\sqrt{2}} + v_0 \tag{2.386}$$

$$\dot{y} = \frac{l\omega}{2\sqrt{2}} - v_0 \tag{2.387}$$

These $\dot{q}$s apply at $t = 0^+$.

The impulse $\hat{R}_1$ is obtained by using the principle of impulse and momentum in the vertical direction.

$$\hat{R}_1 = m\left(\dot{y} + v_0 + \frac{l\omega}{\sqrt{2}}\right) = \frac{3ml\omega}{2\sqrt{2}} \tag{2.388}$$

Similarly, from the principle of impulse and momentum in the horizontal direction, we obtain

$$\hat{R}_2 = m(\dot{r} - v_0) = \frac{ml\omega}{2\sqrt{2}} \tag{2.389}$$

Now let us consider the kinetic energy of this system with a suddenly moving constraint. The increase in kinetic energy is

$$\Delta T = \frac{m}{2}\left[\left(\frac{l\omega}{2\sqrt{2}} + v_0\right)^2 + \left(\frac{l\omega}{2\sqrt{2}} - v_0\right)^2\right] - mv_0^2 = \frac{ml^2\omega^2}{8} \tag{2.390}$$

From (2.369) the kinetic energy of relative motion is

$$K = \frac{m}{2}[(\dot{r} - v_0)^2 + (\dot{y} + v_0)^2] = \frac{ml^2\omega^2}{8} \tag{2.391}$$

We see that

$$\Delta T = K \tag{2.392}$$

for this system with an acatastatic constraint given by (2.380).

2.7 Bibliography

Desloge, E. A. *Classical Mechanics*, Vol. 1. New York: John Wiley and Sons, 1982.
Ginsberg, J. H. *Advanced Engineering Dynamics*, 2nd edn. Cambridge, UK: Cambridge University Press, 1995.
Greenwood, D. T. *Principles of Dynamics*, 2nd edn. Englewood Cliffs, NJ: Prentice-Hall, 1988.
Pars, L. A. *A Treatise on Analytical Dynamics*. London: William Heinemann, 1965.

2.8 Problems

2.1. A string of length $2L$ is attached at the fixed points A and B which are separated by a horizontal distance L. A particle of mass m can slide without friction on the string which remains taut, resulting in an elliptical path. (a) Using Lagrange's equation and (x, y) as the particle coordinates, find the differential equations of motion. (b) Assume the initial conditions $x(0) = 0$, $y(0) = -(\sqrt{3}/2)L$, $\dot{x}(0) = \sqrt{gL}$, $\dot{y}(0) = 0$. Solve for $x_{\max}$ during the motion and the initial string tension P.

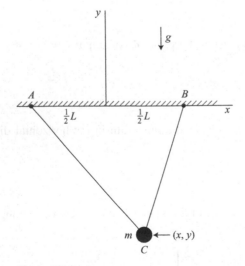

Figure P 2.1.

2.2. Consider the dumbbell system of Example 2.3 on page 83. Suppose there is a Coulomb friction coefficient μ for skidding, that is, sliding in a direction perpendicular to the knife edge, but there is no friction for sliding along the knife edge. (a) What are the differential equations of motion, assuming that skidding occurs? (b) Write an inequality of the form $|f(q, \dot{q})| > C$ which is the necessary and sufficient condition for skidding to occur.

2.3. A particle of mass m can slide on a smooth rigid wire in the form of the parabola $y = x^2$ under the action of gravity. (a) Obtain the x and y differential equations of motion and solve for $\ddot{x}$ as a function of $(x, \dot{x})$. (b) Assuming the initial conditions $x(0) = x_0$, $\dot{x}(0) = 0$, solve for $\dot{x}(x)$.

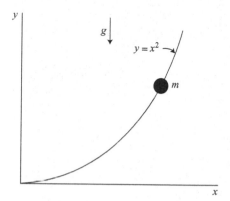

Figure P 2.3.

2.4. A cylinder of radius r rotates at a constant rate ω about its fixed central axis. A particle of mass m is attached to point P on its circumference by a flexible rope of length L. (a) Assume that the initial value $\phi(0) = 0$ and the absolute velocity of the particle is zero at this time. Solve for $\dot{\phi}$ as a function of ϕ for $0 \leq \phi \leq \frac{1}{2}\pi$. (b) After ϕ reaches $\frac{1}{2}\pi$ the rope begins to wrap around the cylinder and its straight portion is of length $l = L - r\theta$. Assuming that the initial value of $\dot{\theta}$ for this phase of the motion is equal to the value of $\dot{\phi}$ at $\phi = \frac{1}{2}\pi$, solve for $\dot{\theta}$ as a function of l and show that $\ddot{\theta} = r^3\omega^2/l^3$.

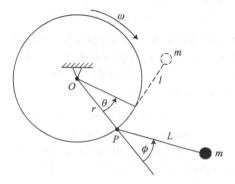

Figure P 2.4.

2.5. A particle P of mass m is attached by a flexible massless string of constant length L to a point A on the rim of a smooth fixed vertical cylinder of radius R. The string can wrap around the cylinder, but leaves at point C, creating a straight portion CP. At any given time, all portions of the string maintain the same slope angle ϕ relative to a horizontal plane. (a) Using (θ, ϕ) as generalized coordinates, obtain the differential equations of motion. (b) Solve for the tension in the string. Assume that the string remains taut and the distance $CP > 0$.

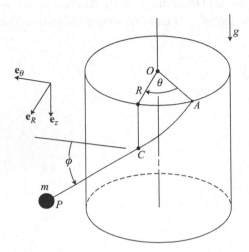

Figure P 2.5.

2.6. A particle of mass m moves on the surface of a spherical earth of radius R which rotates at a constant rate Ω. Linear damping with a damping coefficient c exists for motion of the particle relative to the earth's surface. Choose an inertial frame at the center of the earth and use (θ, ϕ) as generalized coordinates, where θ is the latitude angle (positive north of the equator) and ϕ is the longitude angle (positive toward the east). (a) Obtain the differential equations of motion. (b) Show that the work rate $\dot{W}_p$ of the damper acting on the particle plus its work rate $\dot{W}_e$ acting on the earth plus the energy dissipation rate $\dot{D}$ sums to zero.

2.7. A dumbbell consists of two particles, each of mass m, connected by a massless rod of length l. A knife edge is located at particle 1 and is aligned at $45°$ to the longitudinal direction, as shown in Fig. P 2.7. Using (x, y, ϕ) as generalized coordinates, find the differential equations of motion of the system.

2.8. Two particles, each of mass m, can slide on the horizontal xy-plane (Fig. 2.8). Particle 1 at (x, y) has a knife-edge constraint and is attached to a massless rigid rod. Particle 2 can slide without friction on this rod and is connected to particle 1 by a spring of stiffness k and unstressed length l_0. Using (x, y, ϕ, s) as generalized coordinates, obtain the differential equations of motion.

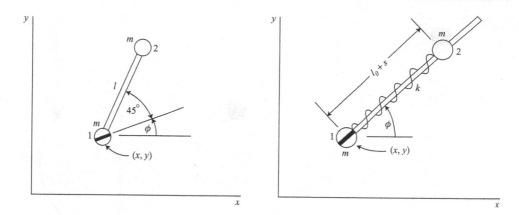

Figure P 2.7.   **Figure P 2.8.**

2.9. Particles B and C, each of mass m, are motionless in the position shown in Fig. P 2.9 when particle B is hit by particle A, also of mass m and moving with speed v_0. The coefficient of restitution for this impact is $e = \frac{2}{3}$. Particles A and B can move without friction on a fixed rigid wire, the displacement of B being x. There is a joint at C and the two rods of length l are each massless. (a) Find the velocity $\dot{x}$ of particle B immediately after impact. (b) What is the impulse $\hat{P}$ transmitted by rod BC?

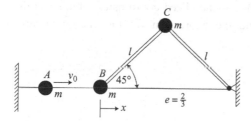

Figure P 2.9.

2.10. Three particles, each of mass m, are connected by strings of length l to form an equilateral triangle (Fig. P 2.10). The system is rotating in planar motion about its fixed center of mass at an angular rate $\dot{\phi} = \omega_0$ when string BC suddenly breaks. (a) Take (ϕ, θ) as generalized coordinates and use the Routhian procedure to obtain the differential equation for θ in the motion which follows. (b) Solve for the value of $\dot{\theta}$ when the system first reaches a straight configuration with $\theta = \frac{1}{2}\pi$, assuming the strings remain taut.

2.11. A rigid wire in the form of a circle of radius r rotates about a fixed vertical diameter at a constant rate Ω (Fig. P 2.11). A particle of mass m can slide on the wire, and there is a coefficient of friction μ between the particle and the wire. Find the differential equation for the position angle θ.

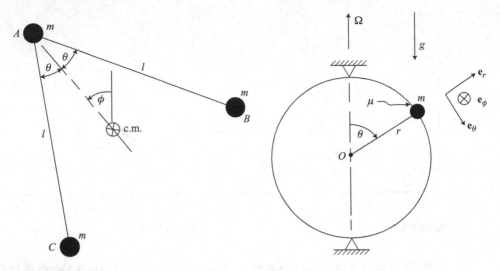

Figure P 2.10.   **Figure P 2.11.**

2.12. A rod of length R rotates in the horizontal xy-plane at a constant rate ω. A pendulum of mass m and length l is attached to point P at the end of the rod. The orientation of the pendulum relative to the rod is given by the angle θ measured from the downward vertical and the angle ϕ between the vertical plane through the pendulum and the direction OP. (a) Using (θ, ϕ) as generalized coordinates, obtain the differential equations of motion. (b) Write the expression for the energy integral of this system.

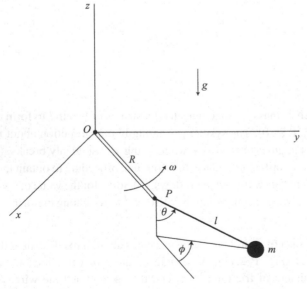

Figure P 2.12.

2.13. An xy Cartesian frame rotates at a constant angular rate Ω about the origin in planar motion. The position of a particle of mass m is (x, y) relative to the rotating frame. There is an impulsive constraint, discontinuous at time t_1, which is given by $\dot{y} = 0$ $(t < t_1)$, $\dot{y} = v_0$ $(t > t_1)$. The position at $t = t_1$ is (x_0, y_0) and $\dot{x}(t_1^-) = \dot{x}_0$. (a) Solve for the values of the relative velocity components $(\dot{x}, \dot{y})$ at time t_1^+. (b) Find ΔT_2 as well as the increase ΔT in the total kinetic energy. (c) Solve for the constraint impulse.

2.14. A particle of mass m can move without friction on a horizontal spiral of the form $r = k\theta$, as in Fig. 1.8 on page 13. (a) Choose (r, θ) as generalized coordinates and write the differential equations of motion using Lagrange's equation. (b) Assuming that the particle starts at the origin with speed v_0, solve for $\dot{r}$ and the normal constraint force N as functions of r. (c) Now add Coulomb friction with coefficient μ and obtain a single equation of motion giving $\ddot{r}$ as a function of r and $\dot{r}$.

2.15. A dumbbell AB can move in the vertical xy-plane. There is a coefficient of friction μ for sliding at A and B. The normal constraint forces are N_x and N_y, as shown. (a) Assuming downward sliding ($\dot{x} > 0$), write the differential equations of motion in the forms $\ddot{x} = \ddot{x}(x, y, \dot{x}, \dot{y})$ and $\ddot{y} = \ddot{y}(x, y, \dot{x}, \dot{y})$, that is, eliminate λ, N_x, and N_y by expressing them in terms of these variables. (b) Use θ as a single generalized coordinate and write the differential equation for θ in the form $\ddot{\theta} = \ddot{\theta}(\theta, \dot{\theta})$.

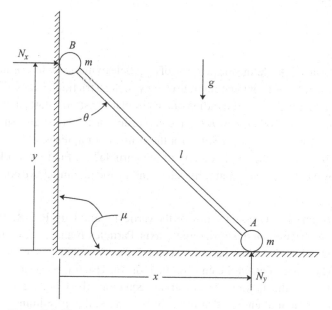

Figure P 2.15.

2.16. Four particles, each of mass m, are connected by massless rods of length l, as shown in Fig. P 2.16. There is a joint at each particle. Assume planar motion with no external forces, so the center of mass can be considered fixed in an inertial frame. (a) Using (r_1, r_2, θ) as generalized coordinates, obtain the differential equations of motion. Reduce these equations to three equations of motion, each involving a single variable and its time derivatives. (b) Assuming an initially square configuration with $\dot{r}_1(0) = v_0$, $\dot{r}_2(0) = -v_0$, $\theta(0) = 0$, $\dot{\theta}(0) = \omega_0$, solve for r_1, r_2, and θ as functions of time.

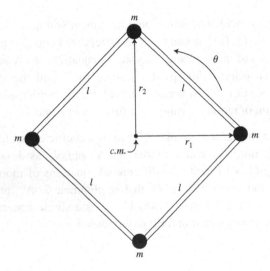

Figure P 2.16.

2.17. The polar coordinates (r, θ) define the position of a particle of mass m which is initially constrained to move at a constant angular rate $\dot{\theta} = \omega_0$ along a frictionless circular path of radius R. Then, at time $t = 0$, the radius of the circular constraint starts increasing at a constant rate $\dot{r} = v_0$. When $t = R/v_0$ the expansion of the constraint suddenly stops, and $r = 2R$ for $t > R/v_0$. (a) Solve for the constraint impulses at times $t = 0$ and $t = R/v_0$. (b) What is the final velocity of the particle? (c) Find the work done on the particle by each of the constraint impulses and by the nonimpulsive constraint force.

2.18. Particles A and B can slide without friction on the vertical y-axis (Fig. P 2.18). Particle C can slide without friction on the horizontal x-axis. Particles B and C are connected by a massless rod of length l while particles A and C are connected by a massless rod of length $\sqrt{3}l$. There are joints at the ends of the rods. (a) Use (θ_1, θ_2) as generalized coordinates and write the holonomic constraint equation. (b) Obtain expressions for the kinetic and potential energies in terms of θ_2 and $\dot{\theta}_2$; that is, eliminate θ_1 and $\dot{\theta}_1$ by using the constraint equation. (c) Assume the initial conditions $\theta_1(0) = 30°$, $\dot{\theta}_1(0) = 0$, $\theta_2(0) = 60°$, $\dot{\theta}_2(0) = 0$. Solve for the velocity of particle B when it first passes through the origin. (d) Solve for the acceleration of particle A at this time.

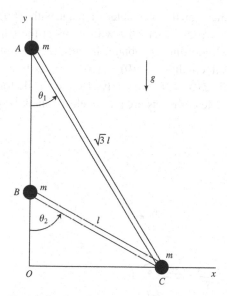

Figure P 2.18.

2.19. Three particles, each of mass m, are connected by massless rods AB and AC with a joint at A. The system is translating in the negative y direction with speed v_0 when particle C hits a fixed smooth wall and there is a coefficient of restitution $e = 1$. Use the generalized coordinates (θ, ϕ, x, y). Assume that the conditions just before impact are $\theta(0^-) = 45°$, $\dot{\theta}(0^-) = 0$, $\phi(0^-) = 45°$, $\dot{\phi}(0^-) = 0$, $\dot{x}(0^-) = 0$, $\dot{y}(0^-) = -v_0$. Find: (a) the constraint impulse $\hat{N}$ during impact and the initial conditions just after impact; (b) the θ and ϕ equations of motion; (c) the minimum value of θ and the value of $\dot{\phi}$ at this time.

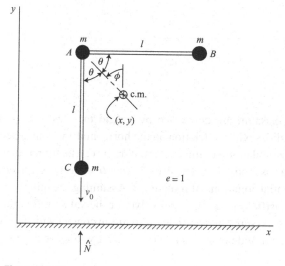

Figure P 2.19.

2.20. Four particles, each of mass m, can move on the horizontal xy-plane without friction. They are connected by four massless rods of length l with joints at the particles. (a) Using (x, y, θ, ϕ) as generalized coordinates, obtain the differential equations of motion. (b) Assuming the initial conditions $x(0) = y(0) = 0$, $\dot{x}(0) = \dot{y}(0) = 0$, $\theta(0) = \phi(0) = 0$, $\dot{\theta}(0) = \dot{\theta}_0 > 0$, $\dot{\phi}(0) = \dot{\phi}_0 > 0$, solve for the tensile forces in the rods. (c) Assuming that any particle collisions are purely elastic, find the period of the motion in ϕ.

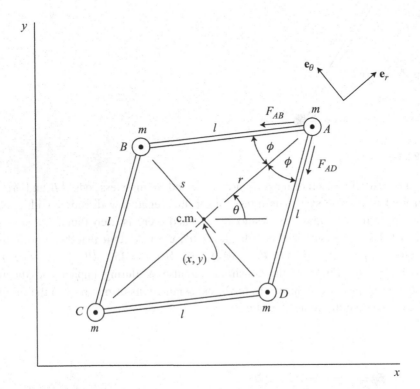

Figure P 2.20.

2.21. Two particles, each of mass m, are connected by a rigid massless rod of length l (Fig. P2.21). Particle 1 slides without friction on the horizontal xy-plane. Because of the conservation of horizontal momentum, an inertial frame can be found such that the motion of the center of mass is strictly vertical. (a) Using (θ, ϕ) as generalized coordinates, obtain the differential equations of motion. (b) Assuming the initial conditions $\theta(0) = \pi/4$, $\dot{\theta}(0) = 0$, $\phi(0) = 0$, $\dot{\phi}(0) = \dot{\phi}_0$, solve for the value of $\dot{\theta}$ at the instant just before θ reaches zero. Assume that particle 1 remains in contact with the xy-plane. (c) Find the minimum magnitude of $\dot{\phi}_0$ such that particle 1 will leave the xy-plane at $t = 0$.

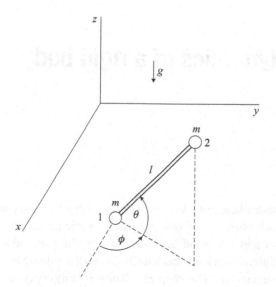

Figure P 2.21.

2.22. Four frictionless uniform spheres, each of mass m and radius r, are placed in the form of a pyramid on a horizontal surface. They are released from rest with the upper sphere moving downward and the lower spheres moving radially outward without rotation. Because of the symmetry, the angle θ to the top sphere is the same for all three lower spheres. (a) Find the differential equation for θ, assuming that sphere 1 remains in contact with the lower spheres. (b) Find the angle θ at which the top sphere loses contact with the others. (c) What is the final velocity of sphere 2?

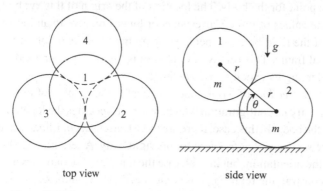

top view   side view

Figure P 2.22.

3   Kinematics and dynamics of a rigid body

A rigid body may be viewed as a special case of a system of particles in which the particles are rigidly interconnected with each other. As a consequence, the basic principles which apply to systems of particles also apply to rigid bodies. However, there are additional kinematical and dynamical formulations which are specifically associated with rigid bodies, and which contribute greatly to their analysis. This chapter will be primarily concerned with a discussion of these topics.

3.1   Kinematical preliminaries

Degrees of freedom

The *configuration* of a rigid body can be expressed by giving the location of an arbitrary point in the body and also giving the orientation of the body in space. This can be accomplished by first considering a Cartesian *body-axis* frame to be fixed in the rigid body with its origin at the arbitrary reference point for the body. The location of the origin of this *xyz* body-axis frame can be given by the values of three Cartesian coordinates relative to an inertial *XYZ* frame. The orientation of the rigid body is specified by giving the orientation of the body axes relative to the inertial frame. This relative orientation is frequently expressed in terms of three independent Euler angles which will be defined.

The number of *degrees of freedom* of a rigid body is equal to the number of independent parameters required to specify its configuration which, in effect, specifies the locations of all the particles comprising the body. In this case, there are three Cartesian coordinates and three Euler angles, so there are *six degrees of freedom* for the rigid body. Alternative methods are available for specifying the orientation, but in each case the number of parameters minus the number of independent constraining relationships is equal to three, the number of degrees of freedom associated with orientation. To summarize, a rigid body has six degrees of freedom, three being translational and three rotational.

Rotation of axes

Let **r** be an *arbitrary vector*. We seek a relationship between its components in the primed and unprimed coordinate systems of Fig. 3.1. This relationship must be linear, so we can

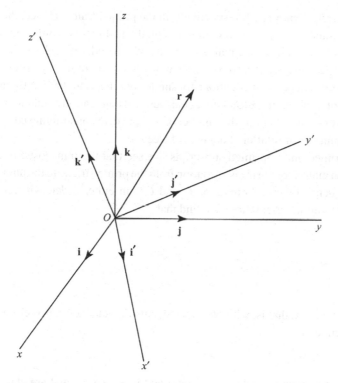

Figure 3.1.

write the matrix equation

$$r' = Cr \tag{3.1}$$

In detail, we have

$$
\begin{bmatrix} x' \\ y' \\ z' \end{bmatrix} = \begin{bmatrix} c_{x'x} & c_{x'y} & c_{x'z} \\ c_{y'x} & c_{y'y} & c_{y'z} \\ c_{z'x} & c_{z'y} & c_{z'z} \end{bmatrix} \begin{bmatrix} x \\ y \\ z \end{bmatrix} \tag{3.2}
$$

where

$$\mathbf{r} = x\mathbf{i} + y\mathbf{j} + z\mathbf{k} = x'\mathbf{i}' + y'\mathbf{j}' + z'\mathbf{k}' \tag{3.3}$$

The matrix C is called a *rotation matrix* and it specifies the relative orientation of the two coordinate systems. An element $c_{i'j}$ is equal to the cosine of the angle between the positive $\mathbf{i}'$ and $\mathbf{j}$ axes, and is called a *direction cosine*. By convention, the first subscript refers to the final coordinate system, and the second subscript refers to the original coordinate system. In order to transform a given set of components to a new coordinate system, premultiply by the corresponding rotation matrix.

In general, the nine elements of a rotation matrix are all different. The first column of the C matrix gives the components of the unit vector $\mathbf{i}$ in the primed frame. Similarly, the second and third columns of C represent $\mathbf{j}$ and $\mathbf{k}$ respectively, in the primed frame. Hence, the sum of the squares of the elements of a single column must equal one. In other words, the scalar product of a unit vector with itself must equal one. On the other hand, the scalar product of any two different columns must equal zero, since the unit vectors are mutually orthogonal. Altogether, there are three independent equations for single columns, and three independent equations for pairs of columns, giving a total of six independent constraining equations. The number of direction cosines (nine) minus the number of independent constraining equations (six) yields three, the number of rotational degrees of freedom.

By interchanging primed and unprimed subscripts, we see that the transposed rotation matrix C^{T} is the rotation matrix for the transformation from the primed frame to the unprimed frame. Hence, the sequence of transformations C and C^{T}, in either order, will return a coordinate frame to its original orientation. We find that

$$r = C^{\mathrm{T}} r' \tag{3.4}$$

and, using (3.1) and (3.4),

$$C^{\mathrm{T}} C = C C^{\mathrm{T}} = U \tag{3.5}$$

where U is a 3×3 unit matrix, that is, with ones on the main diagonal and zeros elsewhere. From (3.5) it is apparent that

$$C^{\mathrm{T}} = C^{-1} \tag{3.6}$$

Matrices such as the rotation matrix whose transpose and inverse are equal are classed as *orthogonal* matrices. The determinant of any rotation matrix is equal to $+1$.

In general, a rotation of axes given by C_a followed by a second rotation C_b is equivalent to a single rotation

$$C = C_b C_a \tag{3.7}$$

The order of matrix multiplications is important, indicating that the order of the corresponding finite rotations is also important. As we shall see, the definitions of the various Euler angle systems used in specifying rigid body orientations always require a particular order in making the rotations.

Euler angles

The rotation of a rigid body from some reference orientation to an arbitrary final orientation can always be accomplished by three rotations in a given sequence about specified body axes. The resulting angles of rotation are known as Euler angles. Many Euler angle systems are possible, but we will consider only two, namely, aircraft Euler angles and classical Euler angles.

Type I (aircraft) Euler angles are shown in Fig. 3.2 and represent an axis-of-rotation order zyx, where each successive rotation is about the latest position of the given body axis. Assume that the xyz body axes and the XYZ inertial axes coincide initially. Then three

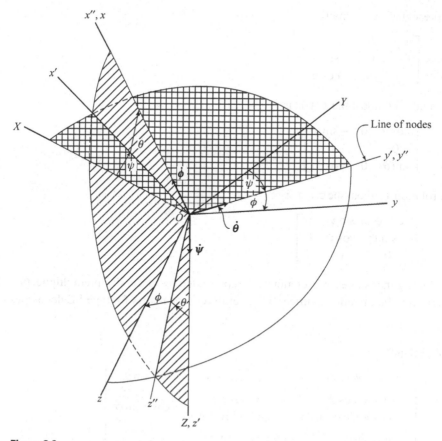

Figure 3.2.

rotations are made in the following order: (1) ψ about the Z-axis, resulting in the primed axis system; (2) θ about the y'-axis (line of nodes) resulting in the double-primed system; (3) ϕ about the x''-axis, resulting in the final xyz body-fixed frame.

Incremental changes in the Euler angles are indicated by the corresponding angular velocity vectors; namely, $\dot{\psi}$ about the Z-axis, $\dot{\theta}$ about the y'-axis (line of nodes), and $\dot{\phi}$ about the x-axis. The absolute angular velocity of the xyz body-axis frame is

$$\boldsymbol{\omega} = \dot{\boldsymbol{\psi}} + \dot{\boldsymbol{\theta}} + \dot{\boldsymbol{\phi}} \tag{3.8}$$

Note that the vectors $\dot{\boldsymbol{\psi}}$ and $\dot{\boldsymbol{\phi}}$ are not orthogonal. In terms of body-axis components,

$$\boldsymbol{\omega} = \omega_x \mathbf{i} + \omega_y \mathbf{j} + \omega_z \mathbf{k} \tag{3.9}$$

For an aircraft, ω_x is the *roll rate*, ω_y is the *pitch rate*, and ω_z is the *yaw rate*. A rotation matrix C, defined in terms of direction cosines, was given by (3.2). For rotations about individual Cartesian axes, we have the following results. A rotation ϕ about the x-axis is represented by the matrix

$$
\Phi = \begin{bmatrix} 1 & 0 & 0 \\ 0 & \cos\phi & \sin\phi \\ 0 & -\sin\phi & \cos\phi \end{bmatrix}
$$

(3.10)

A rotation θ about the y-axis is

$$
\Theta = \begin{bmatrix} \cos\theta & 0 & -\sin\theta \\ 0 & 1 & 0 \\ \sin\theta & 0 & \cos\theta \end{bmatrix}
$$

(3.11)

A rotation ψ about the z-axis is

$$
\Psi = \begin{bmatrix} \cos\psi & \sin\psi & 0 \\ -\sin\psi & \cos\psi & 0 \\ 0 & 0 & 1 \end{bmatrix}
$$

(3.12)

Noting that successive rotations are represented by successive premultiplications by the corresponding rotation matrices, the overall rotation matrix for type I Euler angles is

$$
C = \Phi\Theta\Psi
$$

(3.13)

or, in detail,

$$
C = \begin{bmatrix} \cos\psi\cos\theta & \sin\psi\cos\theta & -\sin\theta \\ \begin{array}{c}(-\sin\psi\cos\phi \\ +\cos\psi\sin\theta\sin\phi)\end{array} & \begin{array}{c}(\cos\psi\cos\phi \\ +\sin\psi\sin\theta\sin\phi)\end{array} & \cos\theta\sin\phi \\ \begin{array}{c}(\sin\psi\sin\phi \\ +\cos\psi\sin\theta\cos\phi)\end{array} & \begin{array}{c}(-\cos\psi\sin\phi \\ +\sin\psi\sin\theta\cos\phi)\end{array} & \cos\theta\cos\phi \end{bmatrix}
$$

(3.14)

Usually, we consider that the ranges of the Euler angles are

$$
0 \le \psi < 2\pi, \qquad -\frac{\pi}{2} \le \theta \le \frac{\pi}{2}, \qquad 0 \le \phi < 2\pi
$$

The body-axis components of the absolute angular velocity in terms of Euler angle rates are

$$
\begin{aligned}
\omega_x &= \dot\phi - \dot\psi\sin\theta \\
\omega_y &= \dot\psi\cos\theta\sin\phi + \dot\theta\cos\phi \\
\omega_z &= \dot\psi\cos\theta\cos\phi - \dot\theta\sin\phi
\end{aligned}
$$

(3.15)

Conversely, the Euler angle rates in terms of angular velocity components are

$$
\begin{aligned}
\dot\psi &= \sec\theta\,(\omega_y\sin\phi + \omega_z\cos\phi) \\
\dot\theta &= \omega_y\cos\phi - \omega_z\sin\phi \\
\dot\phi &= \omega_x + \dot\psi\sin\theta \\
&= \omega_x + \omega_y\tan\theta\sin\phi + \omega_z\tan\theta\cos\phi
\end{aligned}
$$

(3.16)

We see from (3.15) that ω_x, ω_y, ω_z are well-defined for arbitrary Euler angles and Euler angle rates. But if $\theta = \pm\pi/2$, equation (3.16) shows that $\dot\psi$ and $\dot\phi$ can be infinite even if ω is finite. Furthermore, the angles ψ and ϕ are not well-defined if the x-axis of Fig. 3.2 points vertically upward or downward in this singular situation. If $\theta = \pi/2$, for example, ψ and ϕ are undefined, but $(\psi - \phi)$ is the angle between the Y and y axes.

All types of Euler angle systems are subject to singularity problems which occur when two of the Euler angle rates represent rotations about the same axis in space. Thus, in Fig. 3.2 the $\dot\psi$ and $\dot\phi$ rotations occur about the same vertical axis if $\theta = \pm\pi/2$. When combined with a $\dot\theta$ rotation about a horizontal axis, this limits the ω vector to a definite vertical plane when using this Euler angle representation, even though the actual motion may have no such constraint. For these reasons and the associated problems of maintaining accuracy in numerical computations, it is the usual practice to choose an Euler angle system such that the rotational motion of the rigid body does not come near a singular orientation.

Type II Euler angles are shown in Fig. 3.3. The order of rotation is as follows: (1) ϕ about the Z-axis; (2) θ about the x'-axis; (3) ψ about the z''-axis. The usual ranges of the Euler angles are

$$0 \le \phi < 2\pi, \qquad 0 \le \theta \le \pi, \qquad 0 \le \psi < 2\pi$$

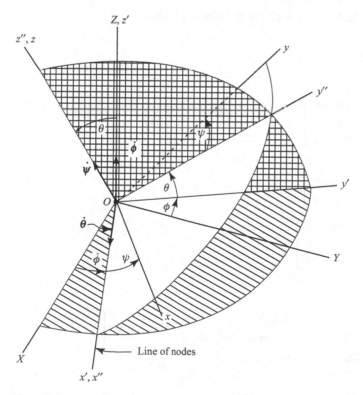

Figure 3.3.

The absolute angular velocity of the *xyz* frame is

$$\boldsymbol{\omega} = \dot{\boldsymbol{\phi}} + \dot{\boldsymbol{\theta}} + \dot{\boldsymbol{\psi}} \tag{3.17}$$

with the body-axis components

$$
\begin{aligned}
\omega_x &= \dot{\phi}\sin\theta\sin\psi + \dot{\theta}\cos\psi \\
\omega_y &= \dot{\phi}\sin\theta\cos\psi - \dot{\theta}\sin\psi \\
\omega_z &= \dot{\psi} + \dot{\phi}\cos\theta
\end{aligned}
\tag{3.18}
$$

Conversely, we obtain

$$
\begin{aligned}
\dot{\phi} &= \csc\theta\,(\omega_x\sin\psi + \omega_y\cos\psi) \\
\dot{\theta} &= \omega_x\cos\psi - \omega_y\sin\psi \\
\dot{\psi} &= \omega_z - \dot{\phi}\cos\theta \\
&= -\omega_x\cot\theta\sin\psi - \omega_y\cot\theta\cos\psi + \omega_z
\end{aligned}
\tag{3.19}
$$

Singular orientations of the body occur for $\theta = 0$ or π, and both $\dot{\phi}$ and $\dot{\psi}$ become infinite even with finite ω_x and ω_y. Furthermore, we note that the $\dot{\phi}$ and $\dot{\psi}$ rotations occur about the same vertical axis in Fig. 3.3.

The overall rotation matrix for Type II Euler angles is

$$C = \Psi\Theta\Phi \tag{3.20}$$

where

$$
\Phi = \begin{bmatrix}
\cos\phi & \sin\phi & 0 \\
-\sin\phi & \cos\phi & 0 \\
0 & 0 & 1
\end{bmatrix}
\tag{3.21}
$$

$$
\Theta = \begin{bmatrix}
1 & 0 & 0 \\
0 & \cos\theta & \sin\theta \\
0 & -\sin\theta & \cos\theta
\end{bmatrix}
\tag{3.22}
$$

$$
\Psi = \begin{bmatrix}
\cos\psi & \sin\psi & 0 \\
-\sin\psi & \cos\psi & 0 \\
0 & 0 & 1
\end{bmatrix}
\tag{3.23}
$$

In detail, we obtain

$$
C = \begin{bmatrix}
(\cos\phi\cos\psi & (\sin\phi\cos\psi & \\
-\sin\phi\cos\theta\sin\psi) & +\cos\phi\cos\theta\sin\psi) & \sin\theta\sin\psi \\
(-\cos\phi\sin\psi & (-\sin\phi\sin\psi & \\
-\sin\phi\cos\theta\cos\psi) & +\cos\phi\cos\theta\cos\psi) & \sin\theta\cos\psi \\
\sin\phi\sin\theta & -\cos\phi\sin\theta & \cos\theta
\end{bmatrix}
\tag{3.24}
$$

Euler angles are true generalized coordinates and, as such, can be used as qs in Lagrange's equations. We shall find, however, that other methods may be more convenient in obtaining the dynamical equations for the rotational motion of a rigid body.

Axis and angle of rotation

An important result in rigid body kinematics is expressed by *Euler's Theorem: The most general displacement of a rigid body with a fixed point is equivalent to a single rotation about some axis through that point.*

Because a fixed point is assumed, any motion involves a change of orientation. Hence, Euler's theorem is concerned with orientation changes. On the basis of this theorem, we conclude that, although any Euler angle system expresses an orientation change in terms of a sequence of three rotations about coordinate axes, the same change of orientation could have been obtained by a single rotation about a fixed axis. Hence, if one can specify the direction of the axis of rotation, as well as the angle of rotation about this axis, it is equivalent to specifying three Euler angles or a rotation matrix consisting of nine direction cosines.

Suppose we let the direction of the axis of rotation be given by the *unit vector*

$$\mathbf{a} = a_x \mathbf{i} + a_y \mathbf{j} + a_z \mathbf{k} \tag{3.25}$$

where $\mathbf{i}, \mathbf{j}, \mathbf{k}$ are Cartesian unit vectors and the scalar as are constrained by the relation

$$a_x^2 + a_y^2 + a_z^2 = 1 \tag{3.26}$$

These as have the *same values in either the inertial or the body-fixed frame*, and remain constant during the rotation because the rotation axis is fixed in both frames. Let ϕ be the angle of rotation of the body-fixed frame relative to the inertial frame, where positive ϕ is measured in a right-hand sense about the unit vector $\mathbf{a}$. The values of the four parameters (a_x, a_y, a_z, ϕ) specify the orientation of the rigid body and its body-fixed frame. These are the axis and angle variables.

Now suppose that we are given a rotation matrix C and we desire the corresponding values of the axis and angle variables. First, the angle of rotation ϕ is found from the trace of the rotation matrix, that is,

$$\mathrm{tr}\, C = c_{xx} + c_{yy} + c_{zz} = 1 + 2\cos\phi, \qquad 0 \le \phi \le \pi \tag{3.27}$$

Then, knowing ϕ, we can find the direction cosines of the axis of rotation from

$$
\begin{aligned}
a_x &= \frac{c_{yz} - c_{zy}}{2\sin\phi} \\
a_y &= \frac{c_{zx} - c_{xz}}{2\sin\phi} \\
a_z &= \frac{c_{xy} - c_{yx}}{2\sin\phi}
\end{aligned}
\tag{3.28}
$$

For convenience, the primes have been dropped from the subscripts of the cs. Note that the limits on ϕ given in (3.27) imply that $\sin\phi$ is positive or zero. This is always possible because a rotation ϕ is equivalent to a rotation $(2\pi - \phi)$ about an oppositely directed **a** axis.

If the value of ϕ is near 0 or π, then $|\sin\phi| << 1$ and (3.28) leads to inaccurate results. In this case, better accuracy can be obtained by using

$$
a_x = \frac{1}{d}(c_{yz} - c_{zy})
$$

$$
a_y = \frac{1}{d}(c_{zx} - c_{xz}) \tag{3.29}
$$

$$
a_z = \frac{1}{d}(c_{xy} - c_{yx})
$$

where

$$
d = [(c_{yz} - c_{zy})^2 + (c_{zx} - c_{xz})^2 + (c_{xy} - c_{yx})^2]^{\frac{1}{2}}
$$

A comparison of (3.28) and (3.29) shows that

$$
\sin\phi = \frac{1}{2}d \tag{3.30}
$$

For values of ϕ near 0 or π, this equation is more accurate than (3.27) in determining ϕ. Note that (3.27) uses the main diagonal terms of C to evaluate ϕ, whereas (3.30) uses the off-diagonal terms. For these small positive values of $\sin\phi$, (3.27) is used to determine the quadrant.

Now consider the case in which the axis and angle variables are given, and we wish to find the corresponding rotation matrix. First, let us write the rotation matrix C as the sum of its symmetric and antisymmetric parts.

$$
C = C_s + C_a \tag{3.31}
$$

The symmetric matrix is

$$
C_s = \frac{1}{2}(C + C^{\mathrm{T}}) \tag{3.32}
$$

and the antisymmetric (or skew-symmetric) matrix is

$$
C_a = \frac{1}{2}(C - C^{\mathrm{T}}) \tag{3.33}
$$

It can be shown that, in terms of axis and angle variables, the symmetric portion of the rotation matrix is

$$
C_s = \cos\phi\, U + (1 - \cos\phi)aa^{\mathrm{T}} \tag{3.34}
$$

or, in detail,

$$
C_s = \cos\phi \begin{bmatrix} 1 & 0 & 0 \\ 0 & 1 & 0 \\ 0 & 0 & 1 \end{bmatrix} + (1 - \cos\phi) \begin{bmatrix} a_x^2 & a_x a_y & a_x a_z \\ a_y a_x & a_y^2 & a_y a_z \\ a_z a_x & a_z a_y & a_z^2 \end{bmatrix} \tag{3.35}
$$

The antisymmetric part of the rotation matrix is

$$C_a = \sin\phi \begin{bmatrix} 0 & a_z & -a_y \\ -a_z & 0 & a_x \\ a_y & -a_x & 0 \end{bmatrix} \tag{3.36}$$

Let us introduce the tilde matrix notation

$$\tilde{a} \equiv \begin{bmatrix} 0 & -a_z & a_y \\ a_z & 0 & -a_x \\ -a_y & a_x & 0 \end{bmatrix} \tag{3.37}$$

Then

$$C_a = -\sin\phi \, \tilde{a} \tag{3.38}$$

and the complete rotation matrix is

$$C = C_s + C_a = \cos\phi \, U + (1 - \cos\phi) a a^{\mathrm{T}} - \sin\phi \, \tilde{a} \tag{3.39}$$

In detail, we find that

$$\begin{aligned}
c_{xx} &= (1 - \cos\phi) a_x^2 + \cos\phi = a_x^2 + \left(a_y^2 + a_z^2\right)\cos\phi \\
c_{yy} &= (1 - \cos\phi) a_y^2 + \cos\phi = a_y^2 + \left(a_z^2 + a_x^2\right)\cos\phi \\
c_{zz} &= (1 - \cos\phi) a_z^2 + \cos\phi = a_z^2 + \left(a_x^2 + a_y^2\right)\cos\phi \\
c_{xy} &= (1 - \cos\phi) a_x a_y + a_z \sin\phi \\
c_{yx} &= (1 - \cos\phi) a_y a_x - a_z \sin\phi \\
c_{xz} &= (1 - \cos\phi) a_x a_z - a_y \sin\phi \\
c_{zx} &= (1 - \cos\phi) a_z a_x + a_y \sin\phi \\
c_{yz} &= (1 - \cos\phi) a_y a_z + a_x \sin\phi \\
c_{zy} &= (1 - \cos\phi) a_z a_y - a_x \sin\phi
\end{aligned} \tag{3.40}$$

Euler parameters

Another method of specifying the orientation of a rigid body is by the use of four *Euler parameters*. The Euler parameters are based on the mathematics of quaternions and are closely related to axis and angle variables. We shall define the Euler parameters in terms of axis and angle variables as follows:

$$\begin{aligned}
\epsilon_x &= a_x \sin\frac{\phi}{2} \\
\epsilon_y &= a_y \sin\frac{\phi}{2} \\
\epsilon_z &= a_z \sin\frac{\phi}{2} \\
\eta &= \cos\frac{\phi}{2}
\end{aligned} \tag{3.41}$$

where we recall that $0 \leq \phi \leq \pi$. These four parameters are constrained by the single equation

$$\epsilon_x^2 + \epsilon_y^2 + \epsilon_z^2 + \eta^2 = 1 \tag{3.42}$$

Euler parameters have the advantages of having no singular orientations and only one equation of constraint. Thus, they exhibit a good balance between overall accuracy and computational efficiency.

Let us define the *Euler 3-vector* as

$$\boldsymbol{\epsilon} = \epsilon_x \mathbf{i} + \epsilon_y \mathbf{j} + \epsilon_z \mathbf{k} = \mathbf{a} \sin \frac{\phi}{2} \tag{3.43}$$

whose magnitude is

$$\epsilon = \left(\epsilon_x^2 + \epsilon_y^2 + \epsilon_z^2 \right)^{\frac{1}{2}} \tag{3.44}$$

Then

$$\epsilon^2 + \eta^2 = 1 \tag{3.45}$$

To convert from the direction cosines of the rotation matrix to Euler parameters, we can use

$$
\begin{aligned}
\epsilon_x &= \frac{1}{4\eta}(c_{yz} - c_{zy}) \\[4pt]
\epsilon_y &= \frac{1}{4\eta}(c_{zx} - c_{xz}) \\[4pt]
\epsilon_z &= \frac{1}{4\eta}(c_{xy} - c_{yx}) \\[4pt]
\eta &= \frac{1}{2}\sqrt{1 + \operatorname{tr} C}, \qquad 0 \leq \phi \leq \pi
\end{aligned}
\tag{3.46}
$$

Conversely, to convert from Euler parameters to the rotation matrix, we obtain

$$
C = \begin{bmatrix}
1 - 2\left(\epsilon_y^2 + \epsilon_z^2\right) & 2(\epsilon_x \epsilon_y + \epsilon_z \eta) & 2(\epsilon_x \epsilon_z - \epsilon_y \eta) \\
2(\epsilon_y \epsilon_x - \epsilon_z \eta) & 1 - 2(\epsilon_z^2 + \epsilon_x^2) & 2(\epsilon_y \epsilon_z + \epsilon_x \eta) \\
2(\epsilon_z \epsilon_x + \epsilon_y \eta) & 2(\epsilon_z \epsilon_y - \epsilon_x \eta) & 1 - 2\left(\epsilon_x^2 + \epsilon_y^2\right)
\end{bmatrix}
\tag{3.47}
$$

Another form of this result is the matrix equation

$$C = (1 - 2\epsilon^{\mathrm{T}}\epsilon)U + 2\epsilon\epsilon^{\mathrm{T}} - 2\eta\tilde{\epsilon} \tag{3.48}$$

where $\epsilon^{\mathrm{T}}\epsilon = \epsilon_x^2 + \epsilon_y^2 + \epsilon_z^2$ and we see that

$$1 - 2\epsilon^{\mathrm{T}}\epsilon = \eta^2 - \left(\epsilon_x^2 + \epsilon_y^2 + \epsilon_z^2\right) = \cos^2 \frac{\phi}{2} - \sin^2 \frac{\phi}{2} = \cos\phi \tag{3.49}$$

Successive rotations

In the earlier discussion of rotation matrices, we found that successive rotations of a co-ordinate frame are accomplished by successive premultiplications by the corresponding rotation matrices. The question arises whether a similar procedure is available for use with Euler parameters. The answer is that such a procedure is indeed available, and consists in premultiplying the current Euler 4-vector by the matrix

$$
R = \begin{bmatrix}
\eta & \epsilon_z & -\epsilon_y & \epsilon_x \\
-\epsilon_z & \eta & \epsilon_x & \epsilon_y \\
\epsilon_y & -\epsilon_x & \eta & \epsilon_z \\
-\epsilon_x & -\epsilon_y & -\epsilon_z & \eta
\end{bmatrix}
\tag{3.50}
$$

where the components of ϵ are taken in the body-fixed frame.

As an example, suppose a first rotation "a", represented by $(\epsilon_{xa}, \epsilon_{ya}, \epsilon_{za}, \eta_a)$, is followed by a second rotation "b" given by $(\epsilon_{xb}, \epsilon_{yb}, \epsilon_{zb}, \eta_b)$. These two rotations are equivalent to a single rotation given by

$$
\begin{bmatrix}
\epsilon_x \\
\epsilon_y \\
\epsilon_z \\
\eta
\end{bmatrix} =
\begin{bmatrix}
\eta_b & \epsilon_{zb} & -\epsilon_{yb} & \epsilon_{xb} \\
-\epsilon_{zb} & \eta_b & \epsilon_{xb} & \epsilon_{yb} \\
\epsilon_{yb} & -\epsilon_{xb} & \eta_b & \epsilon_{zb} \\
-\epsilon_{xb} & -\epsilon_{yb} & -\epsilon_{zb} & \eta_b
\end{bmatrix}
\begin{bmatrix}
\epsilon_{xa} \\
\epsilon_{ya} \\
\epsilon_{za} \\
\eta_a
\end{bmatrix}
\tag{3.51}
$$

An alternate form involving the vector ϵ and the scalar η is given by the two equations

$$
\epsilon = \eta_b \epsilon_a + \eta_a \epsilon_b + \epsilon_a \times \epsilon_b
\tag{3.52}
$$

$$
\eta = \eta_a \eta_b - \epsilon_a \cdot \epsilon_b
\tag{3.53}
$$

where the vectors are again expressed relative to the body-fixed frame. Notice that the overall axis of rotation depends on the order of the individual rotations, but the angle of rotation does not.

In terms of axis and angle variables, the sequence of two rotations is equivalent to a single rotation ϕ about an axis **a** where

$$
\cos \frac{\phi}{2} = \cos \frac{\phi_a}{2} \cos \frac{\phi_b}{2} - \sin \frac{\phi_a}{2} \sin \frac{\phi_b}{2} \mathbf{a}_a \cdot \mathbf{a}_b
\tag{3.54}
$$

and, using this value of ϕ,

$$
\mathbf{a} = \frac{1}{\sin \dfrac{\phi}{2}} \left(\sin \frac{\phi_a}{2} \cos \frac{\phi_b}{2} \mathbf{a}_a + \cos \frac{\phi_a}{2} \sin \frac{\phi_b}{2} \mathbf{a}_b + \sin \frac{\phi_a}{2} \sin \frac{\phi_b}{2} \mathbf{a}_a \times \mathbf{a}_b \right)
\tag{3.55}
$$

in agreement with (3.52) and (3.53). Again, for a sequence of two given rotations about specified axes, the final equivalent angle ϕ is independent of the order of rotation, but the final axis of rotation is not.

Angular velocity

Consider the rotation matrix which gives the orientation of a body-fixed xyz frame relative to an inertial XYZ frame. Let $(\mathbf{i}, \mathbf{j}, \mathbf{k})$ be the unit vectors for the xyz frame and let $(\mathbf{I}, \mathbf{J}, \mathbf{K})$ be the corresponding unit vectors for the inertial frame. Then we can write

$$C = \begin{bmatrix} \mathbf{i}\cdot\mathbf{I} & \mathbf{i}\cdot\mathbf{J} & \mathbf{i}\cdot\mathbf{K} \\ \mathbf{j}\cdot\mathbf{I} & \mathbf{j}\cdot\mathbf{J} & \mathbf{j}\cdot\mathbf{K} \\ \mathbf{k}\cdot\mathbf{I} & \mathbf{k}\cdot\mathbf{J} & \mathbf{k}\cdot\mathbf{K} \end{bmatrix} \tag{3.56}$$

that is, $c_{xx} = \mathbf{i}\cdot\mathbf{I}$, $c_{xy} = \mathbf{i}\cdot\mathbf{J}$, and so on in accordance with the direction cosine definitions.

Now suppose that the xyz frame has an angular velocity

$$\boldsymbol{\omega} = \omega_x \mathbf{i} + \omega_y \mathbf{j} + \omega_z \mathbf{k} \tag{3.57}$$

This will result in changing values of the elements of the rotation matrix. For example,

$$\begin{aligned} \dot{c}_{xx} &= \frac{d\mathbf{i}}{dt}\cdot\mathbf{I} = (\boldsymbol{\omega}\times\mathbf{i})\cdot\mathbf{I} \\ &= (\omega_z \mathbf{j} - \omega_y \mathbf{k})\cdot\mathbf{I} = c_{yx}\omega_z - c_{zx}\omega_y \end{aligned} \tag{3.58}$$

A similar procedure results in the complete set of *Poisson equations*.

$$\begin{aligned} \dot{c}_{xx} &= c_{yx}\omega_z - c_{zx}\omega_y \\ \dot{c}_{xy} &= c_{yy}\omega_z - c_{zy}\omega_y \\ \dot{c}_{xz} &= c_{yz}\omega_z - c_{zz}\omega_y \\ \dot{c}_{yx} &= c_{zx}\omega_x - c_{xx}\omega_z \\ \dot{c}_{yy} &= c_{zy}\omega_x - c_{xy}\omega_z \\ \dot{c}_{yz} &= c_{zz}\omega_x - c_{xz}\omega_z \\ \dot{c}_{zx} &= c_{xx}\omega_y - c_{yx}\omega_x \\ \dot{c}_{zy} &= c_{xy}\omega_y - c_{yy}\omega_x \\ \dot{c}_{zz} &= c_{xz}\omega_y - c_{yz}\omega_x \end{aligned} \tag{3.59}$$

These equations can be written in the matrix form

$$\dot{C} = -\tilde{\omega}C \tag{3.60}$$

where

$$\tilde{\omega} = \begin{bmatrix} 0 & -\omega_z & \omega_y \\ \omega_z & 0 & -\omega_x \\ -\omega_y & \omega_x & 0 \end{bmatrix} \tag{3.61}$$

Upon solving (3.60) for $\tilde{\omega}$, we obtain

$$\tilde{\omega} = -\dot{C}C^{\mathrm{T}} = C\dot{C}^{\mathrm{T}} \tag{3.62}$$

which results in

$$\omega_x = c_{zx}\dot{c}_{yx} + c_{zy}\dot{c}_{yy} + c_{zz}\dot{c}_{yz}$$
$$\omega_y = c_{xx}\dot{c}_{zx} + c_{xy}\dot{c}_{zy} + c_{xz}\dot{c}_{zz} \qquad (3.63)$$
$$\omega_z = c_{yx}\dot{c}_{xx} + c_{yy}\dot{c}_{xy} + c_{yz}\dot{c}_{xz}$$

Now consider the angular velocity of a rigid body in terms of axis and angle variables. It can be shown that

$$\omega = \dot{\phi}\mathbf{a} + \sin\phi\,\dot{\mathbf{a}} + (1 - \cos\phi)\,\mathbf{a}\times\dot{\mathbf{a}} \qquad (3.64)$$

where the three terms represent mutually orthogonal components.

We can solve for $\dot{\phi}$ and $\dot{\mathbf{a}}$ in terms of ω by first taking the dot product of $\mathbf{a}$ with (3.64), obtaining

$$\dot{\phi} = \mathbf{a}\cdot\omega \qquad (3.65)$$

Next, noting that $\mathbf{a}\times(\mathbf{a}\times\dot{\mathbf{a}}) = -\dot{\mathbf{a}}$, we obtain from (3.64) that

$$\mathbf{a}\times\omega = \sin\phi\,\mathbf{a}\times\dot{\mathbf{a}} - (1 - \cos\phi)\,\dot{\mathbf{a}} \qquad (3.66)$$

and

$$\mathbf{a}\times(\mathbf{a}\times\omega) = -\sin\phi\,\dot{\mathbf{a}} - (1 - \cos\phi)\,\mathbf{a}\times\dot{\mathbf{a}} \qquad (3.67)$$

Recall that

$$\cot\frac{\phi}{2} = \frac{\sin\phi}{1 - \cos\phi} = \frac{1 + \cos\phi}{\sin\phi} \qquad (3.68)$$

so

$$\cot\frac{\phi}{2}\,\mathbf{a}\times(\mathbf{a}\times\omega) = -(1 + \cos\phi)\dot{\mathbf{a}} - \sin\phi\,\mathbf{a}\times\dot{\mathbf{a}} \qquad (3.69)$$

Then

$$\mathbf{a}\times\omega + \cot\frac{\phi}{2}\,\mathbf{a}\times(\mathbf{a}\times\omega) = -2\dot{\mathbf{a}} \qquad (3.70)$$

or

$$\dot{\mathbf{a}} = -\frac{1}{2}\left[\mathbf{a}\times\omega + \cot\frac{\phi}{2}\,\mathbf{a}\times(\mathbf{a}\times\omega)\right] \qquad (3.71)$$

It is convenient for computational reasons to find the component rates $(\dot{a}_x, \dot{a}_y, \dot{a}_z)$ of $\mathbf{a}$ relative to the rotating body axes. Using vector notation, we have

$$(\dot{\mathbf{a}})_r = \dot{\mathbf{a}} - \omega\times\mathbf{a} = \frac{1}{2}\left[\mathbf{a}\times\omega - \cot\frac{\phi}{2}\,\mathbf{a}\times(\mathbf{a}\times\omega)\right] \qquad (3.72)$$

where

$$(\dot{\mathbf{a}})_r = \dot{a}_x\mathbf{i} + \dot{a}_y\mathbf{j} + \dot{a}_z\mathbf{k} \qquad (3.73)$$

Similar equations in terms of Euler parameters can be obtained by first recalling that

$$\epsilon = \mathbf{a}\sin\frac{\phi}{2}, \qquad \eta = \cos\frac{\phi}{2} \tag{3.74}$$

and

$$\dot{\eta} = -\frac{1}{2}\dot{\phi}\sin\frac{\phi}{2} \tag{3.75}$$

Then

$$\dot{\eta} = -\frac{1}{2}\mathbf{a}\cdot\boldsymbol{\omega}\sin\frac{\phi}{2} = -\frac{1}{2}\boldsymbol{\epsilon}\cdot\boldsymbol{\omega} \tag{3.76}$$

A differentiation of (3.74) with respect to time and the use of (3.65) results in

$$\dot{\boldsymbol{\epsilon}} = \frac{1}{2}\cos\frac{\phi}{2}(\mathbf{a}\cdot\boldsymbol{\omega})\mathbf{a} + \sin\frac{\phi}{2}\dot{\mathbf{a}} \tag{3.77}$$

Now substitute for $\dot{\mathbf{a}}$ from (3.71) and note that $\mathbf{a}\times(\mathbf{a}\times\boldsymbol{\omega}) = (\mathbf{a}\cdot\boldsymbol{\omega})\mathbf{a} - \boldsymbol{\omega}$. We obtain

$$\begin{aligned}
\dot{\boldsymbol{\epsilon}} &= \frac{1}{2}\cos\frac{\phi}{2}(\mathbf{a}\cdot\boldsymbol{\omega})\mathbf{a} - \frac{1}{2}\sin\frac{\phi}{2}\left[\mathbf{a}\times\boldsymbol{\omega} + \cot\frac{\phi}{2}((\mathbf{a}\cdot\boldsymbol{\omega})\mathbf{a} - \boldsymbol{\omega})\right] \\
&= -\frac{1}{2}\sin\frac{\phi}{2}\mathbf{a}\times\boldsymbol{\omega} + \frac{1}{2}\cos\frac{\phi}{2}\boldsymbol{\omega} \\
&= -\frac{1}{2}\boldsymbol{\epsilon}\times\boldsymbol{\omega} + \frac{1}{2}\eta\boldsymbol{\omega}
\end{aligned} \tag{3.78}$$

The basic Euler parameter rates are given by (3.76) and (3.78). However, for computational purposes we need to find $(\dot{\boldsymbol{\epsilon}})_r$, that is, the values of $\dot{\epsilon}_x$, $\dot{\epsilon}_y$, and $\dot{\epsilon}_z$. We obtain

$$(\dot{\boldsymbol{\epsilon}})_r = \dot{\boldsymbol{\epsilon}} + \boldsymbol{\epsilon}\times\boldsymbol{\omega} = \frac{1}{2}\boldsymbol{\epsilon}\times\boldsymbol{\omega} + \frac{1}{2}\eta\boldsymbol{\omega} \tag{3.79}$$

or, in detail,

$$\begin{aligned}
\dot{\epsilon}_x &= \frac{1}{2}(\omega_z\epsilon_y - \omega_y\epsilon_z + \omega_x\eta) \\
\dot{\epsilon}_y &= \frac{1}{2}(\omega_x\epsilon_z - \omega_z\epsilon_x + \omega_y\eta) \\
\dot{\epsilon}_z &= \frac{1}{2}(\omega_y\epsilon_x - \omega_x\epsilon_y + \omega_z\eta) \\
\dot{\eta} &= -\frac{1}{2}(\omega_x\epsilon_x + \omega_y\epsilon_y + \omega_z\epsilon_z)
\end{aligned} \tag{3.80}$$

These kinematic differential equations are widely used and, upon numerical integration, yield the orientation of a rigid body as a function of time. Note that, in contrast to the comparable $(\dot{\mathbf{a}})_r$ equations, they do not involve trigonometric functions. They require the body-axis components of $\boldsymbol{\omega}$ as inputs, and these are often obtained from the solution of the dynamical equations.

Conversely, the body-axis components of the angular velocity ω can be obtained from (3.80) and the Euler parameter constraint equation (3.42). They are

$$\omega_x = 2(\eta \dot{\epsilon}_x + \epsilon_z \dot{\epsilon}_y - \epsilon_y \dot{\epsilon}_z - \epsilon_x \dot{\eta})$$
$$\omega_y = 2(\eta \dot{\epsilon}_y + \epsilon_x \dot{\epsilon}_z - \epsilon_z \dot{\epsilon}_x - \epsilon_y \dot{\eta}) \tag{3.81}$$
$$\omega_z = 2(\eta \dot{\epsilon}_z + \epsilon_y \dot{\epsilon}_x - \epsilon_x \dot{\epsilon}_y - \epsilon_z \dot{\eta})$$

Infinitesimal rotations

Consider a body-fixed xyz frame and an inertial XYZ frame which initially coincide. Then let the xyz frame undergo a sequence of three *infinitesimal rotations*; namely, θ_x about the x-axis, θ_y about the y-axis, and θ_z about the z-axis. The overall rotation matrix is equal to the product of the three individual rotation matrices. If we retain terms to first order in the small θs, we obtain

$$C = \begin{bmatrix} 1 & \theta_z & 0 \\ -\theta_z & 1 & 0 \\ 0 & 0 & 1 \end{bmatrix} \begin{bmatrix} 1 & 0 & -\theta_y \\ 0 & 1 & 0 \\ \theta_y & 0 & 1 \end{bmatrix} \begin{bmatrix} 1 & 0 & 0 \\ 0 & 1 & \theta_x \\ 0 & -\theta_x & 1 \end{bmatrix}$$

$$= \begin{bmatrix} 1 & \theta_z & -\theta_y \\ -\theta_z & 1 & \theta_x \\ \theta_y & -\theta_x & 1 \end{bmatrix} \tag{3.82}$$

or

$$C = U - \tilde{\theta} \tag{3.83}$$

where

$$\tilde{\theta} = \begin{bmatrix} 0 & -\theta_z & \theta_y \\ \theta_z & 0 & -\theta_x \\ -\theta_y & \theta_x & 0 \end{bmatrix} \tag{3.84}$$

and U is a 3×3 unit matrix.

It is important to realize that the order of the infinitesimal rotations does not matter, in contrast to the situation for *finite rotations*, where the order must be specified. We can express a general *infinitesimal rotation* as a vector

$$\theta = \theta_x \mathbf{I} + \theta_y \mathbf{J} + \theta_z \mathbf{K} \tag{3.85}$$

where $\mathbf{I}, \mathbf{J}, \mathbf{K}$ are inertial unit vectors. To first order in terms of body-fixed unit vectors,

$$\theta = \theta_x \mathbf{i} + \theta_y \mathbf{j} + \theta_z \mathbf{k} \tag{3.86}$$

The component infinitesimal rotations $\theta_x, \theta_y, \theta_z$ can be taken in any order. Hence, a sequence of several infinitesimal rotations is equivalent to a single rotation equal to their vector sum.

The angular velocity of the rigid body is equal to an infinitesimal rotation $\Delta\boldsymbol{\theta}$ divided by the corresponding time interval Δt, in the limit as $\Delta\boldsymbol{\theta}$ and Δt approach zero. Thus, to first order,

$$\boldsymbol{\omega} = \lim_{\Delta t \to 0} \frac{\Delta\boldsymbol{\theta}}{\Delta t} = \dot{\boldsymbol{\theta}} = \dot{\theta}_x \mathbf{i} + \dot{\theta}_y \mathbf{j} + \dot{\theta}_z \mathbf{k} \tag{3.87}$$

where $\boldsymbol{\omega}$ is finite, in general. We see that

$$\omega_x = \dot{\theta}_x, \qquad \omega_y = \dot{\theta}_y, \qquad \omega_z = \dot{\theta}_z \tag{3.88}$$

Upon differentiation of (3.83), we obtain

$$\dot{C} = -\dot{\tilde{\theta}} = -\tilde{\omega} \tag{3.89}$$

for infinitesimal rotations.

If we use axis and angle variables, we find that

$$\boldsymbol{\theta} = \phi\mathbf{a} \tag{3.90}$$

where

$$\phi = \sqrt{\theta_x^2 + \theta_y^2 + \theta_z^2} \tag{3.91}$$

Thus,

$$a_x = \frac{\theta_x}{\phi}, \qquad a_y = \frac{\theta_y}{\phi}, \qquad a_z = \frac{\theta_z}{\phi} \tag{3.92}$$

The angular velocity of the rigid body is

$$\boldsymbol{\omega} = \dot{\boldsymbol{\theta}} = \dot{\phi}\mathbf{a} + \phi\dot{\mathbf{a}} = \dot{\phi}\mathbf{a} \tag{3.93}$$

where we note that $\phi\dot{\mathbf{a}}$ is infinitesimal.

In terms of Euler parameters, we have, from (3.43) and (3.90) for small ϕ,

$$\boldsymbol{\epsilon} = \frac{1}{2}\boldsymbol{\theta} \tag{3.94}$$

or

$$\epsilon_x = \frac{1}{2}\theta_x, \qquad \epsilon_y = \frac{1}{2}\theta_y, \qquad \epsilon_z = \frac{1}{2}\theta_z, \qquad \eta = 1 \tag{3.95}$$

The angular velocity is

$$\boldsymbol{\omega} = \dot{\boldsymbol{\theta}} = 2\dot{\boldsymbol{\epsilon}} \tag{3.96}$$

where

$$\dot{\boldsymbol{\epsilon}} = \dot{\epsilon}_x \mathbf{i} + \dot{\epsilon}_y \mathbf{j} + \dot{\epsilon}_z \mathbf{k} \tag{3.97}$$

Instantaneous axis of rotation

Euler's theorem states that the most general displacement of a rigid body with a fixed base point is equivalent to a single rotation about some axis through that point. A *completely general displacement* of a rigid body, however, would require a displacement of all points in the body. So, if one is given a base point fixed in the body, then the most general displacement is equivalent to (1) a translation of the body resulting in the correct final position of the base point, followed by (2) a rotation about an axis through the final position of the base point which gives the correct final orientation. Now, for a given base point, it is apparent that the translation and rotation are independent and can take place in any order or possibly together. Furthermore, for any nonzero rotation angle ϕ and a given axis orientation, it is always possible to choose a location of the axis of rotation such that any given base point in the body will undergo a prescribed translation in a plane perpendicular to the axis of rotation. Hence, all that remains is to give the base point the required translation parallel to the axis of rotation. Thus we obtain *Chasles' Theorem: The most general displacement of a rigid body is equivalent to a screw displacement consisting of a rotation about a fixed axis plus a translation parallel to that axis.*

Now suppose that a rigid body undergoes a general infinitesimal displacement during the time interval Δt. Then, in accordance with Chasles' theorem, the motion in the limit as Δt approaches zero can be considered as the superposition of an angular velocity ω about some axis plus a translational velocity $\mathbf{v}$ in a direction parallel to ω. All points in the body that lie on this axis have the same velocity $\mathbf{v}$ along the axis. This axis exists for all cases in which $\omega \neq 0$ and is known as the *instantaneous axis of rotation*.

For the special case of planar motion, the velocity $\mathbf{v}$ is equal to zero; and the instantaneous axis becomes the *instantaneous center of rotation*. The instantaneous axis of rotation, or the instantaneous center for planar motion, can move relative to the body and relative to inertial space. However, at the instant under consideration these points in the rigid body have no velocity component normal to the instantaneous axis of rotation.

Example 3.1 A solid cone of semivertex angle 30° rolls without slipping on the inside surface of a conical cavity with a semivertex angle at O equal to 60° (Fig. 3.4). The cone has an angular velocity ω with a constant magnitude ω_0. We wish to determine the values of the Euler angles, axis and angle variables, Euler parameters, and their rates of change when the configuration is as shown in the figure.

Let us use Type II Euler angles. The three successive rotations about the Z, x, and z axes, respectively, are $\phi = 0$, $\theta = 30°$, and $\psi = 0$. From (3.24), or directly by observation of Fig. 3.4, we find that the rotation matrix is

$$C = \begin{bmatrix} 1 & 0 & 0 \\ 0 & \frac{\sqrt{3}}{2} & \frac{1}{2} \\ 0 & -\frac{1}{2} & \frac{\sqrt{3}}{2} \end{bmatrix}$$

(3.98)

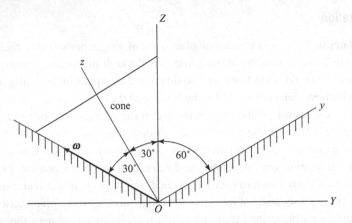

Figure 3.4.

The cone rotates about its line of contact. Therefore, the body-axis components of its angular velocity ω are

$$\omega_x = 0, \qquad \omega_y = -\frac{1}{2}\omega_0, \qquad \omega_z = \frac{\sqrt{3}}{2}\omega_0 \tag{3.99}$$

Then, using (3.19), the Euler angle rates are

$$\dot{\phi} = \csc\theta\,(\omega_x \sin\psi + \omega_y \cos\psi) = -\omega_0$$
$$\dot{\theta} = \omega_x \cos\psi - \omega_y \sin\psi = 0 \tag{3.100}$$
$$\dot{\psi} = \omega_z - \dot{\phi}\cos\theta = \sqrt{3}\,\omega_0$$

Next, consider axis and angle variables, where ϕ is now the rotation angle about the rotation axis **a**. Since the first and third Euler angle rotations were equal to zero, we see that the required rotation is $30°$ about the x-axis. Hence,

$$a_x = 1, \qquad a_y = 0, \qquad a_z = 0, \qquad \phi = 30° \tag{3.101}$$

This is in agreement with (3.27) and (3.28) which use values from the rotation matrix.

The values of the Euler parameters can now be obtained from (3.41). They are

$$\epsilon_x = \sin 15°, \qquad \epsilon_y = 0, \qquad \epsilon_z = 0, \qquad \eta = \cos 15° \tag{3.102}$$

The corresponding Euler parameter rates are found from (3.80).

$$\dot{\epsilon}_x = \frac{1}{2}(\omega_z\epsilon_y - \omega_y\epsilon_z + \omega_x\eta) = 0$$

$$\dot{\epsilon}_y = \frac{1}{2}(\omega_x\epsilon_z - \omega_z\epsilon_x + \omega_y\eta) = -\frac{\omega_0}{2}\sin 45° = -\frac{\omega_0}{2\sqrt{2}}$$

$$\dot{\epsilon}_z = \frac{1}{2}(\omega_y\epsilon_x - \omega_x\epsilon_y + \omega_z\eta) = \frac{\omega_0}{2}\cos 45° = \frac{\omega_0}{2\sqrt{2}} \tag{3.103}$$

$$\dot{\eta} = -\frac{1}{2}(\omega_x\epsilon_x + \omega_y\epsilon_y + \omega_z\epsilon_z) = 0$$

Now consider the rates of the axis and angle variables. From (3.65) we find that

$$\dot{\phi} = \mathbf{a} \cdot \boldsymbol{\omega} = 0 \tag{3.104}$$

Next, recall from (3.71) that

$$\dot{\mathbf{a}} = -\frac{1}{2} \left[\mathbf{a} \times \boldsymbol{\omega} + \cot \frac{\phi}{2} \, \mathbf{a} \times (\mathbf{a} \times \boldsymbol{\omega}) \right] \tag{3.105}$$

where, for this example,

$$\mathbf{a} \times \boldsymbol{\omega} = -\omega_0 \left(\frac{\sqrt{3}}{2} \mathbf{j} + \frac{1}{2} \mathbf{k} \right) \tag{3.106}$$

$$\cot \frac{\phi}{2} \, \mathbf{a} \times (\mathbf{a} \times \boldsymbol{\omega}) = -\omega \cot \frac{\phi}{2} = (2 + \sqrt{3})\omega_0 \left(\frac{1}{2} \mathbf{j} - \frac{\sqrt{3}}{2} \mathbf{k} \right) \tag{3.107}$$

Thus, we obtain

$$\dot{\mathbf{a}} = \omega_0 \left[-\frac{1}{2} \mathbf{j} + \left(1 + \frac{\sqrt{3}}{2} \right) \mathbf{k} \right] \tag{3.108}$$

To find the components of $\dot{\mathbf{a}}$ in the inertial frame, multiply by C^{T} and obtain

$$\dot{\mathbf{a}} = \left(\frac{1 + \sqrt{3}}{2} \right) \omega_0 (-\mathbf{J} + \mathbf{K}) \tag{3.109}$$

The time derivative of $\mathbf{a}$ with respect to the rotating body axes is, from (3.72),

$$(\dot{\mathbf{a}})_r = \frac{1}{2} \left[\mathbf{a} \times \boldsymbol{\omega} - \cot \frac{\phi}{2} \, \mathbf{a} \times (\mathbf{a} \times \boldsymbol{\omega}) \right]$$

$$= \left(\frac{1 + \sqrt{3}}{2} \right) \omega_0 (-\mathbf{j} + \mathbf{k}) \tag{3.110}$$

or

$$\dot{a}_x = 0, \qquad \dot{a}_y = -\left(\frac{1 + \sqrt{3}}{2} \right) \omega_0, \qquad \dot{a}_z = \left(\frac{1 + \sqrt{3}}{2} \right) \omega_0 \tag{3.111}$$

Notice that these components are equal to the corresponding components of $\dot{\mathbf{a}}$ in the inertial frame. This is explained by the fact that the axis $\mathbf{a}$ about which the rotation ϕ takes place is fixed in both frames during the rotation, and these frames coincide initially.

3.2   Dyadic notation

Definition of a dyadic

Dyadics are used in vector equations to represent the second-order tensors that are often expressed as 3×3 matrices. A *dyad* consists of a pair of vectors such as **ab** written in a particular order. A *dyadic* is the sum of dyads.

Each vector in 3-space may be expressed in terms of an orthogonal set of unit vectors. If we choose the same set of unit vectors for $\mathbf{a}$ and $\mathbf{b}$, we obtain

$$\mathbf{a} = a_1\mathbf{e}_1 + a_2\mathbf{e}_2 + a_3\mathbf{e}_3$$
$$\mathbf{b} = b_1\mathbf{e}_1 + b_2\mathbf{e}_2 + b_3\mathbf{e}_3 \tag{3.112}$$

We can define the dyadic

$$\mathbf{A} = \mathbf{ab} = \sum_{i=1}^{3}\sum_{j=1}^{3} A_{ij}\mathbf{e}_i\mathbf{e}_j \tag{3.113}$$

where $A_{ij} = a_ib_j$. Thus $\mathbf{A}$ consists of the sum of nine terms, each containing a pair of unit vectors.

Now let us define the *conjugate dyadic* by interchanging the vectors in each term. We obtain

$$\mathbf{A}^{\mathrm{T}} = \sum_{i=1}^{3}\sum_{j=1}^{3} A_{ij}\mathbf{e}_j\mathbf{e}_i \tag{3.114}$$

A dyadic is *symmetric* if $\mathbf{A} = \mathbf{A}^{\mathrm{T}}$; it is *skew-symmetric* if $\mathbf{A} = -\mathbf{A}^{\mathrm{T}}$. An example of a symmetric dyadic is the *inertia dyadic*.

$$\mathbf{I} = I_{xx}\mathbf{ii} + I_{xy}\mathbf{ij} + I_{xz}\mathbf{ik} + I_{yx}\mathbf{ji} + I_{yy}\mathbf{jj} + I_{yz}\mathbf{jk} + I_{zx}\mathbf{ki} + I_{zy}\mathbf{kj} + I_{zz}\mathbf{kk} \tag{3.115}$$

The *moments of inertia* I_{xx}, I_{yy}, and I_{zz} are defined as follows:

$$I_{xx} = \int_V \rho(y^2 + z^2)dV$$
$$I_{yy} = \int_V \rho(z^2 + x^2)dV \tag{3.116}$$
$$I_{zz} = \int_V \rho(x^2 + y^2)dV$$

where ρ is the mass density and dV is a volume element of the rigid body. The *products of inertia* are defined with a minus sign, that is,

$$I_{xy} = I_{yx} = -\int_V \rho xy\,dV$$
$$I_{xz} = I_{zx} = -\int_V \rho xz\,dV \tag{3.117}$$
$$I_{yz} = I_{zy} = -\int_V \rho yz\,dV$$

Dyadic operations

In general, the order of the dot product of a dyadic and a vector is important, that is,

$$\mathbf{A}\cdot\mathbf{b} \neq \mathbf{b}\cdot\mathbf{A} \tag{3.118}$$

for an arbitrary vector **b**. On the other hand, for a *symmetric dyadic* such as the inertia dyadic **I**, we find that

$$\mathbf{I} \cdot \omega = \omega \cdot \mathbf{I} = \mathbf{H} \tag{3.119}$$

where ω is an arbitrary angular velocity and **H** is the angular momentum about the origin of the body-axis frame. Notice that the dot product of a vector and a dyadic results in a vector having a different magnitude and direction, in general.

The cross product of a dyadic and a vector results in another dyadic. Again, the order is important, that is,

$$\mathbf{A} \times \mathbf{b} \neq \mathbf{b} \times \mathbf{A} \tag{3.120}$$

in general.

The defining equation for a *unit dyadic* is

$$\mathbf{U} = \mathbf{ii} + \mathbf{jj} + \mathbf{kk} \tag{3.121}$$

where one can use any orthogonal set of unit vectors. For an arbitrary vector **b** we see that

$$\mathbf{U} \cdot \mathbf{b} = \mathbf{b} \cdot \mathbf{U} = \mathbf{b} \tag{3.122}$$

A defining equation for the inertia dyadic of a rigid body is

$$\mathbf{I} = \int \left[(\rho \cdot \rho) \mathbf{U} - \rho\rho \right] dm \tag{3.123}$$

where $\rho = x\mathbf{i} + y\mathbf{j} + z\mathbf{k}$ is the position vector of the mass element dm relative to the origin of the body axes.

Consider next the *time derivative* of a dyadic. As an example, suppose a rigid body has an angular velocity ω. Its inertia dyadic is

$$\mathbf{I} = \sum_{i=1}^{3} \sum_{j=1}^{3} I_{ij} \mathbf{e}_i \mathbf{e}_j \tag{3.124}$$

where the unit vectors are mutually orthogonal and rotate with the body. Upon differentiation with respect to time, we obtain

$$\dot{\mathbf{I}} = \sum_{i=1}^{3} \sum_{j=1}^{3} (\dot{I}_{ij} \mathbf{e}_i \mathbf{e}_j + I_{ij} \dot{\mathbf{e}}_i \mathbf{e}_j + I_{ij} \mathbf{e}_i \dot{\mathbf{e}}_j) \tag{3.125}$$

For a rigid body the moments and products of inertia are constant scalar quantities, so $\dot{I}_{ij} = 0$. Also,

$$\dot{\mathbf{e}}_i = \omega \times \mathbf{e}_i \tag{3.126}$$

$$\dot{\mathbf{e}}_j = -\mathbf{e}_j \times \omega \tag{3.127}$$

Hence we see that

$$\dot{\mathbf{I}} = \omega \times \mathbf{I} - \mathbf{I} \times \omega \tag{3.128}$$

As another example of the use of dyadic notation, one can write the rotation matrix C in dyadic form. This *rotation dyadic* is

$$\mathbf{C} = c_{x'x}\mathbf{i'i} + c_{x'y}\mathbf{i'j} + c_{x'z}\mathbf{i'k} + c_{y'x}\mathbf{j'i} + c_{y'y}\mathbf{j'j} + c_{y'z}\mathbf{j'k}$$
$$+ c_{z'x}\mathbf{k'i} + c_{z'y}\mathbf{k'j} + c_{z'z}\mathbf{k'k} \tag{3.129}$$

The first subscript and unit vector in each term refer to an axis of the new primed coordinate system after rotation, while the second subscript and unit vector refer to an axis of the original unprimed system. Then, corresponding to (3.1) and (3.4), we obtain

$$\mathbf{r'} = \mathbf{C} \cdot \mathbf{r} \tag{3.130}$$

and

$$\mathbf{r} = \mathbf{C}^{\mathrm{T}} \cdot \mathbf{r'} \tag{3.131}$$

In order to obtain the transformation equations relating the inertia dyadic of a rigid body with respect to primed and unprimed reference frames, let us begin with (3.123) and note that

$$\rho = \mathbf{C}^{\mathrm{T}} \cdot \rho' = \rho' \cdot \mathbf{C} \tag{3.132}$$
$$\rho \cdot \rho = \rho' \cdot \rho' \tag{3.133}$$
$$\mathbf{U} = \mathbf{C}^{\mathrm{T}} \cdot \mathbf{U'} \cdot \mathbf{C} \tag{3.134}$$

where $\mathbf{U'}$ is the unit dyadic expressed in terms of primed unit vectors. We obtain

$$\mathbf{I} = \mathbf{C}^{\mathrm{T}} \cdot \mathbf{I'} \cdot \mathbf{C} \tag{3.135}$$

or, conversely,

$$\mathbf{I'} = \mathbf{C} \cdot \mathbf{I} \cdot \mathbf{C}^{\mathrm{T}} \tag{3.136}$$

3.3   Basic rigid body dynamics

Kinetic energy

In accordance with *Koenig's theorem*, the kinetic energy of a rigid body is equal to the sum of (1) the kinetic energy due to the translational velocity of the center of mass and (2) the kinetic energy due to rotation about the center of mass. Thus,

$$T = T_{\mathrm{tr}} + T_{\mathrm{rot}} \tag{3.137}$$

where

$$T_{\mathrm{tr}} = \frac{1}{2}m\mathbf{v}_c^2 \tag{3.138}$$

$$T_{\mathrm{rot}} = \frac{1}{2}\boldsymbol{\omega} \cdot \mathbf{I}_c \cdot \boldsymbol{\omega} \tag{3.139}$$

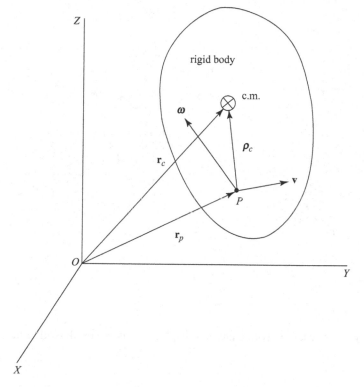

Figure 3.5.

Here $\mathbf{v}_c$ is the velocity of the center of mass, $\mathbf{I}_c$ is the inertia dyadic about the center of mass, and $\boldsymbol{\omega}$ is the angular velocity of the rigid body.

Frequently it is convenient to write the kinetic energy of a rigid body in terms of the velocity of a reference point in the body. Suppose that a point P is fixed in a rigid body and moves with a velocity $\mathbf{v}$. The position vector of the center of mass relative to P is $\boldsymbol{\rho}_c$ (Fig. 3.5). Then, recalling (1.127), the kinetic energy of the rigid body is

$$T = \frac{1}{2}m\mathbf{v}^2 + \frac{1}{2}\boldsymbol{\omega} \cdot \mathbf{I} \cdot \boldsymbol{\omega} + m\mathbf{v} \cdot \dot{\boldsymbol{\rho}}_c \tag{3.140}$$

where the inertia dyadic $\mathbf{I}$ is taken about the reference point P.

Example 3.2 A comparison of methods for obtaining the kinetic energy of a system of rigid bodies can be shown by considering the physical double pendulum of Fig. 3.6. Two identical rigid bodies, supported by joints at A and B, move with planar motion. The bodies each have mass m, length l, and moment of inertia I_c about the center of mass.

In general, Koenig's theorem applied to a system of N rigid bodies gives

$$T = \frac{1}{2}\sum_{i=1}^{N} m_i v_{ci}^2 + \frac{1}{2}\sum_{i=1}^{N} \boldsymbol{\omega}_i \cdot \mathbf{I}_{ci} \cdot \boldsymbol{\omega}_i \tag{3.141}$$

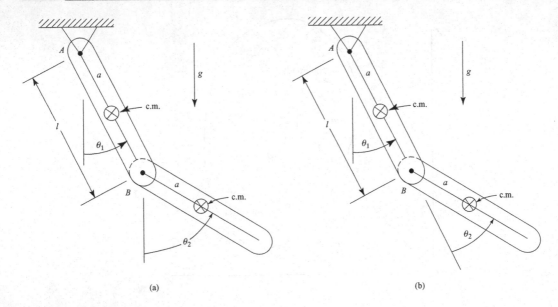

Figure 3.6.

First, consider Fig. 3.6a where absolute angular displacements are used. Noting that

$$m_1 = m_2 = m \tag{3.142}$$

$$I_{c1} = I_{c2} = I_c \tag{3.143}$$

$$v_{c1}^2 = a^2\dot{\theta}_1^2, \qquad v_{c2}^2 = l^2\dot{\theta}_1^2 + a^2\dot{\theta}_2^2 + 2al\dot{\theta}_1\dot{\theta}_2\cos(\theta_2 - \theta_1) \tag{3.144}$$

we obtain the total kinetic energy

$$T = \frac{1}{2}(I_c + ma^2)\dot{\theta}_1^2 + \frac{1}{2}ml^2\dot{\theta}_1^2 + \frac{1}{2}(I_c + ma^2)\dot{\theta}_2^2 + mal\dot{\theta}_1\dot{\theta}_2\cos(\theta_2 - \theta_1) \tag{3.145}$$

Another approach is to invoke (3.140) for a system of N rigid bodies, namely,

$$T = \frac{1}{2}\sum_{i=1}^{N} m_i v_i^2 + \frac{1}{2}\sum_{i=1}^{N} \boldsymbol{\omega}_i \cdot \mathbf{I}_i \cdot \boldsymbol{\omega}_i + \sum_{i=1}^{N} m_i \mathbf{v}_i \cdot \dot{\boldsymbol{\rho}}_{ci} \tag{3.146}$$

where $\mathbf{v}_i$ is the velocity of the reference point of the ith body. In this case, we take reference points at A and B, respectively, for the two bodies. The velocities of the reference points are

$$v_1 = 0, \qquad v_2 = l\dot{\theta}_1 \tag{3.147}$$

and the moments of inertia about the reference points are

$$I_1 = I_2 = I = I_c + ma^2 \tag{3.148}$$

Noting that

$$\dot{\boldsymbol{\rho}}_{c2} = \boldsymbol{\omega}_2 \times \boldsymbol{\rho}_{c2} \tag{3.149}$$

we obtain

$$T = \frac{1}{2}I\left(\dot{\theta}_1^2 + \dot{\theta}_2^2\right) + \frac{1}{2}ml^2\dot{\theta}_1^2 + mal\dot{\theta}_1\dot{\theta}_2\cos(\theta_2 - \theta_1) \qquad (3.150)$$

This result agrees with that obtained earlier in (3.145) but the use of (3.146) is somewhat more direct. Thus, it is sometimes advantageous to use a reference point which is not at the center of mass.

The kinetic energy expression can be used to find the inertia coefficients. They are given by

$$m_{ij} = \frac{\partial^2 T}{\partial \dot{q}_i \partial \dot{q}_j} \qquad (3.151)$$

which results in

$$m_{11} = I + ml^2, \qquad m_{22} = I, \qquad m_{12} = m_{21} = mal\cos(\theta_2 - \theta_1) \qquad (3.152)$$

Here m_{11} is equal to the moment of inertia about A due to the first body plus the mass of the second body considered as a particle at B. Furthermore, m_{22} is the moment of inertia of the second body about its reference point B.

Now let us consider the case illustrated in Fig. 3.6b in which θ_2 is measured relative to the first body rather than having a fixed reference. With the aid of (3.146), the kinetic energy is found to be

$$T = \frac{1}{2}I\dot{\theta}_1^2 + \frac{1}{2}ml^2\dot{\theta}_1^2 + \frac{1}{2}I(\dot{\theta}_1 + \dot{\theta}_2)^2 + mal\dot{\theta}_1(\dot{\theta}_1 + \dot{\theta}_2)\cos\theta_2$$

$$= \frac{1}{2}(2I + ml^2 + 2mal\cos\theta_2)\dot{\theta}_1^2 + \frac{1}{2}I\dot{\theta}_2^2 + (I + mal\cos\theta_2)\dot{\theta}_1\dot{\theta}_2 \qquad (3.153)$$

Then (3.151) yields

$$m_{11} = 2I + ml^2 + 2mal\cos\theta_2, \qquad m_{22} = I, \qquad m_{12} = m_{21} = I + mal\cos\theta_2 \qquad (3.154)$$

In this case, notice that m_{11} is equal to the moment of inertia about A of the rigidly connected bodies. Again, m_{22} is equal to the moment of inertia of the second body about B.

Vectorial dynamics, Euler equations

The translational equation of motion for a rigid body has the vector form

$$m\mathbf{a} = \mathbf{F} \qquad (3.155)$$

where $\mathbf{a}$ is the acceleration of the center of mass and $\mathbf{F}$ is the total force acting on the body, including any constraint forces. For a given total force acting on the body, this equation has the same form as if the total mass m were concentrated as a particle at the center of mass.

Most of our attention, however, will be given to the rotational motion of a rigid body. A fruitful approach is to use the vector equation

$$\dot{\mathbf{H}} = \mathbf{M} \qquad (3.156)$$

which applies if the reference point for the angular momentum **H** and the total moment **M** acting on the body is either (1) fixed or (2) at the center of mass. Let us choose a body-fixed coordinate system and let the reference point be at the origin O. The angular momentum is

$$\mathbf{H} = \mathbf{I} \cdot \boldsymbol{\omega} \tag{3.157}$$

or, in terms of body-axis components,

$$
\begin{aligned}
H_x &= I_{xx}\omega_x + I_{xy}\omega_y + I_{xz}\omega_z \\
H_y &= I_{yx}\omega_x + I_{yy}\omega_y + I_{yz}\omega_z \\
H_z &= I_{zx}\omega_x + I_{zy}\omega_y + I_{zz}\omega_z
\end{aligned}
\tag{3.158}
$$

Upon differentiation with respect to time, and referring to (3.128), we obtain

$$
\begin{aligned}
\dot{\mathbf{H}} &= \mathbf{I} \cdot \dot{\boldsymbol{\omega}} + \dot{\mathbf{I}} \cdot \boldsymbol{\omega} \\
&= \mathbf{I} \cdot \dot{\boldsymbol{\omega}} + (\boldsymbol{\omega} \times \mathbf{I} - \mathbf{I} \times \boldsymbol{\omega}) \cdot \boldsymbol{\omega} \\
&= \mathbf{I} \cdot \dot{\boldsymbol{\omega}} + \boldsymbol{\omega} \times \mathbf{I} \cdot \boldsymbol{\omega}
\end{aligned}
\tag{3.159}
$$

where we note that

$$(\mathbf{I} \times \boldsymbol{\omega}) \cdot \boldsymbol{\omega} = 0 \tag{3.160}$$

From (3.156) and (3.159) we obtain

$$\mathbf{I} \cdot \dot{\boldsymbol{\omega}} + \boldsymbol{\omega} \times \mathbf{I} \cdot \boldsymbol{\omega} = \mathbf{M} \tag{3.161}$$

Written in detail, we have the generalized Euler equations:

$$
\begin{aligned}
I_{xx}\dot{\omega}_x + I_{xy}(\dot{\omega}_y - \omega_z\omega_x) + I_{xz}(\dot{\omega}_z + \omega_x\omega_y) + (I_{zz} - I_{yy})\omega_y\omega_z + I_{yz}\left(\omega_y^2 - \omega_z^2\right) &= M_x \\
I_{xy}(\dot{\omega}_x + \omega_y\omega_z) + I_{yy}\dot{\omega}_y + I_{yz}(\dot{\omega}_z - \omega_x\omega_y) + (I_{xx} - I_{zz})\omega_z\omega_x + I_{xz}\left(\omega_z^2 - \omega_x^2\right) &= M_y \\
I_{xz}(\dot{\omega}_x - \omega_y\omega_z) + I_{yz}(\dot{\omega}_y + \omega_z\omega_x) + I_{zz}\dot{\omega}_z + (I_{yy} - I_{xx})\omega_x\omega_y + I_{xy}\left(\omega_x^2 - \omega_y^2\right) &= M_z
\end{aligned}
\tag{3.162}
$$

These equations apply for an arbitrary orientation of the body axes relative to the rigid body. Equations (3.161) or (3.162) can also be written in the matrix form

$$I\dot{\omega} + \tilde{\omega}I\omega = M \tag{3.163}$$

where $\tilde{\omega}$ is given by (3.61) and represents the cross product of $\boldsymbol{\omega}$.

Usually it is convenient to choose body axes which are also *principal axes*. Then the products of inertia vanish and we obtain the familiar *Euler equations*:

$$
\begin{aligned}
I_{xx}\dot{\omega}_x + (I_{zz} - I_{yy})\omega_y\omega_z &= M_x \\
I_{yy}\dot{\omega}_y + (I_{xx} - I_{zz})\omega_z\omega_x &= M_y \\
I_{zz}\dot{\omega}_z + (I_{yy} - I_{xx})\omega_x\omega_y &= M_z
\end{aligned}
\tag{3.164}
$$

Here the reference point for **H** and **M** is the origin O of the body-fixed frame, and this point is located at the center of mass or at an inertially fixed point. If the applied moments are known, these nonlinear first-order differential equations can be integrated numerically to give the ω components as functions of time. To obtain the orientation as a function of time, the ωs are substituted into the kinematic equations (3.16) or (3.19) or (3.80), and these are integrated numerically to give Euler angles or Euler parameters as functions of time.

As an extension of this vectorial approach to the rotational dynamics of a rigid body, let us consider the case of a reference point P which is fixed in the body, but it is neither at the center of mass nor at a fixed point in space. The angular momentum with respect to P is

$$\mathbf{H}_p = \mathbf{I}_p \cdot \omega \tag{3.165}$$

where $\mathbf{I}_p$ is the inertia dyadic of the rigid body about P. The equation of motion corresponding to (3.156) is

$$\dot{\mathbf{H}}_p = \mathbf{M}_p - \rho_c \times m\dot{\mathbf{v}}_p \tag{3.166}$$

in agreement with (1.115) which was obtained for a system of particles. The last term of (3.166) can be interpreted as the moment about P due to an *inertia force* $-m\dot{\mathbf{v}}_p$ acting through the center of mass, and due to the acceleration of the reference point. Thus we obtain

$$\mathbf{I}_p \cdot \dot{\omega} + \omega \times \mathbf{I}_p \cdot \omega + \rho_c \times m\dot{\mathbf{v}}_p = \mathbf{M}_p \tag{3.167}$$

This result may be regarded as an extension of Euler's equation for the case of an accelerating reference point.

In general, the angular momentum $\mathbf{H}_p$ about an arbitrary reference point P is equal to the angular momentum $\mathbf{H}_c$ about the center of mass plus the angular momentum due to the translational velocity of the center of mass relative to the reference point. For this case in which the reference point P is fixed in the rigid body, we have (see Fig. 3.5)

$$\mathbf{H}_p = \mathbf{H}_c + m\rho_c \times \dot{\rho}_c = \mathbf{H}_c + m\rho_c \times (\omega \times \rho_c) \tag{3.168}$$

Hence, we obtain

$$\begin{aligned}
\mathbf{I}_p \cdot \omega &= \mathbf{I}_c \cdot \omega + m\rho_c^2\omega - m(\rho_c \cdot \omega)\rho_c \\
&= \mathbf{I}_c \cdot \omega + m\left(\rho_c^2\mathbf{U} - \rho_c\rho_c\right) \cdot \omega
\end{aligned} \tag{3.169}$$

Since ω is arbitrary, we see that

$$\mathbf{I}_p = \mathbf{I}_c + m\left(\rho_c^2\mathbf{U} - \rho_c\rho_c\right) \tag{3.170}$$

where, in terms of body-axis components,

$$\rho_c = x_c\mathbf{i} + y_c\mathbf{j} + z_c\mathbf{k} \tag{3.171}$$

In detail, the moments and products of inertia about P due to a *translation of axes* are

$$\begin{aligned}
I_{xx} &= (I_{xx})_c + m\left(y_c^2 + z_c^2\right) \\
I_{yy} &= (I_{yy})_c + m\left(z_c^2 + x_c^2\right) \\
I_{zz} &= (I_{zz})_c + m\left(x_c^2 + y_c^2\right)
\end{aligned} \tag{3.172}$$

and

$$I_{xy} = (I_{xy})_c - m x_c y_c$$
$$I_{xz} = (I_{xz})_c - m x_c z_c \qquad\qquad (3.173)$$
$$I_{yz} = (I_{yz})_c - m y_c z_c$$

We see that a translation of axes away from the center of mass always increases the moments of inertia, but the products of inertia may increase or decrease.

Ellipsoid of inertia

Consider the kinetic energy of a rigid body rotating about a fixed reference point. It is, similar to (3.139),

$$T = \frac{1}{2} \boldsymbol{\omega} \cdot \mathbf{I} \cdot \boldsymbol{\omega} \qquad\qquad (3.174)$$

where $\mathbf{I}$ is the inertia dyadic about the reference point and $\boldsymbol{\omega}$ is the angular velocity vector. An alternate expression for the kinetic energy due to rotation about a fixed reference point is

$$T = \frac{1}{2} I \omega^2 \qquad\qquad (3.175)$$

where the scalar I is the moment of inertia about the axis of rotation and ω is the magnitude of the angular velocity. This is apparent from (3.174) for the case where $\boldsymbol{\omega}$ is directed along one of the coordinate axes.

From (3.174) and (3.175) we obtain

$$\frac{1}{2} \boldsymbol{\omega} \cdot \mathbf{I} \cdot \boldsymbol{\omega} = \frac{1}{2} I \omega^2 \qquad\qquad (3.176)$$

or, dividing by the right-hand side,

$$\boldsymbol{\rho} \cdot \mathbf{I} \cdot \boldsymbol{\rho} = 1 \qquad\qquad (3.177)$$

where

$$\boldsymbol{\rho} = \frac{\boldsymbol{\omega}}{\omega \sqrt{I}} \qquad\qquad (3.178)$$

We see that $\boldsymbol{\rho}$ has the same direction as $\boldsymbol{\omega}$ and a magnitude that is inversely proportional to $\sqrt{I}$ or the radius of gyration. Suppose we consider $\boldsymbol{\rho}$ to be a vector drawn from the origin of the body-fixed frame to the point (x, y, z), that is,

$$\boldsymbol{\rho} = x\mathbf{i} + y\mathbf{j} + z\mathbf{k} \qquad\qquad (3.179)$$

Then, (3.177) takes the form

$$I_{xx}x^2 + I_{yy}y^2 + I_{zz}z^2 + 2I_{xy}xy + 2I_{xz}xz + 2I_{yz}yz = 1 \qquad\qquad (3.180)$$

This is the equation of an ellipsoidal surface and is known as the *ellipsoid of inertia*. The ellipsoid of inertia is fixed relative to the rigid body and the body-fixed frame. It is essentially a three-dimensional plot which gives the value of $1/\sqrt{I}$ for any axis passing through the origin of the *xyz* frame. The major axis is the axis having the minimum moment of inertia. Similarly, the minor axis is the axis having the maximum moment of inertia; it is perpendicular to the major axis. These two axes plus an orthogonal third axis constitute the set of three *principal axes* corresponding to the *principal moments of inertia*. The orientations of the three principal axes are such that all products of inertia are equal to zero. This means that the inertial coupling terms associated with products of inertia dissappear and, for example, the generalized Euler equations (3.162) simplify to the standard equations of (3.164).

Suppose we are given a general inertia dyadic **I** and we wish to calculate the principal moments of inertia of the body. If a rigid body is rotating about a principal axis, then its angular velocity and angular momentum are parallel, that is

$$\mathbf{H} = \mathbf{I} \cdot \boldsymbol{\omega} = I\boldsymbol{\omega} \qquad (3.181)$$

where the proportionality constant I on the right is a principal moment of inertia. Equation (3.181) has the form of an *eigenvalue problem*. In terms of dyadics we have

$$(\mathbf{I} - I\mathbf{U}) \cdot \boldsymbol{\omega} = 0 \qquad (3.182)$$

or, using matrix notation,

$$\begin{bmatrix} (I_{xx} - I) & I_{xy} & I_{xz} \\ I_{yx} & (I_{yy} - I) & I_{yz} \\ I_{zx} & I_{zy} & (I_{zz} - I) \end{bmatrix} \begin{bmatrix} \omega_x \\ \omega_y \\ \omega_z \end{bmatrix} = \begin{bmatrix} 0 \end{bmatrix} \qquad (3.183)$$

We assume that $\boldsymbol{\omega} \neq 0$, so the determinant of the coefficients must equal zero, that is,

$$\begin{vmatrix} (I_{xx} - I) & I_{xy} & I_{xz} \\ I_{yx} & (I_{yy} - I) & I_{yz} \\ I_{zx} & I_{zy} & (I_{zz} - I) \end{vmatrix} = 0 \qquad (3.184)$$

This determinant yields a cubic equation in I known as the *characteristic equation*. Its three real roots are the three principal moments of inertia.

For each root I the direction of the corresponding principal axis is found by solving (3.183) for the ratios of the components of $\boldsymbol{\omega}$. These axes are also the principal axes of the inertia ellipsoid of the body.

Example 3.3   Consider a rigid body having an inertia matrix

$$I = \begin{bmatrix} 28 & -8 & -4 \\ -8 & 28 & -4 \\ -4 & -4 & 24 \end{bmatrix} \text{kg·m}^2 \qquad (3.185)$$

with respect to a body-fixed *xyz* frame. We wish to solve for the principal moments of inertia and the directions of the corresponding principal axes. The characteristic equation is obtained from (3.184). It is

$$
\begin{vmatrix}
(28-I) & -8 & -4 \\
-8 & (28-I) & -4 \\
-4 & -4 & (24-I)
\end{vmatrix} = 0
\tag{3.186}
$$

or

$$
I^3 - 80\,I^2 + 2032\,I - 16\,128 = 0
\tag{3.187}
$$

The roots of this characteristic equation are the values of the principal moments of inertia. In increasing magnitude, they are

$$
I_1 = 16, \qquad I_2 = 28, \qquad I_3 = 36 \text{ kg·m}^2
\tag{3.188}
$$

The corresponding axis directions are obtained from (3.183), namely,

$$
\begin{aligned}
(28-I)\omega_x - 8\omega_y - 4\omega_z &= 0 \\
-8\omega_x + (28-I)\omega_y - 4\omega_z &= 0 \\
-4\omega_x - 4\omega_y + (24-I)\omega_z &= 0
\end{aligned}
\tag{3.189}
$$

All three equations must be satisfied, but usually any two equations are sufficient to solve for the ω ratios.

Assuming that no principal axis lies in the yz-plane, let us set $\omega_x = 1$. Then, for $I_1 = 16$ we obtain $\omega_x = \omega_y = \omega_z = 1$. For $I_2 = 28$ the result is $\omega_x = \omega_y = 1$, $\omega_z = -2$. The result for $I_3 = 36$ is $\omega_x = 1$, $\omega_y = -1$, $\omega_z = 0$. Now let us normalize each set of ωs to give a unit vector magnitude, resulting in a set of direction cosines for each principal axis. The direction cosines can be combined to form the three rows of a rotation matrix C for rotation from an original xyz frame to the principal $x'y'z'$ frame. A system of labeling for the primed axes which requires the least rotation is obtained by placing the most positive terms on the main diagonal of C, possibly by multiplying a row by -1. We obtain

$$
\begin{bmatrix}
\frac{1}{\sqrt{3}} & \frac{1}{\sqrt{3}} & \frac{1}{\sqrt{3}} \\
-\frac{1}{\sqrt{2}} & \frac{1}{\sqrt{2}} & 0 \\
\frac{-1}{\sqrt{6}} & \frac{-1}{\sqrt{6}} & \frac{2}{\sqrt{6}}
\end{bmatrix}
\tag{3.190}
$$

One can check that the principal moments of inertia are obtained from

$$
I' = CIC^{\mathrm{T}}
\tag{3.191}
$$

Also, to be sure of a right-handed principal coordinate system, one should check that the determinant $|C| = 1$.

Modified Euler equations

Sometimes it is convenient to use a rotational equation having a form similar to (3.161) but being different in that strict body axes are not used. In general, suppose that $\mathbf{H}$ and $\mathbf{M}$

are expressed in terms of unit vectors associated with a Cartesian frame having an angular velocity ω_c. Then the rotational equation has the vector form

$$(\dot{\mathbf{H}})_r + \omega_c \times \mathbf{H} = \mathbf{M} \tag{3.192}$$

where $(\dot{\mathbf{H}})_r$ is the time derivative of total angular momentum $\mathbf{H}$, as viewed from the rotating frame and assuming that the unit vectors fixed in this frame have time derivatives equal to zero.

An important application of this approach occurs in the case of an axially symmetric rigid body. Let us assume that the x-axis of an xyz frame is the axis of symmetry, and the reference point is chosen at the center of mass or at an inertially fixed point on the axis of symmetry. Let I_a be the moment of inertia about the symmetry axis, whereas I_t is the moment of inertia about any transverse axis at the reference point.

Now suppose that the body rotates about the axis of symmetry with an angular velocity or spin rate s measured relative to the xyz frame. If we let the angular velocity of the xyz frame be

$$\omega_c = \omega_x \mathbf{i} + \omega_y \mathbf{j} + \omega_z \mathbf{k} \tag{3.193}$$

we see that the angular momentum of the body is

$$\mathbf{H} = I_a(\omega_x + s)\mathbf{i} + I_t \omega_y \mathbf{j} + I_t \omega_z \mathbf{k} \tag{3.194}$$

Also,

$$(\dot{\mathbf{H}})_r = I_a(\dot{\omega}_x + \dot{s})\mathbf{i} + I_t \dot{\omega}_y \mathbf{j} + I_t \dot{\omega}_z \mathbf{k} \tag{3.195}$$

Then, substituting into (3.192), we obtain the *modified Euler equations*:

$$\begin{aligned}
I_a(\dot{\omega}_x + \dot{s}) &= I_a \dot{\omega} = M_x \\
I_t \dot{\omega}_y - (I_t - I_a)\omega_z \omega_x + I_a s \omega_z &= M_y \\
I_t \dot{\omega}_z + (I_t - I_a)\omega_x \omega_y - I_a s \omega_y &= M_z
\end{aligned} \tag{3.196}$$

where the *total spin* Ω, which is the total angular velocity about the symmetry axis, is

$$\Omega = \omega_x + s \tag{3.197}$$

Until now the *relative spin s* has been arbitrary. For example, if we set $s = 0$, the modified Euler equations revert to ordinary Euler equations for an axially symmetric body. But we can choose s in such a way that the kinematics of solving for the orientation of axially symmetric body is relatively easy. Let us use type I Euler angles and let ψ and θ specify the orientation of the xyz frame. The inertial XYZ frame is chosen so that the XY-plane is horizontal and the Z-axis is positive downward. Since the third Euler angle ϕ does not affect the orientation of the xyz frame, its y-*axis remains horizontal.* This implies that the horizontal component of ω_c must lie along the y-axis, and we have the constraint equation (see Fig. 3.7)

$$\omega_x \cos\theta + \omega_z \sin\theta = 0 \tag{3.198}$$

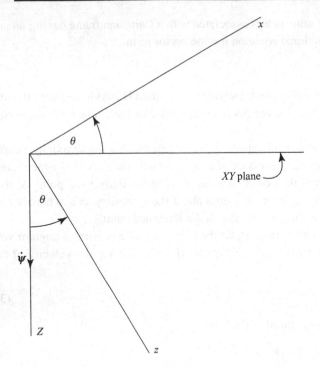

Figure 3.7.

or

$$\omega_x = -\omega_z \tan \theta \tag{3.199}$$

where we note that the xz-plane is vertical. The vertically downward component of ω_c is $\dot{\psi}$, that is,

$$\dot{\psi} = -\omega_x \sin \theta + \omega_z \cos \theta = \omega_z \sec \theta \tag{3.200}$$

Also, we have

$$\dot{\theta} = \omega_y \tag{3.201}$$

$$\dot{\phi} = s = \Omega - \omega_x = \Omega + \omega_z \tan \theta \tag{3.202}$$

Now let us eliminate ω_x and s from (3.196) and the modified Euler equations have the form

$$I_a \dot{\omega} = M_x \tag{3.203}$$

$$I_t \dot{\omega}_y + I_t \omega_z^2 \tan \theta + I_a \Omega \omega_z = M_y \tag{3.204}$$

$$I_t \dot{\omega}_z - I_t \omega_y \omega_z \tan \theta - I_a \Omega \omega_y = M_z \tag{3.205}$$

These three first-order dynamical equations plus the three first-order kinematical equations (3.200)–(3.202) can be integrated numerically to obtain the Euler angles as functions of time. Thus, we have obtained relatively simple equations for the rotational motion of an axially-symmetric body. Note that the third terms in (3.204) and (3.205) are gyroscopic coupling terms.

Another approach is to write the dynamical equations directly in terms of type I Euler angles and their time derivatives. One can use

$$\Omega = \dot{\phi} - \dot{\psi}\sin\theta \tag{3.206}$$

$$\omega_y = \dot{\theta} \tag{3.207}$$

$$\omega_z = \dot{\psi}\cos\theta \tag{3.208}$$

and then (3.203)–(3.205) take the form

$$I_a(\ddot{\phi} - \ddot{\psi}\sin\theta - \dot{\psi}\dot{\theta}\cos\theta) = M_x \tag{3.209}$$

$$I_t(\ddot{\theta} + \dot{\psi}^2\sin\theta\cos\theta) + I_a\dot{\psi}(\dot{\phi} - \dot{\psi}\sin\theta)\cos\theta = M_y \tag{3.210}$$

$$I_t(\ddot{\psi}\cos\theta - 2\dot{\psi}\dot{\theta}\sin\theta) - I_a\dot{\theta}(\dot{\phi} - \dot{\psi}\sin\theta) = M_z \tag{3.211}$$

These three second-order dynamical equations are somewhat more complicated than those obtained earlier in (3.203)–(3.205).

Finally, let us consider the rather common situation in which there is no applied moment about the axis of symmetry, that is,

$$M_x = 0 \tag{3.212}$$

Then there are only two first-order dynamical equations, namely,

$$I_t\dot{\omega}_y + I_t\omega_z^2\tan\theta + I_a\Omega\omega_z = M_y \tag{3.213}$$

$$I_t\dot{\omega}_z - I_t\omega_y\omega_z\tan\theta - I_a\Omega\omega_y = M_z \tag{3.214}$$

where the *total spin* Ω *is constant*. If M_y and M_z are known as functions of (ψ, θ, t), these dynamic equations plus the kinematic equations (3.200) and (3.201) can be integrated numerically to give the orientation of the axis of symmetry as a function of time.

Example 3.4   As an example of an axially symmetric body with no moment applied about its axis of symmetry, consider the motion of a top with a fixed point under the action of gravity (Fig. 3.8). The top has a constant total spin Ω and its center of mass lies at distance l from its fixed point O.

The differential equations of motion, obtained from (3.204) and (3.205), are

$$I_t\dot{\omega}_y + I_t\omega_z^2\tan\theta + I_a\Omega\omega_z = -mgl\cos\theta \tag{3.215}$$

$$I_t\dot{\omega}_z - I_t\omega_y\omega_z\tan\theta - I_a\Omega\omega_y = 0 \tag{3.216}$$

and we recall that

$$\dot{\theta} = \omega_y \tag{3.217}$$

Let us consider the case of a top with a large spin Ω and assume that the axis of symmetry remains nearly horizontal, that is, θ is small. In addition, assume that ω_y and ω_z are small.

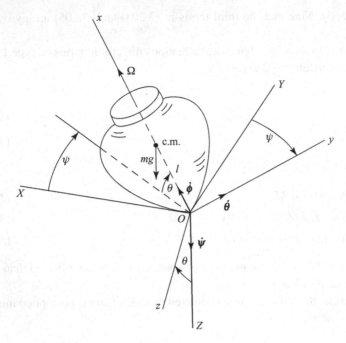

Figure 3.8.

Then, neglecting higher-order terms, (3.215) and (3.216) take the linear forms

$$I_t \dot{\omega}_y + I_a \Omega \omega_z = -mgl \tag{3.218}$$

$$I_t \dot{\omega}_z - I_a \Omega \omega_y = 0 \tag{3.219}$$

Now differentiate (3.219) with respect to time, obtaining

$$\dot{\omega}_y = \frac{I_t}{I_a \Omega} \ddot{\omega}_z \tag{3.220}$$

and substitute into (3.218). The result is

$$\frac{I_t^2}{I_a \Omega} \ddot{\omega}_z + I_a \Omega \omega_z = -mgl \tag{3.221}$$

This differential equation has a solution of the general form

$$\omega_z = \dot{\psi} = -\frac{mgl}{I_a \Omega} + A \cos \frac{I_a \Omega t}{I_t} + B \sin \frac{I_a \Omega t}{I_t} \tag{3.222}$$

Then, from (3.219) we find that

$$\omega_y = \dot{\theta} = -A \sin \frac{I_a \Omega t}{I_t} + B \cos \frac{I_a \Omega t}{I_t} \tag{3.223}$$

where the constants A and B are evaluated from initial conditions.

As a specific example, consider the case of *cuspidal motion* of a fast-spinning top or gyroscope and assume that its symmetry axis is nearly horizontal. We have the initial conditions

$$\omega_y(0) = \omega_z(0) = 0 \tag{3.224}$$

and obtain the solutions

$$\omega_y = \dot{\theta} = -\frac{mgl}{I_a\Omega} \sin \frac{I_a\Omega t}{I_t} \tag{3.225}$$

$$\omega_z = \dot{\psi} = -\frac{mgl}{I_a\Omega} \left(1 - \cos \frac{I_a\Omega t}{I_t}\right) \tag{3.226}$$

Assuming the initial condition $\theta(0) = 0$, (3.225) can be integrated to yield

$$\theta = -\frac{I_t mgl}{I_a^2\Omega^2} \left(1 - \cos \frac{I_a\Omega t}{I_t}\right) \tag{3.227}$$

We have assumed that θ is small and Ω is large. More specifically, let us now assume that $\Omega^2 >> I_t mgl/I_a^2$. Then ω_y and ω_z will also be small, as we have assumed.

A plot of θ versus ψ indicates that a point on the symmetry axis follows a cycloidal path with upward-pointing cusps. The average precession rate is $mgl/I_a\Omega$ and the frequency of the sinusoidal nutational oscillation in θ is $I_a\Omega/I_t$.

Example 3.5 Let us use Euler's equations to solve for the rotational motion of a rigid body under the action of a *constant applied moment* relative to the body axes. Assume that the principal moments of inertia are $I_{xx} < \frac{1}{2}I_{yy} < \frac{1}{2}I_{zz}$. There is a constant applied moment M_z about the z-axis and we assume that the initial conditions and the magnitude of M_z are such that $\omega_x >> (\omega_y, \omega_z)$.

The Euler equations in this instance are

$$I_{xx}\dot{\omega}_x + (I_{zz} - I_{yy})\omega_y\omega_z = 0 \tag{3.228}$$

$$I_{yy}\dot{\omega}_y - (I_{zz} - I_{xx})\omega_x\omega_z = 0 \tag{3.229}$$

$$I_{zz}\dot{\omega}_z + (I_{yy} - I_{xx})\omega_x\omega_y = M_z \tag{3.230}$$

Since ω_y and ω_z are small, (3.228) can be approximated by $\dot{\omega}_x = 0$ or

$$\omega_x = \Omega \tag{3.231}$$

where Ω is a constant. Then a differentiation of (3.229) with respect to time results in

$$\ddot{\omega}_y = \frac{I_{zz} - I_{xx}}{I_{yy}}\Omega\dot{\omega}_z \tag{3.232}$$

where, from (3.230),

$$\dot{\omega}_z = -\frac{I_{yy} - I_{xx}}{I_{zz}}\Omega\omega_y + \frac{M_z}{I_{zz}} \tag{3.233}$$

Thus we obtain

$$\ddot{\omega}_y + \frac{(I_{yy} - I_{xx})(I_{zz} - I_{xx})}{I_{yy}I_{zz}}\Omega^2\omega_y = \frac{I_{zz} - I_{xx}}{I_{yy}I_{zz}}\Omega M_z \tag{3.234}$$

The solution of this linear differential equation has the form

$$\omega_y = C_1 \cos \lambda t + C_2 \sin \lambda t + \frac{M_z}{(I_{yy} - I_{xx})\Omega} \tag{3.235}$$

where the natural frequency is

$$\lambda = \sqrt{\frac{(I_{yy} - I_{xx})(I_{zz} - I_{xx})}{I_{yy}I_{zz}}}\,\Omega \tag{3.236}$$

Then, from (3.229),

$$\omega_z = \frac{I_{yy}}{(I_{zz} - I_{xx})\Omega}\dot{\omega}_y$$

$$= \frac{I_{yy}\lambda}{(I_{zz} - I_{xx})\Omega}(C_2 \cos \lambda t - C_1 \sin \lambda t) \tag{3.237}$$

As a specific example, let us assume the initial conditions $\omega_y(0) = \omega_z(0) = 0$. Then

$$C_1 = -\frac{M_z}{(I_{yy} - I_{xx})\Omega}, \qquad C_2 = 0 \tag{3.238}$$

and we obtain the solutions

$$\omega_y = \frac{M_z}{(I_{yy} - I_{xx})\Omega}(1 - \cos \lambda t) \tag{3.239}$$

$$\omega_z = \frac{M_z}{I_{zz}\lambda} \sin \lambda t \tag{3.240}$$

Since Ω and λ are large, we see that ω_y and ω_z are small, in agreement with the earlier assumption.

It is perhaps surprising that a constant applied moment about the z-axis produces an ω_z which is *periodic* with average value zero rather than constantly increasing. A plot of ω_z versus ω_y is elliptical with its center on the ω_y axis. Furthermore, notice from (3.236) that if I_{xx} is the *intermediate* principal moment of inertia, then the solutions for ω_y and ω_z are no longer sinusoidal and they do not remain small.

Lagrange's equations

The various forms of Lagrange's equations such as (2.34), (2.37) and (2.49) are all useful in obtaining equations of motion for a system involving one or more rigid bodies. There are restrictions, however, on the use of Lagrange's equations.

The first restriction is that a *complete set* of qs and $\dot{q}$s must be used in writing the energy functions T and V. This means that the chosen set of generalized coordinates must be able to specify the configuration of the system, thereby specifying the locations of all particles and mass elements of the system.

As an example, suppose we wish to use type I Euler angles as generalized coordinates in the Lagrangian analysis of the rotational motion of an axially-symmetric body having no applied moment about the axis of symmetry. The rotational kinetic energy can be written in the form

$$T = \frac{1}{2}I_a\Omega^2 + \frac{1}{2}I_t(\dot{\theta}^2 + \dot{\psi}^2\cos^2\theta) \tag{3.241}$$

where the moments of inertia are taken about the center of mass and the total spin Ω is constant. Let us apply Lagrange's equation in the fundamental form

$$\frac{d}{dt}\left(\frac{\partial T}{\partial \dot{q}_i}\right) - \frac{\partial T}{\partial q_i} = Q_i \tag{3.242}$$

where the qs are the Euler angles (ψ, θ, ϕ). If we assume that neither Q_ψ nor Q_θ are functions of ϕ or $\dot{\phi}$, then it appears that the ψ and θ equations can be solved separately, treating Ω as a constant in the differentiations, and effectively reducing the number of degrees of freedom to two.

This application of Lagrange's equation would be *incorrect*, however, because the system actually has three rotational degrees of freedom, but the third Euler angle ϕ has not entered the analysis. Thus, we have not used a complete set of generalized coordinates. A correct approach would be to substitute

$$\Omega = \dot{\phi} - \dot{\psi}\sin\theta \tag{3.243}$$

into (3.241) and obtain

$$T = \frac{1}{2}I_a(\dot{\phi} - \dot{\psi}\sin\theta)^2 + \frac{1}{2}I_t(\dot{\theta}^2 + \dot{\psi}^2\cos^2\theta) \tag{3.244}$$

The use of this kinetic energy function in Lagrange's equation (3.242) leads to correct equations of motion. An alternate approach would be to notice that ϕ is an ignorable coordinate and use the Routhian method.

The second restriction on the use of Lagrange's equation in a form such as (3.242) is that true $\dot{q}$s must be used in writing the kinetic energy $T(q, \dot{q}, t)$, rather than using quasi-velocities as velocity variables. A *quasi-velocity* u_j is equal to a linear function of the $\dot{q}$s and has the form

$$u_j = \sum_{i=1}^{n} \Psi_{ji}(q, t)\dot{q}_i + \Psi_{jt}(q, t) \qquad (j = 1, \ldots, n) \tag{3.245}$$

where the right-hand side is *not integrable*.

A common example of quasi-velocities would be the body-axis components $\omega_x, \omega_y, \omega_z$ of the angular velocity ω of a rigid body. In terms of Euler angles, which are true coordinates, we have

$$\omega_x = \dot{\phi} - \dot{\psi}\sin\theta$$
$$\omega_y = \dot{\psi}\cos\theta\sin\phi + \dot{\theta}\cos\phi \tag{3.246}$$
$$\omega_z = \dot{\psi}\cos\theta\cos\phi - \dot{\theta}\sin\phi$$

and we see that the right-hand sides are not integrable. Assuming principal axes, the kinetic energy due to rotation about the center of mass of a rigid body is

$$T = \frac{1}{2}I_{xx}\omega_x^2 + \frac{1}{2}I_{yy}\omega_y^2 + \frac{1}{2}I_{zz}\omega_z^2 \tag{3.247}$$

However, a substitution into Lagrange's equation (3.242), treating the ωs as $\dot{q}$s, does not result in correct equations of motion. Terms are missing. Thus, the correct Euler equations of motion are not obtained by using Lagrange's equation. It is possible, however, to expand the Lagrange formulation to include the required terms. This is a topic to be discussed in the next chapter.

Example 3.6   Let us obtain the differential equations for the rotational motion of a rigid body, using type I Euler angles as generalized coordinates. We choose the center of mass as the reference point and thereby decouple the translational and rotational motions. Assume a principal axis system and arbitrary applied moments.

The rotational kinetic energy is found by substituting from (3.246) into (3.247) with the result that

$$T = \frac{1}{2}I_{xx}(\dot{\phi} - \dot{\psi}\sin\theta)^2 + \frac{1}{2}I_{yy}(\dot{\psi}\cos\theta\sin\phi + \dot{\theta}\cos\phi)^2$$
$$+ \frac{1}{2}I_{zz}(\dot{\psi}\cos\theta\cos\phi - \dot{\theta}\sin\phi)^2 \tag{3.248}$$

The generalized momenta are

$$p_\psi = \frac{\partial T}{\partial\dot{\psi}} = [I_{xx}\sin^2\theta + (I_{yy}\sin^2\phi + I_{zz}\cos^2\phi)\cos^2\theta]\dot{\psi}$$
$$+ [(I_{yy} - I_{zz})\cos\theta\sin\phi\cos\phi]\dot{\theta} - I_{xx}\dot{\phi}\sin\theta$$
$$p_\theta = \frac{\partial T}{\partial\dot{\theta}} = (I_{yy} - I_{zz})\dot{\psi}\cos\theta\sin\phi\cos\phi + (I_{yy}\cos^2\phi + I_{zz}\sin^2\phi)\dot{\theta} \tag{3.249}$$
$$p_\phi = \frac{\partial T}{\partial\dot{\phi}} = I_{xx}(\dot{\phi} - \dot{\psi}\sin\theta)$$

These equations have the matrix form

$$\mathbf{p} = \mathbf{m}\dot{\mathbf{q}} \tag{3.250}$$

where the mass or inertia matrix is

$$\mathbf{m} = \begin{bmatrix} [I_{xx}\sin^2\theta + (I_{yy}\sin^2\phi \\ + I_{zz}\cos^2\phi)\cos^2\theta] & (I_{yy} - I_{zz})\cos\theta\sin\phi\cos\phi & -I_{xx}\sin\theta \\ (I_{yy} - I_{zz})\cos\theta\sin\phi\cos\phi & I_{yy}\cos^2\phi + I_{zz}\sin^2\phi & 0 \\ -I_{xx}\sin\theta & 0 & I_{xx} \end{bmatrix} \tag{3.251}$$

for the order (ψ, θ, ϕ).

The differential equations of rotational motion are obtained by using Lagrange's equation in the form

$$\frac{d}{dt}\left(\frac{\partial T}{\partial \dot{q}_i}\right) - \frac{\partial T}{\partial q_i} = Q_i \tag{3.252}$$

These equations of motion are

$$[I_{xx}\sin^2\theta + (I_{yy}\sin^2\phi + I_{zz}\cos^2\phi)\cos^2\theta]\ddot{\psi} + (I_{yy} - I_{zz})\ddot{\theta}\cos\theta\sin\phi\cos\phi$$
$$- I_{xx}\ddot{\phi}\sin\theta + 2[I_{xx} - (I_{yy}\sin^2\phi + I_{zz}\cos^2\phi)]\dot{\psi}\dot{\theta}\sin\theta\cos\theta$$
$$+ 2(I_{yy} - I_{zz})\dot{\psi}\dot{\phi}\cos^2\theta\sin\phi\cos\phi + (I_{zz} - I_{yy})\dot{\theta}^2\sin\theta\sin\phi\cos\phi$$
$$+ [-I_{xx} + (I_{yy} - I_{zz})(\cos^2\phi - \sin^2\phi)]\dot{\theta}\dot{\phi}\cos\theta = Q_\psi$$

$$(I_{yy} - I_{zz})\ddot{\psi}\cos\theta\sin\phi\cos\phi + (I_{yy}\cos^2\phi + I_{zz}\sin^2\phi)\ddot{\theta}$$
$$- [I_{xx} - (I_{yy}\sin^2\phi + I_{zz}\cos^2\phi)]\dot{\psi}^2\sin\theta\cos\theta$$
$$+ [I_{xx} + (I_{yy} - I_{zz})(\cos^2\phi - \sin^2\phi)]\dot{\psi}\dot{\phi}\cos\theta + 2(I_{zz} - I_{yy})\dot{\theta}\dot{\phi}\sin\phi\cos\phi = Q_\theta$$

$$- I_{xx}\ddot{\psi}\sin\theta + I_{xx}\ddot{\phi} + (I_{zz} - I_{yy})\dot{\psi}^2\cos^2\theta\sin\phi\cos\phi$$
$$- [I_{xx} + (I_{yy} - I_{zz})(\cos^2\phi - \sin^2\phi)]\dot{\psi}\dot{\theta}\cos\theta$$
$$+ (I_{yy} - I_{zz})\dot{\theta}^2\sin\phi\cos\phi = Q_\phi \tag{3.253}$$

In accordance with the virtual work approach to finding generalized forces, we see that a total applied moment **M** is associated with a virtual work

$$\delta W = \mathbf{M}\cdot\mathbf{e}_\psi\,\delta\psi + \mathbf{M}\cdot\mathbf{e}_\theta\,\delta\theta + \mathbf{M}\cdot\mathbf{e}_\phi\,\delta\phi \tag{3.254}$$

where es are unit vectors in the directions of the respective rotation axes (Fig. 3.2). Thus,

$$Q_\psi = \mathbf{M}\cdot\mathbf{e}_\psi, \qquad Q_\theta = \mathbf{M}\cdot\mathbf{e}_\theta, \qquad Q_\phi = \mathbf{M}\cdot\mathbf{e}_\phi \tag{3.255}$$

The three second-order differential equations given in (3.253) are relatively complicated. Furthermore, they must be solved for the individual accelerations $\ddot{\psi}$, $\ddot{\theta}$, and $\ddot{\phi}$ before being integrated numerically. This involves an inversion of the mass matrix **m**. By comparison, the use of the three first-order Euler equations given in (3.164) to solve for ω_x, ω_y, and ω_z, followed by the three first-order kinematic equations of (3.16) to obtain the Euler angles ψ, θ, and ϕ, appears to be much simpler.

The use of type II Euler angles and Lagrange's equation would result in equally complicated equations of motion.

Angular velocity coefficients

The state of motion of the ith rigid body can be expressed by giving the velocity $\mathbf{v}_i$ of a reference point P_i fixed in the body and also giving its angular velocity ω_i. In terms of $\dot{q}$s, the velocity $\mathbf{v}_i$ can be written in the form

$$\mathbf{v}_i = \sum_{j=1}^{n} \boldsymbol{\gamma}_{ij}(q,t)\dot{q}_j + \boldsymbol{\gamma}_{it}(q,t) \tag{3.256}$$

as in (1.43). The $\boldsymbol{\gamma}$s are called *velocity coefficients* and, in particular, $\boldsymbol{\gamma}_{ij}$ is a vector coefficient representing the sensitivity of the reference point velocity $\mathbf{v}_i$ to changes in $\dot{q}_j$.

In a similar manner, the angular velocity $\boldsymbol{\omega}_i$ of the ith body can be expressed as

$$\boldsymbol{\omega}_i = \sum_{j=1}^{n} \boldsymbol{\beta}_{ij}(q,t)\dot{q}_j + \boldsymbol{\beta}_{it}(q,t) \tag{3.257}$$

where the $\boldsymbol{\beta}$s are called *angular velocity coefficients*.

We see that

$$\boldsymbol{\gamma}_{ij} = \frac{\partial \mathbf{v}_i}{\partial \dot{q}_j} \tag{3.258}$$

and

$$\boldsymbol{\beta}_{ij} = \frac{\partial \boldsymbol{\omega}_i}{\partial \dot{q}_j} \tag{3.259}$$

Furthermore, for the case of a *scleronomic system*, we find that all $\boldsymbol{\gamma}_{it}$ and all $\boldsymbol{\beta}_{it}$ are equal to zero.

Now, let us consider *virtual work* as it applies to a rigid body. An arbitrary set of forces acting on the ith body is dynamically equivalent to a total force $\mathbf{F}_i$ acting at the reference point plus a couple of moment $\mathbf{M}_i$ which may be applied anywhere on the body. The virtual work of this force system due to an arbitrary virtual displacement is

$$\delta W_i = \mathbf{F}_i \cdot \delta \mathbf{r}_i + \mathbf{M}_i \cdot \delta \boldsymbol{\theta}_i \tag{3.260}$$

where $\delta \mathbf{r}_i$ is a small displacement of the reference point and $\delta \boldsymbol{\theta}_i$ represents a small rotation of the body. Now

$$\delta \mathbf{r}_i = \sum_{j=1}^{n} \boldsymbol{\gamma}_{ij}\delta q_j \tag{3.261}$$

and

$$\delta \boldsymbol{\theta}_i = \sum_{j=1}^{n} \boldsymbol{\beta}_{ij}\delta q_i \tag{3.262}$$

Therefore, for the ith body,

$$\delta W_i = \sum_{j=1}^{n} (\mathbf{F}_i \cdot \boldsymbol{\gamma}_{ij} + \mathbf{M}_i \cdot \boldsymbol{\beta}_{ij})\delta q_j \tag{3.263}$$

For a system of N rigid bodies, the total virtual work is

$$\delta W = \sum_{i=1}^{N} \delta W_i = \sum_{i=1}^{N} \sum_{j=1}^{n} (\mathbf{F}_i \cdot \boldsymbol{\gamma}_{ij} + \mathbf{M}_i \cdot \boldsymbol{\beta}_{ij})\delta q_j \tag{3.264}$$

This is of the form

$$\delta W = \sum_{j=1}^{n} Q_j \delta q_j \tag{3.265}$$

where the generalized force associated with q_j is

$$Q_j = \sum_{i=1}^{N} (\mathbf{F}_i \cdot \boldsymbol{\gamma}_{ij} + \mathbf{M}_i \cdot \boldsymbol{\beta}_{ij}) \qquad (j = 1, \ldots, n) \tag{3.266}$$

As an illustration of the meaning of the βs in (3.266), consider a rigid body whose orientation is given by three Euler angles, as in Example 3.6 on page 178. Referring to (3.255), we see that

$$\boldsymbol{\beta}_{11} = \mathbf{e}_\psi, \qquad \boldsymbol{\beta}_{12} = \mathbf{e}_\theta, \qquad \boldsymbol{\beta}_{13} = \mathbf{e}_\phi \tag{3.267}$$

where the **e**s are unit vectors in the directions of the respective rotation axes. In general, however, the angular velocity coefficients need not be of unit magnitude.

We have defined the velocity coefficients and angular velocity coefficients with respect to true velocities ($\dot{q}$s), but one can also define these coefficients with respect to *quasi-velocities*. Thus, when using quasi-velocities (us) as velocity variables, we write

$$\boldsymbol{\gamma}_{ij} = \frac{\partial \mathbf{v}_i}{\partial u_j} \tag{3.268}$$

$$\boldsymbol{\beta}_{ij} = \frac{\partial \boldsymbol{\omega}_i}{\partial u_j} \tag{3.269}$$

Furthermore, we find that

$$\mathbf{v}_i = \sum_{j=1}^{n} \boldsymbol{\gamma}_{ij}(q, t) u_j + \boldsymbol{\gamma}_{it}(q, t) \tag{3.270}$$

$$\boldsymbol{\omega}_i = \sum_{j=1}^{n} \boldsymbol{\beta}_{ij}(q, t) u_j + \boldsymbol{\beta}_{it}(q, t) \tag{3.271}$$

The generalized force Q_j associated with u_j is given by (3.266) where the γs and βs are obtained from (3.268) and (3.269).

If the velocity variables are a mixture of quasi-velocities and true velocities, one can use quasi-velocity notation and identify some of the us with $\dot{q}$s as a special case of (3.245).

Free rotational motion

The rotational motion of a rigid body is *free* if the system of applied forces results in a zero moment about the reference point which is inertially fixed or at the center of mass. Our approach to solving for the free rotational motion will be to use integrals of the motion directly, rather than starting with second-order differential equations.

Let us represent the orientation of the body by type II Euler angles (Fig. 3.3). Since there are no applied moments, the angular momentum is constant in magnitude and direction. Let us define the vertical direction as the direction of the **H** vector, that is, **H** can be considered to lie along the positive Z-axis. Then, assuming that the xyz body axes are principal axes, we see from the geometry that

$$H_x = I_{xx}\omega_x = H\sin\theta\sin\psi \tag{3.272}$$

$$H_y = I_{yy}\omega_y = H\sin\theta\cos\psi \tag{3.273}$$

$$H_z = I_{zz}\omega_z = H\cos\theta \tag{3.274}$$

where H is a positive constant. Thus, we obtain

$$\omega_x = \dot{\phi}\sin\theta\sin\psi + \dot{\theta}\cos\psi = \frac{H}{I_{xx}}\sin\theta\sin\psi \tag{3.275}$$

$$\omega_y = \dot{\phi}\sin\theta\cos\psi - \dot{\theta}\sin\psi = \frac{H}{I_{yy}}\sin\theta\cos\psi \tag{3.276}$$

$$\omega_z = \dot{\phi}\cos\theta + \dot{\psi} = \frac{H}{I_{zz}}\cos\theta \tag{3.277}$$

These three equations represent the integrals of the motion; that is, we can write three independent functions of the qs and $\dot{q}$s which are equal to zero. Finally, solving (3.275)–(3.277) for the type II Euler angle rates, we obtain

$$\dot{\phi} = H\left(\frac{\sin^2\psi}{I_{xx}} + \frac{\cos^2\psi}{I_{yy}}\right) \tag{3.278}$$

$$\dot{\theta} = H\left(\frac{1}{I_{xx}} - \frac{1}{I_{yy}}\right)\sin\theta\sin\psi\cos\psi \tag{3.279}$$

$$\dot{\psi} = H\left(\frac{1}{I_{zz}} - \frac{\sin^2\psi}{I_{xx}} - \frac{\cos^2\psi}{I_{yy}}\right)\cos\theta \tag{3.280}$$

Note that the precession rate $\dot{\phi}$ is always positive, but the nutation rate $\dot{\theta}$ and the relative spin $\dot{\psi}$ may have either sign. If the z-axis is an axis of either maximum or minimum moment of inertia, we find that $0 \le \theta \le \pi/2$ for any physically realizable body. If I_{zz} is the minimum moment of inertia, $\dot{\psi}$ is positive; if I_{zz} is maximum, then $\dot{\psi}$ is negative.

Axial symmetry

Now suppose that the body-fixed z-axis is a symmetry axis. Let

$$I_{xx} = I_{yy} = I_t, \qquad I_{zz} = I_a \tag{3.281}$$

First, we note from (3.279) that

$$\dot{\theta} = 0 \quad \text{or} \quad \theta = \text{const} \tag{3.282}$$

From (3.278), the constant precession rate is

$$\dot{\phi} = \frac{H}{I_t} \tag{3.283}$$

The relative spin rate, from (3.280), is also constant, that is,

$$\dot{\psi} = H \left(\frac{1}{I_a} - \frac{1}{I_t} \right) \cos\theta \tag{3.284}$$

where

$$H = \frac{I_a \Omega}{\cos\theta} \tag{3.285}$$

and the total spin Ω is

$$\Omega = \omega_z = \dot{\phi} \cos\theta + \dot{\psi} \tag{3.286}$$

which is constant. Thus, we find that

$$\dot{\psi} = \left(1 - \frac{I_a}{I_t} \right) \Omega \tag{3.287}$$

For the *prolate* case ($I_a < I_t$) we find that $\dot{\psi}$ is positive; whereas, for the *oblate* case ($I_a > I_t$), $\dot{\psi}$ is negative. Here we take the positive direction of the symmetry axis in this free motion so that the total spin Ω is always positive or zero.

Notice that, since θ is constant, the symmetry axis will sweep out a conical surface as it precesses about the **H** vector which lies on the Z-axis. This is called a *coning motion* and is typical of the free rotational motion of an axially-symmetric body.

The Poinsot method

The Poinsot method is a geometrical method of representing the free rotational motion of a rigid body. It is exact and is based upon the conservation of angular momentum and of kinetic energy.

Let us choose a set of principal axes at the center of mass as the xyz body-fixed frame. Then, if I_1, I_2, and I_3 are the principal moments of inertia, (3.180) representing the ellipsoid of inertia reduces to

$$I_1 x^2 + I_2 y^2 + I_3 z^2 = 1 \tag{3.288}$$

The Poinsot construction (Fig. 3.9) pictures the free rotational motion of a rigid body as the rolling of its ellipsoid of inertia on an invariable plane which is perpendicular to the constant angular momentum vector **H** drawn from the fixed center O. To see how this comes about, recall from (3.178) that the vector ρ from O to P is

$$\rho = \frac{\omega}{\omega \sqrt{I}} \tag{3.289}$$

Also, from (3.176) we note that the rotational kinetic energy is

$$T_{\text{rot}} = \frac{1}{2} \omega \cdot \mathbf{H} = \frac{1}{2} I \omega^2 \tag{3.290}$$

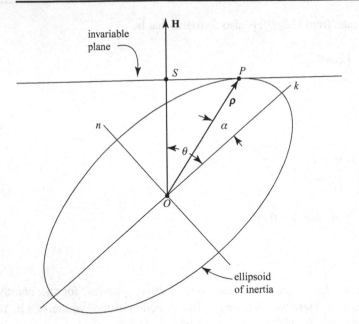

Figure 3.9.

Thus, we find that the length OS is

$$\frac{\rho \cdot \mathbf{H}}{H} = \frac{\omega \cdot \mathbf{H}}{\omega H \sqrt{I}} = \frac{\sqrt{2T_{\text{rot}}}}{H} \tag{3.291}$$

which is constant as P moves. This means that the point P moves in a plane that is perpendicular to the angular momentum vector $\mathbf{H}$ and intersects it at S. This plane is fixed in space and is known as the *invariable plane*.

Next, we need to show that the invariable plane is tangent to the inertia ellipsoid at P. The ellipsoid has the form

$$F(x, y, z) = I_1 x^2 + I_2 y^2 + I_3 z^2 = 1 \tag{3.292}$$

The direction of the normal to the ellipsoid at P is found by evaluating the gradient ∇F at that point. Now

$$\begin{aligned} \nabla F &= \frac{\partial F}{\partial x}\mathbf{i} + \frac{\partial F}{\partial y}\mathbf{j} + \frac{\partial F}{\partial z}\mathbf{k} \\ &= 2I_1 x\mathbf{i} + 2I_2 y\mathbf{j} + 2I_3 z\mathbf{k} \end{aligned} \tag{3.293}$$

We see that ∇F is parallel to $\mathbf{H}$ since

$$\mathbf{H} = I_1 \omega_x \mathbf{i} + I_2 \omega_y \mathbf{j} + I_3 \omega_z \mathbf{k} \tag{3.294}$$

and

$$\omega_x : \omega_y : \omega_z = x : y : z \tag{3.295}$$

that is, ω and ρ have the same direction in space. We conclude that the invariable plane is tangent to the inertia ellipsoid at P. The instantaneous axis of rotation passes through the contact point P, so the ellipsoid of inertia rolls without slipping on the invariable plane. Because ω has a component normal to the invariable plane, there is also some pivoting about P. Of course, the actual rigid body goes through the same rotational motions as the ellipsoid of inertia.

It is interesting to consider the path of the contact point P on both the invariable plane and the inertia ellipsoid. The path of P on the invariable plane is called the *herpolhode*. It is not a closed curve, in general, as P moves continuously between two circles centered on S and corresponding to extreme values of $(\theta - \alpha)$. During successive intervals of $\pi/2$ in ψ, P moves from tangency with one circle to tangency with the other.

The curve traced by P on the ellipsoid of inertia is called a *polhode*. For any rotation not exactly about a principal axis, the polhode curves are closed and encircle either the axis of minimum moment of inertia Ok or the axis of maximum moment of inertia On (Fig. 3.10).

The polhode curves are formed by the intersection of two ellipsoids, namely, the inertia ellipsoid and the momentum ellipsoid. The inertia ellipsoid relative to principal axes was given by (3.288). To obtain the momentum ellipsoid we start with the expression for the square of the angular momentum.

$$I_1^2\omega_x^2 + I_2^2\omega_y^2 + I_3^2\omega_z^2 = H^2 \tag{3.296}$$

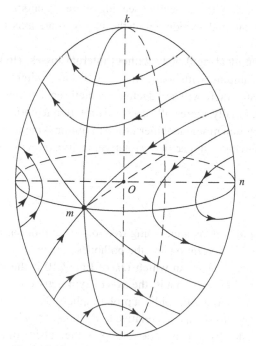

Figure 3.10.

However, from (3.289) and (3.290) we see that

$$
\begin{aligned}
\omega_x &= x\omega\sqrt{I} = x\sqrt{2T_{\text{rot}}} \\
\omega_y &= y\omega\sqrt{I} = y\sqrt{2T_{\text{rot}}} \\
\omega_z &= z\omega\sqrt{I} = z\sqrt{2T_{\text{rot}}}
\end{aligned}
\tag{3.297}
$$

Hence we obtain the *momentum ellipsoid*

$$
I_1^2 x^2 + I_2^2 y^2 + I_3^2 z^2 = \frac{H^2}{2T_{\text{rot}}} = D
\tag{3.298}
$$

where H and T_{rot} are constants which are usually evaluated from the initial conditions.

The polhode curve for a given case of free rotational motion is determined by the values of (x, y, z) satisfying the initial value of ρ as well as (3.288) and (3.298). It can be shown that if $D < I_2$, assuming $I_1 < I_2 < I_3$, the polhode encircles the axis Ok corresponding to the minimum moment of inertia. If $D > I_2$, the polhode encircles the axis On corresponding to the maximum moment of inertia.

Now consider the *stability* of rotational motion about a principal axis. The polhodes in the vicinity of k and n are tiny ellipses indicating stability, that is, a small displacement of the axis of rotation relative to the body will remain small. On the other hand, polhodes near m are *hyperbolic* in nature, indicating instability. The axis of rotation relative to the body will suddenly flip over to nearly the opposite direction and then return back again, only to repeat the cycle. Meanwhile the angular momentum vector **H** remains constant in space. Thus the rotational motion of the body in space is quite irregular. In theory it takes an infinite time for the point P to leave m but, practically speaking, there are enough small disturbances that the instability of rotational motion about the intermediate axis Om is immediately apparent.

The above analysis has assumed an ideal rigid body without internal losses. However, an actual body will have structural damping with some energy loss due to slight elastic deflections during the motion. As a result, there will be a decline in kinetic energy consistent with the constant angular momentum. Finally, there will be a steady rotational motion about the axis On corresponding to the maximum moment of inertia and minimum kinetic energy. This explains why a spin-stabilized rigid body in space must rotate about its axis of maximum moment of inertia.

Axial symmetry

Now assume an axially-symmetric rigid body is undergoing free rotational motion. In this case the inertia ellipsoid is an ellipsoid of revolution and the polhodes are circles centered on the axis of symmetry. First consider the case in which Ok of Fig. 3.10 is the axis of symmetry and $I_1 = I_a$, $I_2 = I_3 = I_t$ and $I_a < I_t$, that is, the axis of symmetry corresponds to the minimum moment of inertia. The inertia ellipsoid is a prolate spheroid and, as it rolls on the invariable plane, the ω vector sweeps out a cone relative to the body and also a cone in space. This is conveniently represented by a body cone rolling on the outside of a fixed space cone (Fig. 3.11a). Notice that the symmetry axis Oz and the angular velocity vector

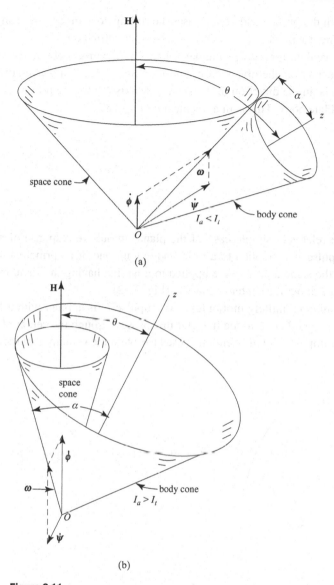

Figure 3.11.

$\boldsymbol{\omega}$ lie on the same side of the angular momentum vector $\mathbf{H}$. Also the precession rate $\dot{\phi}$ is smaller than ω.

If, on the other hand, the axis of symmetry corresponds to the maximum moment of inertia, the inertia ellipsoid is an oblate spheroid and we can take $I_1 = I_2 = I_t, I_3 = I_a$ with $I_a > I_t$. Again the polhodes are circles and, in this case, the axis of symmetry is On. Because of the oblateness, the symmetry axis and the $\boldsymbol{\omega}$ vector lie on opposite sides of the $\mathbf{H}$ vector. This leads to the rolling cone model shown in Fig. 3.11b in which the inside of the body cone rolls on the outside of the fixed space cone. Usually the precession rate $\dot{\phi}$

is larger than ω. In both the prolate and oblate cases the rolling cone model gives an exact representation of the free rotational motion of an axially symmetric body.

From Fig. 3.11 we see that, for a given angular velocity ω, the precession rate $\dot{\phi}$ for the prolate case is smaller than for the oblate case. Furthermore, notice that the relative spin rate $\dot{\psi}$ is positive, that is, in the direction of the symmetry axis Oz for the prolate case but $\dot{\psi}$ is negative for an oblate body. This is in agreement with (3.287).

3.4 Impulsive motion

Planar rigid body motion

Let us begin with the relatively simple case of the planar impulsive response of a rigid body. Suppose an impulse $\hat{F}$ is applied to a rigid body along one of its principal planes. The response will be the same as if $\hat{F}$ were applied to a lamina having the same mass m and moment of inertia I about the center of mass C (Fig. 3.12).

Assume that the lamina is initially motionless. An impulse $\hat{F}$ is applied with a line of action through B and perpendicular to the line AB that passes through the center of mass. The principle of linear impulse and momentum is used to obtain the velocity v of the center

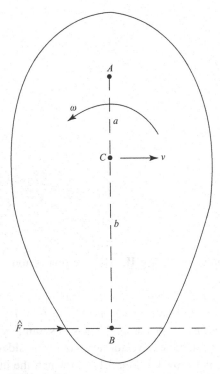

Figure 3.12.

of mass. It is

$$v = \frac{\hat{F}}{m} \qquad (3.299)$$

Similarly, from the principle of angular impulse and momentum, we obtain the angular velocity

$$\omega = \frac{b\hat{F}}{I} \qquad (3.300)$$

where I is the moment of inertia about the center of mass.

   Now consider the conditions on the lengths a and b such that point A is the instantaneous center of rotation immediately after the impulse $\hat{F}$ is applied, that is, its velocity is zero. This requires that

$$v - a\omega = 0$$

or

$$\frac{\hat{F}}{m} = \frac{ab\hat{F}}{I}$$

or

$$ab = \frac{I}{m} = k_c^2 \qquad (3.301)$$

where k_c is the *radius of gyration* about the center of mass.

   If point A is fixed during the impulse, then there will be no impulsive reaction at A if the values of a and b satisfy (3.301). Under these conditions, the point B is known as the *center of percussion* relative to A. Similarly, point A is the center of percussion relative to B.

Example 3.7   Let us solve for the height h at which a billiard ball must be struck by a horizontal impulse in order that the response will be pure rolling with no tendency to slip (Fig. 3.13).

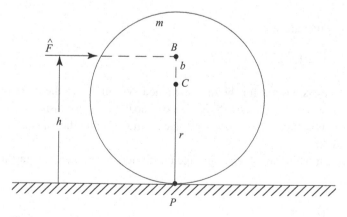

Figure 3.13.

Here we wish to find the location of the center of percussion B relative to the contact point P. We see that $a = r$ and $I = \frac{2}{5} mr^2$. From (3.301) we obtain

$$ab = \frac{I}{m} = \frac{2}{5}r^2 \tag{3.302}$$

so $b = \frac{2}{5}r$ and height $h = r + b = \frac{7}{5}r$.

Constrained impulse response

Consider a system of N rigid bodies and assume that the kinetic energy has the form $T(q, \dot{q}, t)$. In general,

$$T = \sum_{i=1}^{N} \left[\frac{1}{2} m_i v_i^2 + \frac{1}{2} \boldsymbol{\omega}_i \cdot \mathbf{I}_i \cdot \boldsymbol{\omega}_i + m_i \mathbf{v}_i \cdot \dot{\boldsymbol{\rho}}_{ci} \right] \tag{3.303}$$

where $\mathbf{v}_i$ is the velocity of the reference point P_i of the ith body, and $\mathbf{I}_i$ is the inertia dyadic about that reference point (see Fig. 3.5). Assume that there are m nonholonomic constraints of the form

$$\sum_{i=1}^{n} a_{ji}(q, t)\dot{q}_i + a_{jt}(q, t) = 0 \qquad (j = 1, \ldots, m) \tag{3.304}$$

and generalized impulses $\hat{Q}_i$ are applied over an infinitesimal time interval Δt. Then (2.323) applies, that is,

$$\sum_{i=1}^{n} \left[\sum_{j=1}^{n} m_{ij}(\dot{q}_j - \dot{q}_{j0}) - \hat{Q}_i \right] \delta w_i = 0 \tag{3.305}$$

where δws are virtual velocities which satisfy the m instantaneous constraint equations of the form

$$\sum_{i=1}^{n} a_{ji} \delta w_i = 0 \qquad (j = 1, \ldots, m) \tag{3.306}$$

The mass or inertia coefficients are

$$m_{ij} = \frac{\partial^2 T}{\partial \dot{q}_i \partial \dot{q}_j} \qquad (i, j = 1, \ldots, n) \tag{3.307}$$

where the kinetic energy is written for the unconstrained system. Since there are $(n - m)$ independent sets of δws in (3.305) which satisfy (3.306), and there are m constraint equations from (3.304), the result is a total of n equations to solve for the values of the n $\dot{q}$s immediately after the time interval Δt.

An alternate approach which involves Lagrange multipliers is obtained by starting with the basic equation

$$\Delta p_i = \sum_{j=1}^{n} m_{ij} \Delta \dot{q}_j = \hat{Q}_i + \hat{C}_i \qquad (i = 1, \ldots, n) \tag{3.308}$$

where

$$\hat{C}_i = \sum_{k=1}^{m} \hat{\lambda}_k a_{ki} \qquad (i = 1, \ldots, n) \tag{3.309}$$

is the ith generalized constraint impulse. This results in n equations of the form

$$\sum_{j=1}^{n} m_{ij}(\dot{q}_j - \dot{q}_{j0}) = \hat{Q}_i + \sum_{k=1}^{m} \hat{\lambda}_k a_{ki} \qquad (i = 1, \ldots, n) \tag{3.310}$$

These n equations plus the m constraint equations form a total of $(n + m)$ equations from which to solve for the n $\dot{q}$s and m $\hat{\lambda}$s.

Example 3.8   Two thin rods, each of mass m and length l are connected by a pin joint at B (Fig. 3.14). They are falling with velocity v_0 in planar motion when end A strikes the smooth floor at $y = 0$ inelastically. Assuming that $\theta_1 = \theta_2 = 30°$ and $\dot{\theta}_1 = \dot{\theta}_2 = 0$ just before impact, we wish to solve for the $\dot{q}$s immediately after impact.

First method   Let us use the Lagrange multiplier method, as given in (3.310). The generalized coordinates, in order, are $(x, y, \theta_1, \theta_2)$. The initial velocities ($\dot{q}_{j0}$s) are

$$\dot{x}_0 = 0, \qquad \dot{y}_0 = -v_0, \qquad \dot{\theta}_{10} = 0, \qquad \dot{\theta}_{20} = 0 \tag{3.311}$$

The equation of constraint is $y = 0$ or $\dot{y} = 0$, yielding

$$a_{11} = 0, \qquad a_{12} = 1, \qquad a_{13} = 0, \qquad a_{14} = 0 \tag{3.312}$$

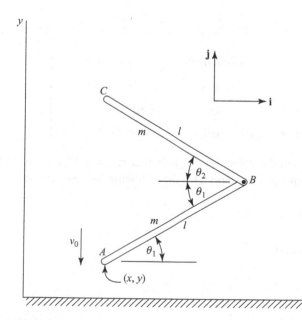

Figure 3.14.

The mass coefficients m_{ij} are found by first obtaining the kinetic energy expression for the unconstrained system. For a system of N rigid bodies, the kinetic energy given by (3.303) is

$$T = \frac{1}{2}\sum_{i=1}^{N} m_i v_i^2 + \frac{1}{2}\sum_{i=1}^{N} \boldsymbol{\omega}_i \cdot \mathbf{I}_i \cdot \boldsymbol{\omega}_i + \sum_{i=1}^{N} m_i \mathbf{v}_i \cdot \dot{\boldsymbol{\rho}}_{ci} \tag{3.313}$$

Let us choose A and B as reference points for the two rods. We see that

$$\mathbf{v}_A = \dot{x}\mathbf{i} + \dot{y}\mathbf{j}, \qquad \mathbf{v}_B = (\dot{x} - l\dot{\theta}_1 \sin\theta_1)\mathbf{i} + (\dot{y} + l\dot{\theta}_1 \cos\theta_1)\mathbf{j} \tag{3.314}$$

$$I_1 = I_2 = \frac{1}{3}ml^2 \tag{3.315}$$

$$\dot{\boldsymbol{\rho}}_{c1} = \frac{1}{2}l\dot{\theta}_1(-\sin\theta_1\,\mathbf{i} + \cos\theta_1\,\mathbf{j}), \qquad \dot{\boldsymbol{\rho}}_{c2} = \frac{1}{2}l\dot{\theta}_2(\sin\theta_2\,\mathbf{i} + \cos\theta_2\,\mathbf{j}) \tag{3.316}$$

Using (3.313), the kinetic energy is

$$T = m(\dot{x}^2 + \dot{y}^2) + \frac{2}{3}ml^2\dot{\theta}_1^2 + \frac{1}{6}ml^2\dot{\theta}_2^2 - \frac{3}{2}ml\dot{x}\dot{\theta}_1 \sin\theta_1$$
$$+ \frac{3}{2}ml\dot{y}\dot{\theta}_1 \cos\theta_1 + \frac{1}{2}ml\dot{x}\dot{\theta}_2 \sin\theta_2 + \frac{1}{2}ml\dot{y}\dot{\theta}_2 \cos\theta_2$$
$$+ \frac{1}{2}ml^2\dot{\theta}_1\dot{\theta}_2 \cos(\theta_1 + \theta_2) \tag{3.317}$$

Using

$$m_{ij} = \frac{\partial^2 T}{\partial \dot{q}_i \partial \dot{q}_j} \tag{3.318}$$

we obtain the mass matrix

$$\mathbf{m} = \begin{bmatrix} 2m & 0 & -\frac{3}{2}ml\sin\theta_1 & \frac{1}{2}ml\sin\theta_2 \\ 0 & 2m & \frac{3}{2}ml\cos\theta_1 & \frac{1}{2}ml\cos\theta_2 \\ -\frac{3}{2}ml\sin\theta_1 & \frac{3}{2}ml\cos\theta_1 & \frac{4}{3}ml^2 & \frac{1}{2}ml^2\cos(\theta_1 + \theta_2) \\ \frac{1}{2}ml\sin\theta_2 & \frac{1}{2}ml\cos\theta_2 & \frac{1}{2}ml^2\cos(\theta_1 + \theta_2) & \frac{1}{3}ml^2 \end{bmatrix} \tag{3.319}$$

Note that all $\hat{Q}_i$ equal zero, and the constraint requires that $\Delta\dot{y} = \dot{y} - \dot{y}_0 = v_0$. Then recalling that $\theta_1 = \theta_2 = 30°$, equation (3.310) yields the following four equations:

$$2m\Delta\dot{x} - \frac{3}{4}ml\Delta\dot{\theta}_1 + \frac{1}{4}ml\Delta\dot{\theta}_2 = 0 \tag{3.320}$$

$$2mv_0 + \frac{3\sqrt{3}}{4}ml\Delta\dot{\theta}_1 + \frac{\sqrt{3}}{4}ml\Delta\dot{\theta}_2 = \hat{\lambda} \tag{3.321}$$

$$-\frac{3}{4}ml\Delta\dot{x} + \frac{3\sqrt{3}}{4}mlv_0 + \frac{4}{3}ml^2\Delta\dot{\theta}_1 + \frac{1}{4}ml^2\Delta\dot{\theta}_2 = 0 \tag{3.322}$$

$$\frac{1}{4}ml\Delta\dot{x} + \frac{\sqrt{3}}{4}mlv_0 + \frac{1}{4}ml^2\Delta\dot{\theta}_1 + \frac{1}{3}ml^2\Delta\dot{\theta}_2 = 0 \tag{3.323}$$

Here are four linear algebraic equations in $\Delta\dot{x}$, $\Delta\dot{\theta}_1$, $\Delta\dot{\theta}_2$ and $\hat{\lambda}$. The solutions are

$$\Delta\dot{x} = \dot{x} = -\frac{6\sqrt{3}}{23}v_0, \qquad \Delta\dot{\theta}_1 = \dot{\theta}_1 = -\frac{81\sqrt{3}}{115}\frac{v_0}{l},$$

$$\Delta\dot{\theta}_2 = \dot{\theta}_2 = -\frac{3\sqrt{3}}{115}\frac{v_0}{l}, \qquad \hat{\lambda} = \frac{91}{230}mv_0 \tag{3.324}$$

We see that the contact point A moves to the left, and both angles decrease, but $\dot{\theta}_1$ is 27 times larger than $\dot{\theta}_2$. From (3.309), we see that $\hat{\lambda}$ is equal to the upward constraint impulse applied to the system at the contact point A.

Second method   Let us now apply the virtual velocity method of (3.305). We will need to find $n - m = 3$ independent sets of δws which satisfy the constraint equation

$$\delta\dot{y} = 0 \tag{3.325}$$

For example, we can use

$$\delta\mathbf{w}_1 = (1, 0, 0, 0), \qquad \delta\mathbf{w}_2 = (0, 0, 1, 0), \qquad \delta\mathbf{w}_3 = (0, 0, 0, 1) \tag{3.326}$$

By applying (3.305), and recalling that $\Delta\dot{y} = v_0$, we obtain (3.320), (3.322), and (3.323). These equations are solved with the result

$$\Delta\dot{x} = -\frac{6\sqrt{3}}{23}v_0, \qquad \Delta\dot{\theta}_1 = -\frac{81\sqrt{3}}{115}\frac{v_0}{l}, \qquad \Delta\dot{\theta}_2 = -\frac{3\sqrt{3}}{115}\frac{v_0}{l} \tag{3.327}$$

in agreement with (3.324).

This second method is, in effect, a method for taking linear combinations of the n equations of the form of (3.310) in such a manner that the $\hat{\lambda}$s are eliminated. Thus, a reduced set of equations is obtained from which one can solve for the final $\dot{q}$s.

Quasi-velocities

In the discussion of the use of Lagrange's equations in rigid body dynamics, it was mentioned that quasi-velocities (us) as defined in (3.245) cannot be used in the expression for kinetic energy. Rather, the kinetic energy, in general, must have the form $T(q, \dot{q}, t)$ where the $\dot{q}$s are true velocities in the sense that they are time derivatives of parameters which specify the configuration of the system. On the other hand, the body-axis components ω_x, ω_y and ω_z are quasi-velocities whose time integrals do not specify the orientation of a rigid body. Thus, kinetic energy written as $T(q, \omega, t)$ cannot be used in Lagrange's equation. This restriction, however, does not apply to impulsive equations. Thus, (3.305) can be generalized to

$$\sum_{i=1}^{n}\left[\sum_{j=1}^{n}m_{ij}(u_j - u_{j0}) - \hat{Q}_i\right]\delta w_i = 0 \tag{3.328}$$

where there are m constraints of the form

$$\sum_{i=1}^{n}a_{ji}(q, t)u_i + a_{jt}(q, t) = 0 \qquad (j = 1, \ldots, m) \tag{3.329}$$

and the δws satisfy

$$\sum_{i=1}^{n} a_{ji}\delta w_i = 0 \tag{3.330}$$

The us are quasi-velocities, in general, and the mass coefficients are

$$m_{ij} = \frac{\partial^2 T(q, u, t)}{\partial u_i \partial u_j} \qquad (i, j = 1, \ldots, n) \tag{3.331}$$

$\hat{Q}_i$ is an applied impulse associated with u_i.

A Lagrange multiplier approach, similar to (3.310), results in equations of the form

$$\Delta p_i = \sum_{j=1}^{n} m_{ij}(u_j - u_{j0}) = \hat{Q}_i + \sum_{k=1}^{m} \hat{\lambda}_k a_{ki} \qquad (i = 1, \ldots, n) \tag{3.332}$$

These n equations plus the m constraint equations can be solved for the n us and m $\hat{\lambda}$s.

A particularly simple case occurs if the us are *independent*. Then we obtain

$$\sum_{j=1}^{n} m_{ij}(u_j - u_{j0}) = \hat{Q}_i \qquad (i = 1, \ldots, n) \tag{3.333}$$

As a specific example, consider the rotational motion of a rigid body about either its center of mass or a fixed point. Its response to an applied angular impulse $\hat{\mathbf{M}}$, in terms of body-axis components, is obtained from

$$\sum_{j=1}^{3} I_{ij}(\omega_j - \omega_{j0}) = \hat{M}_i \qquad (i = 1, 2, 3) \tag{3.334}$$

Here the ωs are quasi-velocities and $m_{ij} \equiv I_{ij}$.

Example 3.9 A rigid body is moving with general motion when suddenly its reference point P (Fig. 3.5) is stopped and remains with zero velocity, although its rotational motion is not impeded. We wish to find the angular velocity of the body immediately after application of this impulsive constraint.

First method Let us choose a body-fixed xyz frame with its origin at the reference point P, and $(\mathbf{i}, \mathbf{j}, \mathbf{k})$ as unit vectors attached to this frame. The velocity of P is

$$\mathbf{v} = v_x \mathbf{i} + v_y \mathbf{j} + v_z \mathbf{k} \tag{3.335}$$

and the angular velocity of the rigid body is

$$\boldsymbol{\omega} = \omega_x \mathbf{i} + \omega_y \mathbf{j} + \omega_z \mathbf{k} \tag{3.336}$$

The position vector of the center of mass relative to the reference point is

$$\boldsymbol{\rho}_c = x_c \mathbf{i} + y_c \mathbf{j} + z_c \mathbf{k} \tag{3.337}$$

and we see that

$$\dot{\rho}_c = \omega \times \rho_c = (\omega_y z_c - \omega_z y_c)\mathbf{i} + (\omega_z x_c - \omega_x z_c)\mathbf{j} + (\omega_x y_c - \omega_y x_c)\mathbf{k} \qquad (3.338)$$

In general, the kinetic energy of a rigid body has the form

$$T = \frac{1}{2}mv^2 + \frac{1}{2}\omega \cdot \mathbf{I} \cdot \omega + m\mathbf{v} \cdot \dot{\rho}_c \qquad (3.339)$$

where $\mathbf{I}$ is the inertia dyadic about the reference point. For this example, we obtain

$$\begin{aligned}
T = \frac{1}{2}m\left(v_x^2 + v_y^2 + v_z^2\right) + \frac{1}{2}\left(I_{xx}\omega_x^2 + I_{yy}\omega_y^2 + I_{zz}\omega_z^2\right. \\
\left. + 2I_{xy}\omega_x\omega_y + 2I_{xz}\omega_x\omega_z + 2I_{yz}\omega_y\omega_z\right) + mv_x(\omega_y z_c - \omega_z y_c) \\
+ mv_y(\omega_z x_c - \omega_x z_c) + mv_z(\omega_x y_c - \omega_y x_c)
\end{aligned} \qquad (3.340)$$

The equations of motion in terms of quasi-velocities are given by (3.328), namely,

$$\sum_{i=1}^{n}\left[\sum_{j=1}^{n}m_{ij}(u_j - u_{j0}) - \hat{Q}_i\right]\delta w_i = 0 \qquad (3.341)$$

where each of the $(n - m)$ independent sets of δws satisfies the instantaneous constraint equations

$$\sum_{i=1}^{n}a_{ji}\delta w_i = 0 \qquad (j = 1, \ldots, m) \qquad (3.342)$$

As quasi-velocities, let us choose $(v_x, v_y, v_z, \omega_x, \omega_y, \omega_z)$ which are subject to the suddenly appearing constraint equations

$$u_1 = v_x = 0 \qquad (3.343)$$
$$u_2 = v_y = 0 \qquad (3.344)$$
$$u_3 = v_z = 0 \qquad (3.345)$$

resulting in

$$a_{11} = a_{22} = a_{33} = 1 \qquad (3.346)$$

All the other as equal zero.

Consider the kinetic energy expression of (3.340). Using

$$m_{ij} = \frac{\partial^2 T}{\partial u_i \partial u_j} \qquad (i, j = 1, \ldots, n) \qquad (3.347)$$

the resulting mass matrix is

$$\mathbf{m} = \begin{bmatrix}
m & 0 & 0 & 0 & mz_c & -my_c \\
0 & m & 0 & -mz_c & 0 & mx_c \\
0 & 0 & m & my_c & -mx_c & 0 \\
0 & -mz_c & my_c & I_{xx} & I_{xy} & I_{xz} \\
mz_c & 0 & -mx_c & I_{yx} & I_{yy} & I_{yz} \\
-my_c & mx_c & 0 & I_{zx} & I_{zy} & I_{zz}
\end{bmatrix} \qquad (3.348)$$

The initial conditions of the quasi-velocities are

$$u_{j0} = (v_{x0}, v_{y0}, v_{z0}, \omega_{x0}, \omega_{y0}, \omega_{z0}) \tag{3.349}$$

and these velocities occur just before the reference point P undergoes inelastic impact with a fixed point. At the time of impact there are constraint impulses acting on the system, but all applied impulses ($\hat{Q}$s) are equal to zero.

The instantaneous constraint equations, in accordance with (3.342), are

$$\delta u_1 = \delta v_x = 0, \qquad \delta u_2 = \delta v_y = 0, \qquad \delta u_3 = \delta v_z = 0 \tag{3.350}$$

Three independent sets of δws which also satisfy the instantaneous constraints are

$$\delta w_1 = (0, 0, 0, 1, 0, 0)$$
$$\delta w_2 = (0, 0, 0, 0, 1, 0)$$
$$\delta w_3 = (0, 0, 0, 0, 0, 1) \tag{3.351}$$

Finally, substituting into (3.341), we obtain the following three equations to be solved for the values of the ωs just after impact.

$$mz_c v_{y0} - my_c v_{z0} + I_{xx}(\omega_x - \omega_{x0}) + I_{xy}(\omega_y - \omega_{y0}) + I_{xz}(\omega_z - \omega_{z0}) = 0 \tag{3.352}$$

$$-mz_c v_{x0} + mx_c v_{z0} + I_{yx}(\omega_x - \omega_{x0}) + I_{yy}(\omega_y - \omega_{y0}) + I_{yz}(\omega_z - \omega_{z0}) = 0 \tag{3.353}$$

$$my_c v_{x0} - mx_c v_{y0} + I_{zx}(\omega_x - \omega_{x0}) + I_{zy}(\omega_y - \omega_{y0}) + I_{zz}(\omega_z - \omega_{z0}) = 0 \tag{3.354}$$

where the Is are taken about the reference point P.

Second method   Let us choose the fixed origin O as the final location of the reference point P. There is conservation of angular momentum about O because the only impulse applied to the rigid body acts through that point.

First, we need to find the angular momentum about O in terms of the angular momentum about P. From Fig. 3.5 we see that, in general,

$$\mathbf{H}_0 = \mathbf{I}_c \cdot \boldsymbol{\omega} + m(\mathbf{r}_p + \boldsymbol{\rho}_c) \times (\dot{\mathbf{r}}_p + \dot{\boldsymbol{\rho}}_c) \tag{3.355}$$

where $\mathbf{I}_c$ is the inertia dyadic about the center of mass. The angular momentum about P is

$$\mathbf{H}_p = \mathbf{I} \cdot \boldsymbol{\omega} = \mathbf{I}_c \cdot \boldsymbol{\omega} + m\boldsymbol{\rho}_c \times \dot{\boldsymbol{\rho}}_c \tag{3.356}$$

Hence, we find that

$$\mathbf{H}_0 = \mathbf{H}_p + m\mathbf{r}_p \times \dot{\mathbf{r}}_p + m\mathbf{r}_p \times \dot{\boldsymbol{\rho}}_c + m\boldsymbol{\rho}_c \times \dot{\mathbf{r}}_p \tag{3.357}$$

This equation would be valid even if the point O were moving since the definition of angular momentum involves *relative* velocities.

For this example, the point O is fixed and, at the time of impact, $\mathbf{r}_p = 0$. Just before impact,

$$\dot{\mathbf{r}}_p = v_{x0}\mathbf{i} + v_{y0}\mathbf{j} + v_{z0}\mathbf{k} \tag{3.358}$$

$$\boldsymbol{\omega} = \omega_{x0}\mathbf{i} + \omega_{y0}\mathbf{j} + \omega_{z0}\mathbf{k} \tag{3.359}$$

Thus, the angular momentum about O at this time, as given by (3.357), is

$$
\begin{aligned}
\mathbf{H}_0 = {} & [I_{xx}\omega_{x0} + I_{xy}\omega_{y0} + I_{xz}\omega_{z0} + m(y_c v_{z0} - z_c v_{y0})]\mathbf{i} \\
& + [I_{yx}\omega_{x0} + I_{yy}\omega_{y0} + I_{yz}\omega_{z0} + m(z_c v_{x0} - x_c v_{z0})]\mathbf{j} \\
& + [I_{zx}\omega_{x0} + I_{zy}\omega_{y0} + I_{zz}\omega_{z0} + m(x_c v_{y0} - y_c v_{x0})]\mathbf{k}
\end{aligned} \tag{3.360}
$$

Immediately after impact, we have $\dot{\mathbf{r}}_p = 0$, so we obtain

$$
\begin{aligned}
\mathbf{H}_0 = \mathbf{I} \cdot \boldsymbol{\omega} = {} & (I_{xx}\omega_x + I_{xy}\omega_y + I_{xz}\omega_z)\mathbf{i} + (I_{yx}\omega_x + I_{yy}\omega_y + I_{yz}\omega_z)\mathbf{j} \\
& + (I_{zx}\omega_x + I_{zy}\omega_y + I_{zz}\omega_z)\mathbf{k}
\end{aligned} \tag{3.361}
$$

Now we use conservation of angular momentum to obtain the following three component equations expressing the final value minus the initial value of $\mathbf{H}_0$.

$$
I_{xx}(\omega_x - \omega_{x0}) + I_{xy}(\omega_y - \omega_{y0}) + I_{xz}(\omega_z - \omega_{z0}) + m(z_c v_{y0} - y_c v_{z0}) = 0 \tag{3.362}
$$

$$
I_{yx}(\omega_x - \omega_{x0}) + I_{yy}(\omega_y - \omega_{y0}) + I_{yz}(\omega_z - \omega_{z0}) + m(x_c v_{z0} - z_c v_{x0}) = 0 \tag{3.363}
$$

$$
I_{zx}(\omega_x - \omega_{x0}) + I_{zy}(\omega_y - \omega_{y0}) + I_{zz}(\omega_z - \omega_{z0}) + m(y_c v_{x0} - x_c v_{y0}) = 0 \tag{3.364}
$$

in agreement with (3.352)–(3.354). Thus, we see that the conservation of angular momentum approach is rather direct in this case.

Input–output methods

For the general case in which the motion of a mechanical system is expressed in terms of quasi-velocities, we found that

$$
\Delta p_i = \sum_{j=1}^{n} m_{ij} \Delta u_j = \hat{Q}_i + \hat{C}_i \qquad (i = 1, \ldots, n) \tag{3.365}
$$

where the $\hat{Q}$s are applied impulses and the $\hat{C}$s are constraint impulses which can be expressed in the Lagrange multiplier form

$$
\hat{C}_i = \sum_{k=1}^{m} \hat{\lambda}_k a_{ki} \tag{3.366}
$$

All impulses occur at the same instant.

A comparison of (3.365) with (3.308) shows that equations of the same form apply whether we use $\dot{q}$s or us as velocity variables. Furthermore, the constraints may be holonomic or nonholonomic and, in either case, it is always possible to find a set of *independent* us. Thus, the impulsive equations of motion tend to be much simpler than the full dynamical equations for a given system. For example, there are no elastic force terms and no terms resulting from centripetal or Coriolis accelerations.

Let us consider a system in which the qs and $\dot{q}$s may be constrained, but the us are *independent*. Then the $\hat{C}$s vanish and we obtain

$$
\Delta p_i = \sum_{j=1}^{n} m_{ij} \Delta u_j = \hat{Q}_i \qquad (i = 1, \ldots, n) \tag{3.367}
$$

The mass coefficients $m_{ij}(q, t)$ are constant during the application of impulses. They form a mass matrix $\mathbf{m}$ which is positive definite and symmetric.

One can solve for the Δus in (3.367) and obtain

$$\Delta u_j = \sum_{i=1}^{n} Y_{ji} \hat{Q}_i \qquad (j = 1, \ldots, n) \tag{3.368}$$

where the matrix $\mathbf{Y} = \mathbf{m}^{-1}$ is symmetric and positive definite.

Equations (3.367) and (3.368) have the same linear form as the equations of an electrical circuit. Hence, one can consider input–output relationships. The $\hat{Q}$s might be taken as input impulses whereas the Δus are output responses. The $\mathbf{m}$ matrix represents an inertial resistance to changes in velocity. Proportional increases in magnitudes of the m_{ij} coefficients result in smaller Δus for given $\hat{Q}$s. On the other hand, the $\mathbf{Y}$ matrix of (3.368) has an inverse effect and acts as an *admittance matrix* for impulsive inputs.

A system characteristic of considerable practical importance is the *input* or *driving-point mass* at a given location. This is the effective mass which an input impulse $\hat{Q}_i$ experiences when all other input impulses are set equal to zero. In the usual case in which u_i represents a linear velocity, the corresponding input mass has the units of mass. But if u_i represents an angular velocity, then the associated input mass is actually a moment of inertia and $\hat{Q}_i$ is an impulsive moment or couple. In any event, we find that the input or driving-point mass for an impulsive input at u_i is

$$\overline{m}_i = \frac{\hat{Q}_i}{\Delta u_i} = \frac{1}{Y_{ii}} \tag{3.369}$$

Note that, in general, this is not equal to m_{ii}.

For the common case in which a linear impulse $\hat{F}$ is applied at a point P, the Δu_i in (3.369) is the velocity change at P in the direction of $\hat{F}$.

Let us consider a system with n degrees of freedom and n us. Choose u_1 as the input and u_2 as the output. Then, in accordance with (3.368) and Fig. 3.15a, we obtain

$$\Delta u_1 = Y_{11} \hat{Q}_1 + Y_{12} \hat{Q}_2 \tag{3.370}$$

$$\Delta u_2 = Y_{21} \hat{Q}_1 + Y_{22} \hat{Q}_2 \tag{3.371}$$

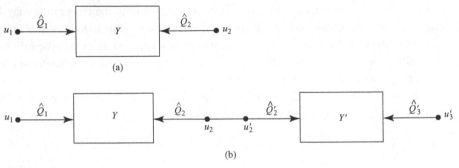

(a)

(b)

Figure 3.15.

and we note that the entire $\mathbf{Y}$ matrix is $n \times n$. The input mass at u_1 for $\hat{Q}_2 = 0$ is

$$\bar{m}_1 = \frac{1}{Y_{11}} \tag{3.372}$$

Now let us consider a second primed system with an input u_2' and an output u_3'. For this system, we have

$$\Delta u_2' = Y_{22}' \hat{Q}_2' + Y_{23}' \hat{Q}_3' \tag{3.373}$$
$$\Delta u_3' = Y_{32}' \hat{Q}_2' + Y_{33}' \hat{Q}_3' \tag{3.374}$$

Finally, let us connect the two systems in accordance with the constraint equation

$$\Delta u_2 = \Delta u_2' \tag{3.375}$$

as shown in Fig. 3.15b. Then the impulse $\hat{Q}_2$ is actually due to the constraint and is not imposed from outside the system. Furthermore, due to Newton's law of action and reaction, we have

$$\hat{Q}_2' = -\hat{Q}_2 \tag{3.376}$$

Equating the right-hand sides of (3.371) and (3.373), and setting $\hat{Q}_3'$ equal to zero, we obtain

$$Y_{21} \hat{Q}_1 + Y_{22} \hat{Q}_2 = Y_{22}' \hat{Q}_2' = -Y_{22}' \hat{Q}_2$$

or

$$\hat{Q}_2 = -\frac{Y_{21}}{Y_{22} + Y_{22}'} \hat{Q}_1 \tag{3.377}$$

Then, using (3.370), we have

$$\Delta u_1 = \left(Y_{11} - \frac{Y_{12} Y_{21}}{Y_{22} + Y_{22}'} \right) \hat{Q}_1 \tag{3.378}$$

Since the $\mathbf{Y}$ matrix is symmetric, the input or driving-point mass for the combined system is

$$\bar{m}_1 = \frac{\hat{Q}_1}{\Delta u_1} = \frac{Y_{22} + Y_{22}'}{Y_{11}(Y_{22} + Y_{22}') - Y_{12}^2} \tag{3.379}$$

In general, the self-admittances Y_{11}, Y_{22} and Y_{22}' are positive and Y_{12} is nonzero. Hence, the input mass at u_1 is increased, in general, by connecting a second system. The exceptions are that $\bar{m}_1$ is unchanged if $Y_{12} = 0$, or if the input mass of the second system is zero, that is, Y_{22}' is infinite.

Notice that the effect of connecting the second system is to attach its input mass

$$\bar{m}_2' = \frac{1}{Y_{22}'} \tag{3.380}$$

at u_2. This results in an impulse $\hat{Q}_2$ acting on the first system, where

$$\hat{Q}_2 = -\hat{Q}_2' = -\bar{m}_2' \Delta u_2 = -\frac{\Delta u_2}{Y_{22}'} \tag{3.381}$$

Then, upon substituting into (3.370) and (3.371), we can solve for Δu_1 in agreement with (3.378) and obtain $\overline{m}_1$ as before.

The energy input to an *initially motionless* system due to a single impulse $\hat{Q}_i$ at u_i is equal to the work

$$W = \frac{1}{2} \hat{Q}_i \Delta u_i \tag{3.382}$$

where, from (3.369),

$$\Delta u_i = \frac{\hat{Q}_i}{\overline{m}_i} \tag{3.383}$$

Thus, we obtain

$$W = \frac{\hat{Q}_i^2}{2\overline{m}_i} = \frac{1}{2}\overline{m}_i \Delta u_i^2 \tag{3.384}$$

This, by the principle of work and kinetic energy, must equal the total kinetic energy of the system. Hence, we can find the kinetic energy without explicitly solving for all the Δus if we use the concept of input mass. Note that the work done by an impulse on an initially motionless system is always positive or zero.

More generally, the work done by a single impulse $\hat{Q}_i$ on a moving system is

$$W = \hat{Q}_i \left(u_{i0} + \frac{1}{2}\Delta u_i \right) = \hat{Q}_i u_{i0} + \frac{\hat{Q}_i^2}{2\overline{m}_i} \tag{3.385}$$

Again, the work is equal to $\hat{Q}_i$ multiplied by the average velocity at u_i. In this case, however, the work may be positive or negative, depending on the value of u_{i0}, the velocity just before the impulse.

An interesting consequence of the symmetry of the $\mathbf{Y}$ matrix is the *reciprocity of impulsive responses*. Thus, the response Δu_j due to a unit impulse $\hat{Q}_i$ is equal to the response Δu_i due to a unit impulse $\hat{Q}_j$, in accordance with (3.368) and $Y_{ij} = Y_{ji}$.

Example 3.10   Two rods, each of mass m and length l, are connected by a pin joint at B (Fig. 3.16). When the system is in a straight configuration, a transverse impulse $\hat{F}_1$ is applied at A. We wish to solve for the responses Δu_1, Δu_2, Δu_3, as well as the driving-point mass at u_1.

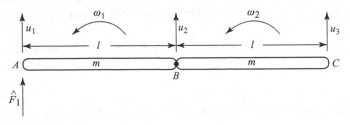

Figure 3.16.

First method   First consider the rod AB separately. The velocity of its center of mass is

$$v_{c1} = \frac{1}{2}(u_1 + u_2) \tag{3.386}$$

and its angular velocity is

$$\omega_1 = \frac{1}{l}(u_2 - u_1) \tag{3.387}$$

Hence, its kinetic energy is

$$T = \frac{m}{8}(u_1 + u_2)^2 + \frac{m}{24}(u_2 - u_1)^2 = \frac{m}{6}\left(u_1^2 + u_2^2 + u_1 u_2\right) \tag{3.388}$$

With the aid of (3.331), namely,

$$m_{ij} = \frac{\partial^2 T}{\partial u_i \partial u_j} \tag{3.389}$$

we obtain the mass matrix

$$\mathbf{m} = \begin{bmatrix} \dfrac{m}{3} & \dfrac{m}{6} \\ \dfrac{m}{6} & \dfrac{m}{3} \end{bmatrix} \tag{3.390}$$

The inverse matrix is

$$\mathbf{Y} = \mathbf{m}^{-1} = \begin{bmatrix} \dfrac{4}{m} & \dfrac{-2}{m} \\ \dfrac{-2}{m} & \dfrac{4}{m} \end{bmatrix} \tag{3.391}$$

These $\mathbf{m}$ and $\mathbf{Y}$ matrices apply to rods AB and BC individually.

If we now connect the two rods, the driving-point mass at u_1 is given by (3.379)

$$\overline{m}_1 = \frac{\hat{F}_1}{\Delta u_1} = \frac{2Y_{22}}{2Y_{11}Y_{22} - Y_{12}^2} = m\left(\frac{8}{32 - 4}\right) = \frac{2}{7}m \tag{3.392}$$

Furthermore,

$$\Delta u_1 = \frac{\hat{F}_1}{\overline{m}_1} = \frac{7\hat{F}_1}{2m} \tag{3.393}$$

Let us return to the basic equation

$$\sum_{j=1}^{n} m_{ij}\Delta u_j = \hat{Q}_i \tag{3.394}$$

and apply it to rod AB in the combined system. We obtain

$$\frac{m}{3}\Delta u_1 + \frac{m}{6}\Delta u_2 = \hat{F}_1 \tag{3.395}$$

Knowing Δu_1, we find that

$$\Delta u_2 = -\frac{\hat{F}_1}{m} \tag{3.396}$$

In addition, for rod AB,

$$\hat{Q}_2 = \frac{m}{6}\Delta u_1 + \frac{m}{3}\Delta u_2 = \frac{1}{4}\hat{F}_1 \tag{3.397}$$

where $\hat{Q}_2$ is actually the constraint impulse applied to rod AB at B.

Finally, for rod BC, noting the law of action and reaction at B, we have

$$\frac{m}{3}\Delta u_2 + \frac{m}{6}\Delta u_3 = -\hat{Q}_2 = -\frac{1}{4}\hat{F}_1 \tag{3.398}$$

from which we obtain

$$\Delta u_3 = \frac{\hat{F}_1}{2m} \tag{3.399}$$

This result is in agreement with the second motion equation for rod BC, namely,

$$\frac{m}{6}\Delta u_2 + \frac{m}{3}\Delta u_3 = 0 \tag{3.400}$$

Second method   Let us consider the complete system from the beginning. Referring to (3.388) and adding over the two rods, the kinetic energy of the system is

$$\begin{aligned}
T &= \frac{m}{6}\left(u_1^2 + u_2^2 + u_1 u_2\right) + \frac{m}{6}\left(u_2^2 + u_3^2 + u_2 u_3\right) \\
&= \frac{m}{6}\left(u_1^2 + 2u_2^2 + u_3^2 + u_1 u_2 + u_2 u_3\right)
\end{aligned} \tag{3.401}$$

Then, using (3.389), the mass matrix is

$$\mathbf{m} = \begin{bmatrix} \dfrac{m}{3} & \dfrac{m}{6} & 0 \\[2mm] \dfrac{m}{6} & \dfrac{2m}{3} & \dfrac{m}{6} \\[2mm] 0 & \dfrac{m}{6} & \dfrac{m}{3} \end{bmatrix} \tag{3.402}$$

and, upon inversion,

$$\mathbf{Y} = \begin{bmatrix} \dfrac{7}{2m} & -\dfrac{1}{m} & \dfrac{1}{2m} \\[2mm] -\dfrac{1}{m} & \dfrac{2}{m} & -\dfrac{1}{m} \\[2mm] \dfrac{1}{2m} & -\dfrac{1}{m} & \dfrac{7}{2m} \end{bmatrix} \tag{3.403}$$

Now we can use the general equation

$$\Delta u_j = \sum_{i=1}^{n} Y_{ji} \hat{Q}_i \qquad (3.404)$$

and note that

$$\hat{Q}_1 = \hat{F}_1, \qquad \hat{Q}_2 = \hat{Q}_3 = 0 \qquad (3.405)$$

Here $\hat{Q}_2$ is equal to zero because no external impulse is applied at B in the u_2 direction. The constraint impulses at B are internal to the complete system.

From the first column of $\mathbf{Y}$ and (3.404), we obtain

$$\Delta u_1 = \frac{7\hat{F}_1}{2m}, \qquad \Delta u_2 = -\frac{\hat{F}_1}{m}, \qquad \Delta u_3 = \frac{\hat{F}_1}{2m} \qquad (3.406)$$

in agreement with our previous results.

The second method appears to be simpler, but it involves the inversion of a larger matrix. The second method avoids calculating constraint impulses, but they are thereby not directly available. Thus, the choice of method depends upon the nature of the desired results.

Example 3.11   As an example of reciprocity, let us consider the angular impulse response of a rigid body. Let xyz be a principal axis system at the center of mass of the body. Suppose an angular impulse $\hat{M}$ is applied about an arbitrary 1-axis which passes through the center of mass. We wish to find the change $\Delta \omega_2$ in the angular velocity component about a second arbitrary 2-axis through the center of mass.

Let (c_{1x}, c_{1y}, c_{1z}) be the direction cosines of the positive 1-axis relative to the xyz frame. Similarly, let (c_{2x}, c_{2y}, c_{2z}) specify the direction of the positive 2-axis. The response components in the body-fixed Cartesian frame are

$$\Delta \omega_x = \frac{c_{1x} \hat{M}}{I_{xx}}, \qquad \Delta \omega_y = \frac{c_{1y} \hat{M}}{I_{yy}}, \qquad \Delta \omega_z = \frac{c_{1z} \hat{M}}{I_{zz}} \qquad (3.407)$$

Now take the component of ω in the direction of the positive 2-axis. We obtain

$$\Delta \omega_2 = c_{2x} \Delta \omega_x + c_{2y} \Delta \omega_y + c_{2z} \Delta \omega_z = \left(\frac{c_{1x} c_{2x}}{I_{xx}} + \frac{c_{1y} c_{2y}}{I_{yy}} + \frac{c_{1z} c_{2z}}{I_{zz}} \right) \hat{M} \qquad (3.408)$$

This is the angular velocity about the 2-axis due to an angular impulse $\hat{M}$ about the 1-axis.

If, on the other hand, we wish to find the response $\Delta \omega_1$ about the 1-axis due to an angular impulse $\hat{M}$ about the 2-axis, we merely interchange the 1 and 2 subscripts in (3.407) and (3.408). But this leaves the right-hand side of (3.408) unchanged. Hence,

$$\Delta \omega_1 = \Delta \omega_2 \qquad (3.409)$$

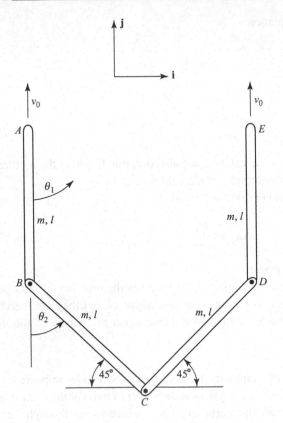

Figure 3.17.

and we confirm the reciprocity relation for the responses to angular impulses about two arbitrary axes through the center of mass of a rigid body.

Example 3.12   A system consists of four rods, each of mass m and length l, which are connected by pin joints, as shown in Fig. 3.17. The system is motionless until ends A and E are suddenly given equal vertical velocities v_0. We wish to find the resulting angular velocities of the rods.

First, notice that the system is symmetrical about a vertical line through C. Thus, we can consider only rods AB and BC which are the left half of the system. From the symmetry, any motion of point C must be vertical and any interaction impulse at C must be horizontal. The velocities of points A and B are

$$\mathbf{v}_A = v_0\mathbf{j} \tag{3.410}$$

$$\mathbf{v}_B = l\dot{\theta}_1\mathbf{i} + v_0\mathbf{j} \tag{3.411}$$

The kinetic energy of ABC is

$$T = T_{AB} + T_{BC} \tag{3.412}$$

where, using (3.339), we find that

$$T_{AB} = \frac{1}{2}mv_0^2 + \frac{1}{6}ml^2\dot{\theta}_1^2 \tag{3.413}$$

$$T_{BC} = \frac{1}{2}m\left(v_0^2 + l^2\dot{\theta}_1^2\right) + \frac{1}{6}ml^2\dot{\theta}_2^2 + \frac{ml}{2\sqrt{2}}\dot{\theta}_2(l\dot{\theta}_1 + v_0) \tag{3.414}$$

Thus,

$$T = mv_0^2 + \frac{2}{3}ml^2\dot{\theta}_1^2 + \frac{1}{6}ml^2\dot{\theta}_2^2 + \frac{ml^2}{2\sqrt{2}}\dot{\theta}_1\dot{\theta}_2 + \frac{mlv_0}{2\sqrt{2}}\dot{\theta}_2 \tag{3.415}$$

The half-system has only one degree of freedom because of the constraint that the horizontal velocity at C is equal to zero. The holonomic constraint equation is

$$l\dot{\theta}_1 + \frac{1}{\sqrt{2}}l\dot{\theta}_2 = 0$$

or

$$\dot{\theta}_2 = -\sqrt{2}\dot{\theta}_1 \tag{3.416}$$

Hence, the kinetic energy is

$$T = \frac{1}{2}ml^2\dot{\theta}_1^2 - \frac{1}{2}mlv_0\dot{\theta}_1 + mv_0^2 \tag{3.417}$$

Note that $\hat{Q}_1$ equals zero and use (3.367) to obtain

$$\Delta p_1 = p_1 = \frac{\partial T}{\partial \dot{\theta}_1} = ml^2\dot{\theta}_1 - \frac{1}{2}mlv_0 = 0 \tag{3.418}$$

which results in

$$\dot{\theta}_1 = \frac{v_0}{2l}, \qquad \dot{\theta}_2 = -\frac{v_0}{\sqrt{2}l} \tag{3.419}$$

By symmetry, rods DE and CD move as mirror images of AB and BC, respectively.

It is interesting to note that by setting $\partial T/\partial \dot{\theta}_1$ equal to zero in (3.418) we apply a stationarity condition on T with respect to changes in the value of $\dot{\theta}_1$. Actually, the kinetic energy is a minimum consistent with the prescribed velocity v_0, in accordance with Kelvin's theorem.

In general, *Kelvin's theorem* states that if a system, initially at rest, is suddenly set into motion by prescribed velocities at some of its points, the actual kinetic energy of the resulting motion is a minimum compared to other kinematically possible motions.

3.5   Bibliography

Ginsberg, J. H. *Advanced Engineering Dynamics*, 2nd edn. Cambridge, UK: Cambridge University Press, 1995.

Greenwood, D. T., *Principles of Dynamics*, 2nd edn. Englewood Cliffs, NJ: Prentice-Hall, 1988.

Kane, T. R. and Levinson, D. A. *Dynamics: Theory and Applications*. New York: McGraw-Hill, 1985.

3.6   Problems

3.1. Let XYZ be an inertially-fixed frame and let xyz be the body-fixed frame of a rigid body. Suppose the initial orientation of the body is given by the axis and angle variables $\mathbf{a} = \mathbf{I} = \mathbf{i}$ and $\phi = 60°$. (a) Assuming that $\dot{\mathbf{a}} = 0.1\mathbf{J}$ and $\dot{\phi} = 0$, find the initial angular velocity $\boldsymbol{\omega}$ of the body, expressed in inertial and in body-fixed components. (b) Find the initial values of C and $\dot{C}$.

3.2. Obtain the rotational equations of motion for a general rigid body, assuming principal axes and using type II Euler angles as generalized coordinates.

3.3. Two spheres, each of mass m, radius r, and moment of inertia I undergo impact, as shown. Just before impact sphere 1 has a velocity v_0 at $45°$ from the x-axis. Sphere 2 is initially motionless and neither sphere is rotating. (a) Assuming smooth spheres and a coefficient of restitution e, solve for the velocities $\mathbf{v}_1$ and $\mathbf{v}_2$ of the two spheres just after impact. (b) Repeat for the case in which there is a coefficient of friction μ between the spheres, as well as a coefficient of restitution e. Also solve for the resulting angular velocities $\boldsymbol{\omega}_1$ and $\boldsymbol{\omega}_2$.

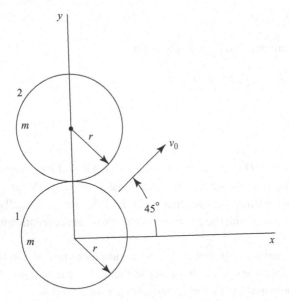

Figure P 3.3.

3.4. An axially-symmetric rocket has constant axial and transverse moments of inertia with $I_t = 5I_a$. Initially the rocket is rotating about the symmetry axis (z-axis) at Ω rad/s. Due to a small thrust misalignment, a constant body-fixed moment M_y is applied to the rocket, where $M_y << I_a\Omega^2$. (a) Assume the initial conditions $\omega_x(0) = \omega_y(0) = 0$ and solve for ω_x and ω_y as functions of time. (b) At what time does the symmetry axis first return to its original orientation?

3.5. Use type I Euler angles to analyze the free rotational motion of a rigid body. (a) Obtain equations for $\dot{\psi}, \dot{\theta}, \dot{\phi}$ in terms of ψ, θ, ϕ, assuming that the **H** vector points upward in the direction of the negative Z-axis. (b) What do these equations reduce to if the rigid body is symmetrical about its x-axis?

3.6. Two slabs, each of mass m_0 can translate on rollers, as shown. The rollers are uniform cylinders, each of mass m and radius r, and they roll without slipping. The applied force F is constant. (a) Obtain the differential equations of motion, using (x_1, x_2) as generalized coordinates. (b) Suppose the system starts from rest with $x_1(0) = x_2(0) = 0$. Assuming that $m_0 = 2m$, solve for x_2 when $x_1 = 1$. (c) Solve for the horizontal forces applied to the upper and lower faces of the lower slab.

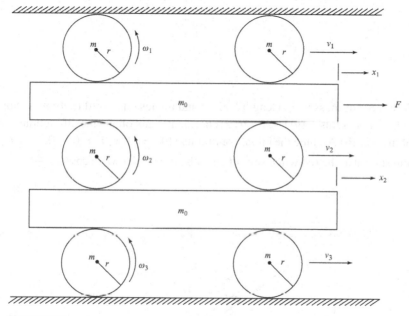

Figure P 3.6.

3.7. A rigid body of mass m has principal moments of inertia $I_{xx} = 3ma^2$, $I_{yy} = 4ma^2$, $I_{zz} = 5ma^2$ about its center of mass. Initially the velocity of the center of mass is zero and the angular velocity components are $\omega_x(0) = \omega_0$, $\omega_y(0) = \omega_z(0) = 0$. Suddenly a point P at $x = y = z = a$ is fixed, although there is free rotation about P. Find (a) the values of $\omega_x, \omega_y, \omega_z$ immediately after impact; (b) the final kinetic energy.

3.8. The center of mass of a thin circular disk of radius r, and with central moments of inertia I_a and I_t, moves with constant speed around a circular path of radius R. The plane of the disk is inclined at a constant angle θ from the vertical. Solve for the required precession rate $\dot{\psi}$.

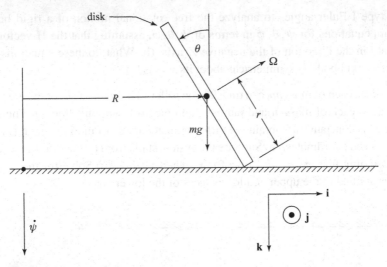

Figure P 3.8.

3.9. A thin rod of mass m and length l slides and rotates on a fixed frictionless horizontal cylinder of radius r. (a) Use (x, θ) as coordinates and obtain the differential equations of motion. (b) Assume the initial conditions $x(0) = \frac{1}{5}l$, $\dot{x}(0) = 0$, $\theta(0) = 0$, $\dot{\theta}(0) = 0$ and show that the rod will lose contact with the cylinder when $\cos \theta = \frac{1}{g}(r\dot{\theta}^2 + 2\dot{x}\dot{\theta})$.

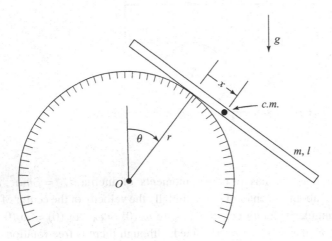

Figure P 3.9.

3.10. An axially symmetric rigid body of mass m has axial and transverse moments of inertia I_a and I_t, where I_t is taken about the fixed point O. The xyz coordinate system has its origin at O and the x-axis is the axis of symmetry. The z-axis remains horizontal and the xy-plane is vertical. (a) Find the θ and ϕ differential equations of motion. (b) Assume the initial conditions $\theta(0) = 60°, \dot{\theta}(0) = \dot{\phi}(0) = 0, \omega_x(0) = \Omega$ and solve for the minimum value of θ during the motion. Assume that $I_a^2 \Omega^2 = 2I_t mgl$.

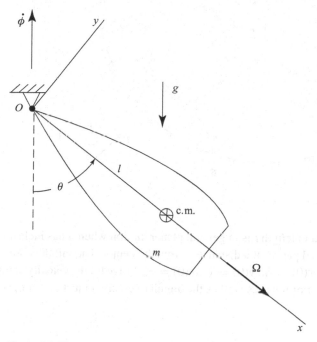

Figure P 3.10.

3.11. A rigid body with xyz principal body axes is suspended by a massless gimbal system associated with type I Euler angles. The outer gimbal can rotate freely about the vertical ψ-axis. Assume that the inner gimbal is locked relative to the outer gimbal, and the Euler angle θ is constant. A motor drives the rigid body relative to the inner gimbal and produces a constant moment Q_ϕ about the x-axis. Assume axial symmetry, that is, $I_{xx} = I_a$, $I_{yy} = I_{zz} = I_t$. (a) Write the differential equations of motion. (b) Show that the vertical angular momentum is constant, even though Q_ϕ has a vertical component. Explain.

3.12. A dumbbell consists of particles A and B, each of mass m, connected by a massless rigid rod of length l (Fig. P 3.12). It can slide without friction on a fixed spherical depression of radius r under the influence of gravity. The configuration of the system is given by the type II Euler angles (ϕ, θ, ψ), where ψ is the angle between the rod AB and the vertical plane which includes the center O and the center of mass. Note that the center of mass of the dumbbell moves on a sphere of radius $R = \sqrt{r^2 - \frac{1}{4}l^2}$. Obtain the equations of motion using the Lagrangian method.

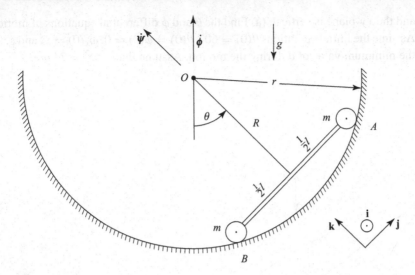

Figure P 3.12.

3.13. A rod of mass m and length l is in general planar motion when it has inelastic impact with a smooth fixed peg P at a distance a from the center. The initial conditions just before impact are $\dot{x}(0) = \dot{x}_0, \dot{y}(0) = \dot{y}_0, \omega(0) = \omega_0$. Solve for the velocity components $(\dot{x}, \dot{y})$ of the center of mass as well as the angular velocity ω just after impact.

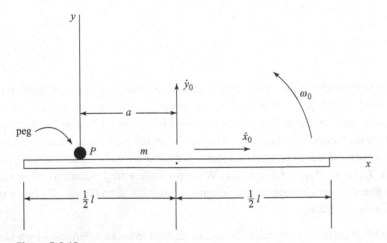

Figure P 3.13.

3.14. The angles (θ, ϕ) describe the orientation of a uniform rod of mass m and length l which is spinning with one end in contact with a smooth floor (Fig. P 3.14). (a) Use the Routhian method to obtain a differential equation of the form $\ddot{\theta}(\theta, \dot{\theta}) = 0$. (b) Assume the initial conditions $\theta(0) = 45°, \dot{\theta}(0) = 0, \dot{\phi}(0) = \omega_0$ and solve for the initial value of the floor reaction N.

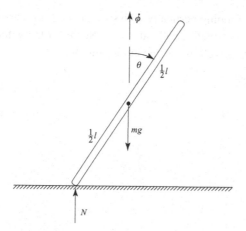

Figure P 3.14.

3.15. An axially-symmetric top with mass m, and with axial and transverse moments of inertia I_a and I_t, respectively, about the center of mass has a total spin rate Ω about its symmetry axis. Its center of mass is located at a distance l from its point O which can slide without friction on a horizontal floor. (a) Use type I Euler angles as coordinates and the modified Euler equations to obtain the θ and ψ equations of motion. (b) Let $I_a = ml^2/12$, $I_t = ml^2/6$, and $\Omega^2 = 72g/l$. Assume that the top is initially vertical and is disturbed slightly. Solve for θ_{min} in the motion which follows. (c) What is the precession rate $\dot\psi$ when $\theta = \theta_{min}$?

3.16. Two thin disks, each of mass m and radius r, are connected by a thin rigid axle of mass m_0 and length L, and roll without slipping on the horizontal xy-plane. (a) Using $(v, \dot\phi)$ as velocity variables, find the kinetic energy and the mass matrix. (b) If the system is initially at rest, solve for v and $\dot\phi$ due to a transverse horizontal impulse $\hat{F}$ which is applied to the axle at a distance c from its center.

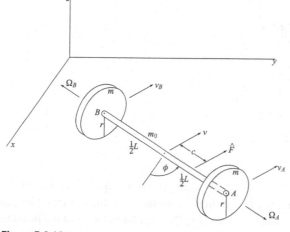

Figure P 3.16.

3.17. A rod of mass m and length l is dropping vertically with speed v_0 and angular velocity ω_0, when end A hits a smooth floor inelastically at $y = 0$. Solve for the values of $\dot{x}$ and $\dot{\theta}$ immediately after impact, as well as the constraint impulse.

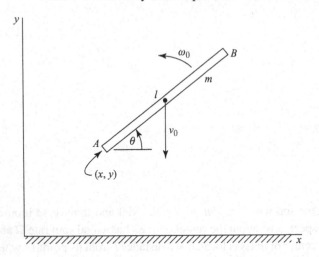

Figure P 3.17.

3.18. A uniform rod of mass m and length l is at rest with $\theta = 45°$ and end B touching a smooth plane surface. Then a transverse impulse $\hat{F}$ is applied at end A. (a) Solve for u_1 and $u_2 \equiv \dot{\theta}$ immediately after the impulse. (b) Find the value of the constraint impulse $\hat{C}$. End B remains in contact with the plane.

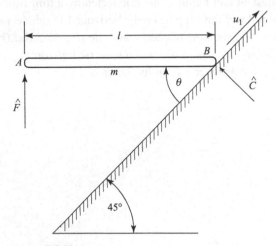

Figure P 3.18.

3.19. A dumbbell is released from the position shown in Fig. P 3.19 and slides without friction down a circular track of radius R. Assume that the dumbbell length $l \ll R$ and keep terms to first order in l/R. Find the angular velocity of the dumbbell when: (a) particle 1 leaves the track; (b) particle 2 leaves the track.

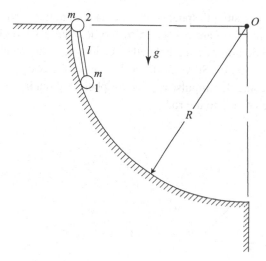

Figure P 3.19.

3.20. A thin disk of mass m and radius r is spinning at $\omega_0 = \sqrt{g/r}$ about a vertical diameter on a frictionless floor. Then it is disturbed slightly. (a) Find the maximum value of the angle θ between the plane of the disk and the vertical in the motion which follows. (b) What is the minimum value of ω_0 for stability of the initial motion?

3.21. A thin horizontal rectangular plate is translating downward (into page) with a velocity v_0 when its corner A is suddenly stopped at a fixed point. Assuming the rigid plate can rotate freely about A, solve for its angular velocity ω and the velocity v_c of its center immediately after impact.

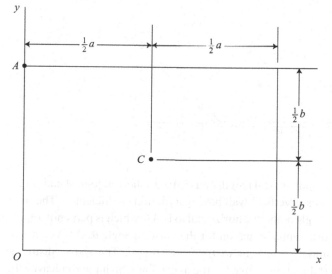

Figure P 3.21.

3.22. A dumbbell is composed of two smooth spheres, each of mass m, radius r, and central moment of inertia $I_c = \frac{2}{5}mr^2$. They are connected by a rigid massless rod of length L. Just before perfectly elastic impact the dumbbell has a longitudinal velocity v_0 at an angle 45° relative to the floor. (a) Solve for the velocities of the two spheres just after impact. (b) What is the angular impulse applied to sphere 1 about its center? (c) Solve for the shear impulse normal to the rod.

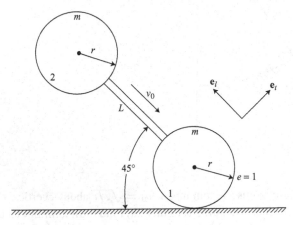

Figure P 3.22.

3.23. A system consists of three rods, each of mass m and length l, which are connected by joints at B and C. The system is initially motionless with $\theta_1 = \theta_3 = 45°$ and $\theta_2 = 0$ when the velocities v_1 and v_2 are suddenly applied, as shown. Solve for the resulting immediate values of $\dot{\theta}_1$, $\dot{\theta}_2$ and $\dot{\theta}_3$.

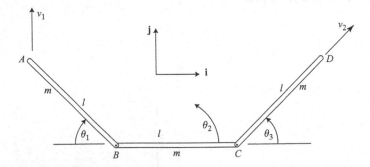

Figure P 3.23.

3.24. A uniform rod of mass m and length l has a fixed spherical joint at end B (Fig. P 3.24). End C rests against a vertical wall having a friction coefficient μ. The rod can rotate with a constant angle α about a horizontal axis AB which is perpendicular to the wall. (a) Obtain the differential equation for the position angle θ. (b) Assume $\mu = 0$, $\alpha = 30°$ with the initial conditions $\theta(0) = 0$, $\dot{\theta}(0) = 0$. The rod is disturbed slightly in the positive θ direction. Solve for the angle θ at which the rod leaves the wall.

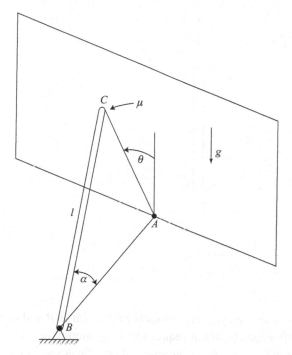

Figure P 3.24.

3.25. End B of a uniform rod of mass m and length l is initially at rest and then is hit by an impulse $\hat{\mathbf{F}}$ at an angle α from the transverse direction, as shown. (a) Find the resulting velocity at B and the driving-point mass $\overline{m}$ associated with $\hat{\mathbf{F}}$. (b) Show that the kinetic energy is $T = \frac{1}{2\overline{m}} \hat{F}^2$.

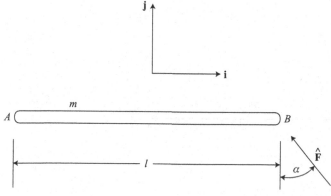

Figure P 3.25.

3.26. Find the input mass at A for the directions u_1 and u_2 (Fig. P 3.26). There is a pin joint at B connecting the two rods.

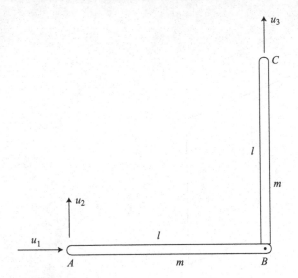

Figure P 3.26.

3.27. Two rods, each of mass m and length l are connected by a joint at B and move in a horizontal plane. A knife-edge constraint requires that any motion at A be perpendicular to the rod. Choose $(x, y, \theta_1, \theta_2)$ as generalized coordinates and obtain three second-order differential equations of motion by using Jourdain's principle.

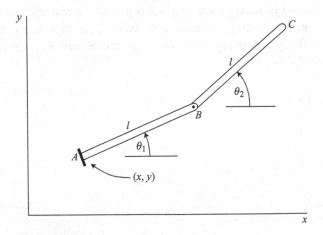

Figure P 3.27.

4 Equations of motion: differential approach

In the previous chapters, we have considered relatively familiar methods of obtaining the differential equations of motion for a mechanical system. In this chapter, we shall introduce a number of other methods, partly in order to give the student a broader view of dynamics, and partly to present some practical and efficient approaches to obtaining the differential equations of motion. The results presented here are applicable to holonomic and nonholonomic systems alike, but the emphasis will be on nonholonomic systems because greater insight is needed in finding applicable theoretical approaches for these systems. As we proceed, we will show the advantages of using quasi-velocities in the analysis of nonholonomic systems.

4.1 Quasi-coordinates and quasi-velocities

Transformation equations

It is often possible to simplify the analysis of dynamical systems by using quasi-velocities (us) rather than $\dot{q}$s as velocity variables. An example is the use of Euler's equations for the rotational motion of a rigid body. Here the velocity variables are the body-axis components of the angular velocity ω rather than Euler angle rates. The ω components are quasi-velocities, whereas the Euler angle rates are true $\dot{q}$s whose time integrals result in generalized coordinates. As we showed in Chapter 3, the Euler equations are simpler than the corresponding Lagrange equations written in terms of Euler angles and Euler angle rates.

The transformation equations relating the us and $\dot{q}$s can be obtained by starting with the differential form

$$d\theta_j = \sum_{i=1}^{n} \Psi_{ji}(q, t)\, dq_i + \Psi_{jt}(q, t)\, dt \qquad (j = 1, \ldots, n) \tag{4.1}$$

where the θs are called *quasi-coordinates* and $d\theta_j = u_j dt$ or $u_j = \dot{\theta}_j$. In general, the right-hand expressions in (4.1) are *not integrable*; if they were integrable, the θs would be true generalized coordinates.

If (4.1) is divided by dt, it has the form

$$u_j = \sum_{i=1}^{n} \Psi_{ji}(q, t)\dot{q}_i + \Psi_{jt}(q, t) \qquad (j = 1, \ldots, n) \tag{4.2}$$

as in (3.245). We assume that these n equations can be solved for the $\dot{q}$s, resulting in

$$\dot{q}_i = \sum_{j=1}^{n} \Phi_{ij}(q, t)u_j + \Phi_{it}(q, t) \qquad (i = 1, \ldots, n) \tag{4.3}$$

Notice that for this unconstrained case, there are the same number of us and $\dot{q}$s. For example, the rotational dynamics of a rigid body might have ω_x, ω_y, ω_z as us and the Euler angles ψ, θ, ϕ as qs. The dynamical equations would be written in terms of us and then the qs would be generated by integrating (4.3) which are kinematic equations.

Constraints

Now let us impose m nonholonomic constraints of the linear form

$$\sum_{i=1}^{n} a_{ji}(q, t)\dot{q}_i + a_{jt}(q, t) = 0 \qquad (j = 1, \ldots, m) \tag{4.4}$$

as in (1.15). This will be accomplished by changing the notation and setting the last m us in (4.2) equal to zero. Thus, we can write

$$u_j = \sum_{i=1}^{n} \Psi_{ji}(q, t)\dot{q}_i + \Psi_{jt}(q, t) \qquad (j = 1, \ldots, n - m) \tag{4.5}$$

$$u_j = \sum_{i=1}^{n} \Psi_{ji}(q, t)\dot{q}_i + \Psi_{jt}(q, t) = 0 \qquad (j = n - m + 1, \ldots, n) \tag{4.6}$$

The first $(n - m)$ us are *independent* quasi-velocities, while the last m us are set equal to zero to enforce the m nonholonomic constraints.

As before, we assume that the $(n \times n)$ Ψ matrix is invertible and we can solve for the $\dot{q}$s in the form

$$\dot{q}_i = \sum_{j=1}^{n-m} \Phi_{ij}(q, t)u_j + \Phi_{it}(q, t) \qquad (i = 1, \ldots, n) \tag{4.7}$$

The upper limit on the summation is $(n - m)$ because the last m us are equal to zero. Note that the $(n \times n)$ matrix Φ is

$$\Phi = \Psi^{-1} \tag{4.8}$$

Furthermore,

$$\Phi_{it} = -\sum_{j=1}^{n} \Phi_{ij}\Psi_{jt} \qquad (i = 1, \ldots, n) \tag{4.9}$$

If there are any holonomic constraints, we have

$$\phi_j(q,t) = 0 \tag{4.10}$$

and (4.6) applies, where

$$\Psi_{ji} = \frac{\partial \phi_j}{\partial q_i}, \qquad \Psi_{jt} = \frac{\partial \phi_j}{\partial t} \tag{4.11}$$

Using the notation of (4.5) and (4.6), virtual displacements are constrained in accordance with

$$\delta \theta_j = \sum_{i=1}^{n} \Psi_{ji}(q,t)\delta q_i = 0 \qquad (j = n - m + 1, \ldots, n) \tag{4.12}$$

In velocity space, we have

$$\sum_{i=1}^{n} \Psi_{ji}(q,t)\delta w_i = 0 \qquad (j = n - m + 1, \ldots, n) \tag{4.13}$$

implying that any virtual velocity $\delta \mathbf{w}$ lies in the common intersection of the constraint planes in velocity space.

4.2   Maggi's equation

Derivation of Maggi's equation

Consider a mechanical system whose configuration is described by n qs and which has m nonholonomic constraints with the linear form of (4.6). One approach to the problem of obtaining differential equations of motion is to use Lagrange's equation in the multiplier form of (2.49), namely,

$$\frac{d}{dt}\left(\frac{\partial T}{\partial \dot{q}_i}\right) - \frac{\partial T}{\partial q_i} = Q_i + \sum_{j=n-m+1}^{n} \lambda_j \Psi_{ji} \qquad (i = 1, \ldots, n) \tag{4.14}$$

These n second-order equations plus the m first-order constraint equations comprise a total of $(n + m)$ equations to solve for the n qs and m λs as functions of time. Frequently, however, one is not interested in the λ solutions, and would prefer a method which eliminates the λs from the beginning. The use of Maggi's equation is one such method.

Let us begin with Lagrange's principle, that is,

$$\sum_{i=1}^{n} \left[\frac{d}{dt}\left(\frac{\partial T}{\partial \dot{q}_i}\right) - \frac{\partial T}{\partial q_i} - Q_i \right] \delta q_i = 0 \tag{4.15}$$

where the δqs satisfy the virtual constraints of (4.12) and we note that the kinetic energy $T(q, \dot{q}, t)$ is written for the *unconstrained* system. The virtual displacements of quasi-coordinates and true coordinates are related by

$$\delta \theta_j = \sum_{i=1}^{n} \Psi_{ji}(q,t)\delta q_i \qquad (j = 1, \ldots, n) \tag{4.16}$$

or, upon inversion,

$$\delta q_i = \sum_{j=1}^{n-m} \Phi_{ij}(q,t)\delta\theta_j \qquad (i=1,\ldots,n) \tag{4.17}$$

where, from (4.12), the last m $\delta\theta$s are equal to zero.

Now substitute this expression for δq_i into (4.15). We obtain

$$\sum_{j=1}^{n-m}\sum_{i=1}^{n}\left[\frac{d}{dt}\left(\frac{\partial T}{\partial \dot{q}_i}\right) - \frac{\partial T}{\partial q_i} - Q_i\right]\Phi_{ij}\delta\theta_j = 0 \tag{4.18}$$

The first $(n-m)$ $\delta\theta$s are *independent* so each coefficient must equal zero. The result is *Maggi's equation* (1896).

$$\sum_{i=1}^{n}\left[\frac{d}{dt}\left(\frac{\partial T}{\partial \dot{q}_i}\right) - \frac{\partial T}{\partial q_i}\right]\Phi_{ij} = \sum_{i=1}^{n}Q_i\Phi_{ij} \qquad (j=1,\ldots,n-m) \tag{4.19}$$

The term on the right represents the generalized applied force associated with the quasi-coordinate $\delta\theta_j$. The set of Maggi equations are $(n-m)$ second-order differential equations in $(q,\dot{q},\ddot{q},t)$. In addition, there are m equations of constraint, making a total of n equations to solve for the n qs as functions of time.

The Ψ_{ji} coefficients are not unique but are chosen in such a way that the first $(n-m)$ $\delta\theta$s are independent and are consistent with the constraints. In the resulting Φ matrix, each column represents the amplitude ratios of an independent set of δqs which could be used in Lagrange's principle, equation (4.15). Maggi's equation uses the first $(n-m)$ columns of Φ to obtain the $(n-m)$ differential equations of motion.

Example 4.1 Two particles, each of mass m, are connected by a massless rod of length l (Fig. 4.1). There is a knife-edge constraint at particle 1 which prevents a velocity component perpendicular to the rod at that point. We wish to use Maggi's equation to find the differential equations of motion.

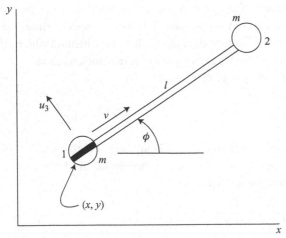

Figure 4.1.

Let (x, y, ϕ) be chosen as generalized coordinates. As quasi-velocities (us) let us take

$$u_1 = v = \dot{x} \cos \phi + \dot{y} \sin \phi \qquad (4.20)$$

$$u_2 = \dot{\phi} \qquad (4.21)$$

$$u_3 = -\dot{x} \sin \phi + \dot{y} \cos \phi = 0 \qquad (4.22)$$

where $u_3 = 0$ is the nonholonomic constraint equation and (u_1, u_2) are independent. Thus, we obtain the coefficient matrix

$$\mathbf{\Psi} = \begin{bmatrix} \cos \phi & \sin \phi & 0 \\ 0 & 0 & 1 \\ -\sin \phi & \cos \phi & 0 \end{bmatrix}$$

$$(4.23)$$

for the unconstrained system. The inverse matrix is

$$\mathbf{\Phi} = \mathbf{\Psi}^{-1} = \begin{bmatrix} \cos \phi & 0 & -\sin \phi \\ \sin \phi & 0 & \cos \phi \\ 0 & 1 & 0 \end{bmatrix}$$

$$(4.24)$$

We shall use Maggi's equation, (4.19), in the form

$$\sum_{i=1}^{3} \left[\frac{d}{dt} \left(\frac{\partial T}{\partial \dot{q}_i} \right) - \frac{\partial T}{\partial q_i} \right] \Phi_{ij} = 0 \qquad (j = 1, 2) \qquad (4.25)$$

because all Qs are equal to zero. The unconstrained kinetic energy is obtained by using (3.140). It is

$$T = m \left(\dot{x}^2 + \dot{y}^2 + \frac{1}{2} l^2 \dot{\phi}^2 - l\dot{x}\dot{\phi} \sin \phi + l\dot{y}\dot{\phi} \cos \phi \right) \qquad (4.26)$$

We find that

$$\frac{d}{dt} \left(\frac{\partial T}{\partial \dot{x}} \right) = m(2\ddot{x} - l\ddot{\phi} \sin \phi - l\dot{\phi}^2 \cos \phi), \qquad \frac{\partial T}{\partial x} = 0 \qquad (4.27)$$

$$\frac{d}{dt} \left(\frac{\partial T}{\partial \dot{y}} \right) = m(2\ddot{y} + l\ddot{\phi} \cos \phi - l\dot{\phi}^2 \sin \phi), \qquad \frac{\partial T}{\partial y} = 0 \qquad (4.28)$$

$$\frac{d}{dt} \left(\frac{\partial T}{\partial \dot{\phi}} \right) = m(l^2 \ddot{\phi} - l\ddot{x} \sin \phi + l\ddot{y} \cos \phi - l\dot{x}\dot{\phi} \cos \phi - l\dot{y}\dot{\phi} \sin \phi) \qquad (4.29)$$

$$\frac{\partial T}{\partial \phi} = m(-l\dot{x}\dot{\phi} \cos \phi - l\dot{y}\dot{\phi} \sin \phi) \qquad (4.30)$$

The first equation of motion, using Maggi's equation and involving (4.27), (4.28) and the first column of the $\mathbf{\Phi}$ matrix, is

$$m(2\ddot{x} - l\ddot{\phi} \sin \phi - l\dot{\phi}^2 \cos \phi) \cos \phi + m(2\ddot{y} + l\ddot{\phi} \cos \phi - l\dot{\phi}^2 \sin \phi) \sin \phi = 0$$

or

$$m(2\ddot{x}\cos\phi + 2\ddot{y}\sin\phi - l\dot{\phi}^2) = 0 \tag{4.31}$$

The second equation of motion, involving the second column of the Φ matrix, is

$$\frac{d}{dt}\left(\frac{\partial T}{\partial\dot{\phi}}\right) - \frac{\partial T}{\partial\phi} = m(l^2\ddot{\phi} - l\ddot{x}\sin\phi + l\ddot{y}\cos\phi) = 0 \tag{4.32}$$

In addition, (4.22), the constraint equation, can be differentiated with respect to time to yield,

$$\ddot{x}\sin\phi - \ddot{y}\cos\phi + \dot{x}\dot{\phi}\cos\phi + \dot{y}\dot{\phi}\sin\phi = 0 \tag{4.33}$$

In equations (4.31)–(4.33) we have three equations linear in the $\ddot{q}$s. Thus, we can solve for $\ddot{x}$, $\ddot{y}$, $\ddot{\phi}$ which are then integrated numerically to obtain the response as a function of time.

Example 4.2 Now consider a more complex problem, the rolling motion of a disk on a horizontal plane. Suppose that the classical Euler angles (ϕ, θ, ψ) are used to specify the orientation of a disk of mass m and radius r, whose contact point C has Cartesian coordinates (x, y), (Fig. 4.2).

Let us choose the independent us to be the Euler angle rates. Thus,

$$u_1 = \dot{\phi}, \qquad u_2 = \dot{\theta}, \qquad u_3 = \dot{\psi} \tag{4.34}$$

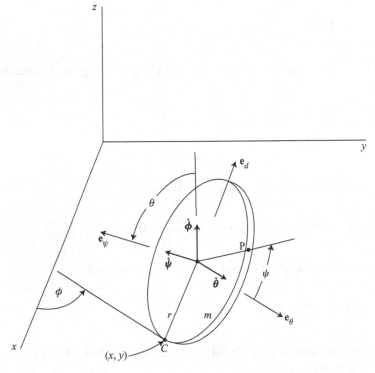

Figure 4.2.

The constraint equations, which enforce no slipping at the contact point, are

$$u_4 = \dot{x} + r\dot{\psi} \cos\phi = 0 \tag{4.35}$$

$$u_5 = \dot{y} + r\dot{\psi} \sin\phi = 0 \tag{4.36}$$

Maggi's equation is

$$\sum_{i=1}^{n} \left[\frac{d}{dt} \left(\frac{\partial T}{\partial \dot{q}_i} \right) - \frac{\partial T}{\partial q_i} \right] \Phi_{ij} = \sum_{i=1}^{n} Q_i \Phi_{ij} \quad (j = 1, \ldots, n - m) \tag{4.37}$$

where, in this case, the only nonzero Q is

$$Q_2 = -mgr \cos\theta \tag{4.38}$$

The coefficients in (4.34)–(4.36) result in

$$\Psi = \begin{bmatrix} 1 & 0 & 0 & 0 & 0 \\ 0 & 1 & 0 & 0 & 0 \\ 0 & 0 & 1 & 0 & 0 \\ 0 & 0 & r\cos\phi & 1 & 0 \\ 0 & 0 & r\sin\phi & 0 & 1 \end{bmatrix} \tag{4.39}$$

for the order $(\dot{\phi}, \dot{\theta}, \dot{\psi}, \dot{x}, \dot{y})$. The inverse matrix is

$$\Phi = \Psi^{-1} = \begin{bmatrix} 1 & 0 & 0 & 0 & 0 \\ 0 & 1 & 0 & 0 & 0 \\ 0 & 0 & 1 & 0 & 0 \\ 0 & 0 & -r\cos\phi & 1 & 0 \\ 0 & 0 & -r\sin\phi & 0 & 1 \end{bmatrix} \tag{4.40}$$

In terms of the Cartesian unit vectors $\mathbf{i}, \mathbf{j}, \mathbf{k}$, the velocity of the center of the unconstrained disk is

$$\mathbf{v} = (\dot{x} - r\dot{\phi}\cos\phi\cos\theta + r\dot{\theta}\sin\phi\sin\theta)\mathbf{i}$$
$$+ (\dot{y} - r\dot{\phi}\sin\phi\cos\theta - r\dot{\theta}\cos\phi\sin\theta)\mathbf{j} + r\dot{\theta}\cos\theta\,\mathbf{k} \tag{4.41}$$

The angular velocity is

$$\omega = (\dot{\phi}\cos\theta + \dot{\psi})\mathbf{e}_\psi + \dot{\theta}\mathbf{e}_\theta + \dot{\phi}\sin\theta\,\mathbf{e}_d \tag{4.42}$$

In accordance with Koenig's theorem, the kinetic energy is

$$T = \frac{1}{2}mv^2 + \frac{1}{2}\omega \cdot \mathbf{I} \cdot \omega \tag{4.43}$$

where the axial and transverse moments of inertia are

$$I_a = \frac{1}{2}mr^2, \quad I_t = \frac{1}{4}mr^2 \tag{4.44}$$

about the center of the disk. Thus, the total unconstrained kinetic energy is

$$\begin{aligned}
T = &\frac{1}{2}m(\dot{x}^2 + \dot{y}^2) + \frac{1}{8}mr^2\dot{\phi}^2(1 + 5\cos^2\theta) + \frac{5}{8}mr^2\dot{\theta}^2 + \frac{1}{4}mr^2\dot{\psi}^2 \\
&+ \frac{1}{2}mr^2\dot{\phi}\dot{\psi}\cos\theta - mr\dot{x}\dot{\phi}\cos\phi\cos\theta + mr\dot{x}\dot{\theta}\sin\phi\sin\theta \\
&- mr\dot{y}\dot{\phi}\sin\phi\cos\theta - mr\dot{y}\dot{\theta}\cos\phi\sin\theta
\end{aligned} \tag{4.45}$$

Next, let us apply the Lagrangian operator with respect to each of the generalized coordinates. We obtain

$$\begin{aligned}
\frac{d}{dt}\left(\frac{\partial T}{\partial \dot{\phi}}\right) - \frac{\partial T}{\partial \phi} = &\frac{1}{4}mr^2\ddot{\phi}(1 + 5\cos^2\theta) + \frac{1}{2}mr^2\ddot{\psi}\cos\theta \\
&- mr\ddot{x}\cos\phi\cos\theta - mr\ddot{y}\sin\phi\cos\theta \\
&- \frac{5}{2}mr^2\dot{\phi}\dot{\theta}\sin\theta\cos\theta - \frac{1}{2}mr^2\dot{\theta}\dot{\psi}\sin\theta
\end{aligned} \tag{4.46}$$

$$\begin{aligned}
\frac{d}{dt}\left(\frac{\partial T}{\partial \dot{\theta}}\right) - \frac{\partial T}{\partial \theta} = &\frac{5}{4}mr^2\ddot{\theta} + mr\ddot{x}\sin\phi\sin\theta - mr\ddot{y}\cos\phi\sin\theta \\
&+ \frac{5}{4}mr^2\dot{\phi}^2\sin\theta\cos\theta + \frac{1}{2}mr^2\dot{\phi}\dot{\psi}\sin\theta
\end{aligned} \tag{4.47}$$

$$\frac{d}{dt}\left(\frac{\partial T}{\partial \dot{\psi}}\right) - \frac{\partial T}{\partial \psi} = \frac{1}{2}mr^2\ddot{\psi} + \frac{1}{2}mr^2\ddot{\phi}\cos\theta - \frac{1}{2}mr^2\dot{\phi}\dot{\theta}\sin\theta \tag{4.48}$$

$$\begin{aligned}
\frac{d}{dt}\left(\frac{\partial T}{\partial \dot{x}}\right) - \frac{\partial T}{\partial x} = &m\ddot{x} - mr\ddot{\phi}\cos\phi\cos\theta \\
&+ mr\ddot{\theta}\sin\phi\sin\theta + mr\dot{\phi}^2\sin\phi\cos\theta \\
&+ mr\dot{\theta}^2\sin\phi\cos\theta + 2mr\dot{\phi}\dot{\theta}\cos\phi\sin\theta
\end{aligned} \tag{4.49}$$

$$\begin{aligned}
\frac{d}{dt}\left(\frac{\partial T}{\partial \dot{y}}\right) - \frac{\partial T}{\partial y} = &m\ddot{y} - mr\ddot{\phi}\sin\phi\cos\theta \\
&- mr\ddot{\theta}\cos\phi\sin\theta - mr\dot{\phi}^2\cos\phi\cos\theta \\
&- mr\dot{\theta}^2\cos\phi\cos\theta + 2mr\dot{\phi}\dot{\theta}\sin\phi\sin\theta
\end{aligned} \tag{4.50}$$

A substitution into (4.37), using the first three columns of the $\mathbf{\Phi}$ matrix, yields the Maggi equations. The ϕ equation is

$$\begin{aligned}
&\frac{1}{4}mr^2\ddot{\phi}(1 + 5\cos^2\theta) + \frac{1}{2}mr^2\ddot{\psi}\cos\theta - mr\ddot{x}\cos\phi\cos\theta \\
&- mr\ddot{y}\sin\phi\cos\theta - \frac{5}{2}mr^2\dot{\phi}\dot{\theta}\sin\theta\cos\theta - \frac{1}{2}mr^2\dot{\theta}\dot{\psi}\sin\theta = 0
\end{aligned} \tag{4.51}$$

The θ equation is

$$\frac{5}{4}mr^2\ddot{\theta} + mr\ddot{x}\sin\phi\sin\theta - mr\ddot{y}\cos\phi\sin\theta$$

$$+ \frac{5}{4}mr^2\dot{\phi}^2\sin\theta\cos\theta + \frac{1}{2}mr^2\dot{\phi}\dot{\psi}\sin\theta = -mgr\cos\theta \tag{4.52}$$

The ψ equation is

$$\frac{1}{2}mr^2\ddot{\psi} + \frac{3}{2}mr^2\ddot{\phi}\cos\theta - \frac{5}{2}mr^2\dot{\phi}\dot{\theta}\sin\theta - mr(\ddot{x}\cos\phi + \ddot{y}\sin\phi) = 0 \tag{4.53}$$

There are two additional equations obtained by differentiating the constraint equations (4.35) and (4.36) with respect to time.

$$\ddot{x} + r\ddot{\psi}\cos\phi - r\dot{\phi}\dot{\psi}\sin\phi = 0 \tag{4.54}$$
$$\ddot{y} + r\ddot{\psi}\sin\phi + r\dot{\phi}\dot{\psi}\cos\phi = 0 \tag{4.55}$$

In (4.51)–(4.55) we have five differential equations which are linear in the $\ddot{q}$s. They can be solved for the $\ddot{q}$s and then integrated to yield the motion as given by ϕ, θ, ψ, x and y as functions of time.

A simplification can be made if we solve (4.54) for $\ddot{x}$ and (4.55) for $\ddot{y}$, and then substitute into the Maggi equations. In (4.51) we find that

$$-mr\cos\theta(\ddot{x}\cos\phi + \ddot{y}\sin\phi) = mr^2\ddot{\psi}\cos\theta \tag{4.56}$$

and the ϕ equation reduces to

$$\frac{1}{4}mr^2\ddot{\phi}(1 + 5\cos^2\theta) + \frac{3}{2}mr^2\ddot{\psi}\cos\theta$$

$$- \frac{5}{2}mr^2\dot{\phi}\dot{\theta}\sin\theta\cos\theta - \frac{1}{2}mr^2\dot{\theta}\dot{\psi}\sin\theta = 0 \tag{4.57}$$

Similarly,

$$mr\sin\theta(\ddot{x}\sin\phi - \ddot{y}\cos\phi) = mr^2\dot{\phi}\dot{\psi}\sin\theta \tag{4.58}$$

and the θ equation becomes

$$\frac{5}{4}mr^2\ddot{\theta} + \frac{5}{4}mr^2\dot{\phi}^2\sin\theta\cos\theta + \frac{3}{2}mr^2\dot{\phi}\dot{\psi}\sin\theta = -mgr\cos\theta \tag{4.59}$$

Finally, from (4.53) and (4.56), we find that the ψ equation is

$$\frac{3}{2}mr^2\ddot{\psi} + \frac{3}{2}mr^2\ddot{\phi}\cos\theta - \frac{5}{2}mr^2\dot{\phi}\dot{\theta}\sin\theta = 0 \tag{4.60}$$

In (4.57), (4.59), and (4.60) we have reduced the original set of five differential equations to three dynamical equations in the Euler angles. This is a minimum set for this system.

4.3   The Boltzmann–Hamel equation

We have seen that the Lagrangian approach does not produce correct equations of motion if the kinetic energy is expressed in terms of quasi-velocities rather than true velocities. The question arises whether we can add terms to Lagrange's equation to produce correct equations in this more general case. The answer is yes, but at the cost of some additional complications.

Derivation

Let us consider an *unconstrained* system whose kinetic energy is expressed in the form $T(q, \dot{q}, t)$, that is, using true velocities ($\dot{q}$s). We know that the fundamental form of Lagrange's equation applies, namely,

$$\frac{d}{dt}\left(\frac{\partial T}{\partial \dot{q}_i}\right) - \frac{\partial T}{\partial q_i} = Q_i \qquad (i = 1, \ldots, n) \tag{4.61}$$

We wish to introduce quasi-velocities (us) in place of the $\dot{q}$s. To accomplish this aim, let us first assume *scleronomic* transformation equations of the form

$$u_j = \sum_{i=1}^{n} \Psi_{ji}(q)\dot{q}_i \qquad (j = 1, \ldots, n) \tag{4.62}$$

or, conversely,

$$\dot{q}_i = \sum_{j=1}^{n} \Phi_{ij}(q)u_j \qquad (i = 1, \ldots, n) \tag{4.63}$$

The kinetic energy will be the same in value, whether expressed in terms of $\dot{q}$s or us, but will be different in form. So let us write

$$T^*(q, u, t) = T(q, \dot{q}, t) \tag{4.64}$$

where T^* is the kinetic energy expressed in terms of quasi-velocities.

First, let us consider

$$\frac{\partial T}{\partial \dot{q}_i} = \sum_{j=1}^{n} \frac{\partial T^*}{\partial u_j}\frac{\partial u_j}{\partial \dot{q}_i} = \sum_{j=1}^{n} \frac{\partial T^*}{\partial u_j}\Psi_{ji} \tag{4.65}$$

Thus, we obtain

$$\frac{d}{dt}\left(\frac{\partial T}{\partial \dot{q}_i}\right) = \sum_{j=1}^{n} \frac{d}{dt}\left(\frac{\partial T^*}{\partial u_j}\right)\Psi_{ji} + \sum_{j=1}^{n}\sum_{k=1}^{n} \frac{\partial T^*}{\partial u_j}\frac{\partial \Psi_{ji}}{\partial q_k}\dot{q}_k \tag{4.66}$$

Then, using (4.63),

$$\frac{d}{dt}\left(\frac{\partial T}{\partial \dot{q}_i}\right) = \sum_{j=1}^{n} \frac{d}{dt}\left(\frac{\partial T^*}{\partial u_j}\right)\Psi_{ji} + \sum_{j=1}^{n}\sum_{k=1}^{n}\sum_{l=1}^{n} \frac{\partial T^*}{\partial u_j}\frac{\partial \Psi_{ji}}{\partial q_k}\Phi_{kl}u_l \tag{4.67}$$

Furthermore, we see that

$$
\frac{\partial T}{\partial q_i} = \frac{\partial T^*}{\partial q_i} + \sum_{j=1}^{n} \frac{\partial T^*}{\partial u_j} \frac{\partial u_j}{\partial q_i} = \frac{\partial T^*}{\partial q_i} + \sum_{j=1}^{n} \sum_{k=1}^{n} \frac{\partial T^*}{\partial u_j} \frac{\partial \Psi_{jk}}{\partial q_i} \dot{q}_k
$$

$$
= \frac{\partial T^*}{\partial q_i} + \sum_{j=1}^{n} \sum_{k=1}^{n} \sum_{l=1}^{n} \frac{\partial T^*}{\partial u_j} \frac{\partial \Psi_{jk}}{\partial q_i} \Phi_{kl} u_l \tag{4.68}
$$

Thus, starting with Lagrange's equation (4.61), and using (4.67) and (4.68), we obtain

$$
\sum_{j=1}^{n} \frac{d}{dt}\left(\frac{\partial T}{\partial u_j}\right) \Psi_{ji} - \frac{\partial T^*}{\partial q_i} + \sum_{j=1}^{n} \sum_{k=1}^{n} \sum_{l=1}^{n} \frac{\partial T^*}{\partial u_j} \left(\frac{\partial \Psi_{ji}}{\partial q_k} - \frac{\partial \Psi_{jk}}{\partial q_i}\right) \Phi_{kl} u_l = Q_i \tag{4.69}
$$

where Q_i is the generalized applied force associated with q_i.

Next, multiply (4.69) by Φ_{ir} and sum over i. Now

$$
\sum_{i=1}^{n} \Psi_{ji} \Phi_{ir} = \delta_{jr} \tag{4.70}
$$

since $\Phi = \Psi^{-1}$ and δ_{jr} is the Kronecker delta. Thus we obtain the result that

$$
\frac{d}{dt}\left(\frac{\partial T^*}{\partial u_r}\right) - \sum_{i=1}^{n} \frac{\partial T^*}{\partial q_i} \Phi_{ir} + \sum_{i=1}^{n} \sum_{j=1}^{n} \sum_{k=1}^{n} \sum_{l=1}^{n} \frac{\partial T^*}{\partial u_j} \left(\frac{\partial \Psi_{ji}}{\partial q_k} - \frac{\partial \Psi_{jk}}{\partial q_i}\right) \Phi_{kl} \Phi_{ir} u_l = Q_r^*
$$
$$
(r = 1, \ldots, n) \tag{4.71}
$$

where

$$
Q_r^* = \sum_{i=1}^{n} Q_i \frac{\partial \dot{q}_i}{\partial u_r} = \sum_{i=1}^{n} Q_i \Phi_{ir} \tag{4.72}
$$

is the generalized applied force associated with the quasi-velocity u_r.

Equation (4.71) is the basic result, giving the n differential equations of motion in terms of quasi-velocities. For convenience, however, we shall introduce notation which will give it a simpler appearance. First, we define the *Hamel coefficients*

$$
\gamma_{rl}^{j}(q) = -\gamma_{lr}^{j} \equiv \sum_{i=1}^{n} \sum_{k=1}^{n} \left(\frac{\partial \Psi_{ji}}{\partial q_k} - \frac{\partial \Psi_{jk}}{\partial q_i}\right) \Phi_{kl} \Phi_{ir} \tag{4.73}
$$

Let us use the notation

$$
\frac{\partial T^*}{\partial \theta_r} = \sum_{i=1}^{n} \frac{\partial T^*}{\partial q_i} \frac{\partial q_i}{\partial \theta_r} = \sum_{i=1}^{n} \frac{\partial T^*}{\partial q_i} \frac{\partial \dot{q}_i}{\partial u_r} = \sum_{i=1}^{n} \frac{\partial T^*}{\partial q_i} \Phi_{ir} \tag{4.74}
$$

Finally, for convenience, let us drop the asterisks, but assume that $T = T(q, u, t)$. Then (4.71) has the form

$$
\frac{d}{dt}\left(\frac{\partial T}{\partial u_r}\right) - \frac{\partial T}{\partial \theta_r} + \sum_{j=1}^{n} \sum_{l=1}^{n} \frac{\partial T}{\partial u_j} \gamma_{rl}^{j} u_l = Q_r \qquad (r = 1, \ldots, n) \tag{4.75}
$$

This is the *Boltzmann–Hamel equation*, published in 1904, for systems described in terms of quasi-velocities, and assuming independent qs and us. Notice that the added term, compared to Lagrange's equation, actually involves a quadruple summation, implying increased complexity. On the other hand, if the us are true velocities, then (4.62) is integrable for all j and thus all the γ_{rl}^{j} parameters vanish. Then (4.75) reduces to Lagrange's equation with the θs representing true coordinates.

Now consider a *scleronomic* system with m *nonholonomic* constraints. The last m us are chosen such that the constraints are applied by setting these m us equal to zero, that is,

$$u_j = \sum_{i=1}^{n} \Psi_{ji}(q)\,\dot{q}_i = 0 \quad (j = n - m + 1, \ldots, n) \tag{4.76}$$

Equation (4.75) is still valid, except that there are now only $(n - m)$ independent us, corresponding to the $(n - m)$ degrees of freedom. Thus, we have

$$\frac{d}{dt}\left(\frac{\partial T}{\partial u_r}\right) - \frac{\partial T}{\partial \theta_r} + \sum_{j=1}^{n}\sum_{l=1}^{n-m} \frac{\partial T}{\partial u_j}\gamma_{rl}^{j} u_l = Q_r \quad (r = 1, \ldots, n - m) \tag{4.77}$$

In addition to these $(n - m)$ first-order dynamical equations, there are n first-order kinematical equations of the form

$$\dot{q}_i = \sum_{j=1}^{n-m} \Phi_{ij}(q)u_j \quad (i = 1, \ldots, n) \tag{4.78}$$

Thus, there are a total of $(2n - m)$ first-order differential equations to solve for the n qs and the $(n - m)$ nonzero us.

This is a minimal set of equations. One should note that the kinetic energy $T(q, u, t)$ must be written for the full, *unconstrained* set of n us. All n partial derivatives of the form $\partial T/\partial u_j$ must be calculated. After this has been completed, the last m us can be set equal to zero.

A generalization of the Boltzmann–Hamel equation can be obtained for systems in which the equations for the us have the *rheonomic* form

$$u_j = \sum_{i=1}^{n} \Psi_{ji}(q, t)\,\dot{q}_i + \Psi_{jt}(q, t) \quad (j = 1, \ldots, n - m) \tag{4.79}$$

$$u_j = \sum_{i=1}^{n} \Psi_{ji}(q, t)\,\dot{q}_i + \Psi_{jt}(q, t) = 0 \quad (j = n - m + 1, \ldots, n) \tag{4.80}$$

Again, the last m equations represent *nonholonomic* constraints, in general. In addition, we have

$$\dot{q}_i = \sum_{j=1}^{n-m} \Phi_{ij}(q, t)u_j + \Phi_{it}(q, t) \quad (i = 1, \ldots, n) \tag{4.81}$$

A derivation of the dynamical equations for this more general case, using procedures similar to those employed in obtaining (4.71), results in

$$
\frac{d}{dt}\left(\frac{\partial T}{\partial u_r}\right) - \frac{\partial T}{\partial \theta_r} + \sum_{i=1}^{n}\sum_{j=1}^{n}\sum_{k=1}^{n}\sum_{l=1}^{n-m} \frac{\partial T}{\partial u_j}\left(\frac{\partial \Psi_{ji}}{\partial q_k} - \frac{\partial \Psi_{jk}}{\partial q_i}\right)\Phi_{kl}\Phi_{ir}u_l
$$

$$
+ \sum_{i=1}^{n}\sum_{j=1}^{n}\sum_{k=1}^{n} \frac{\partial T}{\partial u_j}\left(\frac{\partial \Psi_{ji}}{\partial q_k} - \frac{\partial \Psi_{jk}}{\partial q_i}\right)\Phi_{kt}\Phi_{ir}
$$

$$
+ \sum_{i=1}^{n}\sum_{j=1}^{n} \frac{\partial T}{\partial u_j}\left(\frac{\partial \Psi_{ji}}{\partial t} - \frac{\partial \Psi_{jt}}{\partial q_i}\right)\Phi_{ir} = Q_r \qquad (r = 1,\ldots,n-m) \tag{4.82}
$$

Let us use the notation

$$
\gamma_{rl}^{j}(q,t) = -\gamma_{lr}^{j} = \sum_{i=1}^{n}\sum_{k=1}^{n}\left(\frac{\partial \Psi_{ji}}{\partial q_k} - \frac{\partial \Psi_{jk}}{\partial q_i}\right)\Phi_{kl}\Phi_{ir} \tag{4.83}
$$

$$
\gamma_{r}^{j}(q,t) = \sum_{i=1}^{n}\sum_{k=1}^{n}\left(\frac{\partial \Psi_{ji}}{\partial q_k} - \frac{\partial \Psi_{jk}}{\partial q_i}\right)\Phi_{kt}\Phi_{ir} + \sum_{i=1}^{n}\left(\frac{\partial \Psi_{ji}}{\partial t} - \frac{\partial \Psi_{jt}}{\partial q_i}\right)\Phi_{ir} \tag{4.84}
$$

Then (4.82) takes the form

$$
\frac{d}{dt}\left(\frac{\partial T}{\partial u_r}\right) - \frac{\partial T}{\partial \theta_r} + \sum_{j=1}^{n}\sum_{l=1}^{n-m} \frac{\partial T}{\partial u_j}\gamma_{rl}^{j}u_l + \sum_{j=1}^{n} \frac{\partial T}{\partial u_j}\gamma_{r}^{j} = Q_r \qquad (r = 1,\ldots,n-m) \tag{4.85}
$$

where $T = T(q,u,t)$ is the unconstrained kinetic energy and Q_r is the generalized applied force associated with u_r. Additionally, recall that we use the notation

$$
\frac{\partial T}{\partial \theta_r} = \sum_{i=1}^{n}\frac{\partial T}{\partial q_i}\Phi_{ir} \tag{4.86}
$$

Equation (4.85) is the *generalized form of the Boltzmann–Hamel equation*. It represents a minimum set of $(n-m)$ first-order dynamical equations in which the velocity variables are quasi-velocities. This allows all the us in the final equations to be independent and consistent with the nonholonomic constraints on the $\dot{q}$s.

Note that $\partial T/\partial u_j$ appears several times in (4.85). Its physical significance is that it represents the generalized momentum associated with u_j.

$$
p_j = \frac{\partial T}{\partial u_j} = \sum_{i=1}^{n} m_{ji}(q,t)u_i + a_j(q,t) \qquad (j = 1,\ldots,n) \tag{4.87}
$$

Thus, all the generalized momenta can enter the equations of motion, including those for suppressed us, that is, for $j = n-m+1,\ldots,n$. Furthermore, note that the first term of (4.85) is the source of all the $\dot{u}$ terms in the equations of motion. The equations are linear in the $\dot{u}$s and have inertia coefficients given by

$$
m_{ij} = m_{ji} = \frac{\partial^2 T}{\partial u_i \partial u_j} \tag{4.88}
$$

Example 4.3 Let us derive the Euler equations of rotational motion for a rigid body using the Boltzmann–Hamel equation. Here we have an unconstrained system described in terms of body-axis angular velocity components which are quasi-velocities.

Assume an xyz principal axis system at the center of mass. Choose type I Euler angles (ψ, θ, ϕ) as qs. The quasi-velocities are

$$
\begin{aligned}
u_1 &= \omega_x = -\dot{\psi} \sin \theta + \dot{\phi} \\
u_2 &= \omega_y = \dot{\psi} \cos \theta \sin \phi + \dot{\theta} \cos \phi \\
u_3 &= \omega_z = \dot{\psi} \cos \theta \cos \phi - \dot{\theta} \sin \phi
\end{aligned}
\tag{4.89}
$$

This results in a coefficient matrix

$$
\Psi = \begin{bmatrix} -\sin \theta & 0 & 1 \\ \cos \theta \sin \phi & \cos \phi & 0 \\ \cos \theta \cos \phi & -\sin \phi & 0 \end{bmatrix}
\tag{4.90}
$$

and the inverse matrix

$$
\Phi = \Psi^{-1} = \begin{bmatrix} 0 & \sec \theta \sin \phi & \sec \theta \cos \phi \\ 0 & \cos \phi & -\sin \phi \\ 1 & \tan \theta \sin \phi & \tan \theta \cos \phi \end{bmatrix}
\tag{4.91}
$$

Let us use the scleronomic form of the Boltzmann–Hamel equation for unconstrained systems, namely,

$$
\frac{d}{dt}\left(\frac{\partial T}{\partial u_r}\right) - \frac{\partial T}{\partial \theta_r} + \sum_{j=1}^{n}\sum_{l=1}^{n} \frac{\partial T}{\partial u_j} \gamma_{rl}^{j} u_l = Q_r \qquad (r = 1, \ldots, n)
\tag{4.92}
$$

where the Qs are moments about the body axes.

The rotational kinetic energy is

$$
T = \frac{1}{2}\left(I_{xx}\omega_x^2 + I_{yy}\omega_y^2 + I_{zz}\omega_z^2\right)
\tag{4.93}
$$

and thus we obtain

$$
\frac{\partial T}{\partial \omega_x} = I_{xx}\omega_x, \qquad \frac{\partial T}{\partial \omega_y} = I_{yy}\omega_y, \qquad \frac{\partial T}{\partial \omega_z} = I_{zz}\omega_z
\tag{4.94}
$$

In the process of evaluating the γ_{rl}^{j} coefficients, let us employ the notation

$$
C_{ik}^{j} = \frac{\partial \Psi_{ji}}{\partial q_k} - \frac{\partial \Psi_{jk}}{\partial q_i}
$$

Then, from (4.73), we find that

$$
\gamma_{rl}^{j} = \sum_{i=1}^{n}\sum_{k=1}^{n} C_{ik}^{j} \Phi_{ir} \Phi_{kl}
\tag{4.95}
$$

or, using matrix notation,

$$\gamma^j = \Phi^T C^j \Phi \tag{4.96}$$

For the order (ψ, θ, ϕ), and with $j = 1$, we find that

$$C^1 = \begin{bmatrix} 0 & -\cos\theta & 0 \\ \cos\theta & 0 & 0 \\ 0 & 0 & 0 \end{bmatrix} \tag{4.97}$$

For $j = 2$, we have

$$C^2 = \begin{bmatrix} 0 & -\sin\theta\sin\phi & \cos\theta\cos\phi \\ \sin\theta\sin\phi & 0 & -\sin\phi \\ -\cos\theta\cos\phi & \sin\phi & 0 \end{bmatrix} \tag{4.98}$$

For $j = 3$,

$$C^3 = \begin{bmatrix} 0 & -\sin\theta\cos\phi & -\cos\theta\sin\phi \\ \sin\theta\cos\phi & 0 & -\cos\phi \\ \cos\theta\sin\phi & \cos\phi & 0 \end{bmatrix} \tag{4.99}$$

Then, upon performing the matrix multiplications, we obtain

$$\gamma^1 = \begin{bmatrix} 0 & 0 & 0 \\ 0 & 0 & 1 \\ 0 & -1 & 0 \end{bmatrix} \tag{4.100}$$

$$\gamma^2 = \begin{bmatrix} 0 & 0 & -1 \\ 0 & 0 & 0 \\ 1 & 0 & 0 \end{bmatrix} \tag{4.101}$$

$$\gamma^3 = \begin{bmatrix} 0 & 1 & 0 \\ -1 & 0 & 0 \\ 0 & 0 & 0 \end{bmatrix} \tag{4.102}$$

In γ_{rl}^j, r is the row and l is the column. Notice that the γ matrices are skew-symmetric, and the resulting coupling is gyroscopic in nature.

Now, recalling (4.94), we can evaluate the matrix

$$\sum_{j=1}^{3} \frac{\partial T}{\partial u_j} \gamma^j = \begin{bmatrix} 0 & I_{zz}\omega_z & -I_{yy}\omega_y \\ -I_{zz}\omega_z & 0 & I_{xx}\omega_x \\ I_{yy}\omega_y & -I_{xx}\omega_x & 0 \end{bmatrix}$$

(4.103)

Then, upon substituting into (4.92), we obtain

$$I_{xx}\dot{\omega}_x + (I_{zz} - I_{yy})\omega_y\omega_z = M_x$$
$$I_{yy}\dot{\omega}_y + (I_{xx} - I_{zz})\omega_z\omega_x = M_y$$
$$I_{zz}\dot{\omega}_z + (I_{yy} - I_{xx})\omega_x\omega_y = M_z$$

(4.104)

These are the Euler rotational equations for a rigid body. Thus, a relatively complicated derivation has a rather simple result.

Example 4.4 Consider again the nonholonomic system shown in Fig. 4.1, consisting of a dumbbell with a knife-edge constraint at one of its particles. As quasi-velocities, we choose

$$u_1 = v = \dot{x}\cos\phi + \dot{y}\sin\phi$$

(4.105)

$$u_2 = \dot{\phi}$$

(4.106)

$$u_3 = -\dot{x}\sin\phi + \dot{y}\cos\phi = 0$$

(4.107)

The last equation is the constraint equation. Notice that the first two us can vary independently without violating the constraint.

The kinetic energy of the unconstrained system in terms of quasi-velocities is

$$T = \frac{1}{2}m\left(u_1^2 + u_3^2\right) + \frac{1}{2}m\left[u_1^2 + (u_3 + lu_2)^2\right]$$
$$= m\left(u_1^2 + u_3^2 + \frac{1}{2}l^2u_2^2 + lu_2u_3\right)$$

(4.108)

The Boltzmann–Hamel equation for a nonholonomic scleronomic system has the general form

$$\frac{d}{dt}\left(\frac{\partial T}{\partial u_r}\right) - \frac{\partial T}{\partial \theta_r} + \sum_{j=1}^{n}\sum_{l=1}^{n-m} \frac{\partial T}{\partial u_j}\gamma_{rl}^j u_l = Q_r \qquad (r = 1, \ldots, n-m)$$

(4.109)

Assuming an order (x, y, ϕ) for the qs, the coefficient matrix for (4.105)–(4.107) is

$$\Psi = \begin{bmatrix} \cos\phi & \sin\phi & 0 \\ 0 & 0 & 1 \\ -\sin\phi & \cos\phi & 0 \end{bmatrix}$$

(4.110)

Its inverse is

$$\mathbf{\Phi} = \mathbf{\Psi}^{-1} = \begin{bmatrix} \cos\phi & 0 & -\sin\phi \\ \sin\phi & 0 & \cos\phi \\ 0 & 1 & 0 \end{bmatrix} \tag{4.111}$$

In order to evaluate the γ_{rl}^{j} coefficients, let us again use the notation

$$C_{ik}^{j} = \frac{\partial \Psi_{ji}}{\partial q_k} - \frac{\partial \Psi_{jk}}{\partial q_i} \tag{4.112}$$

Then

$$\gamma_{rl}^{j} = \sum_{i=1}^{n} \sum_{k=1}^{n} C_{ik}^{j} \Phi_{ir} \Phi_{kl} \tag{4.113}$$

or, in matrix notation,

$$\gamma^{j} = \mathbf{\Phi}^{T} \mathbf{C}^{j} \mathbf{\Phi} \tag{4.114}$$

From (4.112), we see that $\mathbf{C}^2 = 0$ because (4.106) is integrable. Also,

$$\mathbf{C}^1 = \begin{bmatrix} 0 & 0 & -\sin\phi \\ 0 & 0 & \cos\phi \\ \sin\phi & -\cos\phi & 0 \end{bmatrix} \tag{4.115}$$

$$\mathbf{C}^3 = \begin{bmatrix} 0 & 0 & -\cos\phi \\ 0 & 0 & -\sin\phi \\ \cos\phi & \sin\phi & 0 \end{bmatrix} \tag{4.116}$$

The γs are obtained from (4.114). After the matrix multiplications, we obtain

$$\gamma^1 = \begin{bmatrix} 0 & 0 & 0 \\ 0 & 0 & -1 \\ 0 & 1 & 0 \end{bmatrix} \tag{4.117}$$

$$\gamma^2 = 0 \tag{4.118}$$

$$\gamma^3 = \begin{bmatrix} 0 & -1 & 0 \\ 1 & 0 & 0 \\ 0 & 0 & 0 \end{bmatrix} \tag{4.119}$$

Using the kinetic energy expression of (4.108), we obtain

$$\frac{\partial T}{\partial u_1} = 2mu_1, \qquad \frac{\partial T}{\partial u_2} = m(l^2 u_2 + l u_3), \qquad \frac{\partial T}{\partial u_3} = m(2u_3 + l u_2) \qquad (4.120)$$

Now we can apply the constraint by setting $u_3 = 0$. Then

$$\sum_{j=1}^{3} \frac{\partial T}{\partial u_j} \gamma^j = \begin{bmatrix} 0 & -ml u_2 & 0 \\ ml u_2 & 0 & -2mu_1 \\ 0 & 2mu_1 & 0 \end{bmatrix} \qquad (4.121)$$

For this example, all Qs equal zero.

Finally, the Boltzmann–Hamel equation (4.109) is used to obtain the two differential equations of motion. They are

$$2m\dot{u}_1 - ml u_2^2 = 0 \quad \text{or} \quad 2m\dot{v} - ml\dot{\phi}^2 = 0 \qquad (4.122)$$

$$ml^2 \dot{u}_2 + ml u_1 u_2 = 0 \quad \text{or} \quad ml^2 \ddot{\phi} + ml v \dot{\phi} = 0 \qquad (4.123)$$

Referring to (4.109), we see that the indices r and l each take values from 1 to $(n - m)$. Thus, only the first $(n - m)$ columns of the $\mathbf{\Phi}$ matrix enter into the equations of motion, and, similarly, only the first $(n - m)$ rows and columns of the γ^j matrices are involved.

We have emphasized that the *unconstrained* kinetic energy is used in the Boltzmann–Hamel equation. The constraints are applied after the differentiations by setting the last m us equal to zero. It turns out, however, that the first two terms of the Boltzmann–Hamel equation are unchanged if the constrained kinetic energy is used in obtaining the $(n-m)$ equations. In other words, for these terms, the order of differentiation and the application of constraints does not matter, but for the remaining terms, the unconstrained kinetic energy must be used.

4.4   The general dynamical equation

In our study of methods for obtaining the differential equations describing the motions of nonholonomical systems, we have considered the multiplier form of Lagrange's equation, as well as the Maggi and Boltzmann–Hamel equations. All of these classical methods have their shortcomings.

Ideally, we would like to be able to use quasi-velocities (us) as velocity variables, and obtain a minimum set of $(n - m)$ first-order dynamical equations for a system with n qs and m independent nonholonomic constraints. In addition, there are n first-order kinematical equations, of the form of (4.81). Thus, ideally we would have $(2n - m)$ first-order differential equations to solve for the n qs and $(n - m)$ us as functions of time. Furthermore, the procedures should not be overly complicated.

The Lagrangian approach results in a full set of n second-order dynamical equations plus the m equations of constraint. Furthermore, the kinetic energy cannot be expressed in terms of quasi-velocities, so us are not present in the equations of motion. This lack of flexibility frequently means that the equations of motion are more complicated than necessary.

The Maggi equations lead to a reduced set of $(n - m)$ second-order dynamical equations compared to the Lagrangian approach, but quasi-velocities do not appear in these equations. In addition, m differentiated constraint equations are needed, making a total of n second-order differential equations to be solved for the n qs as functions of time.

On the other hand, the Boltzmann–Hamel equation produces a minimum set of $(n - m)$ first-order dynamical equations which are written in terms of quasi-velocities. Thus, the final equations have the ideal form. The procedure suffers, however, because the kinetic energy must be written for the unconstrained system having n degrees of freedom rather than the constrained system with $(n - m)$ degrees of freedom, and in addition, the basic equation with its multiple summations is complicated.

In the remaining portion of this chapter, we shall introduce several additional methods of obtaining the dynamical equations of motion, and will look into their computational efficiency, as illustrated by example problems.

D'Alembert's principle

Let us begin with the Lagrangian form of d'Alembert's principle for a system of N particles, as given by (2.5).

$$\sum_{i=1}^{N}(\mathbf{F}_i - m_i\ddot{\mathbf{r}}_i) \cdot \delta\mathbf{r}_i = 0 \tag{4.124}$$

Here $\mathbf{r}_i$ is the position vector of the ith particle and $\mathbf{F}_i$ is the applied force acting on that particle. The virtual displacement $\delta\mathbf{r}_i$ is consistent with the m instantaneous constraints.

Let us assume that the particle velocities are given in terms of $(n - m)$ independent quasi-velocities in accordance with

$$\mathbf{v}_i = \sum_{j=1}^{n-m}\gamma_{ij}(q, t)u_j + \gamma_{it}(q, t) \qquad (i = 1, \ldots, N) \tag{4.125}$$

where the γs are *velocity coefficients*. The virtual displacement $\delta\mathbf{r}_i$ is

$$\delta\mathbf{r}_i = \sum_{j=1}^{n-m}\gamma_{ij}(q, t)\delta\theta_j \tag{4.126}$$

where θ_j is a quasi-coordinate and $u_j = \dot{\theta}_j$. Then (4.124) becomes

$$\sum_{j=1}^{n-m}\sum_{i=1}^{N}(\mathbf{F}_i - m_i\dot{\mathbf{v}}_i) \cdot \gamma_{ij}\delta\theta_j = 0 \tag{4.127}$$

The virtual work of the applied forces is

$$\delta W = \sum_{i=1}^{N}\mathbf{F}_i \cdot \delta\mathbf{r}_i = \sum_{j=1}^{n-m}Q_j\delta\theta_j \tag{4.128}$$

so, using (4.126), we find that the generalized applied force corresponding to u_j or $\delta\theta_j$ is

$$Q_j = \sum_{i=1}^{N} \mathbf{F}_i \cdot \boldsymbol{\gamma}_{ij} \qquad (j = 1, \ldots, n - m) \tag{4.129}$$

The corresponding generalized inertia force is

$$Q_j^* = -\sum_{i=1}^{N} m_i \dot{\mathbf{v}}_i \cdot \boldsymbol{\gamma}_{ij} \qquad (j = 1, \ldots, n - m) \tag{4.130}$$

Then (4.127) can be written in the form

$$\sum_{j=1}^{n-m} (Q_j + Q_j^*)\delta\theta_j = 0 \tag{4.131}$$

Since the $\delta\theta$s are independent for $j = 1, \ldots, n - m$, we obtain

$$Q_j + Q_j^* = 0 \qquad (j = 1, \ldots, n - m) \tag{4.132}$$

These $(n - m)$ equations, written in terms of us and $\dot{u}$s, are sometimes known as *Kane's equations*.

For our purposes, we can write

$$\sum_{i=1}^{N} m_i \dot{\mathbf{v}}_i \cdot \boldsymbol{\gamma}_{ij} = Q_j \qquad (j = 1, \ldots, n - m) \tag{4.133}$$

We shall call this the *general dynamical equation for a system of particles*. As we have seen, it derives directly from d'Alembert's principle. It consists of a minimum set of $(n - m)$ first-order differential equations in the us, since $\dot{\mathbf{v}}_i$, in general, will be a function of $(q, u, \dot{u}, t)$ and is linear in the $\dot{u}$s. In addition, there are n first-order kinematical equations of the form

$$\dot{q}_i = \sum_{j=1}^{n-m} \Phi_{ij}(q, t)u_j + \Phi_{it}(q, t) \qquad (i = 1, \ldots, n) \tag{4.134}$$

Thus, there are a total of $(2n - m)$ first-order equations to solve for the n qs and $(n - m)$ us as functions of time. Notice that the constraint equations do not enter explicitly, but rather implicitly through the choice of independent us. Furthermore, one does not need to solve for the constraint forces.

Rigid body equations

Equation (4.133) can be generalized for the case of a system of N rigid bodies (Fig. 4.3). Suppose that the ith rigid body has a reference point P_i, fixed in the body, a mass m_i, and an inertia dyadic $\mathbf{I}_i$ about P_i. The applied forces acting on the ith body are equivalent to a force $\mathbf{F}_i$ acting at P_i, plus a couple of moment $\mathbf{M}_i$. In terms of quasi-velocities, we can write the velocity of the reference point P_i as

$$\mathbf{v}_i = \sum_{j=1}^{n-m} \boldsymbol{\gamma}_{ij}(q, t)u_j + \boldsymbol{\gamma}_{it}(q, t) \tag{4.135}$$

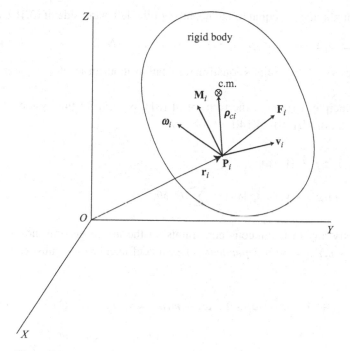

Figure 4.3.

The angular velocity of the ith body is

$$\omega_i = \sum_{j=1}^{n-m} \beta_{ij}(q,t)u_j + \beta_{it}(q,t) \tag{4.136}$$

where the βs are *angular velocity coefficients*. The generalized force associated with u_j is

$$Q_j = \sum_{i=1}^{N}(\mathbf{F}_i \cdot \boldsymbol{\gamma}_{ij} + \mathbf{M}_i \cdot \beta_{ij}) \qquad (j = 1, \ldots, n-m) \tag{4.137}$$

Note that constraint forces do not enter into Q_j if the $(n-m)$ us are independent.

The differential equations of motion are obtained from (4.131) where, for this system of rigid bodies, the generalized inertia force is

$$Q_j^* = \sum_{i=1}^{N}(\mathbf{F}_i^* \cdot \boldsymbol{\gamma}_{ij} + \mathbf{M}_i^* \cdot \beta_{ij}) \qquad (j = 1, \ldots, n-m) \tag{4.138}$$

The inertia force for the ith body is equal to the negative of the mass times the acceleration of the center of mass, that is,

$$\mathbf{F}_i^* = -m_i(\dot{\mathbf{v}}_i + \ddot{\boldsymbol{\rho}}_{ci}) \qquad (i = 1, \ldots, N) \tag{4.139}$$

The inertial moment about P_i is equal to the negative of the left-hand side of (3.167).

$$\mathbf{M}_i^* = -(\mathbf{I}_i \cdot \dot{\boldsymbol{\omega}}_i + \boldsymbol{\omega}_i \times \mathbf{I}_i \cdot \boldsymbol{\omega}_i + m_i \boldsymbol{\rho}_{ci} \times \dot{\mathbf{v}}_i) \qquad (i = 1, \dots, N) \tag{4.140}$$

Recall that (3.167) is essentially Euler's equation for rotation about an accelerating reference point.

D'Alembert's principle, written in the form of (4.131), is valid for this system of rigid bodies, so with the aid of (4.138)–(4.140) we obtain

$$\sum_{j=1}^{n-m} \sum_{i=1}^{N} [m_i(\dot{\mathbf{v}}_i + \ddot{\boldsymbol{\rho}}_{ci}) \cdot \boldsymbol{\gamma}_{ij} + (\mathbf{I}_i \cdot \dot{\boldsymbol{\omega}}_i$$
$$+ \boldsymbol{\omega}_i \times \mathbf{I}_i \cdot \boldsymbol{\omega}_i + m_i \boldsymbol{\rho}_{ci} \times \dot{\mathbf{v}}_i) \cdot \boldsymbol{\beta}_{ij}] \delta\theta_j = \sum_{j=1}^{n-m} Q_j \delta\theta_j \tag{4.141}$$

where the $\delta\theta$s satisfy any instantaneous constraints on the us. In general, however, we assume that the $(n - m)$ $\delta\theta$s are *independent*, so each coefficient of $\delta\theta_j$ must equal zero. Thus, we obtain

$$\sum_{i=1}^{N} [m_i(\dot{\mathbf{v}}_i + \ddot{\boldsymbol{\rho}}_{ci}) \cdot \boldsymbol{\gamma}_{ij} + (\mathbf{I}_i \cdot \dot{\boldsymbol{\omega}}_i + \boldsymbol{\omega}_i \times \mathbf{I}_i \cdot \boldsymbol{\omega}_i + m_i \boldsymbol{\rho}_{ci} \times \dot{\mathbf{v}}_i) \cdot \boldsymbol{\beta}_{ij}] = Q_j \tag{4.142}$$

$$(j = 1, \dots, n - m)$$

where $\boldsymbol{\gamma}_{ij}$ is the velocity coefficient for the reference point P_i and $\boldsymbol{\beta}_{ij}$ is the angular velocity coefficient associated with $\boldsymbol{\omega}_i$. This is the *general dynamical equation for a system of rigid bodies*. It represents a minimum set of $(n - m)$ first-order dynamical equations of the general form

$$m(q, t)\dot{u} + f(q, t) = Q \tag{4.143}$$

In addition, there are n first-order kinematical equations given by (4.134). Thus, there are a total of $(2n - m)$ first-order differential equations to solve for the n qs and $(n - m)$ us as functions of time.

An *alternate form* of the general dynamical equation is

$$\sum_{i=1}^{N} [\dot{\mathbf{p}}_i \cdot \boldsymbol{\gamma}_{ij} + (\dot{\mathbf{H}}_i + m_i \boldsymbol{\rho}_{ci} \times \dot{\mathbf{v}}_i) \cdot \boldsymbol{\beta}_{ij}] = Q_j \qquad (j = 1, \dots, n - m) \tag{4.144}$$

where, in (4.142), the linear momentum rate of the ith body is

$$\dot{\mathbf{p}}_i = m_i(\dot{\mathbf{v}}_i + \ddot{\boldsymbol{\rho}}_{ci}) \qquad (i = 1, \dots, N) \tag{4.145}$$

and its angular momentum rate about P_i is

$$\dot{\mathbf{H}}_i = \mathbf{I}_i \cdot \dot{\boldsymbol{\omega}}_i + \boldsymbol{\omega}_i \times \mathbf{I}_i \cdot \boldsymbol{\omega}_i \qquad (i = 1, \dots, N) \tag{4.146}$$

In (4.146), we assume that the inertia dyadic $\mathbf{I}_i$ is written in terms of body-fixed unit vectors. However, (4.144) is generally valid, so the vectors may be written in terms of any suitable set of unit vectors. For example, if the body is axially symmetric, it is usually advantageous to choose a reference frame other than body axes.

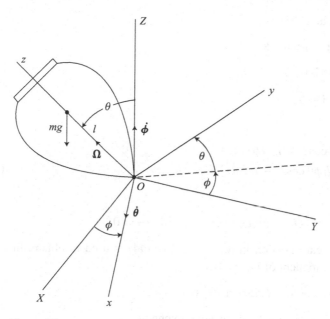

Figure 4.4.

Example 4.5   We can illustrate the use of (4.144) by considering the motion of a top with a fixed point O (Fig. 4.4). Let us use the type II Euler angles ϕ and θ to specify the orientation of the xyz reference frame and its corresponding unit vectors $\mathbf{i}, \mathbf{j}, \mathbf{k}$.

For quasi-velocities, let us choose

$$u_1 = \dot{\phi}, \qquad u_2 = \dot{\theta}, \qquad u_3 = \Omega \tag{4.147}$$

where the first two are actually the time derivatives of true coordinates, and Ω is the total angular velocity about the axis of symmetry z. The angular velocity of the xyz reference frame is

$$\boldsymbol{\omega}_c = \dot{\theta}\mathbf{i} + \dot{\phi}\sin\theta\,\mathbf{j} + \dot{\phi}\cos\theta\,\mathbf{k} \tag{4.148}$$

whereas the angular velocity of the body is

$$\boldsymbol{\omega} = \dot{\theta}\mathbf{i} + \dot{\phi}\sin\theta\,\mathbf{j} + \Omega\mathbf{k} \tag{4.149}$$

The angular velocity of the body relative to the xyz frame is the Euler angle rate $\dot{\psi}$, that is,

$$\Omega = \dot{\phi}\cos\theta + \dot{\psi} \tag{4.150}$$

All the γ_{ij} velocity coefficients are zero because the reference point O is fixed. The βs are

$$\boldsymbol{\beta}_{11} = \frac{\partial\boldsymbol{\omega}}{\partial\dot{\phi}} = \sin\theta\,\mathbf{j}, \qquad \boldsymbol{\beta}_{12} = \frac{\partial\boldsymbol{\omega}}{\partial\dot{\theta}} = \mathbf{i}, \qquad \boldsymbol{\beta}_{13} = \frac{\partial\boldsymbol{\omega}}{\partial\Omega} = \mathbf{k} \tag{4.151}$$

The angular momentum about O is

$$\mathbf{H} = I_t\omega_x\mathbf{i} + I_t\omega_y\mathbf{j} + I_a\omega_z\mathbf{k} = I_t\dot{\theta}\mathbf{i} + I_t\dot{\phi}\sin\theta\,\mathbf{j} + I_a\Omega\mathbf{k} \tag{4.152}$$

The xyz frame rotates at $\boldsymbol{\omega}_c$ so

$$\dot{\mathbf{i}} = \boldsymbol{\omega}_c \times \mathbf{i} = \dot{\phi} \cos\theta \, \mathbf{j} - \dot{\phi} \sin\theta \, \mathbf{k}$$

$$\dot{\mathbf{j}} = \boldsymbol{\omega}_c \times \mathbf{j} = -\dot{\phi} \cos\theta \, \mathbf{i} + \dot{\theta}\mathbf{k} \tag{4.153}$$

$$\dot{\mathbf{k}} = \boldsymbol{\omega}_c \times \mathbf{k} = \dot{\phi} \sin\theta \, \mathbf{i} - \dot{\theta}\mathbf{j}$$

Thus, we find that

$$\dot{\mathbf{H}} = (I_t\ddot{\theta} - I_t\dot{\phi}^2 \sin\theta\cos\theta + I_a\Omega\dot{\phi}\sin\theta)\mathbf{i}$$
$$+ (I_t\ddot{\phi}\sin\theta + 2I_t\dot{\phi}\dot{\theta}\cos\theta - I_a\Omega\dot{\theta})\mathbf{j} + I_a\dot{\Omega}\mathbf{k} \tag{4.154}$$

The Qs are

$$Q_1 = Q_\phi = 0, \qquad Q_2 = Q_\theta = mgl\sin\theta, \qquad Q_3 = Q_\Omega = 0 \tag{4.155}$$

The general dynamical equation in the form of (4.144) is used to obtain, first, the ϕ equation from the $\mathbf{j}$ component of $\dot{\mathbf{H}}$.

$$I_t\ddot{\phi}\sin^2\theta + 2I_t\dot{\phi}\dot{\theta}\sin\theta\cos\theta - I_a\Omega\dot{\theta}\sin\theta = 0 \tag{4.156}$$

Assuming that $\sin\theta \neq 0$, we can divide by $\sin\theta$ to obtain

$$I_t\ddot{\phi}\sin\theta + 2I_t\dot{\phi}\dot{\theta}\cos\theta - I_a\Omega\dot{\theta} = 0 \tag{4.157}$$

Similarly, the θ equation is

$$I_t\ddot{\theta} - I_t\dot{\phi}^2 \sin\theta\cos\theta + I_a\Omega\dot{\phi}\sin\theta = mgl\sin\theta \tag{4.158}$$

Finally, the Ω equation is

$$I_a\dot{\Omega} = 0 \tag{4.159}$$

indicating that Ω is constant in the previous two equations of motion.

The motion of the xyz reference frame has physical significance. The precession rate of the top is $\dot{\phi}$ and its nutation rate is $\dot{\theta}$. The rotation rate of the top relative to the xyz frame is the third Euler angle rate $\dot{\psi}$, given by (4.150). However, because we chose the quasi-velocity Ω rather than $\dot{\psi}$ as u_3, the final equations of motion turned out to be simpler in form than if we had used the Lagrangian approach with $\dot{q}$s as velocity variables. Thus, the flexibility in the choice of us can often lead to a simplified analysis.

Example 4.6 Let us apply the general dynamical equation to the problem of the motion of a thin uniform disk rolling on a horizontal plane (Fig. 4.2). We shall use the general form

$$\sum_{i=1}^{N}[\dot{\mathbf{p}}_i \cdot \boldsymbol{\gamma}_{ij} + (\dot{\mathbf{H}}_i + m_i\boldsymbol{\rho}_{ci} \times \mathbf{v}_i) \cdot \boldsymbol{\beta}_{ij}] = Q_j \qquad (j = 1, \ldots, n-m) \tag{4.160}$$

Choose the reference point of the disk at its center, implying that $\boldsymbol{\rho}_{ci} = 0$. Type II Euler angles specify the orientation of the disk. Its angular velocity is

$$\boldsymbol{\omega} = \dot{\theta}\mathbf{e}_\theta + \omega_d\mathbf{e}_d + \Omega\mathbf{e}_\psi \tag{4.161}$$

where $\omega_d = \dot{\phi} \sin \theta$ and where $\mathbf{e}_\theta$, $\mathbf{e}_d$, $\mathbf{e}_\psi$ are mutually orthogonal unit vectors. Noting that $\mathbf{e}_\theta$ remains horizontal, we find that the angular velocity of the unit vectors is $\dot{\phi} + \dot{\theta}$ or

$$\boldsymbol{\omega}_c = \dot{\phi}(\sin \theta \, \mathbf{e}_d + \cos \theta \, \mathbf{e}_\psi) + \dot{\theta} \mathbf{e}_\theta$$

$$= \dot{\theta}\mathbf{e}_\theta + \omega_d \mathbf{e}_d + \omega_d \cot \theta \, \mathbf{e}_\psi \qquad (4.162)$$

Assuming no slipping, the velocity of the center of mass is

$$\mathbf{v} = -r\Omega \mathbf{e}_\theta + r\dot{\theta}\mathbf{e}_\psi \qquad (4.163)$$

Let the independent us be

$$u_1 = \dot{\theta}, \qquad u_2 = \omega_d, \qquad u_3 = \Omega \qquad (4.164)$$

The γs are

$$\gamma_{11} = \frac{\partial \mathbf{v}}{\partial \dot{\theta}} = r\mathbf{e}_\psi, \qquad \gamma_{12} = \frac{\partial \mathbf{v}}{\partial \omega_d} = 0, \qquad \gamma_{13} = \frac{\partial \mathbf{v}}{\partial \Omega} = -r\mathbf{e}_\theta \qquad (4.165)$$

From (4.161), the βs are

$$\beta_{11} = \frac{\partial \boldsymbol{\omega}}{\partial \dot{\theta}} = \mathbf{e}_\theta, \qquad \beta_{12} = \frac{\partial \boldsymbol{\omega}}{\partial \omega_d} = \mathbf{e}_d, \qquad \beta_{13} = \frac{\partial \boldsymbol{\omega}}{\partial \Omega} = \mathbf{e}_\psi \qquad (4.166)$$

The translational momentum of the disk is

$$\mathbf{p} = m\mathbf{v} = -mr\Omega \mathbf{e}_\theta + mr\dot{\theta}\mathbf{e}_\psi \qquad (4.167)$$

Its moments of inertia are

$$I_a = \frac{1}{2}mr^2, \qquad I_t = \frac{1}{4}mr^2 \qquad (4.168)$$

and the angular momentum about the center is

$$\mathbf{H} = \frac{1}{4}mr^2\dot{\theta}\mathbf{e}_\theta + \frac{1}{4}mr^2\omega_d \mathbf{e}_d + \frac{1}{2}mr^2\Omega \mathbf{e}_\psi \qquad (4.169)$$

In order to find $\dot{\mathbf{p}}$ and $\dot{\mathbf{H}}$, we need first to evaluate the time derivatives of the unit vectors.

$$\dot{\mathbf{e}}_\theta = \boldsymbol{\omega}_c \times \mathbf{e}_\theta = \omega_d \cot \theta \mathbf{e}_d - \omega_d \mathbf{e}_\psi$$

$$\dot{\mathbf{e}}_d = \boldsymbol{\omega}_c \times \mathbf{e}_d = -\omega_d \cot \theta \mathbf{e}_\theta + \dot{\theta}\mathbf{e}_\psi \qquad (4.170)$$

$$\dot{\mathbf{e}}_\psi = \boldsymbol{\omega}_c \times \mathbf{e}_\psi = \omega_d \mathbf{e}_\theta - \dot{\theta}\mathbf{e}_d$$

Then we obtain

$$\dot{\mathbf{p}} = mr[(-\dot{\Omega} + \dot{\theta}\omega_d)\mathbf{e}_\theta - (\dot{\theta}^2 + \Omega\omega_d \cot \theta)\mathbf{e}_d + (\ddot{\theta} + \Omega\omega_d)\mathbf{e}_\psi] \qquad (4.171)$$

$$\dot{\mathbf{H}} = \frac{mr^2}{4}\left[(\ddot{\theta} - \omega_d^2 \cot \theta + 2\Omega\omega_d)\mathbf{e}_\theta + (\dot{\omega}_d + \dot{\theta}\omega_d \cot \theta - 2\Omega\dot{\theta})\mathbf{e}_d + 2\dot{\Omega}\mathbf{e}_\psi\right] \qquad (4.172)$$

The generalized applied forces are

$$Q_1 = -mgr \cos \theta, \qquad Q_2 = 0, \qquad Q_3 = 0 \qquad (4.173)$$

Then, using (4.160), the θ equation is

$$\frac{5}{4}mr^2\ddot{\theta} - \frac{1}{4}mr^2\omega_d^2\cot\theta + \frac{3}{2}mr^2\omega_d\Omega = -mgr\cos\theta \tag{4.174}$$

Similarly, the ω_d equation is

$$\frac{1}{4}mr^2\dot{\omega}_d + \frac{1}{4}mr^2\dot{\theta}\omega_d\cot\theta - \frac{1}{2}mr^2\dot{\theta}\Omega = 0 \tag{4.175}$$

and the Ω equation is

$$\frac{3}{2}mr^2\dot{\Omega} - mr^2\dot{\theta}\omega_d = 0 \tag{4.176}$$

Equations (4.174)–(4.176) are a minimum set of first-order dynamical equations in the us. They are identical with the equations obtained by the Boltzmann–Hamel method, but here they are found by a much simpler procedure. Notice that the constraint equations do not enter explicitly.

The coefficients of the us in the dynamical equations are functions of θ only; and θ can be generated by integrating u_1 with respect to time. So, one can obtain θ as a function of time by integrating four first-order equations. However, to obtain the complete configuration as a function of time, we need a total of five first-order kinematical equations for the $\dot{q}$s. First, we see from (4.162) that $\omega_d = \dot{\phi}\sin\theta$ or

$$\dot{\phi} = \frac{u_2}{\sin\theta} \tag{4.177}$$

Also,

$$\dot{\theta} = u_1 \tag{4.178}$$

and

$$\dot{\psi} = \Omega - \dot{\phi}\cos\theta = -u_2\cot\theta + u_3 \tag{4.179}$$

Finally, the contact point C moves with a speed $r\dot{\psi}$ having the components

$$\dot{x} = -r\dot{\psi}\cos\phi = ru_2\cos\phi\cot\theta - ru_3\cos\phi \tag{4.180}$$
$$\dot{y} = -r\dot{\psi}\sin\phi = ru_2\sin\phi\cot\theta - ru_3\sin\phi \tag{4.181}$$

These five kinematical equations plus the three dynamical equations completely determine the motion of the disk.

For scleronomic systems such as this, the kinematical equations have the matrix form

$$\dot{\mathbf{q}} = \mathbf{\Phi u} \tag{4.182}$$

where $\mathbf{\Phi} = \mathbf{\Psi}^{-1}$ and where $\mathbf{\Psi}$ is the coefficient matrix in

$$\mathbf{u} = \mathbf{\Psi}\dot{\mathbf{q}} \tag{4.183}$$

For this example, we have

$$u_1 = \dot{\theta} \tag{4.184}$$
$$u_2 = \dot{\phi}\sin\theta \tag{4.185}$$

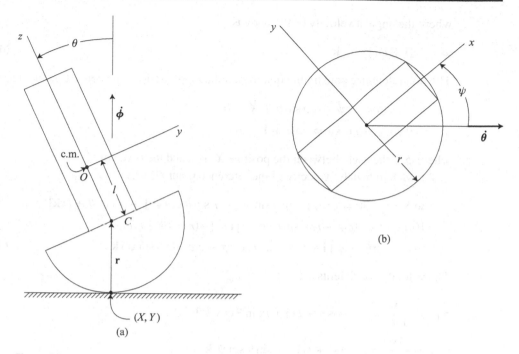

Figure 4.5.

$$u_3 = -\dot{\phi}\cos\theta + \dot{\psi} \tag{4.186}$$

$$u_4 = r\dot{\psi}\cos\phi + \dot{x} = 0 \tag{4.187}$$

$$u_5 = r\dot{\psi}\sin\phi + \dot{y} = 0 \tag{4.188}$$

where the last two equations are the nonholonomic constraint equations, representing the nonslip condition at the contact point C.

Example 4.7 As an example of a more complex nonholonomic system, consider the motion of an unsymmetrical top with a hemispherical base, which rolls without slipping on a horizontal plane (Fig. 4.5). Let us use the general dynamical equation in the form

$$\sum_{i=1}^{N}[m_i(\dot{\mathbf{v}}_i + \ddot{\boldsymbol{\rho}}_{ci})\cdot\boldsymbol{\gamma}_{ij} + (\mathbf{I}_i\cdot\dot{\boldsymbol{\omega}}_i + \boldsymbol{\omega}_i\times\mathbf{I}_i\cdot\boldsymbol{\omega}_i + m_i\boldsymbol{\rho}_{ci}\times\dot{\mathbf{v}}_i)\cdot\boldsymbol{\beta}_{ij}] = Q_j$$

$$(j = 1, \ldots, n-m) \tag{4.189}$$

Choose the reference point O at the center of mass, implying that $\boldsymbol{\rho}_{ci} = 0$. The xyz body axes are principal axes at the center of mass. The orientation of the body is given by type II Euler angles (ϕ, θ, ψ). These three Euler angles plus the inertial location (X, Y) of the contact point on the horizontal plane constitute the five qs.

As independent us, let us choose the quasi-velocities

$$u_1 = \omega_x, \qquad u_2 = \omega_y, \qquad u_3 = \omega_z \tag{4.190}$$

where the angular velocity of the body is

$$\boldsymbol{\omega} = \omega_x \mathbf{i} + \omega_y \mathbf{j} + \omega_z \mathbf{k} \tag{4.191}$$

The nonholonomic constraint equations, indicating no slip at the contact point, are

$$u_4 = -r\dot{\theta}\sin\phi + r\dot{\psi}\cos\phi\sin\theta + \dot{X} = 0 \tag{4.192}$$
$$u_5 = r\dot{\theta}\cos\phi + r\dot{\psi}\sin\phi\sin\theta + \dot{Y} = 0 \tag{4.193}$$

where ϕ is the angle between the positive X-axis and the $\dot{\theta}$ vector.

We wish to find the velocity of the reference point O. It is

$$\begin{aligned}\mathbf{v} = \boldsymbol{\omega} \times (\mathbf{r} + l\mathbf{k}) &= \boldsymbol{\omega} \times [r\sin\theta\sin\psi\,\mathbf{i} + r\sin\theta\cos\psi\,\mathbf{j} + (r\cos\theta + l)\mathbf{k}]\\ &= [(r\cos\theta + l)\omega_y - r\omega_z\sin\theta\cos\psi]\mathbf{i} + [-(r\cos\theta + l)\omega_x\\ &\quad + r\omega_z\sin\theta\sin\psi]\mathbf{j} + (r\omega_x\sin\theta\cos\psi - r\omega_y\sin\theta\sin\psi)\mathbf{k}\end{aligned} \tag{4.194}$$

The velocity coefficients are

$$\gamma_{11} = \frac{\partial\mathbf{v}}{\partial\omega_x} = -(r\cos\theta + l)\mathbf{j} + r\sin\theta\cos\psi\,\mathbf{k}$$

$$\gamma_{12} = \frac{\partial\mathbf{v}}{\partial\omega_y} = (r\cos\theta + l)\mathbf{i} - r\sin\theta\sin\psi\,\mathbf{k} \tag{4.195}$$

$$\gamma_{13} = \frac{\partial\mathbf{v}}{\partial\omega_z} = -r\sin\theta\cos\psi\,\mathbf{i} + r\sin\theta\sin\psi\,\mathbf{j}$$

Similarly, the angular velocity coefficients are

$$\beta_{11} = \frac{\partial\boldsymbol{\omega}}{\partial\omega_x} = \mathbf{i}, \qquad \beta_{12} = \frac{\partial\boldsymbol{\omega}}{\partial\omega_y} = \mathbf{j}, \qquad \beta_{13} = \frac{\partial\boldsymbol{\omega}}{\partial\omega_z} = \mathbf{k} \tag{4.196}$$

We wish to find $\dot{\mathbf{v}}$ and, in the process, we note that

$$\dot{\mathbf{i}} = \omega_z\mathbf{j} - \omega_y\mathbf{k}, \qquad \dot{\mathbf{j}} = -\omega_z\mathbf{i} + \omega_x\mathbf{k}, \qquad \dot{\mathbf{k}} = \omega_y\mathbf{i} - \omega_x\mathbf{j} \tag{4.197}$$

The Euler angle rates are

$$\dot{\phi} = \csc\theta\,(\omega_x\sin\psi + \omega_y\cos\psi) \tag{4.198}$$
$$\dot{\theta} = \omega_x\cos\psi - \omega_y\sin\psi \tag{4.199}$$
$$\dot{\psi} = -\omega_x\cot\theta\sin\psi - \omega_y\cot\theta\cos\psi + \omega_z \tag{4.200}$$

Then a differentiation of (4.194) results in

$$\begin{aligned}\dot{\mathbf{v}} &= [(r\cos\theta + l)\dot{\omega}_y - r\dot{\omega}_z\sin\theta\cos\psi + l\omega_x\omega_z]\mathbf{i}\\ &\quad + [-(r\cos\theta + l)\dot{\omega}_x + r\dot{\omega}_z\sin\theta\sin\psi + l\omega_y\omega_z]\mathbf{j}\\ &\quad + \left[r\dot{\omega}_x\sin\theta\cos\psi - r\dot{\omega}_y\sin\theta\sin\psi - l(\omega_x^2 + \omega_y^2)\right]\mathbf{k}\end{aligned} \tag{4.201}$$

after the cancellation of numerous terms.

In the evaluation of strictly rotational terms, we find that

$$\mathbf{I}_i \cdot \dot{\boldsymbol{\omega}}_i + \boldsymbol{\omega}_i \times \mathbf{I}_i \cdot \boldsymbol{\omega}_i = [I_{xx}\dot{\omega}_x + (I_{zz} - I_{yy})\omega_y\omega_z]\mathbf{i}$$
$$+ [I_{yy}\dot{\omega}_y + (I_{xx} - I_{zz})\omega_z\omega_x]\mathbf{j} + [I_{zz}\dot{\omega}_z + (I_{yy} - I_{xx})\omega_x\omega_y]\mathbf{k} \quad (4.202)$$

Furthermore, the gravitational moment about the $\dot{\theta}$ axis is

$$\mathbf{M}_\theta = mgl\sin\theta \quad (4.203)$$

Hence, we obtain

$$Q_1 = mgl\sin\theta\cos\psi, \qquad Q_2 = -mgl\sin\theta\sin\psi, \qquad Q_3 = 0 \quad (4.204)$$

which are the generalized forces corresponding to ω_x, ω_y, and ω_z, respectively.

Now, we are ready to find the dynamical equations by substituting into (4.189). The ω_x equation is

$$[I_{xx} + mr^2\sin^2\theta\cos^2\psi + m(r\cos\theta + l)^2]\dot{\omega}_x - mr^2\dot{\omega}_y\sin^2\theta\sin\psi\cos\psi$$
$$- mr(r\cos\theta + l)\dot{\omega}_z\sin\theta\sin\psi + [I_{zz} - I_{yy} - ml(r\cos\theta + l)]\omega_y\omega_z$$
$$- mrl\left(\omega_x^2 + \omega_y^2\right)\sin\theta\cos\psi = mgl\sin\theta\cos\psi \quad (4.205)$$

Similarly, the ω_y equation is

$$- mr^2\dot{\omega}_x\sin^2\theta\sin\psi\cos\psi + [I_{yy} + mr^2\sin^2\theta\sin^2\psi + m(r\cos\theta + l)^2]\dot{\omega}_y$$
$$- mr(r\cos\theta + l)\dot{\omega}_z\sin\theta\cos\psi + [I_{xx} - I_{zz} + ml(r\cos\theta + l)]\omega_x\omega_z$$
$$+ mrl\left(\omega_x^2 + \omega_y^2\right)\sin\theta\sin\psi = -mgl\sin\theta\sin\psi \quad (4.206)$$

The ω_z equation is

$$- mr(r\cos\theta + l)\dot{\omega}_x\sin\theta\sin\psi - mr(r\cos\theta + l)\dot{\omega}_y\sin\theta\cos\psi$$
$$+ (I_{zz} + mr^2\sin^2\theta)\dot{\omega}_z + (I_{yy} - I_{xx})\omega_x\omega_y - mrl\omega_x\omega_z\sin\theta\cos\psi$$
$$+ mrl\omega_y\omega_z\sin\theta\sin\psi = 0 \quad (4.207)$$

Equations (4.205)–(4.207) are the three dynamical equations. In addition, we have three kinematical equations for the Euler angle rates given by (4.198)–(4.200). These six equations can be solved for $(\omega_x, \omega_y, \omega_z, \phi, \theta, \psi)$ as functions of time. If one desires to solve for the path of the contact point, one can write the constraint equations in the form

$$\dot{X} = r\dot{\theta}\sin\phi - r\dot{\psi}\cos\phi\sin\theta \quad (4.208)$$
$$\dot{Y} = -r\dot{\theta}\cos\phi - r\dot{\psi}\sin\phi\sin\theta \quad (4.209)$$

These equations are integrated to obtain $X(t)$ and $Y(t)$.

In this example, we have shown how the general dynamical equation can be applied using quasi-velocities to obtain a minimum set of differential equations describing the motion of a relatively complex nonholonomic system. This was accomplished without solving for the constraint forces.

4.5   A fundamental equation

System of particles

Let us consider a system of N particles that are described by d'Alembert's principle in the form

$$\sum_{i=1}^{N} (\mathbf{F}_i - m_i \dot{\mathbf{v}}_i) \cdot \delta \mathbf{r}_i = 0 \tag{4.210}$$

where $\mathbf{r}_i$ is the inertial position vector of the ith particle, $\mathbf{v}_i \equiv \dot{\mathbf{r}}_i$, and $\mathbf{F}_i$ is the applied force acting on that particle. The $\delta \mathbf{r}$s are consistent with any constraints.

Now suppose we express the configuration of the system by using n qs with m linear nonholonomic equations of constraint. The particle velocities are given in terms of $(n - m)$ *independent* quasi-velocities, that is,

$$\mathbf{v}_i = \sum_{j=1}^{n-m} \boldsymbol{\gamma}_{ij}(q, t) u_j + \boldsymbol{\gamma}_{it}(q, t) \tag{4.211}$$

and we find that

$$\delta \mathbf{r}_i = \sum_{j=1}^{n-m} \boldsymbol{\gamma}_{ij}(q, t) \delta \theta_j \tag{4.212}$$

where θ_j is a quasi-coordinate associated with u_j. The corresponding generalized force is

$$Q_j = \sum_{i=1}^{N} \mathbf{F}_i \cdot \boldsymbol{\gamma}_{ij} \qquad (j = 1, \ldots, n - m) \tag{4.213}$$

so (4.210) takes the form

$$\sum_{j=1}^{n-m} \left(Q_j - \sum_{i=1}^{N} m_i \dot{\mathbf{v}}_i \cdot \boldsymbol{\gamma}_{ij} \right) \delta \theta_j = 0 \tag{4.214}$$

The $\delta \theta$s are independent, so we obtain

$$\sum_{i=1}^{N} m_i \dot{\mathbf{v}}_i \cdot \boldsymbol{\gamma}_{ij} = Q_j \qquad (j = 1, \ldots, n - m) \tag{4.215}$$

in agreement with (4.133).

Next, consider the constrained kinetic energy, written in terms of *independent* quasi-velocities.

$$T(q, u, t) = \frac{1}{2} \sum_{i=1}^{N} m_i \mathbf{v}_i \cdot \mathbf{v}_i \tag{4.216}$$

where $\mathbf{v}_i = \mathbf{v}_i(q, u, t)$. We find that

$$\frac{\partial T}{\partial u_j} = \sum_{i=1}^{N} m_i \mathbf{v}_i \cdot \frac{\partial \mathbf{v}_i}{\partial u_j} = \sum_{i=1}^{N} m_i \mathbf{v}_i \cdot \boldsymbol{\gamma}_{ij} \qquad (j = 1, \ldots, n - m) \tag{4.217}$$

Then

$$\frac{d}{dt}\left(\frac{\partial T}{\partial u_j}\right) = \sum_{i=1}^{N} m_i(\dot{\mathbf{v}}_i \cdot \boldsymbol{\gamma}_{ij} + \mathbf{v}_i \cdot \dot{\boldsymbol{\gamma}}_{ij}) \tag{4.218}$$

or

$$\sum_{i=1}^{N} m_i \dot{\mathbf{v}}_i \cdot \boldsymbol{\gamma}_{ij} = \frac{d}{dt}\left(\frac{\partial T}{\partial u_j}\right) - \sum_{i=1}^{N} m_i \mathbf{v}_i \cdot \dot{\boldsymbol{\gamma}}_{ij} \qquad (j = 1, \ldots, n - m) \tag{4.219}$$

Finally, from (4.215) and (4.219) we obtain

$$\frac{d}{dt}\left(\frac{\partial T}{\partial u_j}\right) - \sum_{i=1}^{N} m_i \mathbf{v}_i \cdot \dot{\boldsymbol{\gamma}}_{ij} = Q_j \qquad (j = 1, \ldots, n - m) \tag{4.220}$$

This is a *fundamental dynamical equation for a system of particles*. It results in $(n - m)$ first-order dynamical equations which are supplemented by the n first-order kinematical equations for the $\dot{q}$s. These $(2n - m)$ equations are solved for the $(n - m)$ us and n qs as functions of time.

Notice that the dynamical equations are identical to those obtained by the Boltzmann–Hamel equation but the procedure here is far simpler. Furthermore, the kinetic energy $T(q, u, t)$ is written for the *constrained* rather than the unconstrained system.

System of rigid bodies

Let us generalize the results for a system of N particles to a system of N rigid bodies. Let us choose the reference point for each rigid body at the center of mass. Then in accordance with Koenig's theorem, the kinetic energy is

$$T(q, u, t) = \frac{1}{2}\sum_{i=1}^{N}(m_i \mathbf{v}_{ci} \cdot \mathbf{v}_{ci} + \boldsymbol{\omega}_i \cdot \mathbf{I}_{ci} \cdot \boldsymbol{\omega}_i) \tag{4.221}$$

where $\mathbf{v}_{ci}$ is the velocity of the center of mass of the ith body and $\mathbf{I}_{ci}$ is its inertia dyadic about the center of mass. The $(n - m)$ us are independent.

We see that

$$\frac{\partial T}{\partial u_j} = \sum_{i=1}^{N}(m_i \mathbf{v}_{ci} \cdot \boldsymbol{\gamma}_{ij} + \boldsymbol{\omega}_i \cdot \mathbf{I}_{ci} \cdot \boldsymbol{\beta}_{ij}) \qquad (j = 1, \ldots, n - m) \tag{4.222}$$

where

$$\boldsymbol{\gamma}_{ij} = \frac{\partial \mathbf{v}_{ci}}{\partial u_j}, \qquad \boldsymbol{\beta}_{ij} = \frac{\partial \boldsymbol{\omega}_i}{\partial u_j} \tag{4.223}$$

Hence, we obtain

$$\frac{d}{dt}\left(\frac{\partial T}{\partial u_j}\right) = \sum_{i=1}^{N}[m_i \dot{\mathbf{v}}_{ci} \cdot \boldsymbol{\gamma}_{ij} + (\mathbf{I}_{ci} \cdot \dot{\boldsymbol{\omega}}_i + \boldsymbol{\omega}_i \times \mathbf{I}_{ci} \cdot \boldsymbol{\omega}_i) \cdot \boldsymbol{\beta}_{ij}]$$

$$+ \sum_{i=1}^{N}(m_i \mathbf{v}_{ci} \cdot \dot{\boldsymbol{\gamma}}_{ij} + \boldsymbol{\omega}_i \cdot \mathbf{I}_{ci} \cdot \dot{\boldsymbol{\beta}}_{ij}) \qquad (j = 1, \ldots, n-m) \qquad (4.224)$$

Now recall the general dynamical equation for center-of-mass reference points, that is, for $\rho_{ci} = 0$. From (4.142), we have

$$\sum_{i=1}^{N}[m_i \dot{\mathbf{v}}_{ci} \cdot \boldsymbol{\gamma}_{ij} + (\mathbf{I}_{ci} \cdot \dot{\boldsymbol{\omega}}_i + \boldsymbol{\omega}_i \times \mathbf{I}_{ci} \cdot \boldsymbol{\omega}_i) \cdot \boldsymbol{\beta}_{ij}] = Q_j \qquad (j = 1, \ldots, n-m) \quad (4.225)$$

Finally, from (4.224) and (4.225), we obtain

$$\frac{d}{dt}\left(\frac{\partial T}{\partial u_j}\right) - \sum_{i=1}^{N}(m_i \mathbf{v}_{ci} \cdot \dot{\boldsymbol{\gamma}}_{ij} + \boldsymbol{\omega}_i \cdot \mathbf{I}_{ci} \cdot \dot{\boldsymbol{\beta}}_{ij}) = Q_j \qquad (j = 1, \ldots, n-m) \qquad (4.226)$$

This is a *fundamental equation for a system of rigid bodies*. Using the kinetic energy $T(q, u, t)$ for the *constrained system*, it results in a minimum set of $(n - m)$ first-order dynamical equations.

An alternate form of (4.226) is

$$\frac{d}{dt}\left(\frac{\partial T}{\partial u_j}\right) - \sum_{i=1}^{N}(\mathbf{p}_i \cdot \dot{\boldsymbol{\gamma}}_{ij} + \mathbf{H}_{ci} \cdot \dot{\boldsymbol{\beta}}_{ij}) = Q_j \qquad (j = 1, \ldots, n-m) \qquad (4.227)$$

where $\mathbf{p}_i$ is the linear momentum of the ith body and $\mathbf{H}_{ci}$ is its angular momentum about its center of mass. This form of the equation is more flexible in that the angular momentum may be expressed in terms of arbitrary unit vectors rather than being tied to the body-fixed unit vectors of the inertia dyadic. This is particularly useful in the dynamic analysis of axially symmetric bodies.

If we compare (4.227) with the Boltzmann–Hamel equation, as given by (4.82), we find that

$$\sum_{i=1}^{N}(\mathbf{p}_i \cdot \dot{\boldsymbol{\gamma}}_{ir} + \mathbf{H}_{ci} \cdot \dot{\boldsymbol{\beta}}_{ir}) = \sum_{i=1}^{n}\frac{\partial T}{\partial q_i}\Phi_{ir}$$

$$-\sum_{i=1}^{N}\sum_{j=1}^{n}\sum_{k=1}^{n}\sum_{l=1}^{n-m}\frac{\partial T}{\partial u_j}\left(\frac{\partial \Psi_{ji}}{\partial q_k} - \frac{\partial \Psi_{jk}}{\partial q_i}\right)\Phi_{kl}\Phi_{ir}u_l$$

$$-\sum_{i=1}^{n}\sum_{j=1}^{n}\sum_{k=1}^{n}\frac{\partial T}{\partial u_j}\left(\frac{\partial \Psi_{ji}}{\partial q_k} - \frac{\partial \Psi_{jk}}{\partial q_i}\right)\Phi_{kt}\Phi_{ir}$$

$$-\sum_{i=1}^{n}\sum_{j=1}^{n}\frac{\partial T}{\partial u_j}\left(\frac{\partial \Psi_{ji}}{\partial t} - \frac{\partial \Psi_{jt}}{\partial q_i}\right)\Phi_{ir} \qquad (r = 1, \ldots, n-m) \qquad (4.228)$$

where $T(q, u, t)$ is written for the unconstrained system. Since the differential equations of motion produced by the two methods are identical, we conclude that the fundamental

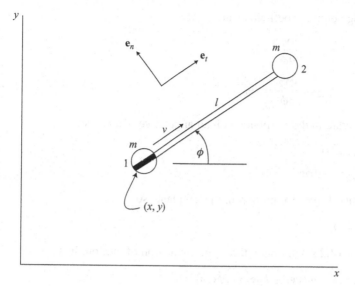

Figure 4.6.

equation (4.226) or (4.227) is more efficient than the Boltzmann–Hamel equation in this general case.

Example 4.8 Let us consider once again the motion of a dumbbell (Fig. 4.6) as it moves on the horizontal xy-plane. There is a knife-edge constraint acting on a particle whose location is (x, y). Let the generalized coordinates be (x, y, ϕ) and choose

$$u_1 = v = \dot{x}\cos\phi + \dot{y}\sin\phi \qquad (4.229)$$

$$u_2 = \dot{\phi} \qquad (4.230)$$

$$u_3 = -\dot{x}\sin\phi + \dot{y}\cos\phi = 0 \qquad (4.231)$$

The first two us are independent and define the motion, while u_3 represents the constraint function which is set equal to zero, indicating no slipping perpendicular to the knife edge.

We shall use the fundamental equation in the form applying to a system of particles.

$$\frac{d}{dt}\left(\frac{\partial T}{\partial u_j}\right) - \sum_{i=1}^{N} m_i \mathbf{v}_i \cdot \dot{\boldsymbol{\gamma}}_{ij} = Q_j \qquad (j = 1, \ldots, n - m) \qquad (4.232)$$

The constrained kinetic energy is

$$T = mu_1^2 + \frac{1}{2}ml^2 u_2^2 = mv^2 + \frac{1}{2}ml^2\dot{\phi}^2 \qquad (4.233)$$

since the particle velocities are

$$\mathbf{v}_1 = v\mathbf{e}_t, \qquad \mathbf{v}_2 = v\mathbf{e}_t + l\dot{\phi}\mathbf{e}_n \qquad (4.234)$$

The corresponding velocity coefficients are

$$\gamma_{11} = \frac{\partial \mathbf{v}_1}{\partial v} = \mathbf{e}_t \qquad \gamma_{12} = \frac{\partial \mathbf{v}_1}{\partial \dot{\phi}} = 0$$

$$\gamma_{21} = \frac{\partial \mathbf{v}_2}{\partial v} = \mathbf{e}_t \qquad \gamma_{22} = \frac{\partial \mathbf{v}_2}{\partial \dot{\phi}} = l\mathbf{e}_n \qquad (4.235)$$

The unit vectors rotate in the xy-plane with an angular velocity $\dot{\phi}$, so

$$\dot{\gamma}_{11} = \dot{\phi}\mathbf{e}_n \qquad \dot{\gamma}_{12} = 0$$
$$\dot{\gamma}_{21} = \dot{\phi}\mathbf{e}_n \qquad \dot{\gamma}_{22} = -l\dot{\phi}\mathbf{e}_t \qquad (4.236)$$

There are no applied forces for motion in the xy-plane, so

$$Q_1 = 0, \qquad Q_2 = 0 \qquad (4.237)$$

Now we can use (4.232) to obtain the u_1 or v equation of motion. It is

$$2m\dot{v} - [mv\mathbf{e}_t \cdot (\dot{\phi}\mathbf{e}_n) + m(v\mathbf{e}_t + l\dot{\phi}\mathbf{e}_n) \cdot (\dot{\phi}\mathbf{e}_n)] = 0$$

or

$$2m\dot{v} - ml\dot{\phi}^2 = 0 \qquad (4.238)$$

In a similar manner, one obtains the u_2 equation, namely,

$$ml^2\ddot{\phi} + mlv\dot{\phi} = 0 \qquad (4.239)$$

These equations of motion are the same as were obtained in Example 4.4, on page 232, by using the Boltzmann–Hamel equation. Here, however, their derivation is more direct.

Example 4.9 Consider the rolling disk problem of Example 4.6 on page 240. Let us use the fundamental equation in the form given by (4.227), namely,

$$\frac{d}{dt}\left(\frac{\partial T}{\partial u_j}\right) - \sum_{i=1}^{N}(\mathbf{p}_i \cdot \gamma_{ij} + \mathbf{H}_{ci} \cdot \beta_{ij}) = Q_j \qquad (j = 1, \ldots, n - m) \qquad (4.240)$$

As in Fig. 4.2, let us choose the Euler angles (ϕ, θ, ψ) as qs and let the independent us be

$$u_1 = \dot{\theta}, \qquad u_2 = \omega_d = \dot{\phi}\sin\theta, \qquad u_3 = \Omega = \dot{\phi}\cos\theta + \dot{\psi} \qquad (4.241)$$

The velocity of the center of the disk is

$$\mathbf{v} = -r\Omega\mathbf{e}_\theta + r\dot{\theta}\mathbf{e}_\psi \qquad (4.242)$$

and the angular velocity is

$$\boldsymbol{\omega} = \dot{\theta}\mathbf{e}_\theta + \omega_d\mathbf{e}_d + \Omega\mathbf{e}_\psi \qquad (4.243)$$

The moments of inertia about the center are

$$I_a = \frac{1}{2}mr^2, \qquad I_t = \frac{1}{4}mr^2 \qquad (4.244)$$

Thus, we find that the *constrained* kinetic energy is

$$T = \frac{1}{2}m(r^2\Omega^2 + r^2\dot{\theta}^2) + \frac{1}{4}mr^2\Omega^2 + \frac{1}{8}mr^2\left(\dot{\theta}^2 + \omega_d^2\right)$$

$$= \frac{5}{8}mr^2\dot{\theta}^2 + \frac{1}{8}mr^2\omega_d^2 + \frac{3}{4}mr^2\Omega^2 \tag{4.245}$$

The velocity coefficients are

$$\boldsymbol{\gamma}_{11} = \frac{\partial\mathbf{v}}{\partial\dot{\theta}} = r\mathbf{e}_\psi, \qquad \boldsymbol{\gamma}_{12} = \frac{\partial\mathbf{v}}{\partial\omega_d} = 0, \qquad \boldsymbol{\gamma}_{13} = \frac{\partial\mathbf{v}}{\partial\Omega} = -r\mathbf{e}_\theta \tag{4.246}$$

and the angular velocity coefficients are

$$\boldsymbol{\beta}_{11} = \frac{\partial\boldsymbol{\omega}}{\partial\dot{\theta}} = \mathbf{e}_\theta, \qquad \boldsymbol{\beta}_{12} = \frac{\partial\boldsymbol{\omega}}{\partial\omega_d} = \mathbf{e}_d, \qquad \boldsymbol{\beta}_{13} = \frac{\partial\boldsymbol{\omega}}{\partial\Omega} = \mathbf{e}_\psi \tag{4.247}$$

The $\mathbf{e}_\theta$, $\mathbf{e}_d$, $\mathbf{e}_\psi$ unit vector triad rotates due to $\dot{\phi}$ and $\dot{\theta}$, resulting in an angular velocity

$$\boldsymbol{\omega}_c = \dot{\theta}\mathbf{e}_\theta + \dot{\phi}(\sin\theta\,\mathbf{e}_d + \cos\theta\,\mathbf{e}_\psi)$$

$$= \dot{\theta}\mathbf{e}_\theta + \omega_d\mathbf{e}_d + \omega_d\cot\theta\,\mathbf{e}_\psi \tag{4.248}$$

where we notice that $\omega_d = \dot{\phi}\,\sin\,\theta$. Thus, we find that

$$\dot{\mathbf{e}}_\theta = \boldsymbol{\omega}_c \times \mathbf{e}_\theta = \omega_d\cot\theta\,\mathbf{e}_d - \omega_d\mathbf{e}_\psi \tag{4.249}$$

$$\dot{\mathbf{e}}_d = \boldsymbol{\omega}_c \times \mathbf{e}_d = -\omega_d\cot\theta\,\mathbf{e}_\theta + \dot{\theta}\mathbf{e}_\psi \tag{4.250}$$

$$\dot{\mathbf{e}}_\psi = \boldsymbol{\omega}_c \times \mathbf{e}_\psi = \omega_d\mathbf{e}_\theta - \dot{\theta}\mathbf{e}_d \tag{4.251}$$

Now we can evaluate

$$\dot{\boldsymbol{\gamma}}_{11} = r\dot{\mathbf{e}}_\psi = r\omega_d - r\dot{\theta}\mathbf{e}_d \tag{4.252}$$

$$\dot{\boldsymbol{\gamma}}_{12} = 0 \tag{4.253}$$

$$\dot{\boldsymbol{\gamma}}_{13} = -r\dot{\mathbf{e}}_\theta = -r\omega_d\cot\theta\,\mathbf{e}_d + r\omega_d\mathbf{e}_\psi \tag{4.254}$$

and we note that

$$\dot{\boldsymbol{\beta}}_{11} = \dot{\mathbf{e}}_\theta, \qquad \dot{\boldsymbol{\beta}}_{12} = \dot{\mathbf{e}}_d, \qquad \dot{\boldsymbol{\beta}}_{13} = \dot{\mathbf{e}}_\psi \tag{4.255}$$

The linear momentum is

$$\mathbf{p} = m\mathbf{v} = -mr\Omega\mathbf{e}_\theta + mr\dot{\theta}\mathbf{e}_\psi \tag{4.256}$$

and the angular momentum about the center of mass is

$$\mathbf{H}_c = I_t\dot{\theta}\mathbf{e}_\theta + I_t\omega_d\mathbf{e}_d + I_a\Omega\mathbf{e}_\psi$$

$$= \frac{1}{4}mr^2\dot{\theta}\mathbf{e}_\theta + \frac{1}{4}mr^2\omega_d\mathbf{e}_d + \frac{1}{2}mr^2\Omega\mathbf{e}_\psi \tag{4.257}$$

The generalized applied forces due to gravity, obtained by using potential energy or virtual work, are

$$Q_1 = -mgr\cos\theta, \qquad Q_2 = 0, \qquad Q_3 = 0 \tag{4.258}$$

Now we are prepared to use (4.240) to obtain the differential equations of motion. The θ equation is

$$\frac{5}{4}mr^2\ddot{\theta} + \frac{3}{2}mr^2\omega_d\Omega - \frac{1}{4}mr^2\omega_d^2\cot\theta = -mgr\cos\theta \qquad (4.259)$$

Similarly, the ω_d equation is

$$\frac{1}{4}mr^2\dot{\omega}_d + \frac{1}{4}mr^2\dot{\theta}\omega_d\cot\theta - \frac{1}{2}mr^2\dot{\theta}\Omega = 0 \qquad (4.260)$$

The Ω equation is

$$\frac{3}{2}mr^2\dot{\Omega} - mr^2\dot{\theta}\omega_d = 0 \qquad (4.261)$$

These equations are identical with (4.174)–(4.176) obtained earlier by the general dynamical equation. Both methods involve about the same amount of effort.

Volterra's equation

It is of some interest to note that Volterra's equation for a system of particles, published in 1898, is equivalent to a special case of the fundamental equation in the form of (4.220).

Consider a system of N particles whose configuration is given by the $3N$ inertial Cartesian coordinates $(x_1, \ldots, x_{3N})$. Suppose there are m constraint equations which are linear in the $\dot{x}$s. Choose $\nu = 3N - m$ independent us which are related to the $\dot{x}$s by the scleronomic equations

$$\dot{x}_i = \sum_{s=1}^{\nu} \Phi_{is}(x)u_s \qquad (i = 1, \ldots, 3N) \qquad (4.262)$$

The kinetic energy of the *constrained* system is

$$T(x, u) = \frac{1}{2}\sum_{i=1}^{3N} m_i\dot{x}_i^2 = \frac{1}{2}\sum_{i=1}^{3N}\sum_{k=1}^{\nu}\sum_{r=1}^{\nu} m_i\Phi_{ik}\Phi_{ir}u_ku_r \qquad (4.263)$$

Volterra's equation for this system of N particles is

$$\frac{d}{dt}\left(\frac{\partial T}{\partial u_s}\right) + \sum_{k=1}^{\nu}\sum_{r=1}^{\nu}\left(b_{rs}^k - b_{sr}^k\right)u_ku_r - \sum_{l=1}^{3N}\frac{\partial T}{\partial x_l}\Phi_{ls} = Q_s \qquad (s = 1, \ldots, \nu) \qquad (4.264)$$

where

$$b_{rs}^k = \sum_{i=1}^{3N}\sum_{j=1}^{3N} m_i\Phi_{ik}\frac{\partial\Phi_{ir}}{\partial x_j}\Phi_{js} \qquad (k, r, s = 1, \ldots, \nu) \qquad (4.265)$$

We note that

$$\frac{\partial T}{\partial x_l} = \sum_{i=1}^{3N} m_i\dot{x}_i\frac{\partial\dot{x}_i}{\partial x_l} = \sum_{i=1}^{3N}\sum_{k=1}^{\nu}\sum_{r=1}^{\nu} m_i\Phi_{ik}\frac{\partial\Phi_{ir}}{\partial x_l}u_ku_r \qquad (4.266)$$

Hence, in (4.264), the term

$$-\sum_{l=1}^{3N}\frac{\partial T}{\partial x_l}\Phi_{ls}=-\sum_{i=1}^{3N}\sum_{l=1}^{3N}\sum_{k=1}^{\nu}\sum_{r=1}^{\nu}m_i\Phi_{ik}\frac{\partial\Phi_{ir}}{\partial x_l}\Phi_{ls}u_k u_r=-\sum_{k=1}^{\nu}\sum_{r=1}^{\nu}b_{rs}^k u_k u_r \qquad (4.267)$$

and the *Volterra equation* can be written in the simpler form

$$\frac{d}{dt}\left(\frac{\partial T}{\partial u_s}\right)-\sum_{k=1}^{\nu}\sum_{r=1}^{\nu}b_{sr}^k u_k u_r=Q_s \qquad (s=1,\ldots,\nu) \qquad (4.268)$$

or

$$\frac{d}{dt}\left(\frac{\partial T}{\partial u_s}\right)-\sum_{i=1}^{3N}\sum_{j=1}^{3N}\sum_{k=1}^{\nu}\sum_{r=1}^{\nu}m_i\Phi_{ik}\frac{\partial\Phi_{is}}{\partial x_j}\Phi_{jr}u_k u_r=Q_s \qquad (s=1,\ldots,\nu) \qquad (4.269)$$

We wish to express this result using vector notation. The velocity of the first particle is

$$\mathbf{v}_1=\sum_{i=1}^{3}\sum_{k=1}^{\nu}\Phi_{ik}u_k\mathbf{e}_i \qquad (4.270)$$

where $\mathbf{e}_1,\mathbf{e}_2,\mathbf{e}_3$ are inertially-fixed orthogonal unit vectors. The velocity coefficients for the first particle are

$$\boldsymbol{\gamma}_{1s}=\frac{\partial\mathbf{v}_1}{\partial u_s}=\sum_{i=1}^{3}\Phi_{is}(x)\mathbf{e}_i \qquad (s=1,\ldots,\nu) \qquad (4.271)$$

Hence,

$$\dot{\boldsymbol{\gamma}}_{1s}=\sum_{i=1}^{3}\sum_{j=1}^{3N}\sum_{r=1}^{\nu}\frac{\partial\Phi_{is}}{\partial x_j}\Phi_{jr}u_r\mathbf{e}_i \qquad (s=1,\ldots,\nu) \qquad (4.272)$$

For the first particle, noting that $m_1=m_2=m_3$, we have

$$m_1\mathbf{v}_1\cdot\dot{\boldsymbol{\gamma}}_{1s}=\sum_{i=1}^{3}\sum_{j=1}^{3N}\sum_{k=1}^{\nu}\sum_{r=1}^{\nu}m_i\Phi_{ik}\frac{\partial\Phi_{is}}{\partial x_j}\Phi_{jr}u_k u_r \qquad (s=1,\ldots,\nu) \qquad (4.273)$$

Finally, sum (4.273) over all N particles. Then using (4.269), we see that Volterra's equation in vector form is

$$\frac{d}{dt}\left(\frac{\partial T}{\partial u_s}\right)-\sum_{l=1}^{N}m_l\mathbf{v}_l\cdot\dot{\boldsymbol{\gamma}}_{ls}=Q_s \qquad (s=1,\ldots,\nu) \qquad (4.274)$$

But this is just the fundamental dynamical equation for a system of particles.

We conclude that Volterra's equation is a special case of the fundamental equation for a scleronomic system of particles, a case in which Cartesian coordinates are used as qs. When Volterra's equation was published, the concept of quasi-velocities was unknown, so the us were regarded as $\dot{q}$s. Equation (4.274), however, is more general and can be used with rheonomic systems that are described using n generalized coordinates with m nonholonomic constraints. The $\nu = n - m$ quasi-velocities are independent and the kinetic energy is written for the constrained system.

4.6   The Gibbs–Appell equation

System of particles

Let us consider a system of N particles whose configuration is specified by n qs. The particle velocities for a system with m nonholonomic constraints are given by $(n - m)$ independent us. Thus, the absolute velocity of the ith particle is

$$\mathbf{v}_i = \sum_{j=1}^{n-m} \gamma_{ij}(q, t) u_j + \gamma_{it}(q, t) \qquad (i = 1, \ldots, N) \tag{4.275}$$

and the *constrained* kinetic energy is

$$T(q, u, t) = \frac{1}{2} \sum_{i=1}^{N} m_i \mathbf{v}_i^2 \tag{4.276}$$

The absolute particle accelerations, obtained by differentiating (4.275), are

$$\dot{\mathbf{v}}_i = \sum_{j=1}^{n-m} (\gamma_{ij} \dot{u}_j + \dot{\gamma}_{ij} u_j) + \dot{\gamma}_{it} \qquad (i = 1, \ldots, N) \tag{4.277}$$

where the $\dot{\gamma}$s are linear functions of the us due to the equations

$$\dot{q}_k = \sum_{l=1}^{n-m} \Phi_{kl} u_l + \Phi_{kt} \qquad (k = 1, \ldots, n) \tag{4.278}$$

Thus, we see that the acceleration $\dot{\mathbf{v}}_i$ is linear in the $\dot{u}$s and quadratic in the us.

Now let us introduce the Gibbs–Appell function

$$S(q, u, \dot{u}, t) = \frac{1}{2} \sum_{i=1}^{N} m_i \dot{\mathbf{v}}_i^2 \tag{4.279}$$

which is obtained by substituting $\dot{\mathbf{v}}_i$ from (4.277) for $\mathbf{v}_i$ in the kinetic energy expression of (4.276). We see that

$$\frac{\partial S}{\partial \dot{u}_j} = \sum_{i=1}^{N} m_i \dot{\mathbf{v}}_i \cdot \frac{\partial \dot{\mathbf{v}}_i}{\partial \dot{u}_j} = \sum_{i=1}^{N} m_i \dot{\mathbf{v}}_i \cdot \gamma_{ij} \qquad (j = 1, \ldots, n - m) \tag{4.280}$$

But, from the general dynamical equation (4.133), we have

$$\sum_{i=1}^{N} m_i \dot{\mathbf{v}}_i \cdot \gamma_{ij} = Q_j \qquad (j = 1, \ldots, n - m) \tag{4.281}$$

Finally, from (4.280) and (4.281) we obtain

$$\frac{\partial S}{\partial \dot{u}_j} = Q_j \qquad (j = 1, \ldots, n - m) \tag{4.282}$$

which is the *Gibbs–Appell equation* for a system of particles. This equation was discovered by Gibbs in 1879 and was studied in detail by Appell in an 1899 publication. It provides a minimal set of dynamical equations which are applicable to systems with quasi-velocities and nonholonomic constraints.

To emphasize an important point, recall that the Gibbs–Appell function is obtained by substituting $\dot{\mathbf{v}}_i$ for $\mathbf{v}_i$ in the kinetic energy expression, where $\dot{\mathbf{v}}_i$ is the absolute acceleration of the ith particle. One cannot in general, obtain S by writing $T(q, \dot{q}, t)$ and then substituting $\ddot{q}$s for $\dot{q}$s. Furthermore, since (4.282) involves differentiations with respect to the $\dot{u}$s, any terms in $S(q, u, \dot{u}, t)$ which do not contain $\dot{u}$s can be omitted.

Example 4.10 Let us return to the dumbbell problem of Fig. 4.6. As independent quasi-velocities consistent with the knife-edge constraint we choose

$$u_1 = v, \qquad u_2 = \dot{\phi} \tag{4.283}$$

The particle velocities are

$$\mathbf{v}_1 = v\mathbf{e}_t, \qquad \mathbf{v}_2 = v\mathbf{e}_t + l\dot{\phi}\mathbf{e}_n \tag{4.284}$$

and the corresponding accelerations are

$$\dot{\mathbf{v}}_1 = \dot{v}\mathbf{e}_t + v\dot{\phi}\mathbf{e}_n \tag{4.285}$$

$$\dot{\mathbf{v}}_2 = (\dot{v} - l\dot{\phi}^2)\mathbf{e}_t + (l\ddot{\phi} + v\dot{\phi})\mathbf{e}_n \tag{4.286}$$

The resulting Gibbs–Appell function is

$$S = \frac{1}{2}m\left(\dot{\mathbf{v}}_1^2 + \dot{\mathbf{v}}_2^2\right) = \frac{1}{2}m[\dot{v}^2 + v^2\dot{\phi}^2 + (\dot{v} - l\dot{\phi}^2)^2 + (l\ddot{\phi} + v\dot{\phi})^2] \tag{4.287}$$

The generalized applied forces are

$$Q_1 = 0, \qquad Q_2 = 0 \tag{4.288}$$

Now we can apply (4.282) and obtain the following equations of motion:

$$\frac{\partial S}{\partial \dot{u}_1} = \frac{\partial S}{\partial \dot{v}} = m(2\dot{v} - l\dot{\phi}^2) = 2m\dot{v} - ml\dot{\phi}^2 = 0 \tag{4.289}$$

$$\frac{\partial S}{\partial \dot{u}_2} = \frac{\partial S}{\partial \ddot{\phi}} = ml(l\ddot{\phi} + v\dot{\phi}) = ml^2\ddot{\phi} + mlv\dot{\phi} = 0 \tag{4.290}$$

It is apparent that, for this problem, the Gibbs–Appell method is quite efficient in producing the differential equations of motion.

System of rigid bodies

Now let us generalize the Gibbs–Appell function to give correct equations of motion for a system of N rigid bodies when (4.282) is used. Let $\mathbf{v}_i$ be the velocity of the reference point of the ith body, and let $\mathbf{I}_i$ be the inertia dyadic about this reference point. The total kinetic energy is

$$T = \frac{1}{2}\sum_{i=1}^{N} m_i v_i^2 + \frac{1}{2}\sum_{i=1}^{N} \boldsymbol{\omega}_i \cdot \mathbf{I}_i \cdot \boldsymbol{\omega}_i + \sum_{i=1}^{N} m_i \mathbf{v}_i \cdot \dot{\boldsymbol{\rho}}_{ci} \tag{4.291}$$

A Gibbs–Appell function which yields correct equations of motion for this system of rigid bodies is

$$S(q, u, \dot{u}, t) = \frac{1}{2} \sum_{i=1}^{N} m_i \dot{\mathbf{v}}_i^2 + \sum_{i=1}^{N} \left[\frac{1}{2} \dot{\boldsymbol{\omega}}_i \cdot \mathbf{I}_i \cdot \dot{\boldsymbol{\omega}}_i + \boldsymbol{\omega}_i \times (\mathbf{I}_i \cdot \boldsymbol{\omega}_i) \cdot \dot{\boldsymbol{\omega}}_i \right] + \sum_{i=1}^{N} m_i \dot{\mathbf{v}}_i \cdot \ddot{\boldsymbol{\rho}}_{ci}$$

(4.292)

where

$$\ddot{\boldsymbol{\rho}}_{ci} = \dot{\boldsymbol{\omega}}_i \times \boldsymbol{\rho}_{ci} + \boldsymbol{\omega}_i \times (\boldsymbol{\omega}_i \times \boldsymbol{\rho}_{ci})$$

(4.293)

and we note that $\mathbf{I}_i$ is symmetric. In the evaluation of $\partial S / \partial \dot{u}_j$, only terms involving $\dot{\mathbf{v}}_i$, $\dot{\boldsymbol{\omega}}_i$, or $\ddot{\boldsymbol{\rho}}_{ci}$ need be considered. Recall that

$$\frac{\partial \dot{\mathbf{v}}_i}{\partial \dot{u}_j} = \frac{\partial \mathbf{v}_i}{\partial u_j} = \boldsymbol{\gamma}_{ij}$$

(4.294)

$$\frac{\partial \dot{\boldsymbol{\omega}}_i}{\partial \dot{u}_j} = \frac{\partial \boldsymbol{\omega}_i}{\partial u_j} = \boldsymbol{\beta}_{ij}$$

(4.295)

Also,

$$\dot{\mathbf{v}}_i \cdot \frac{\partial \dot{\boldsymbol{\omega}}_i}{\partial \dot{u}_j} \times \boldsymbol{\rho}_{ci} = \boldsymbol{\rho}_{ci} \times \dot{\mathbf{v}}_i \cdot \boldsymbol{\beta}_{ij}$$

(4.296)

Then, from (4.282) and (4.292), we obtain

$$\frac{\partial S}{\partial \dot{u}_j} = \sum_{i=1}^{N} m_i (\dot{\mathbf{v}}_i + \ddot{\boldsymbol{\rho}}_{ci}) \cdot \boldsymbol{\gamma}_{ij} + \sum_{i=1}^{N} [\mathbf{I}_i \cdot \dot{\boldsymbol{\omega}}_i + \boldsymbol{\omega}_i \times (\mathbf{I}_i \cdot \boldsymbol{\omega}_i) + m_i \boldsymbol{\rho}_{ci} \times \dot{\mathbf{v}}_i] \cdot \boldsymbol{\beta}_{ij} = Q_j$$

$$(j = 1, \ldots, n - m) \quad (4.297)$$

which is the general dynamical equation for a system of rigid bodies. It results in $(n - m)$ first-order dynamical equations.

If one takes the reference point of each rigid body at its center of mass, then $\boldsymbol{\rho}_{ci} = 0$ and the Gibbs–Appell function becomes

$$S = \frac{1}{2} \sum_{i=1}^{N} m_i \dot{\mathbf{v}}_i^2 + \sum_{i=1}^{N} \left[\frac{1}{2} \dot{\boldsymbol{\omega}}_i \cdot \mathbf{I}_{ci} \cdot \dot{\boldsymbol{\omega}}_i + \boldsymbol{\omega}_i \times (\mathbf{I}_{ci} \cdot \boldsymbol{\omega}_i) \cdot \dot{\boldsymbol{\omega}}_i \right]$$

(4.298)

where $\mathbf{I}_{ci}$ is the inertia dyadic about the center of mass. An equivalent form is

$$S = \frac{1}{2} \sum_{i=1}^{N} \left(m_i \dot{\mathbf{v}}_i^2 + \dot{\mathbf{H}}_{ci} \cdot \dot{\boldsymbol{\omega}}_i + \boldsymbol{\omega}_i \times \mathbf{H}_{ci} \cdot \dot{\boldsymbol{\omega}}_i \right)$$

(4.299)

where $\mathbf{H}_{ci}$ is the angular momentum of the ith body about its center of mass.

Example 4.11 Consider the rolling disk problem of Fig. 4.2. We again choose the independent quasi-velocities

$$u_1 = \dot{\theta}, \qquad u_2 = \omega_d, \qquad u_3 = \Omega$$

(4.300)

The angular velocity of the disk is

$$\omega = \dot{\theta}\mathbf{e}_\theta + \omega_d\mathbf{e}_d + \Omega\mathbf{e}_\psi \tag{4.301}$$

and the angular velocity of the $\mathbf{e}_\theta\mathbf{e}_d\mathbf{e}_\psi$ unit vector triad is equal to $\dot{\phi} + \dot{\theta}$ or

$$\omega_c = \dot{\theta}\mathbf{e}_\theta + \omega_d\mathbf{e}_d + \omega_d\cot\theta\ \mathbf{e}_\psi \tag{4.302}$$

and we note that $\mathbf{e}_\theta$ remains horizontal. Thus, we find that

$$\dot{\mathbf{e}}_\theta = \omega_c \times \mathbf{e}_\theta = \omega_d\cot\theta\ \mathbf{e}_d - \omega_d\mathbf{e}_\psi \tag{4.303}$$
$$\dot{\mathbf{e}}_d = \omega_c \times \mathbf{e}_d = -\omega_d\cot\theta\ \mathbf{e}_\theta + \dot{\theta}\mathbf{e}_\psi \tag{4.304}$$
$$\dot{\mathbf{e}}_\psi = \omega_d\mathbf{e}_\theta - \dot{\theta}\mathbf{e}_d \tag{4.305}$$

and

$$\dot{\omega} = \left(\ddot{\theta} - \omega_d^2\cot\theta + \omega_d\Omega\right)\mathbf{e}_\theta + \left(\dot{\omega}_d + \dot{\theta}\omega_d\cot\theta - \dot{\theta}\Omega\right)\mathbf{e}_d + \dot{\Omega}\mathbf{e}_\psi \tag{4.306}$$

The velocity of the center of the disk is

$$\mathbf{v} = -r\Omega\mathbf{e}_\theta + r\dot{\theta}\mathbf{e}_\psi \tag{4.307}$$

and the corresponding acceleration is

$$\dot{\mathbf{v}} = (-r\dot{\Omega} + r\dot{\theta}\omega_d)\mathbf{e}_\theta - (r\omega_d\Omega\cot\theta + r\dot{\theta}^2)\mathbf{e}_d + (r\ddot{\theta} + r\omega_d\Omega)\mathbf{e}_\psi \tag{4.308}$$

Hence, we see that

$$\dot{\mathbf{v}}^2 = (r\dot{\Omega} - r\dot{\theta}\omega_d)^2 + (r\omega_d\Omega\cot\theta + r\dot{\theta}^2)^2 + (r\ddot{\theta} + r\omega_d\Omega)^2 \tag{4.309}$$

The disk has moments of inertia

$$I_a = \frac{1}{2}mr^2, \qquad I_t = \frac{1}{4}mr^2 \tag{4.310}$$

Thus, the angular momentum about the center is

$$\mathbf{H}_c = \frac{1}{4}mr^2\dot{\theta}\mathbf{e}_\theta + \frac{1}{4}mr^2\omega_d\mathbf{e}_d + \frac{1}{2}mr^2\Omega\mathbf{e}_\psi \tag{4.311}$$

Upon differentiating with respect to time, we find that

$$\dot{\mathbf{H}}_c = \frac{1}{4}mr^2\left[(\ddot{\theta} - \omega_d^2\cot\theta + 2\omega_d\Omega)\mathbf{e}_\theta \right.$$
$$\left. + (\dot{\omega}_d + \dot{\theta}\omega_d\cot\theta - 2\dot{\theta}\Omega)\mathbf{e}_d + 2\dot{\Omega}\mathbf{e}_\psi\right] \tag{4.312}$$

In addition,

$$\omega \times \mathbf{H}_c = \frac{1}{4}mr^2(\omega_d\Omega\mathbf{e}_\theta - \dot{\theta}\Omega\mathbf{e}_d) \tag{4.313}$$

Now we can use (4.299) to obtain the Gibbs–Appell function. If we omit terms not containing $\dot{u}$s, it is

$$S = \frac{1}{2}mr^2[(\dot{\Omega} - \dot{\theta}\omega_d)^2 + (\ddot{\theta} + \omega_d\Omega)^2]$$
$$+ \frac{1}{8}mr^2\left[\left(\ddot{\theta} - \omega_d^2\cot\theta + 2\omega_d\Omega\right)\left(\ddot{\theta} - \omega_d^2\cot\theta + \omega_d\Omega\right)\right.$$
$$\left. + (\dot{\omega}_d + \dot{\theta}\omega_d\cot\theta - 2\dot{\theta}\Omega)(\dot{\omega}_d + \dot{\theta}\omega_d\cot\theta - \dot{\theta}\Omega)\right]$$
$$+ \frac{1}{4}mr^2\dot{\Omega}^2 + \frac{1}{8}mr^2(\omega_d\Omega\ddot{\theta} - \dot{\theta}\Omega\dot{\omega}_d) \tag{4.314}$$

The generalized applied forces are

$$Q_1 = -mgr\cos\theta, \qquad Q_2 = 0, \qquad Q_3 = 0 \tag{4.315}$$

The differential equations of motion are obtained from

$$\frac{\partial S}{\partial \dot{u}_j} = Q_j \quad (j = 1, 2, 3) \tag{4.316}$$

The θ equation is

$$\frac{\partial S}{\partial \ddot{\theta}} = mr^2(\ddot{\theta} + \omega_d\Omega) + \frac{1}{8}mr^2\left(2\ddot{\theta} - 2\omega_d^2\cot\theta + 4\omega_d\Omega\right)$$
$$= \frac{5}{4}mr^2\ddot{\theta} - \frac{1}{4}mr^2\omega_d^2\cot\theta + \frac{3}{2}mr^2\omega_d\Omega = -mgr\cos\theta \tag{4.317}$$

The ω_d equation is

$$\frac{\partial S}{\partial \dot{\omega}_d} = \frac{1}{4}mr^2(\dot{\omega}_d + \dot{\theta}\omega_d\cot\theta - 2\dot{\theta}\Omega) = 0 \tag{4.318}$$

Finally, the Ω equation is

$$\frac{\partial S}{\partial \dot{\Omega}} = \frac{3}{2}mr^2\dot{\Omega} - mr^2\dot{\theta}\omega_d = 0 \tag{4.319}$$

These three dynamical equations constitute a minimum set for this system. The effort required in their derivation is about the same as for the general dynamical equation.

Principle of least constraint

The principle of least constraint was discovered by Gauss in 1829, and thereby preceded the Gibbs–Appell equations, to which it is related, by half a century. The principle of least constraint is an algebraic minimization principle which leads to the differential equations of motion. Briefly, it states that a certain function of the constraint forces is minimized by the actual motion, as compared with other motions which, at any given time, have the same configuration and velocities, but have small variations in the accelerations.

Consider a system of N particles whose configuration is given by $3N$ Cartesian coordinates relative to an inertial frame. Suppose there are m equations of constraint, holonomic or nonholonomic, which can be written in the form

$$\sum_{k=1}^{3N} a_{jk}(x,t)\dot{x}_k + a_{jt}(x,t) = 0 \qquad (j = 1, \ldots, m) \tag{4.320}$$

Newton's law, applied to individual particles, results in

$$m_k \ddot{x}_k = F_k + R_k \qquad (k = 1, \ldots, 3N) \tag{4.321}$$

where F_k is an applied force component and R_k is the corresponding constraint force component. From (4.321),

$$R_k = m_k \left(\ddot{x}_k - \frac{F_k}{m_k} \right) \tag{4.322}$$

Now define the function

$$C = \frac{1}{2} \sum_{k=1}^{3N} \frac{R_k^2}{m_k} = \frac{1}{2} \sum_{k=1}^{3N} m_k \left(\ddot{x}_k - \frac{F_k}{m_k} \right)^2 \tag{4.323}$$

which represents the weighted sum of the squares of the constraint force magnitudes. The principle of least constraint states that C is minimized with respect to variations in the $\ddot{x}$s by the actual motion at each instant of time. It is assumed that the $\ddot{x}$s and $\delta\ddot{x}$s satisfy the constraints.

Thus, noting that F_k is not varied, we obtain

$$\delta C = \sum_{k=1}^{3N} m_k \left(\ddot{x}_k - \frac{F_k}{m_k} \right) \delta\ddot{x}_k = 0 \tag{4.324}$$

where

$$\sum_{k=1}^{3N} a_{jk}\delta\ddot{x}_k = 0 \qquad (j = 1, \ldots, m) \tag{4.325}$$

The constraints are incorporated into the analysis by using Lagrange multipliers. Multiply (4.325) by λ_j and sum over j. Then, upon adding this result to (4.324), we obtain

$$\sum_{k=1}^{3N} \left(m_k \ddot{x}_k - F_k + \sum_{j=1}^{m} \lambda_j a_{jk} \right) \delta\ddot{x}_k = 0 \tag{4.326}$$

where the $\delta\ddot{x}$s are now regarded as independent. Hence, each coefficient must be zero, or

$$m_k \ddot{x}_k = F_k - \sum_{j=1}^{m} \lambda_j a_{jk} \qquad (k = 1, \ldots, 3N) \tag{4.327}$$

These are the equations of motion in terms of Cartesian coordinates.

Now let us broaden the analysis by transforming to quasi-velocities and generalized coordinates. Write (4.323) in the form

$$C = \frac{1}{2} \sum_{k=1}^{3N} m_k \ddot{x}_k^2 - \sum_{k=1}^{3N} F_k \ddot{x}_k + \sum_{k=1}^{3N} \frac{F_k^2}{2m_k} \tag{4.328}$$

The Gibbs–Appell function is, from (4.279),

$$S = \frac{1}{2} \sum_{k=1}^{3N} m_k \ddot{x}_k^2 \tag{4.329}$$

Then, noting that F_k is not varied, we can write

$$C = S - \sum_{k=1}^{3N} F_k \ddot{x}_k + \text{const} \tag{4.330}$$

The transformation equations are

$$x_k = x_k(q, t) \qquad (k = 1, \ldots, 3N) \tag{4.331}$$
$$\dot{x}_k = \dot{x}_k(q, u, t) \qquad (k = 1, \ldots, 3N) \tag{4.332}$$

and therefore

$$\ddot{x}_k = \sum_{j=1}^{n-m} \frac{\partial \dot{x}_k}{\partial u_j} \dot{u}_j + f_k(q, u, t) \qquad (k = 1, \ldots, 3N) \tag{4.333}$$

where the $(n - m)$ us are independent and consistent with any constraints on the xs. Note that $\dot{x}_k$ is a linear function of the us, and $\ddot{x}_k$ is a linear function of the $\dot{u}$s. Also, the generalized force associated with u_j is

$$Q_j = \sum_{k=1}^{3N} F_k \frac{\partial \dot{x}_k}{\partial u_j} \qquad (j = 1, \ldots, n - m) \tag{4.334}$$

Then we obtain

$$C(q, u, \dot{u}, t) = S(q, u, \dot{u}, t) - \sum_{j=1}^{n-m} Q_j \dot{u}_j - \sum_{k=1}^{3N} F_k f_k + \text{const} \tag{4.335}$$

Now consider a variation δC due to small variations in the $\dot{u}$s, with the qs and us held fixed.

$$\delta C = \sum_{j=1}^{n-m} \frac{\partial S}{\partial \dot{u}_j} \delta \dot{u}_j - \sum_{j=1}^{n-m} Q_j \delta \dot{u}_j \tag{4.336}$$

Thus, for a stationary value of C, we have

$$\delta C = \sum_{j=1}^{n-m} \left(\frac{\partial S}{\partial \dot{u}_j} - Q_j \right) \delta \dot{u}_j = 0 \tag{4.337}$$

for arbitrary $\delta \dot{u}$s. This requires that each coefficient be zero, and we obtain

$$\frac{\partial S}{\partial \dot{u}_j} = Q_j \qquad (j = 1, \ldots, n - m) \tag{4.338}$$

Thus, the principle of least constraint applied to a system of particles results in the Gibbs–Appell equation.

It has been shown that the Gibbs–Appell equation follows from the requirement that C be stationary with respect to variations of the $\dot{u}$s. That this stationary point is also a minimum can be shown by using (4.323) to evaluate

$$
\frac{\partial^2 C}{\partial \ddot{x}_i \partial \ddot{x}_j} = \begin{cases} m_i & i = j \\ 0 & i \neq j \end{cases}
$$

(4.339)

The corresponding matrix is positive definite, that is,

$$
\delta^2 C = \sum_{i=1}^{3N} \sum_{j=1}^{3N} \left(\frac{\partial^2 C}{\partial \ddot{x}_i \partial \ddot{x}_j} \right) \delta \ddot{x}_i \delta \ddot{x}_j \geq 0
$$

(4.340)

and the value zero occurs only when all the $\delta \ddot{x}$s are zero. Since the second variation of C is a positive-definite function of the $\delta \ddot{x}$s, the stationary point is also a minimum.

4.7   Constraints and energy rates

Ideal and conservative constraints

Consider a dynamical system having m constraints of the general nonholonomic form

$$
f_j(q, \dot{q}, t) = 0 \qquad (j = 1, \ldots, m)
$$

(4.341)

This general form includes the usual nonholonomic constraints which are linear in the velocities, that is,

$$
\sum_{i=1}^{n} a_{ji}(q, t) \dot{q}_i + a_{jt}(q, t) = 0 \qquad (j = 1, \ldots, m)
$$

(4.342)

Furthermore, holonomic constraints of the form

$$
\phi_j(q, t) = 0 \qquad (j = 1, \ldots, m)
$$

(4.343)

can be expressed in the linear form of (4.342) after differentiation with respect to time.

$$
\dot{\phi}_j(q, \dot{q}, t) = \sum_{i=1}^{n} \frac{\partial \phi_j}{\partial q_i} \dot{q}_i + \frac{\partial \phi_j}{\partial t} = 0 \qquad (j = 1, \ldots, m)
$$

(4.344)

Of course, this linear form is integrable.

Let us define an *ideal constraint* as a workless kinematic constraint which may be either scleronomic or rheonomic. By *workless*, we mean that no work is done by the constraint forces in an arbitrary reversible virtual displacement that is consistent with the instantaneous constraints. For example, an ideal constraint might be a frictionless surface on which sliding occurs, or it might involve rolling contact without slipping. Another example is a knife-edge constraint with no frictional resistance for motion along the knife edge, but with no slipping allowed perpendicular to it.

Let $\mathbf{C}_j$ be the generalized ideal constraint force corresponding to the jth constraint. The virtual work of $\mathbf{C}_j$ in an arbitrary virtual displacement consistent with the instantaneous constraints is

$$\delta W = \mathbf{C}_j \cdot \delta \mathbf{q} = \sum_{i=1}^{n} C_{ji} \delta q_i = 0 \qquad (j = 1, \ldots, m) \tag{4.345}$$

where, for the general nonholonomic case,

$$\sum_{i=1}^{n} \frac{\partial f_j}{\partial \dot{q}_i} \delta q_i = 0 \qquad (j = 1, \ldots, m) \tag{4.346}$$

Assuming the usual case of nonholonomic constraints which are linear in the $\dot{q}$s, we have

$$\sum_{i=1}^{n} a_{ji}(q, t) \delta q_i = 0 \qquad (j = 1, \ldots, m) \tag{4.347}$$

If the constraints are holonomic, the δqs satisfy

$$\sum_{i=1}^{n} \frac{\partial \phi_j}{\partial q_i} \delta q_i = 0 \qquad (j = 1, \ldots, m) \tag{4.348}$$

Equation (4.345) states that the ideal constraint force vector $\mathbf{C}_j$ and an allowable virtual displacement $\delta \mathbf{q}$ are orthogonal in n-dimensional configuration space. A comparison of (4.345) and (4.348) shows that C_{ji} and $\partial \phi_j / \partial q_i$ are proportional for any given j, so $\mathbf{C}_j$ is directed normal to the constraint surface; that is, in the direction of the gradient of $\phi_j(q, t)$ in q-space. Similarly, a comparison of (4.345) and (4.346) indicates that $\mathbf{C}_j$ is directed normal to the constraint surface in velocity space.

A virtual velocity $\delta \mathbf{w}$ is subject to instantaneous constraint equations of the form

$$\sum_{i=1}^{n} \frac{\partial f_j}{\partial \dot{q}_i} \delta w_i = 0 \qquad (j = 1, \ldots, m) \tag{4.349}$$

or

$$\sum_{i=1}^{n} a_{ji} \delta w_i = 0 \qquad (j = 1, \ldots, m) \tag{4.350}$$

A comparison of (4.346) and (4.349) or (4.347) and (4.350) shows that the permitted directions of $\delta \mathbf{q}$ and $\delta \mathbf{w}$ are the same, namely, in the tangent plane at the operating point P in velocity space (Fig. 4.7). The direction of an ideal constraint force $\mathbf{C}_j$ is perpendicular to this tangent plane. Note that, for a holonomic constraint, the tangent plane at the operating point in configuration space has the same orientation as the corresponding constraint plane in velocity space.

Now let us define a *conservative constraint* to be an ideal constraint which meets the additional condition that

$$\mathbf{C}_j \cdot \dot{\mathbf{q}} = 0 \tag{4.351}$$

that is, the generalized constraint force $\mathbf{C}_j$ does no work in any possible actual motion of the system. We found earlier from (4.345) and (4.346) that the components C_{ji} and $\partial f_j / \partial \dot{q}_i$

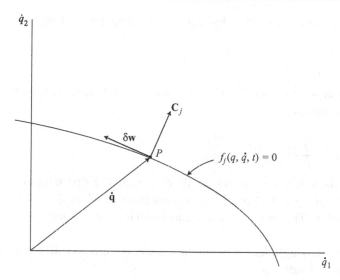

Figure 4.7.

are proportional. Hence,

$$\sum_{i=1}^{n} \frac{\partial f_j}{\partial \dot{q}_i} \dot{q}_i = 0 \tag{4.352}$$

This is just the condition that $f_j(q, \dot{q}, t)$ is a *homogeneous function* of the $\dot{q}$s. In other words, if a generalized velocity $\dot{\mathbf{q}}$ satisfies the constraint, then that velocity multiplied by an arbitrary scalar constant will also satisfy the constraint. This implies that the corresponding constraint surface in velocity space can be generated by sweeping a straight line passing through the origin. It is clear that the common case of a plane passing through the origin is included, but other possibilities exist, such as, for example, a conical surface with its vertex at the origin.

The homogeneity condition for a conservative constraint requires that a_{jt} be zero for the common case of a linear nonholonomic constraint, that is, it must be *catastatic*. For a *holonomic constraint* to be conservative, it must be *scleronomic*. Thus, coefficients of the form $a_{ji}(q, t)$ are acceptable in the nonholonomic case, but any holonomic constraint function must be of the form $\phi_j(q)$.

It is possible that a constraint which does not meet the homogeneity condition for a set of generalized coordinates may be homogeneous in form or may disappear entirely for another choice of generalized coordinates. Thus, whether a constraint is classed as conservative or not may depend upon the choice of coordinates.

Conservative system

A conservative system can be defined as a dynamical system for which an energy integral can be found. As an example, let us consider a dynamical system having m nonholonomic constraints of the general form

$$f_j(q, \dot{q}, t) = 0 \qquad (j = 1, \ldots, m) \tag{4.353}$$

Assume that the system can be described by Lagrange's equation in the form

$$\frac{d}{dt}\left(\frac{\partial L}{\partial \dot{q}_i}\right) - \frac{\partial L}{\partial q_i} = \sum_{j=1}^{m} \lambda_j \frac{\partial f_j}{\partial \dot{q}_i} \qquad (i = 1, \ldots, n) \tag{4.354}$$

Now let us use the same procedure that we employed previously in (2.149)–(2.155). The result in this more general case is

$$\dot{E} = \dot{T}_2 - \dot{T}_0 + \dot{V} = \sum_{i=1}^{n} \sum_{j=1}^{m} \lambda_j \frac{\partial f_j}{\partial \dot{q}_i} \dot{q}_i - \frac{\partial L}{\partial t} \tag{4.355}$$

where the λs are usually nonzero and not easily evaluated. The first term on the right will be zero, however, if $f_j(q, \dot{q}, t)$ is a homogeneous function of the $\dot{q}$s, that is, if each constraint is conservative. The term $\partial L/\partial t$ will equal zero if neither T nor V is an explicit function of time.

In summary, a system having holonomic or nonholonomic constraints will be *conservative* if it meets the following conditions:

1. The standard form of Lagrange's equation, as given by (4.354), applies.
2. All constraints are conservative.
3. The Lagrangian function $L = T - V$ is not an explicit function of time.

These are *sufficient conditions* for a conservative system. Note that the conserved integral of the motion

$$E(q, \dot{q}) = T_2 - T_0 + V \tag{4.356}$$

is equal in value to the Hamiltonian function $H(q, p)$.

Work and energy rates

Consider a system of N rigid bodies. The forces acting on the ith body are equivalent to a force $\mathbf{F}_i$ acting at a reference point P_i plus a couple $\mathbf{M}_i$. Let $\mathbf{v}_i$ be the velocity of point P_i, fixed in the ith body, and let ω_i be the angular velocity of the body. Assume linear nonholonomic constraints.

From the principle of work and kinetic energy, and summing over the N bodies,

$$\dot{T} = \dot{W} = \sum_{i=1}^{N} (\mathbf{F}_i \cdot \mathbf{v}_i + \mathbf{M}_i \cdot \omega_i) \tag{4.357}$$

where $\mathbf{F}_i$ and $\mathbf{M}_i$ may include constraint forces as well as applied forces.

Now let us assume that a portion of $\mathbf{F}_i$ and $\mathbf{M}_i$ arise from a potential energy function $V(q, t)$, but the remaining primed quantities $\mathbf{F}'_i$ and $\mathbf{M}'_i$ do not. Thus, we have

$$\mathbf{F}_i = \mathbf{F}'_i - \frac{\partial V}{\partial \mathbf{r}_i} \qquad (i = 1, \ldots, N) \tag{4.358}$$

$$\mathbf{M}_i = \mathbf{M}'_i - \frac{\partial V}{\partial \theta_i} \qquad (i = 1, \ldots, N) \tag{4.359}$$

where $\mathbf{r}_i$ is the position vector of P_i and $\omega_i = d\theta_i/dt$. In terms of Euler angles, we use the notation

$$\frac{\partial V}{\partial \theta} = \frac{\partial V}{\partial \psi}\mathbf{e}_\psi + \frac{\partial V}{\partial \theta}\mathbf{e}_\theta + \frac{\partial V}{\partial \phi}\mathbf{e}_\phi \qquad (4.360)$$

$$\omega = \dot\psi\mathbf{e}_\psi + \dot\theta\mathbf{e}_\theta + \dot\phi\mathbf{e}_\phi \qquad (4.361)$$

A similar notation is used for $\partial V/\partial \mathbf{r}$. From (4.357)–(4.359), we obtain

$$\dot T = \sum_{i=1}^{N}(\mathbf{F}'_i \cdot \mathbf{v}_i + \mathbf{M}'_i \cdot \omega_i) - \sum_{i=1}^{N}\left(\frac{\partial V}{\partial \mathbf{r}_i} \cdot \mathbf{v}_i + \frac{\partial V}{\partial \theta_i} \cdot \omega_i\right) \qquad (4.362)$$

But,

$$\dot V = \sum_{i=1}^{N}\left(\frac{\partial V}{\partial \mathbf{r}_i} \cdot \mathbf{v}_i + \frac{\partial V}{\partial \theta_i} \cdot \omega_i\right) + \frac{\partial V}{\partial t} \qquad (4.363)$$

Hence, we find that the rate of change of the total energy is

$$\dot T + \dot V = \sum_{i=1}^{N}(\mathbf{F}'_i \cdot \mathbf{v}_i + \mathbf{M}'_i \cdot \omega_i) + \frac{\partial V}{\partial t} \qquad (4.364)$$

Note that constraint forces are included in $\mathbf{F}'_i$ and $\mathbf{M}'_i$, but forces derived from potential energy are not included.

Now let us consider a system whose motion is described in terms of independent quasi-velocities. Assume a system of N rigid bodies and start with the general dynamical equation in the form

$$\sum_{i=1}^{N}[m_i\dot{\mathbf{v}}_i \cdot \gamma_{ij} + (\mathbf{I}_{ci} \cdot \dot\omega_i + \omega_i \times \mathbf{I}_{ci} \cdot \omega_i) \cdot \beta_{ij}] = Q_j \qquad (j = 1, \dots, n-m) \qquad (4.365)$$

where the *center of mass* is chosen as the reference point for each body.

Multiplying (4.365) by u_j and summing over j, we obtain

$$\sum_{i=1}^{N}\sum_{j=1}^{n-m}[m_i\dot{\mathbf{v}}_i \cdot \gamma_{ij}u_j + (\mathbf{I}_{ci} \cdot \dot\omega_i + \omega_i \times \mathbf{I}_{ci} \cdot \omega_i) \cdot \beta_{ij}u_j] = \sum_{j=1}^{n-m}Q_ju_j \qquad (4.366)$$

Recall that $\gamma_{ij} = \partial\mathbf{v}_i/\partial u_j$ and $\beta_{ij} = \partial\omega_i/\partial u_j$. The kinetic energy is

$$T = \frac{1}{2}\sum_{i=1}^{N}m_i\mathbf{v}_i \cdot \mathbf{v}_i + \frac{1}{2}\sum_{i=1}^{N}\omega_i \cdot \mathbf{I}_{ci} \cdot \omega_i \qquad (4.367)$$

Then, using Euler's theorem on homogeneous functions, we see that

$$\sum_{j=1}^{n-m}\frac{\partial T}{\partial u_j}u_j = \sum_{i=1}^{N}\sum_{j=1}^{n-m}\left(m_i\mathbf{v}_i \cdot \frac{\partial\mathbf{v}_i}{\partial u_j}u_j + \omega_i \cdot \mathbf{I}_{ci} \cdot \frac{\partial\omega_i}{\partial u_j}u_j\right)$$

$$= \sum_{i=1}^{N}\sum_{j=1}^{n-m}(m_i\mathbf{v}_i \cdot \gamma_{ij}u_j + \omega_i \cdot \mathbf{I}_{ci} \cdot \beta_{ij}u_j)$$

$$= 2T_2 + T_1 \qquad (4.368)$$

where T_2 is quadratic and T_1 is linear in the us.

Now let us differentiate (4.368) with respect to time and recall that

$$\sum_{j=1}^{n-m} \gamma_{ij} u_j = \mathbf{v}_i - \gamma_{it} \tag{4.369}$$

$$\sum_{j=1}^{n-m} \beta_{ij} u_j = \omega_i - \beta_{it} \tag{4.370}$$

We obtain

$$\frac{d}{dt}\left[\sum_{i=1}^{N}\sum_{j=1}^{n-m}(m_i\mathbf{v}_i\cdot\gamma_{ij}u_j + \omega_i\cdot\mathbf{I}_{ci}\cdot\beta_{ij}u_j)\right] = 2\dot{T}_2 + \dot{T}_1$$

$$= \sum_{i=1}^{N}\sum_{j=1}^{n-m}[m_i\dot{\mathbf{v}}_i\cdot\gamma_{ij}u_j + (\mathbf{I}_{ci}\cdot\dot{\omega}_i + \omega_i\times\mathbf{I}_{ci}\cdot\omega_i)\cdot\beta_{ij}u_j]$$

$$+ \sum_{i=1}^{N}[m_i\mathbf{v}_i\cdot(\dot{\mathbf{v}}_i - \dot{\gamma}_{it}) + \omega_i\cdot\mathbf{I}_{ci}\cdot(\dot{\omega}_i - \dot{\beta}_{it})] \tag{4.371}$$

Differentiating (4.367) with respect to time, we obtain

$$\dot{T} = \sum_{i=1}^{N}(m_i\mathbf{v}_i\cdot\dot{\mathbf{v}}_i + \omega_i\cdot\mathbf{I}_{ci}\cdot\dot{\omega}_i) = \dot{T}_2 + \dot{T}_1 + \dot{T}_0 \tag{4.372}$$

where we note that

$$\omega_i\cdot\dot{\mathbf{I}}_{ci}\cdot\omega_i = 0 \tag{4.373}$$

Next, subtract (4.372) from (4.371). The result is

$$\dot{T}_2 - \dot{T}_0 = \sum_{i=1}^{N}\sum_{j=1}^{n-m}[m_i\dot{\mathbf{v}}_i\cdot\gamma_{ij}u_j + (\mathbf{I}_{ci}\cdot\dot{\omega}_i + \omega_i\times\mathbf{I}_{ci}\cdot\omega_i)\cdot\beta_{ij}u_j]$$

$$- \sum_{i=1}^{N}(m_i\mathbf{v}_i\cdot\dot{\gamma}_{it} + \omega_i\cdot\mathbf{I}_{ci}\cdot\dot{\beta}_{it}) \tag{4.374}$$

Then, using (4.366), we obtain

$$\dot{T}_2 - \dot{T}_0 = \sum_{j=1}^{n-m}Q_j u_j - \sum_{i=1}^{N}m_i\mathbf{v}_i\cdot\dot{\gamma}_{it} - \sum_{i=1}^{N}\omega_i\cdot\mathbf{I}_{ci}\cdot\dot{\beta}_{it} \tag{4.375}$$

Let us assume that a portion of Q_j is obtained from a potential function $V(q, t)$. Thus, we can write

$$Q_j = Q_j' - \sum_{k=1}^{n}\frac{\partial V}{\partial q_k}\frac{\partial\dot{q}_k}{\partial u_j} = Q_j' - \sum_{k=1}^{n}\frac{\partial V}{\partial q_k}\Phi_{kj} \tag{4.376}$$

where Q'_j is that portion which is not obtainable from a potential energy function. Also note that

$$\dot{q}_k = \sum_{j=1}^{n-m} \Phi_{kj} u_j + \Phi_{kt} \tag{4.377}$$

and therefore

$$\dot{V} - \frac{\partial V}{\partial t} = \sum_{j=1}^{n-m} \sum_{k=1}^{n} \frac{\partial V}{\partial q_k} \Phi_{kj} u_j + \sum_{k=1}^{n} \frac{\partial V}{\partial q_k} \Phi_{kt} \tag{4.378}$$

Finally, adding (4.375) and (4.378), and using (4.376), we obtain

$$\dot{E} = \dot{T}_2 - \dot{T}_0 + \dot{V} = \sum_{j=1}^{n-m} Q'_j u_j + \frac{\partial V}{\partial t} + \sum_{k=1}^{n} \frac{\partial V}{\partial q_k} \Phi_{kt}$$
$$- \sum_{i=1}^{N} m_i \mathbf{v}_i \cdot \dot{\boldsymbol{\gamma}}_{it} - \sum_{i=1}^{N} \boldsymbol{\omega}_i \cdot \mathbf{I}_{ci} \cdot \dot{\boldsymbol{\beta}}_{it} \tag{4.379}$$

This is the general energy rate equation for a system of rigid bodies. An alternate form is

$$\dot{E} = \dot{T}_2 - \dot{T}_0 + \dot{V} = \sum_{j=1}^{n-m} Q'_j u_j + \frac{\partial V}{\partial t} + \sum_{k=1}^{n} \frac{\partial V}{\partial q_k} \Phi_{kt} - \sum_{i=1}^{N} \mathbf{p}_i \cdot \dot{\boldsymbol{\gamma}}_{it} - \sum_{i=1}^{N} \mathbf{H}_{ci} \cdot \dot{\boldsymbol{\beta}}_{it}$$
$$\tag{4.380}$$

where

$$\mathbf{p}_i = m_i \mathbf{v}_i \tag{4.381}$$

$$\mathbf{H}_{ci} = \mathbf{I}_{ci} \cdot \boldsymbol{\omega}_i \tag{4.382}$$

and we use a center of mass reference point on each body.

The meaning of these energy rate equations can be clarified by noting that $\boldsymbol{\gamma}_{it}$ represents the velocity of the ith reference point when all the us are set equal to zero. Similarly, $\boldsymbol{\beta}_{it}$ is equal to the angular velocity of the ith body if all us equal zero. Additionally, note that Φ_{kt} is equal to the value of $\dot{q}_k$ when all the us are set equal to zero.

Another approach to energy rate calculations is to begin with the Boltzmann–Hamel equation in the general form of (4.85)

$$\frac{d}{dt}\left(\frac{\partial T}{\partial u_r}\right) - \sum_{i=1}^{n} \frac{\partial T}{\partial q_i} \Phi_{ir} + \sum_{j=1}^{m} \sum_{l=1}^{n-m} \frac{\partial T}{\partial u_j} \gamma_{rl}^j u_l + \sum_{j=1}^{n} \frac{\partial T}{\partial u_j} \gamma_r^j = Q_r$$
$$(r = 1, \ldots, n-m) \tag{4.383}$$

Multiply by u_r and sum over r, using (4.376). The result is

$$\sum_{r=1}^{n-m} \frac{d}{dt}\left(\frac{\partial T}{\partial u_r}\right) u_r - \sum_{i=1}^{n} \sum_{r=1}^{n-m} \frac{\partial T}{\partial q_i} \Phi_{ir} u_r + \sum_{j=1}^{n} \sum_{l=1}^{n-m} \sum_{r=1}^{n-m} \frac{\partial T}{\partial u_j} \gamma_{rl}^j u_l u_r$$
$$+ \sum_{j=1}^{n} \sum_{r=1}^{n-m} \frac{\partial T}{\partial u_j} \gamma_r^j u_r = \sum_{r=1}^{n-m} Q'_r u_r - \sum_{i=1}^{n} \sum_{r=1}^{n-m} \frac{\partial V}{\partial q_i} \Phi_{ir} u_r \tag{4.384}$$

Using Euler's theorem, as in (4.368), we find that

$$\frac{d}{dt}\left(\sum_{r=1}^{n-m}\frac{\partial T}{\partial u_r}u_r\right) = \sum_{r=1}^{n-m}\frac{d}{dt}\left(\frac{\partial T}{\partial u_r}\right)u_r + \sum_{r=1}^{n-m}\frac{\partial T}{\partial u_r}\dot{u}_r = 2\dot{T}_2 + \dot{T}_1 \tag{4.385}$$

Now

$$\dot{T} = \sum_{i=1}^{n}\frac{\partial T}{\partial q_i}\dot{q}_i + \sum_{r=1}^{n-m}\frac{\partial T}{\partial u_r}\dot{u}_r + \frac{\partial T}{\partial t} = \dot{T}_2 + \dot{T}_1 + \dot{T}_0 \tag{4.386}$$

so, from (4.385) and (4.386), we obtain

$$\sum_{r=1}^{n-m}\frac{d}{dt}\left(\frac{\partial T}{\partial u_r}\right)u_r = \dot{T}_2 - \dot{T}_0 + \sum_{i=1}^{n}\frac{\partial T}{\partial q_i}\dot{q}_i + \frac{\partial T}{\partial t} \tag{4.387}$$

Furthermore, we see that

$$\dot{V} = \sum_{i=1}^{n}\frac{\partial V}{\partial q_i}\dot{q}_i + \frac{\partial V}{\partial t} \tag{4.388}$$

and

$$\sum_{r=1}^{n-m}\Phi_{ir}u_r = \dot{q}_i - \Phi_{it} \tag{4.389}$$

From (4.387) and (4.388), we see that

$$\dot{T}_2 - \dot{T}_0 + \dot{V} = \sum_{r=1}^{n-m}\frac{d}{dt}\left(\frac{\partial T}{\partial u_r}\right)u_r - \sum_{i=1}^{n}\frac{\partial L}{\partial q_i}\dot{q}_i - \frac{\partial L}{\partial t} \tag{4.390}$$

where $L(q, u, t) = T(q, u, t) - V(q, t)$. Because of the skew symmetry of γ_{rl}^{j} with respect to r and l, we note that

$$\sum_{l=1}^{n-m}\sum_{r=1}^{n-m}\gamma_{rl}^{j}u_r u_l = 0 \tag{4.391}$$

Then, from (4.384) and (4.389), we obtain

$$\sum_{r=1}^{n-m}\frac{d}{dt}\left(\frac{\partial T}{\partial u_r}\right)u_r = \sum_{r=1}^{n-m}Q_r'u_r + \sum_{i=1}^{n}\frac{\partial L}{\partial q_i}(\dot{q}_i - \Phi_{it}) - \sum_{j=1}^{n}\sum_{r=1}^{n-m}\frac{\partial T}{\partial u_j}\gamma_r^{j}u_r \tag{4.392}$$

Finally, using (4.390) and (4.392), we have the energy rate expression

$$\dot{E} = \dot{T}_2 - \dot{T}_0 + \dot{V} = \sum_{r=1}^{n-m}Q_r'u_r - \frac{\partial L}{\partial t} - \sum_{i=1}^{n}\frac{\partial L}{\partial q_i}\Phi_{it} - \sum_{j=1}^{n}\sum_{r=1}^{n-m}\frac{\partial T}{\partial u_j}\gamma_r^{j}u_r \tag{4.393}$$

Let us compare the energy rate equations (4.380) and (4.393). For a general system of N rigid bodies, we have the corresponding terms

$$\sum_{i=1}^{N}(\mathbf{p}_i \cdot \dot{\gamma}_{it} + \mathbf{H}_{ci} \cdot \dot{\beta}_{it}) = \frac{\partial T}{\partial t} + \sum_{k=1}^{n}\frac{\partial T}{\partial q_k}\Phi_{kt} + \sum_{j=1}^{n}\sum_{r=1}^{n-m}\frac{\partial T}{\partial u_j}\gamma_r^{j}u_r \tag{4.394}$$

where

$$\gamma_r^j = \sum_{i=1}^{n} \sum_{k=1}^{n} \left(\frac{\partial \Psi_{ji}}{\partial q_k} - \frac{\partial \Psi_{jk}}{\partial q_i} \right) \Phi_{kt} \Phi_{ir} + \sum_{i=1}^{n} \left(\frac{\partial \Psi_{ji}}{\partial t} - \frac{\partial \Psi_{jt}}{\partial q_i} \right) \Phi_{ir} \tag{4.395}$$

It appears that, in the general nonholonomic case, the left-hand side of (4.394) is easier to evaluate that is its right-hand side. Thus, (4.380) is more direct than (4.393) in the general case.

From (4.380), we see that sufficient conditions for a *conservative system*, implying a constant value of E, are:

1. $Q'_j = 0$ for all j, that is, all the generalized forces Q_j are derivable from a potential energy function of the form $V(q)$.
2. The functions Φ_{kt}, $\mathbf{p}_i \cdot \dot{\gamma}_{it}$, and $\mathbf{H}_{ci} \cdot \dot{\beta}_{it}$ are all continuously equal to zero.

Example 4.12 A particle of mass m can slide on a wire in the form of a circle of radius r which rotates about a vertical diameter with a variable angular velocity $\Omega(t)$ (Fig. 4.8). We wish to determine the energy rate $\dot{E}$.

This is a rheonomic holonomic system with one generalized coordinate and no constraints. Lagrange's equation applies and (4.355) reduces to

$$\dot{E} = \dot{T}_2 - \dot{T}_0 + \dot{V} = -\frac{\partial L}{\partial t} \tag{4.396}$$

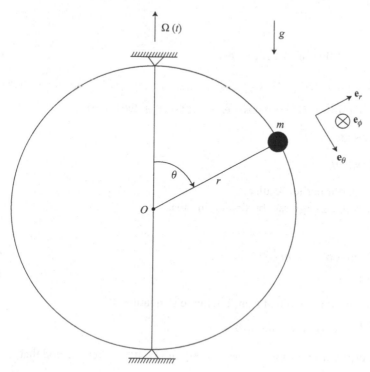

Figure 4.8.

We see that

$$T = \frac{1}{2}mr^2(\dot{\theta}^2 + \Omega^2 \sin^2\theta) \tag{4.397}$$

$$V = mgr\cos\theta \tag{4.398}$$

Thus, we obtain

$$\dot{E} = -\frac{\partial T}{\partial t} = -mr^2\Omega\dot{\Omega}\sin^2\theta \tag{4.399}$$

For the particular case in which Ω is constant, we see that $E = T_2 - T_0 + V$ has a constant value and the system is conservative. The total energy $T + V$ is not constant, however, because there is a torque about the vertical axis that is required to keep the angular velocity Ω constant even though θ is varying.

If we use the Boltzmann–Hamel approach, as given in (4.393), the same result occurs. Here we can take $u_1 = \dot{\theta}$ and note that both Φ_{it} and γ_r^j vanish.

Finally, if we use the general energy rate expression in (4.379), we find that it reduces to

$$\dot{E} = -m\mathbf{v}_1 \cdot \dot{\boldsymbol{\gamma}}_{1t} \tag{4.400}$$

where

$$\mathbf{v}_1 = r\dot{\theta}\mathbf{e}_\theta + r\Omega\sin\theta\,\mathbf{e}_\phi \tag{4.401}$$

$$\boldsymbol{\gamma}_{1t} = r\Omega\sin\theta\,\mathbf{e}_\phi \tag{4.402}$$

Now

$$\dot{\mathbf{e}}_\phi = \boldsymbol{\Omega} \times \mathbf{e}_\phi = -\Omega\sin\theta\,\mathbf{e}_r - \Omega\cos\theta\,\mathbf{e}_\theta \tag{4.403}$$

so

$$\dot{\boldsymbol{\gamma}}_{1t} = -r\Omega^2\sin^2\theta\,\mathbf{e}_r - r\Omega^2\sin\theta\cos\theta\,\mathbf{e}_\theta + r(\dot{\Omega}\sin\theta + \Omega\dot{\theta}\cos\theta)\mathbf{e}_\phi \tag{4.404}$$

Thus, (4.400) results in

$$\dot{E} = -mr^2\Omega\dot{\Omega}\sin^2\theta \tag{4.405}$$

in agreement with our earlier results.

One can check the energy rate by first noting that

$$\dot{T}_2 = mr^2\dot{\theta}\ddot{\theta} \tag{4.406}$$

$$\dot{T}_0 = mr^2\Omega^2\dot{\theta}\sin\theta\cos\theta + mr^2\Omega\dot{\Omega}\sin^2\theta \tag{4.407}$$

$$\dot{V} = -mgr\dot{\theta}\sin\theta \tag{4.408}$$

The equation of motion, obtained from Lagrange's equation, is

$$mr^2\ddot{\theta} - mr^2\Omega^2\sin\theta\cos\theta - mgr\sin\theta = 0 \tag{4.409}$$

Substitute the expression for $mr^2\ddot{\theta}$ from (4.409) into (4.406). Then we find that

$$\dot{E} = \dot{T}_2 - \dot{T}_0 + \dot{V} = -mr^2\Omega\dot{\Omega}\sin^2\theta \tag{4.410}$$

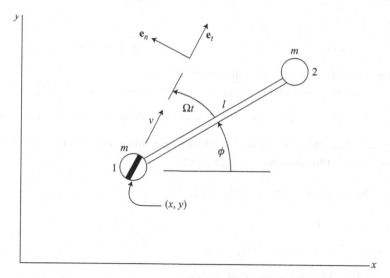

Figure 4.9.

which checks with (4.405).

Example 4.13 A rheonomic nonholonomic system consists of two particles, each of mass m which are connected by a massless rod of length l, as shown in Fig. 4.9. Particle 1 has a nonholonomic constraint in the form of a knife-edge which rotates at a constant rate Ω relative to the rod. Let us choose (x, y, ϕ) as qs, and let

$$u_1 = v, \qquad u_2 = \dot{\phi} \tag{4.411}$$

where v is a quasi-velocity. We wish to find the differential equations of motion and to evaluate the energy rate $\dot{E}$.

First, the differential equations of motion are obtained from the fundamental equation for a system of particles, namely,

$$\frac{d}{dt}\left(\frac{\partial T}{\partial u_j}\right) - \sum_{i=1}^{N} m_i \mathbf{v}_i \cdot \dot{\gamma}_{ij} = Q_j \tag{4.412}$$

Assume that the xy-plane is horizontal. The only constraint force is perpendicular to the knife edge and does no work in an arbitrary virtual displacement. Therefore, each generalized applied force Q_j is zero. The velocities of the two particles are

$$\mathbf{v}_1 = v\mathbf{e}_t, \qquad \mathbf{v}_2 = (v + l\dot{\phi}\sin\Omega t)\mathbf{e}_t + l\dot{\phi}\cos\Omega t \, \mathbf{e}_n \tag{4.413}$$

Thus, the constrained kinetic energy is

$$T = \frac{1}{2}m\left(\mathbf{v}_1^2 + \mathbf{v}_2^2\right) = \frac{1}{2}m(2v^2 + l^2\dot{\phi}^2 + 2lv\dot{\phi}\sin\Omega t) \tag{4.414}$$

The velocity coefficients are

$$\gamma_{ij} = \frac{\partial \mathbf{v}_i}{\partial u_j} \tag{4.415}$$

resulting in

$$\gamma_{11} = \mathbf{e}_t, \qquad \gamma_{12} = 0, \qquad \gamma_{21} = \mathbf{e}_t, \qquad \gamma_{22} = l(\sin \Omega t \, \mathbf{e}_t + \cos \Omega t \, \mathbf{e}_n) \tag{4.416}$$

Also, we note that the γ_{it} coefficients are equal to zero.

A differentiation of (4.416) with respect to time results in

$$\dot{\gamma}_{12} = 0, \qquad \dot{\gamma}_{11} = \dot{\gamma}_{21} = (\dot{\phi} + \Omega)\mathbf{e}_n, \qquad \dot{\gamma}_{22} = l\dot{\phi}(-\cos \Omega t \, \mathbf{e}_t + \sin \Omega t \, \mathbf{e}_n) \tag{4.417}$$

where we note that $\mathbf{e}_t$ and $\mathbf{e}_n$ rotate counterclockwise with an angular velocity $(\dot{\phi} + \Omega)$.

The u_1 equation is obtained by first evaluating

$$\frac{d}{dt}\left(\frac{\partial T}{\partial v}\right) = 2m\dot{v} + ml\ddot{\phi} \sin \Omega t + ml\Omega\dot{\phi} \cos \Omega t \tag{4.418}$$

In addition, we find that

$$m(\mathbf{v}_1 \cdot \dot{\gamma}_{11} + \mathbf{v}_2 \cdot \dot{\gamma}_{21}) = ml\dot{\phi}(\dot{\phi} + \Omega) \cos \Omega t \tag{4.419}$$

Then, using (4.412), the first equation of motion is

$$2m\dot{v} + ml\ddot{\phi} \sin \Omega t - ml\dot{\phi}^2 \cos \Omega t = 0 \tag{4.420}$$

In a similar manner, we obtain

$$\frac{d}{dt}\left(\frac{\partial T}{\partial \dot{\phi}}\right) = ml^2\ddot{\phi} + ml\dot{v} \sin \Omega t + ml\Omega v \cos \Omega t \tag{4.421}$$

$$m(\mathbf{v}_1 \cdot \dot{\gamma}_{12} + \mathbf{v}_2 \cdot \dot{\gamma}_{22}) = -mlv\dot{\phi} \cos \Omega t \tag{4.422}$$

The second equation of motion is

$$ml^2\ddot{\phi} + ml\dot{v} \sin \Omega t + mlv(\dot{\phi} + \Omega) \cos \Omega t = 0 \tag{4.423}$$

These two equations of motion constitute a minimum set for this system which has two degrees of freedom.

Equation (4.420) can be interpreted as stating that the rate of change of linear momentum in the direction of the knife edge is equal to zero. The second equation of motion, (4.423), states that, if particle 1 is chosen as a noninertial reference point, the rate of change of angular momentum is equal to the inertial moment due to the acceleration of particle 1. Thus, it is convenient to think in terms of an accelerating but nonrotating reference frame in this instance.

The energy rate $\dot{E}$ for a system of N particles can be written in the form

$$\dot{E} = \sum_{r=1}^{n-m} Q_r' u_r + \frac{\partial V}{\partial t} + \sum_{k=1}^{n} \frac{\partial V}{\partial q_k} \Phi_{kt} - \sum_{i=1}^{N} m_i \mathbf{v}_i \cdot \dot{\gamma}_{it} \tag{4.424}$$

in accordance with (4.379). We find that Q_r', V, Φ_{kt} and γ_{it} are all equal to zero, so $\dot{E} = 0$. Thus, the energy function E, which in this case is the total kinetic energy, is constant during the motion.

This is an example of a system for which the kinetic energy is an explicit function of time and yet it is conservative.

Second method   If the Lagrangian method is used, we have three differential equations of motion involving Lagrange multipliers. Also, there is the nonholonomic constraint equation

$$-\dot{x}\sin(\phi + \Omega t) + \dot{y}\cos(\phi + \Omega t) = 0 \tag{4.425}$$

which states that the velocity of particle 1 normal to the knife edge is zero.

The kinetic energy of the unconstrained system is

$$T = \frac{1}{2}m[2\dot{x}^2 + 2\dot{y}^2 + l^2\dot{\phi}^2 + 2l\dot{\phi}(-\dot{x}\sin\phi + \dot{y}\cos\phi)] \tag{4.426}$$

We note that T is not an explicit function of time, and we can take the potential energy function V equal to zero; hence $\partial L/\partial t = 0$. Furthermore, the nonholonomic constraint is homogeneous and linear in the $\dot{q}$s, and is therefore conservative. Thus, the sufficient conditions for a conservative system are satisfied, implying, in this case, that the total kinetic energy is a constant of the motion.

Third method   Let us use the Boltzmann–Hamel approach. The equations for a complete set of three us in terms of $\dot{q}$s are

$$u_1 = v = \dot{x}\cos(\phi + \Omega t) + \dot{y}\sin(\phi + \Omega t) \tag{4.427}$$

$$u_2 = \dot{\phi} \tag{4.428}$$

$$u_3 = -\dot{x}\sin(\phi + \Omega t) + \dot{y}\cos(\phi + \Omega t) \tag{4.429}$$

The nonholonomic constraint is applied by setting u_3 equal to zero.

The unconstrained kinetic energy is

$$T = m\left[u_1^2 + \frac{1}{2}l^2u_2^2 + u_3^2 + lu_2(u_1\sin\Omega t + u_3\cos\Omega t)\right] \tag{4.430}$$

and the general Boltzmann–Hamel equation (4.85) reduces to

$$\frac{d}{dt}\left(\frac{\partial T}{\partial u_r}\right) + \sum_{j=1}^{3}\sum_{l=1}^{2}\frac{\partial T}{\partial u_j}\gamma_{rl}^j u_l + \sum_{j=1}^{3}\frac{\partial T}{\partial u_j}\gamma_r^j = 0 \qquad (r = 1, 2) \tag{4.431}$$

In evaluating γ_{rl}^j and γ_r^j, we note that the Ψ_{ji} coefficients are explicit functions of time, in general, but all the Ψ_{jt} coefficients vanish, as well as the Φ_{kt}.

After a rather lengthy calculation, the equations of motion found earlier in (4.420) and (4.423) are obtained. The Boltzmann–Hamel energy rate expression in (4.393) reduces for this example to

$$\dot{E} = -\frac{\partial T}{\partial t} - \sum_{j=1}^{3}\sum_{r=1}^{2}\frac{\partial T}{\partial u_j}\gamma_r^j u_r \tag{4.432}$$

We find that

$$\frac{\partial T}{\partial t} = ml\Omega v\dot{\phi}\cos\Omega t \tag{4.433}$$

and

$$\sum_{j=1}^{3}\sum_{r=1}^{2}\frac{\partial T}{\partial u_j}\gamma_r^j u_r = -ml\Omega v\phi\cos\Omega t \tag{4.434}$$

Hence, $\dot{E} = 0$ and $E = T$ is a constant of the motion even though T is an explicit function of time. Note that the kinetic energy is a homogeneous quadratic function of the us.

Comparing the three methods which were presented for analyzing this rheonomic non-holonomic system, the first method using the fundamental dynamical equation for a system of particles would seem to be preferable. It provides a minimum set of equations of motion without Lagrange multipliers. Furthermore, the energy rate is found to be zero by inspection.

4.8   Summary of differential methods

In the study of differential methods in the dynamics of systems of particles or rigid bodies, it is well to begin with Newton's law of motion. Angular momentum methods can also be employed, resulting in Euler's equations for the rotational motion of rigid bodies. These elementary approaches frequently require the introduction of constraint forces and moments as additional variables in the dynamical equations, thereby complicating the analysis.

Lagrange's equations, when applied to holonomic systems with independent qs, result in a minimum set of equations of motion without the necessity of solving for constraint forces. In other words, although one set of qs may be subject to holonomic constraints, another set of qs can be found which are independent and are consistent with the previous constraints. No generalized constraint forces enter into the Lagrange equations of motion for this system.

On the other hand, if there are nonholonomic constraints, then more qs are required than the number of degrees of freedom. The use of the Lagrangian procedure involves Lagrange multipliers which are associated with generalized constraint forces. If there are n qs and m constraint equations, one obtains n second-order dynamical equations in addition to the m constraint equations.

The use of Maggi's equation eliminates the Lagrange multipliers and results in $(n - m)$ second-order equations of motion plus the m constraint equations. The Lagrange and Maggi methods have the disadvantages, however, that the kinetic energy cannot be written in terms of quasi-velocities, and must be written for the unconstrained system.

In the efficient representation of dynamical systems, it is desirable to obtain a minimal set of $(n - m)$ first-order differential equations of motion. This is possible in the general nonholonomic case if one uses independent quasi-velocities as velocity variables. The remaining four differential methods discussed in this chapter all result in a minimal set of dynamical equations. The use of the Boltzmann–Hamel equation is the most complicated of these methods and requires that the kinetic energy be written for the unconstrained system having n degrees of freedom. The other three methods allow one to assume a constrained system with $(n - m)$ degrees of freedom from the beginning.

Energy rate expressions can be obtained in several forms, depending upon the type of dynamical equations used in their derivation. Each of these expressions can be used to obtain a set of sufficient conditions for a conservative system, that is, a system having a constant energy function E. In the usual case, $E = T_2 - T_0 + V$ rather than the total energy $T + V$.

If the equations of motion for a given system are being integrated numerically, the energy rate $\dot{E}$ can be used as a check on the calculations. This is accomplished by integrating $\dot{E}$ separately with respect to time, and comparing the change in the energy E with that obtained from the integrated solutions to the equations of motion.

A note on quasi-velocities

We have defined quasi-velocities (us) in accordance with the equations

$$u_j = \sum_{i=1}^{n} \Psi_{ji}(q, t)\dot{q}_i + \Psi_{jt}(q, t) \qquad (j = 1, \ldots, n) \tag{4.435}$$

where the right-hand sides are not integrable, in general. Moreover, we assumed that these equations can be solved for the $\dot{q}$s, resulting in

$$\dot{q}_i = \sum_{j=1}^{n} \Phi_{ij}(q, t)u_j + \Phi_{it}(q, t) \qquad (i = 1, \ldots, n) \tag{4.436}$$

Thus, we have assumed that the us and $\dot{q}$s are related linearly. If there are m linear non-holonomic constraints, these are represented by setting the last m us equal to zero. The remaining $(n - m)$ us are independent.

Furthermore, we have expressed velocities and angular velocities in accordance with the linear relations

$$\mathbf{v}_i = \sum_{j=1}^{n} \gamma_{ij}(q, t)u_j + \gamma_{it}(q, t) \tag{4.437}$$

$$\omega_i = \sum_{j=1}^{n} \beta_{ij}(q, t)u_j + \beta_{it}(q, t) \tag{4.438}$$

and we note that the γs and βs can be used in writing the differential equations of motion.

It is possible, however, to define the us in a way such that, in general, the $\dot{q}$s are nonlinear functions of the us. For a system with n qs and m independent nonholonomic constraints, one can define the $(n - m)$ independent us as *a set of parameters which specify the operating point in velocity space*, that is, $\dot{q}$-space. This operating point must lie on each of the m constraint surfaces in velocity space, and therefore on their common intersection.

We need to find expressions for the γs and βs in writing the equations of motion, but the linear equations (4.437) and (4.438) are no longer valid for the more general us. Nevertheless, we can use

$$\gamma_{ij}(q, u, t) = \frac{\partial \mathbf{v}_i(q, u, t)}{\partial u_j} \tag{4.439}$$

$$\beta_{ij}(q, u, t) = \frac{\partial \omega_i(q, u, t)}{\partial u_j} \tag{4.440}$$

A given set of us may not have uniform dimensions, and so the question now arises concerning how the corresponding Qs are to be found. One can use virtual work or, in this case, *virtual power* to evaluate the Qs. For example, if a system of N particles has a force $\mathbf{F}_i$ applied to the ith particle ($i = 1, \ldots, N$), then, using virtual velocities,

$$\sum_{i=1}^{N} \mathbf{F}_i \cdot \delta\mathbf{v}_i = \sum_{j=1}^{n-m} Q_j \delta u_j \qquad (4.441)$$

where

$$\delta\mathbf{v}_i = \sum_{j=1}^{n-m} \frac{\partial \mathbf{v}_i}{\partial u_j} \delta u_j = \sum_{j=1}^{n-m} \gamma_{ij} \delta u_j \qquad (4.442)$$

Then, since the δus are independent, we obtain

$$Q_j = \sum_{i=1}^{N} \mathbf{F}_i \cdot \gamma_{ij} \qquad (j = 1, \ldots, n - m) \qquad (4.443)$$

Example 4.14   Consider the simple case of a particle of mass m which has a force $\mathbf{F}$ applied to it. Let us choose (v, θ, ϕ) as us to represent the velocity vector $\mathbf{v}$ of the particle in three-dimensional velocity space (Fig. 4.10).

This is similar to the use of spherical coordinates to designate a position in ordinary 3-space. The unit vectors $\mathbf{e}_v, \mathbf{e}_\theta, \mathbf{e}_\phi$ form an orthogonal triad with $\mathbf{e}_\phi = \mathbf{e}_v \times \mathbf{e}_\theta$. In place of (4.436) we have the equations

$$\dot{x} = v \sin\theta \cos\phi \qquad (4.444)$$
$$\dot{y} = v \sin\theta \sin\phi \qquad (4.445)$$
$$\dot{z} = v \cos\theta \qquad (4.446)$$

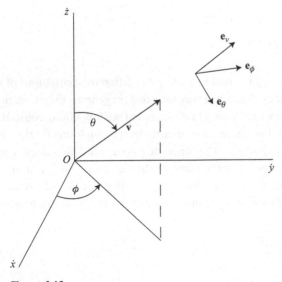

Figure 4.10.

which are nonlinear in the us. Conversely, we find that

$$u_1 = v = \sqrt{\dot{x}^2 + \dot{y}^2 + \dot{z}^2} \tag{4.447}$$

$$u_2 = \theta = \cos^{-1} \frac{\dot{z}}{\sqrt{\dot{x}^2 + \dot{y}^2 + \dot{z}^2}} \tag{4.448}$$

$$u_3 = \phi = \tan^{-1} \frac{\dot{y}}{\dot{x}} \tag{4.449}$$

which are nonlinear and replace (4.435).

First method   Let us use the general dynamical equation in the form

$$\sum_{i=1}^{N} m_i \dot{\mathbf{v}}_i \cdot \boldsymbol{\gamma}_{ij} = Q_j \tag{4.450}$$

The velocity is

$$\mathbf{v} = v \mathbf{e}_v(\theta, \phi) \tag{4.451}$$

and the velocity coefficients, obtained by using (4.439), are

$$\boldsymbol{\gamma}_{11} = \mathbf{e}_v, \qquad \boldsymbol{\gamma}_{12} = v \frac{\partial \mathbf{e}_v}{\partial \theta} = v \mathbf{e}_\theta, \qquad \boldsymbol{\gamma}_{13} = v \frac{\partial \mathbf{e}_v}{\partial \phi} = v \sin \theta \, \mathbf{e}_\phi \tag{4.452}$$

Note that the γs can be functions of the us in contrast to (4.437).

The particle acceleration is

$$\dot{\mathbf{v}} = \dot{v} \mathbf{e}_v + v \dot{\mathbf{e}}_v = \dot{v} \mathbf{e}_v + v \dot{\theta} \mathbf{e}_\theta + v \dot{\phi} \sin \theta \, \mathbf{e}_\phi \tag{4.453}$$

and the force $\mathbf{F}$, in terms of its components, is

$$\mathbf{F} = F_v \mathbf{e}_v + F_\theta \mathbf{e}_\theta + F_\phi \mathbf{e}_\phi \tag{4.454}$$

Hence, using (4.443), we find that

$$Q_1 = F_v, \qquad Q_2 = F_\theta v, \qquad Q_3 = F_\phi v \sin \theta \tag{4.455}$$

Finally, using (4.450), the v equation is

$$m\dot{v} = F_v \tag{4.456}$$

Similarly, the θ equation is

$$mv^2 \dot{\theta} = F_\theta v$$

or

$$mv\dot{\theta} = F_\theta \tag{4.457}$$

The ϕ equation is

$$mv^2 \dot{\phi} \sin^2 \theta = F_\phi v \sin \theta$$

or

$$mv\dot{\phi}\sin\theta = F_\phi \tag{4.458}$$

Actually, knowing $\dot{\mathbf{v}}$, these equations of motion could have been obtained directly by applying Newton's law in the three orthogonal directions.

Second method   Let us consider the Gibbs–Appell equation

$$\frac{\partial S}{\partial \dot{u}_j} = Q_j \tag{4.459}$$

where, in this case,

$$S = \frac{1}{2}m\dot{\mathbf{v}}^2 = \frac{1}{2}m[\dot{v}^2 + (v\dot{\theta})^2 + (v\dot{\phi}\sin\theta)^2] \tag{4.460}$$

The v equation of motion is

$$\frac{\partial S}{\partial \dot{v}} = m\dot{v} = F_v \tag{4.461}$$

The θ equation is

$$\frac{\partial S}{\partial \dot{\theta}} = mv^2\dot{\theta} = F_\theta v \tag{4.462}$$

or

$$mv\dot{\theta} = F_\theta \tag{4.463}$$

The ϕ equation is

$$\frac{\partial S}{\partial \dot{\phi}} = mv^2\dot{\phi}\sin^2\theta = F_\phi v \sin\theta \tag{4.464}$$

or

$$mv\dot{\phi}\sin\theta = F_\phi \tag{4.465}$$

We see that the Gibbs–Appell method is quite direct when applied to this example.

4.9   Bibliography

Desloge, E. A. *Classical Mechanics*, Vol. 2. New York: John Wiley and Sons, 1982.

Greenwood, D. T. *Principles of Dynamics*, 2nd edn. Englewood Cliffs, NJ: Prentice-Hall, 1988.

Kane, T. R. and Levinson, D. A. *Dynamics: Theory and Applications*. New York: McGraw-Hill, 1985.

Neimark, Ju. I. and Fufaev, N. A. *Dynamics of Nonholonomic Systems*. Translations of Mathematical Monographs, Vol. 33. Providence, RI: American Mathematical Society, 1972.

Papastavridis, J. G. *Analytical Mechanics*. Oxford: Oxford University Press, 2002.

Pars, L. A. *A Treatise on Analytical Dynamics*. London: William Heinemann, 1965.

4.10 Problems

4.1. Suppose we wish to analyze the unsymmetrical top of Example 4.7, using Maggi's equation. Referring to Fig. 4.5 on page 243, let the generalized coordinates be $(\phi, \theta, \psi, X, Y)$ where ϕ is the angle between the positive X-axis and the $\dot{\theta}$ vector. Let the independent us be $u_1 = \dot{\phi}$, $u_2 = \dot{\theta}$, and $u_3 = \dot{\psi}$. (a) Obtain the unconstrained kinetic energy $T(q, \dot{q})$ and the potential energy $V(q)$. (b) Write the constraint equations. (c) Find Q_i and Φ_{ij} for $i = 1, \ldots, 5$ and $j = 1, 2, 3$.

4.2. Two particles, each of mass m, are connected by a massless rod of length l. They can move on a horizontal turntable which rotates at a constant rate Ω. There is a knife-edge constraint at particle 1 which allows no velocity relative to the turntable in a direction normal to the knife edge. Use (r, θ, ϕ) as qs and $(v_r, \dot{\phi})$ as independent us, where v_r is the speed of particle 1 relative to the turntable. (a) Find the differential equations of motion. (b) Evaluate the energy rate $\dot{E}$.

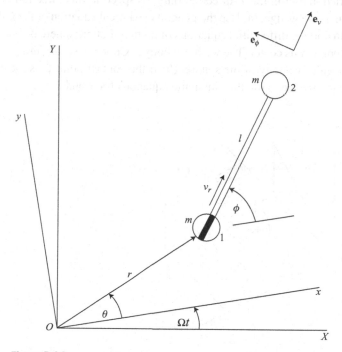

Figure P 4.2.

4.3. A sphere of mass m, radius r, and central moment of inertia I rolls without slipping in a fixed horizontal cylinder (Fig. P 4.3). Choose $(\omega_1, \omega_2, \omega_3)$ as quasi-velocities, where ω is the angular velocity of the sphere. Use the Gibbs–Appell equation to obtain the differential equations of motion.

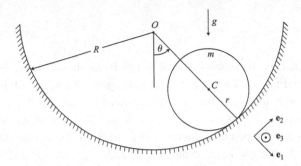

Figure P 4.3.

4.4. A sphere of radius r, mass m, and central moment of inertia I can roll without slipping on a wedge of mass m_0 and wedge angle α. The XYZ frame is inertial with the Z-axis vertical. The xyz frame is fixed in the wedge; the xy-plane forms the slanted surface on which the contact point C moves. The wedge can slide on the horizontal XY-plane without friction, but in the Y-direction only. Its speed in this direction is v. There is no rotation of the wedge. (a) Use the general dynamical equation or the Gibbs–Appell equation to find the differential equations of motion. Let the sphere be body 1 with the reference point at its center. The wedge is body 2. Choose $(\omega_x, \omega_y, \omega_z, v)$ as us, where ω is the angular velocity of the sphere. (b) If the contact point C is located at (x, y) relative to the wedge, write the kinematic equations for $\dot{x}$ and $\dot{y}$.

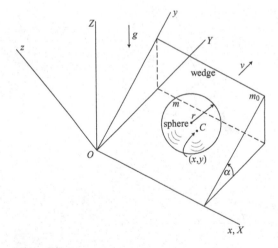

Figure P 4.4.

4.5. Consider a sphere of mass m, radius r, and central moment of inertia I which rolls without slipping on the inside surface of a fixed vertical cylinder of radius R (Fig. P4.5). (a) Choose the angular velocity components $(\omega_r, \omega_\phi, \omega_z)$ as quasi-velocities and obtain the differential equations of motion. (b) Show that the vertical motion of the sphere is sinusoidal.

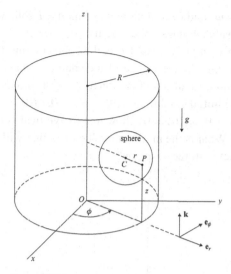

Figure P 4.5.

4.6. An axially-symmetric top with a spherical point moves without slipping on a horizontal plane. The xy-plane remains horizontal and there is no rotation of the xyz frame about its vertical z-axis. Classical Euler angles (ϕ, θ, ψ) are used to specify the orientation of the top. The orthogonal $\mathbf{e}_1\mathbf{e}_2\mathbf{e}_3$ unit vector triad has $\mathbf{e}_1$ along the symmetry axis and $\mathbf{e}_3$ remains horizontal. The center of mass C is located at a distance l from P, the center of the spherical point. The top has mass m and moments of inertia I_a about the symmetry axis and I_p about a transverse axis through P. Choose angular velocity components (ω_1, ω_2, ω_3) of the top as quasi-velocities and obtain the differential equations of motion by using the general dynamical equation.

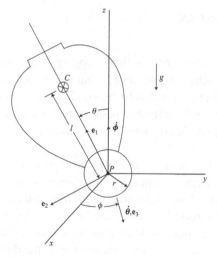

Figure P 4.0.

4.7. A uniform sphere of radius r, mass m, and central moment of inertia I rolls without slipping on the vertical xz-plane which rotates about the fixed z-axis at a constant rate $\Omega > 0$. The sphere has a center C at (x, y, z) and P is the contact point. Choose the components $(\omega_x, \omega_y, \omega_z)$ of the angular velocity ω of the sphere as independent quasi-velocities. (a) Find the differential equations of motion by using the fundamental dynamical equation. (b) Assume the initial conditions $x(0) = x_0 > 0$, $\dot{x}(0) = v_0 > 0$, $y(0) = r$, $\dot{y}(0) = 0$, $z(0) = z_0$, $\dot{z}(0) = 0$, $\omega_y(0) = 0$. Show that the vertical velocity $\dot{z}$ is a sinusoidal function of time. (c) What is the minimum value of v_0 that will ensure that the sphere will remain in contact with the xz-plane?

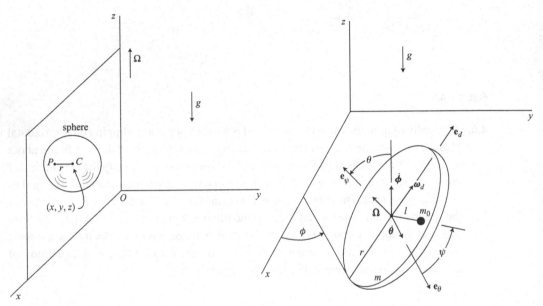

Figure P 4.7.   **Figure P 4.8.**

4.8. A thin uniform disk of mass m and radius r rolls without slipping on the horizontal xy-plane. A particle of mass m_0 is attached at a distance l from its center. Choose the classical Euler angles (ϕ, θ, ψ) as qs and let the us be $(\dot{\theta}, \omega_d, \Omega)$. Consider the disk and particle as separate bodies and take the reference point of the disk at its center. Use the general dynamical equation to obtain the equations of motion.

4.9. A two-wheeled cart consists of body 1 of mass m_0 and moment of inertia I_c about its center of mass C, plus wheels 2 and 3, each a thin uniform disk of mass m and radius r. The wheels roll without slipping, and the body slides without friction on the horizontal XY-plane. The center of mass C of the body is at a distance l from P, where P is the midpoint of the rigidly attached axle of length $2b$. Obtain the differential equations of motion using $(v, \dot{\theta})$ as quasi-velocities, where v is the speed of P in the direction of the x-axis.

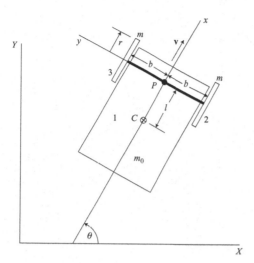

Figure P 4.9.

4.10. A uniform sphere of mass m, radius r, and central moment of inertia I rolls without slipping on the inside surface of a cone that has a vertical axis and a vertex angle α. Let $\dot{\phi}$ be the angular velocity of the center C about the vertical axis, and let ω be the angular velocity of the sphere. Use the components $(\omega_1, \omega_2, \omega_3)$ as quasi-velocities and obtain the differential equations of motion. To simplify the algebra, use the notation $b = r/(R\sin\alpha - r\cos\alpha)$ which is ratio of r to the distance of the center C from the vertical axis.

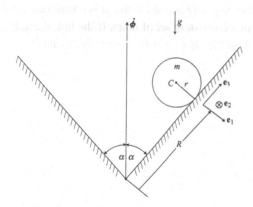

Figure P 4.10.

4.11. The system of three particles (Fig. P 4.11) approximates the dynamical characteristics of a cart with castered wheels. Particle 1 with its knife edge represents the rear wheels. Particle 3 with its knife edge represents the front wheels that have an offset l. There is a joint at particle 2. The force F is constant in magnitude and is directed longitudinally along $\mathbf{e}_1$. Use $(v, \dot{\theta})$ as quasi-velocities and obtain the differential equations of motion.

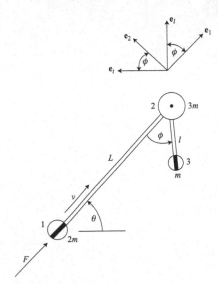

Figure P 4.11.

4.12. A sphere of mass m, radius r, and central moment of inertia I rolls on the horizontal xy-plane which has a system of concentric knife edges embedded in it. At the contact point (R, θ) between the sphere and the horizontal plane, there is no slipping in the radial direction, but slipping can occur without friction in the $\mathbf{e}_\theta$ direction. (a) Obtain the differential equations of motion using $(v_\theta, \omega_r, \omega_\theta, \omega_z)$ as quasi-velocities, where $\mathbf{v}$ is the velocity of the center of the sphere and $\boldsymbol{\omega}$ is its angular velocity. (b) Find the angular momentum about the origin O and show that it is conserved. (c) Assume that $m = 1, r = 1$, and $I = \frac{2}{5}$ in a consistent set of units. If the initial conditions are $R(0) = 10, \theta(0) = 0, v_\theta(0) = 1$ and $\boldsymbol{\omega}(0) = 0$, solve for the final values of θ, v_θ, ω_r, and ω_θ.

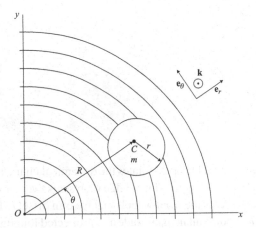

Figure P 4.12.

4.13. A dumbbell consists of two particles, each of mass m, connected by a massless rod of length l. It moves on a smooth spherical depression of radius r. The configuration of the system is given by the generalized coordinates (α, θ, ϕ), where $\theta > 0$ and where the positive $\dot{\phi}$ vector is in the $\mathbf{e}_3$ direction. Let $\phi = 0$ when the rod is horizontal. The unit vectors $\mathbf{e}_2$ and $\mathbf{e}_3$ lie in the plane of triangle $O12$. The center of mass C moves on a sphere of radius $R = (r^2 - \frac{1}{4}l^2)^{1/2}$. A knife-edge constraint allows no velocity component of particle 1 in the $\mathbf{e}_1$ direction. Choose (v, ω_3) as independent us, where v is the velocity component of C in the $\mathbf{e}_2$ direction and ω_3 is the $\mathbf{e}_3$ component of the angular velocity $\boldsymbol{\omega}$ of $O12$. (a) Find the differential equations of motion. (b) Obtain the kinematic expressions for $\dot{\alpha}$, $\dot{\theta}$, and $\dot{\phi}$ as functions of the qs and us.

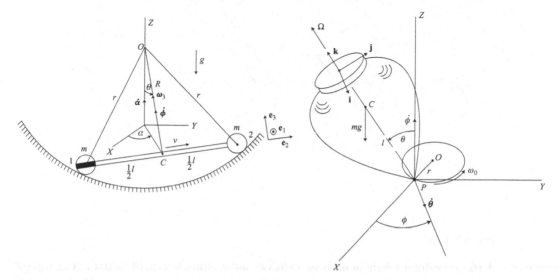

Figure P 4.13.　　　　　　　　**Figure P 4.14.**

4.14. The point P of an axially symmetric top moves with a constant angular velocity ω_0 around a horizontal circular path of radius r. At time $t = 0$, the line OP has the direction of the positive X-axis. The frame XYZ is nonrotating and translates with P. Use type II Euler angles as qs and $(\dot{\phi}, \dot{\theta}, \Omega)$ as us, where Ω is the total angular velocity about the symmetry axis. (a) Find the differential equations of motion. (b) Obtain the energy rate $\dot{E}$ as a function of (q, u, t), assuming that the kinetic energy is calculated relative to the XYZ frame. (c) Find the possible steady precession rates, that is, $\dot{\phi} = \omega_0$ with the top leaning outward at a constant angle θ_0 and with $\phi - \omega_0 t = \pi/2$. Assume that Ω^2 is sufficiently large to ensure stability.

4.15. A rectangular block of mass m has a massless axle and wheels attached along its base, and can roll without slipping on the horizontal XY-plane (Fig. P 4.15). The origin O of the xyz body-fixed principal axis system is the reference point at the midpoint of the axle. The y-axis is horizontal and the x-axis makes an angle θ with the horizontal plane. Choose (X, Y, ϕ, θ) as generalized coordinates and $(v, \dot{\phi}, \dot{\theta})$ as us. (a) Obtain the differential equations of motion. (b) Find the kinematic equations for $\dot{X}$ and $\dot{Y}$.

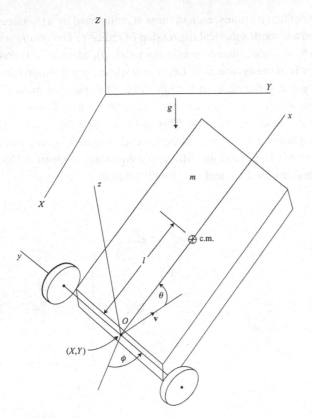

Figure P 4.15.

4.16. A uniform sphere of mass m, radius r, and central moment of inertia I rolls without slipping on a fixed sphere of radius R. Let the coordinates (ϕ, θ) specify the location of the center C and choose $(\omega_1, \omega_2, \omega_3)$ as quasi-velocities, where ω is the angular velocity of the sphere. (a) Obtain the differential equations of motion. (b) Show that the motion of the line OC is similar to that of the symmetry axis of a top with a fixed point.

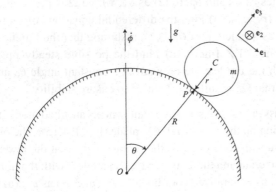

Figure P 4.16.

4.17. A particle of mass m moves relative to a fixed Cartesian xyz frame under the action of a constant force F_x that is directed parallel to the positive x-axis. There is a nonholonomic constraint of the form

$$\dot{z} - k\sqrt{\dot{x}^2 + \dot{y}^2} = 0$$

where k is a positive constant. Note that $k = \tan \beta$ where β is the constant angle of the velocity vector $\mathbf{v}$ above the horizontal xy-plane. Let (x, y, z) denote the position of the particle and choose (v, ϕ) as us, where $v = (\dot{x}^2 + \dot{y}^2 + \dot{z}^2)^{1/2}$ and $\tan \phi = \dot{y}/\dot{x}$. Obtain the differential equations of motion.

4.18. A uniform sphere of mass m, radius r, and central moment of inertia I rolls without slipping on a horizontal turntable which rotates at a constant rate Ω about its center O. The xyz frame is inertial and the coordinates of the center C of the sphere are (x, y, r). Let $\boldsymbol{\omega}$ be the angular velocity of the sphere and choose $(\omega_x, \omega_y, \omega_z)$ as quasi-velocities. (a) Obtain the differential equations of motion. (b) Assuming general initial conditions $x(0) = x_0$, $y(0) = y_0$, $\dot{x}(0) = \dot{x}_0$, $\dot{y}(0) = \dot{y}_0$, show that the path of C is circular by solving for $x(t)$ and $y(t)$. (c) Show that the horizontal component of the angular momentum about O is constant.

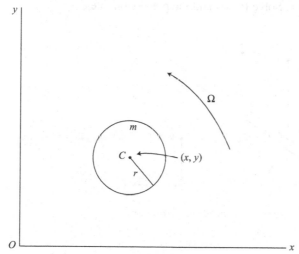

Figure P 4.18.

4.19. A uniform sphere of mass m, radius r, and central moment of inertia I rolls without slipping on a fixed wavy surface described by

$$z = \frac{h}{2}\left(1 - \cos\frac{2\pi y}{L}\right)$$

where $h/L \ll 1$ (Fig. P 4.19). (a) Choose the angular velocity components $(\omega_x, \omega_y, \omega_z)$ of the sphere as quasi-velocities and use the general dynamical equation to obtain the differential equations of motion. Retain terms to first order in the small parameter h/L. (b) Give the kinematic differential equation for $\dot{y}$.

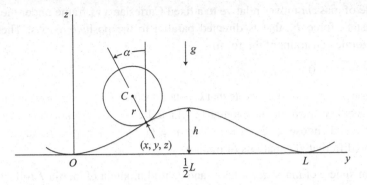

Figure P 4.19.

4.20. A thin uniform disk of mass m and radius r moves on a frictionless horizontal plane. Choose type II Euler angles (ϕ, θ, ψ) as generalized coordinates. The us are $(\dot{\theta}, \omega_d, \Omega)$ where Ω is the total angular velocity about the symmetry axis and ω_d is the angular velocity about a diameter through the contact point. (a) Obtain the differential equations of motion. (b) Consider the case of motion with small positive θ, positive ω_d, and let $\Omega = -\sqrt{2g/r}$. Solve for ω_d and the precession rate $\dot{\phi}$.

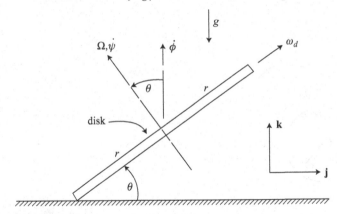

Figure P 4.20.

5 Equations of motion: integral approach

Integral principles and, in particular, Hamilton's principle, have long occupied a prominent position in analytical mechanics. Hamilton's principle, first announced in 1834, presents a *variational principle* as the basis for the dynamical description of a holonomic system. This approach tends to view the motion as a whole and involves a search for the path in configuration space which yields a stationary value for a certain integral. As a result, one obtains the differential equations of motion.

The requirement of stationarity does not apply to nonholonomic systems. Nevertheless, one can use integral methods to obtain the equations of motion for nonholonomic systems. Here we use the integral of the variation rather than the variation of the integral. In this chapter, we shall discuss the derivation and application of these methods, particularly with respect to nonholonomic systems.

5.1   Hamilton's principle

Holonomic system

Consider a dynamical system whose motion satisfies *Lagrange's principle*, namely,

$$\sum_{i=1}^{n} \left[\frac{d}{dt} \left(\frac{\partial T}{\partial \dot{q}_i} \right) - \frac{\partial T}{\partial q_i} - Q_i \right] \delta q_i = 0 \tag{5.1}$$

There are n generalized coordinates and the δqs satisfy the instantaneous constraints. The kinetic energy $T(q, \dot{q}, t)$ is written for the unconstrained system, and is assumed to have at least two continuous derivatives in each of its arguments. Q_i is the generalized applied force associated with q_i.

Now integrate (5.1) with respect to time over the fixed interval t_1 to t_2. Using integration by parts, we find that

$$\int_{t_1}^{t_2} \sum_{i=1}^{n} \frac{d}{dt} \left(\frac{\partial T}{\partial \dot{q}_i} \right) \delta q_i dt = - \int_{t_1}^{t_2} \sum_{i=1}^{n} \frac{\partial T}{\partial \dot{q}_i} \frac{d}{dt} (\delta q_i) dt + \left[\sum_{i=1}^{n} \frac{\partial T}{\partial \dot{q}_i} \delta q_i \right]_{t_1}^{t_2} \tag{5.2}$$

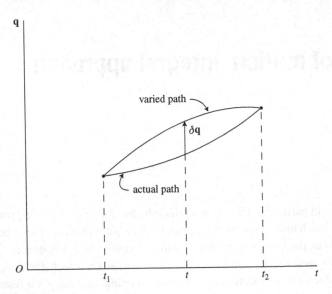

Figure 5.1.

Hence, we obtain

$$\int_{t_1}^{t_2} \left[\sum_{i=1}^{n} \frac{\partial T}{\partial q_i} \delta q_i + \sum_{i=1}^{n} \frac{\partial T}{\partial \dot{q}_i} \frac{d}{dt}(\delta q_i) + \sum_{i=1}^{n} Q_i \delta q_i \right] dt = \left[\sum_{i=1}^{n} \frac{\partial T}{\partial \dot{q}_i} \delta q_i \right]_{t_1}^{t_2} \tag{5.3}$$

The δqs satisfy the m instantaneous constraint equations

$$\sum_{i=1}^{n} a_{ji}(q, t) \delta q_i = 0 \qquad (j = 1, \ldots, m) \tag{5.4}$$

and are assumed to equal zero at the fixed end points t_1 and t_2. Thus, the right-hand side of (5.3) vanishes.

The actual and varied paths in extended configuration space are shown in Fig. 5.1. The δqs are *contemporaneous* variations, that is, they take place with time held fixed. Note that, for a given actual path, the varied path is specified by the δqs which satisfy (5.4).

Let us assume that the qs and $\dot{q}$s are continuous functions of time along the actual and varied paths. Then we can write

$$\frac{d}{dt}(\delta q_i) = \delta \dot{q}_i \qquad (i = 1, \ldots, n) \tag{5.5}$$

The transposition of d and δ operators will be discussed later in the chapter.

Referring again to (5.3), note that

$$\delta T = \sum_{i=1}^{n} \frac{\partial T}{\partial q_i} \delta q_i + \sum_{i=1}^{n} \frac{\partial T}{\partial \dot{q}_i} \delta \dot{q}_i \tag{5.6}$$

and the virtual work is

$$\delta W = \sum_{i=1}^{n} Q_i \delta q_i \tag{5.7}$$

Thus, we obtain

$$\int_{t_1}^{t_2} (\delta T + \delta W) dt = 0 \tag{5.8}$$

This important result applies to the same wide variety of dynamical systems as does Lagrange's principle as given by (5.1). We shall return to this equation when we consider nonholonomic systems.

Now let us assume that all the applied forces are associated with a potential energy function $V(q, t)$. Then $\delta W = -\delta V$ and we can write (5.8) in the form

$$\int_{t_1}^{t_2} \delta L \, dt = 0 \tag{5.9}$$

where the Lagrangian function $L(q, \dot{q}, t) = T - V$. Assuming a *holonomic system*, the operations of integration and variation can be interchanged. Thus, we obtain

$$\delta \int_{t_1}^{t_2} L \, dt = 0 \tag{5.10}$$

which is the usual form of *Hamilton's principle*.

The variation of the integral in (5.10) implies that the actual and varied paths satisfy the m constraint equations of the form

$$\sum_{i-1}^{n} q_{ji}(q, t)\dot{q}_i + a_{jt}(q, t) = 0 \qquad (j = 1, \ldots, m) \tag{5.11}$$

where these expressions are integrable in this holonomic case. On the other hand, the integral of the variation, as in (5.9), implies that the δqs in the expression for δL must satisfy the instantaneous constraints of (5.4). For holonomic systems, the varied paths satisfy both the actual and instantaneous constraint equations. The solutions of (5.10) have the property of *stationarity*; whereas the solutions of (5.9) may or may not have this property, depending on the nature of the constraints.

Now let us restate Hamilton's principle, as it applies to holonomic systems, as follows: *The actual path in configuration space followed by a holonomic dynamical system between the fixed times t_1 and t_2 is such that the integral*

$$I = \int_{t_1}^{t_2} L \, dt \tag{5.12}$$

is stationary with respect to path variations which vanish at the end-points.

To reiterate, the primary assumptions in the derivation of Hamilton's principle are that: (1) the variations δq_i satisfy the instantaneous constraint equations; (2) the end-points are fixed in configuration space and time; and (3) all the applied forces are derivable from a potential energy function $V(q, t)$. Note that the system need not be conservative.

An alternate approach to obtaining the equations of motion for a holonomic system is to begin with Hamilton's principle as a stationarity principle. With this as a starting point, and using the same assumptions as before, we can derive Lagrange's principle. To see how this develops, let us begin with

$$\delta \int_{t_1}^{t_2} L\,dt = \int_{t_1}^{t_2} \delta L\,dt = \int_{t_1}^{t_2} \sum_{i=1}^{n} \left(\frac{\partial L}{\partial q_i} \delta q_i + \frac{\partial L}{\partial \dot{q}_i} \delta \dot{q}_i \right) dt = 0 \tag{5.13}$$

Then, using (5.5) and integrating by parts, we obtain

$$\int_{t_1}^{t_2} \sum_{i=1}^{n} \left[\frac{d}{dt} \left(\frac{\partial L}{\partial \dot{q}_i} \right) - \frac{\partial L}{\partial q_i} \right] \delta q_i\,dt = \sum_{i=1}^{n} \left[\frac{\partial L}{\partial \dot{q}_i} \delta q_i \right]_{t_1}^{t_2} = 0 \tag{5.14}$$

If the δqs are unconstrained, and therefore arbitrary, each coefficient must equal zero, yielding

$$\frac{d}{dt} \left(\frac{\partial L}{\partial \dot{q}_i} \right) - \frac{\partial L}{\partial q_i} = 0 \qquad (i = 1, \ldots, n) \tag{5.15}$$

which is Lagrange's equation. This is also the Euler–Lagrange equation of the calculus of variations.

On the other hand, if the δqs are constrained by (5.4), then the integrand must equal zero at each instant of time since the limits t_1 and t_2 are arbitrary. Thus, we obtain Lagrange's principle, namely,

$$\sum_{i=1}^{n} \left[\frac{d}{dt} \left(\frac{\partial L}{\partial \dot{q}_i} \right) - \frac{\partial L}{\partial q_i} \right] \delta q_i = 0 \tag{5.16}$$

for this case where all the generalized applied forces are obtained from the potential energy $V(q, t)$.

Nonholonomic system

Although stationarity, as expressed in Hamilton's principle, is central to the dynamical theory of holonomic systems, it does not apply to nonholonomic systems. To see how this comes about, let us consider a nonholonomic system and require that each varied path must satisfy the actual constraint equations of the general form

$$f_j(q, \dot{q}, t) = 0 \qquad (j = 1, \ldots, m) \tag{5.17}$$

These constraints are enforced by invoking the *multiplier rule*. The multiplier rule states that the constrained stationary values of the integral of (5.12) are found by considering the *free variations* of

$$I = \int_{t_1}^{t_2} \Lambda\,dt \tag{5.18}$$

where $\Lambda(q, \dot{q}, \mu, t)$ is the *augmented Lagrangian function* which is formed by adjoining the constraint functions to the Lagrangian function by using Lagrange multipliers. Thus,

$$\Lambda = L(q, \dot{q}, t) + \sum_{j=1}^{m} \mu_j f_j(q, \dot{q}, t) \tag{5.19}$$

where the Lagrange multipliers $\mu_j(t)$ are treated as additional variables to be determined.

The stationarity of the free variations of the integral of (5.18) results in the Euler-Lagrange equations

$$\frac{d}{dt}\left(\frac{\partial \Lambda}{\partial \dot{q}_i}\right) - \frac{\partial \Lambda}{\partial q_i} = 0 \qquad (i = 1, \dots, n) \tag{5.20}$$

$$\frac{\partial \Lambda}{\partial \mu_j} = f_j(q, \dot{q}, t) = 0 \qquad (j = 1, \dots, m) \tag{5.21}$$

Note that (5.21) merely restates the constraint equations.

Now let us apply (5.20) to a nonholonomic system in which the constraint functions are linear in the $\dot{q}$s. Thus, the constraint equations have the familiar form

$$f_j(q, \dot{q}, t) = \sum_{i=1}^{n} a_{ji}(q, t)\dot{q}_i + a_{jt}(q, t) = 0 \qquad (j = 1, \dots, m) \tag{5.22}$$

Then, using (5.19) and (5.20), we obtain

$$\frac{d}{dt}\left(\frac{\partial L}{\partial \dot{q}_i}\right) - \frac{\partial L}{\partial q_i} = -\sum_{j=1}^{m} \frac{d}{dt}(\mu_j a_{ji}) + \sum_{j=1}^{m}\sum_{k=1}^{n} \mu_j \frac{\partial a_{jk}}{\partial q_i}\dot{q}_k + \sum_{j=1}^{m} \mu_j \frac{\partial a_{jt}}{\partial q_i}$$

$$= -\sum_{j=1}^{m} \dot{\mu}_j a_{ji} + \sum_{j=1}^{m}\sum_{k=1}^{n} \mu_j \left(\frac{\partial a_{jk}}{\partial q_i} - \frac{\partial a_{ji}}{\partial q_k}\right)\dot{q}_k$$

$$+ \sum_{j=1}^{m} \mu_j \left(\frac{\partial a_{jt}}{\partial q_i} - \frac{\partial a_{ji}}{\partial t}\right) \qquad (i = 1, \dots, n) \tag{5.23}$$

These are the Euler–Lagrange equations for finding the solution path leading to a stationary value of the integral I of (5.12), where both the actual and varied paths satisfy the constraints given by (5.22). Comparing (5.23) with the known form of Lagrange's equation for this nonholonomic system, namely,

$$\frac{d}{dt}\left(\frac{\partial L}{\partial \dot{q}_i}\right) - \frac{\partial L}{\partial q_i} = \sum_{j=1}^{m} \lambda_j a_{ji} \qquad (i = 1, \dots, n) \tag{5.24}$$

we see that, in general, the equations are different. We conclude that the requirement of stationarity leads to *incorrect dynamical equations* for the general case of *nonholonomic* constraints. Conversely, the solution path of a nonholonomic system will not, in general, result in a stationary value of the integral in (5.12).

On the other hand, if we equate $-\dot{\mu}_j$ with λ_j and set

$$\frac{\partial a_{jk}}{\partial q_i} - \frac{\partial a_{ji}}{\partial q_k} = 0 \quad \text{and} \quad \frac{\partial a_{jt}}{\partial q_i} - \frac{\partial a_{ji}}{\partial t} = 0 \qquad \begin{pmatrix} i, k = 1, \dots, n \\ j = 1, \dots, m \end{pmatrix} \tag{5.25}$$

which are the exactness conditions, then the system is *holonomic* and (5.23) reduces to the correct equation (5.24).

The correct equations of motion of a nonholonomic system can be obtained from

$$\int_{t_1}^{t_2} \delta L \, dt = 0 \tag{5.26}$$

which may be considered to be the *nonholonomic form of Hamilton's principle*. It is not a stationarity principle, however, and thereby differs fundamentally from its usual holonomic form given in (5.10). Equation (5.26) assumes that: (1) the actual and varied paths are continuous functions of time and their difference $\delta \mathbf{q}$ satisfies the instantaneous constraint equations; (2) the δqs equal zero at the fixed end-points t_1 and t_2; and (3) all the applied forces arise from a potential energy function $V(q, t)$.

More generally, when the applied forces do not arise from a potential energy function, one can use

$$\int_{t_1}^{t_2} (\delta T + \delta W) \, dt = 0 \tag{5.27}$$

where the virtual work is

$$\delta W = \sum_{i=1}^{n} Q_i \delta q_i \tag{5.28}$$

and the δqs satisfy (5.4). As we found in the derivation of (5.8), this result is essentially an integrated form of Lagrange's principle, (5.1).

Example 5.1 A flat rigid body of mass m moves in the horizontal xy-plane (Fig. 5.2). There is a knife-edge constraint at the reference point P, about which the moment of inertia

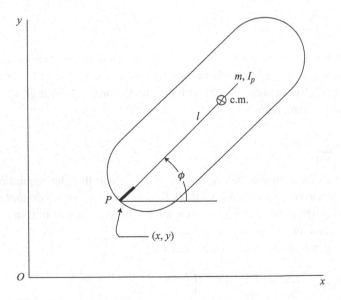

Figure 5.2.

is I_p. Assuming the center of mass is located at a distance l from P, and using (x, y, ϕ) as generalized coordinates, let us find the differential equations of motion.

We can take $V = 0$, so Hamilton's principle has the nonholonomic form

$$\int_{t_1}^{t_2} \delta T \, dt = 0 \tag{5.29}$$

The unconstrained kinetic energy is, from (3.146),

$$T = \frac{1}{2} m(\dot{x}^2 + \dot{y}^2) + \frac{1}{2} I_p \dot{\phi}^2 + ml\dot{\phi}(-\dot{x} \sin \phi + \dot{y} \cos \phi) \tag{5.30}$$

and we obtain

$$\delta T = m\dot{x}\delta\dot{x} + m\dot{y}\delta\dot{y} + I_p\dot{\phi}\delta\dot{\phi} - ml\dot{\phi} \sin \phi \, \delta\dot{x} + ml\dot{\phi} \cos \phi \, \delta\dot{y}$$
$$+ ml(-\dot{x} \sin \phi + \dot{y} \cos \phi)\delta\dot{\phi} + ml\dot{\phi}(-\dot{x} \cos \phi - \dot{y} \sin \phi)\delta\phi \tag{5.31}$$

Noting that

$$\delta\dot{x} = \frac{d}{dt}(\delta x), \qquad \delta\dot{y} = \frac{d}{dt}(\delta y), \qquad \delta\dot{\phi} = \frac{d}{dt}(\delta\phi) \tag{5.32}$$

we find that

$$\delta T = (m\dot{x} - ml\dot{\phi} \sin \phi)\frac{d}{dt}(\delta x) + (m\dot{y} + ml\dot{\phi} \cos \phi)\frac{d}{dt}(\delta y)$$
$$+ (I_p\dot{\phi} - ml\dot{x} \sin \phi + ml\dot{y} \cos \phi)\frac{d}{dt}(\delta\phi) - ml\dot{\phi}(\dot{x} \cos \phi + \dot{y} \sin \phi)\delta\phi \tag{5.33}$$

Now integrate by parts, noting that the δqs equal zero at the end-points. We obtain

$$\int_{t_1}^{t_2} \delta T \, dt = -\int_{t_1}^{t_2} \left\{ \frac{d}{dt}(m\dot{x} - ml\dot{\phi} \sin \phi)\delta x + \frac{d}{dt}(m\dot{y} + ml\dot{\phi} \cos \phi)\delta y \right.$$
$$+ \left[\frac{d}{dt}(I_p\dot{\phi} - ml\dot{x} \sin \phi + ml\dot{y} \cos \phi) \right.$$
$$\left. + ml\dot{\phi}(\dot{x} \cos \phi + \dot{y} \sin \phi) \right] \delta\phi \bigg\} \, dt = 0 \tag{5.34}$$

The end-points t_1 and t_2 are arbitrary, so the integrand must be zero continuously. Thus, we obtain

$$(m\ddot{x} - ml\ddot{\phi} \sin \phi - ml\dot{\phi}^2 \cos \phi)\delta x + (m\ddot{y} + ml\ddot{\phi} \cos \phi - ml\dot{\phi}^2 \sin \phi)\delta y$$
$$+ (I_p\ddot{\phi} - ml\ddot{x} \sin \phi + ml\ddot{y} \cos \phi)\delta\phi = 0 \tag{5.35}$$

which is essentially Lagrange's principle of (5.1).

The nonholonomic constraint equation is

$$\dot{x} \sin \phi - \dot{y} \cos \phi = 0 \tag{5.36}$$

which states that the velocity of point P perpendicular to the knife edge is zero. The corresponding instantaneous constraint equation is

$$\sin \phi \, \delta x - \cos \phi \, \delta y = 0 \tag{5.37}$$

We can choose two independent sets of $(\delta x, \delta y, \delta \phi)$ which satisfy (5.37). Let us choose virtual displacements proportional to $(\cos \phi, \sin \phi, 0)$ and $(0, 0, 1)$. Then, from (5.35) we obtain the following two differential equations of motion:

$$m\ddot{x} \cos \phi + m\ddot{y} \sin \phi - ml\dot{\phi}^2 = 0 \tag{5.38}$$
$$I_p \ddot{\phi} - ml\ddot{x} \sin \phi + ml\ddot{y} \cos \phi = 0 \tag{5.39}$$

Alternatively, we could have noted that

$$\delta y = \tan \phi \, \delta x \tag{5.40}$$

and then considered δx and $\delta \phi$ to be independent.

We need a third differential equation which is obtained by differentiating (5.36) with respect to time.

$$\ddot{x} \sin \phi - \ddot{y} \cos \phi + \dot{x}\dot{\phi} \cos \phi + \dot{y}\dot{\phi} \sin \phi = 0 \tag{5.41}$$

Equations (5.38), (5.39), and (5.41) are linear in the $\ddot{q}$s and can be solved for $\ddot{x}$, $\ddot{y}$, and $\ddot{\phi}$, which are integrated to obtain the motion as a function of time.

The approach used in this example has yielded three second-order equations, namely, two dynamical equations and one kinematical equation. Notice that $\dot{q}$s have been used as velocity variables in the kinetic energy function; quasi-velocities should be avoided because equations similar to (5.32) will not apply.

5.2   Transpositional relations

Now let us examine the kinematical effects due to the nonintegrability of the quasi-velocity expressions and constraint equations often associated with nonholonomic systems. As before, we shall ultimately be concerned with time integrals of variational expressions, although the transpositional relations under consideration here are differential in nature. Hence, we will actually study the kinematics of differential paths in configuration space.

The d and δ operators

Let us begin with the differential form

$$d\theta_j = \sum_{i=1}^{n} \Psi_{ji}(q, t) dq_i + \Psi_{jt}(q, t) dt \qquad (j = 1, \dots, n) \tag{5.42}$$

In general, the differential form is not integrable, so θ_j is a quasi-coordinate. The operator d, as in dq_i, represents an infinitesimal change in the variable q_i which occurs during the infinitesimal time interval dt. If one divides (5.42) by dt, the result is

$$u_j = \sum_{i=1}^{n} \Psi_{ji}(q,t)\dot{q}_i + \Psi_{jt}(q,t) \qquad (j = 1, \ldots, n) \tag{5.43}$$

which, again, is not integrable, in general. Here the quasi-velocity $u_j = d\theta_j/dt$.

The variational equation corresponding to (5.42) is

$$\delta\theta_j = \sum_{i=1}^{n} \Psi_{ji}(q,t)\delta q_i \qquad (j = 1, \ldots, n) \tag{5.44}$$

The operator δ in δq_i represents an infinitesimal change in q_i which is assumed to occur with time held fixed.

Now let us consider the operations δ and d taken in sequence. Taking the total differential of (5.44), we obtain

$$d\delta\theta_j = \sum_{i=1}^{n}\sum_{k=1}^{n} \frac{\partial\Psi_{ji}}{\partial q_k}dq_k\delta q_i + \sum_{i=1}^{n} \frac{\partial\Psi_{ji}}{\partial t}dt\delta q_i + \sum_{i=1}^{n} \Psi_{ji}d\delta q_i \tag{5.45}$$

Similarly, taking the first variation of (5.42), and changing the summing index, we obtain

$$\delta d\theta_j = \sum_{i=1}^{n}\sum_{k=1}^{n} \frac{\partial\Psi_{jk}}{\partial q_i}dq_k\delta q_i + \sum_{i=1}^{n} \frac{\partial\Psi_{jt}}{\partial q_i}dt\delta q_i + \sum_{i=1}^{n} \Psi_{ji}\delta dq_i \tag{5.46}$$

where $\delta t = 0$ and $\delta dt = 0$ since the variations are *contemporaneous*.

Next, subtract (5.46) from (5.45). The result is the important transpositional relation

$$d\delta\theta_j - \delta d\theta_j = \sum_{i=1}^{n}\sum_{k=1}^{n} \left(\frac{\partial\Psi_{ji}}{\partial q_k} - \frac{\partial\Psi_{jk}}{\partial q_i}\right)dq_k\delta q_i + \sum_{i=1}^{n} \left(\frac{\partial\Psi_{ji}}{\partial t} - \frac{\partial\Psi_{jt}}{\partial q_i}\right)dt\delta q_i$$
$$+ \sum_{i=1}^{n} \Psi_{ji}(d\delta q_i - \delta dq_i) \qquad (j = 1, \ldots, n) \tag{5.47}$$

The differential form of (5.42) is linear in the infinitesimal quantities dq_i and dt, but we notice that (5.47) is of second degree in these small quantities. Assuming the nonintegrability of (5.42) and (5.44), the coefficients of $dq_k\delta q_i$ and $dt\delta q_i$ in (5.47) are generally nonzero. To simplify the notation, we can let

$$F_j = \sum_{i=1}^{n}\sum_{k=1}^{n} \left(\frac{\partial\Psi_{ji}}{\partial q_k} - \frac{\partial\Psi_{jk}}{\partial q_i}\right)dq_k\delta q_i + \sum_{i=1}^{n} \left(\frac{\partial\Psi_{ji}}{\partial t} - \frac{\partial\Psi_{jt}}{\partial q_i}\right)dt\delta q_i$$
$$(j = 1, \ldots, n) \tag{5.48}$$

Then we obtain

$$d\delta\theta_j - \delta d\theta_j = F_j + \sum_{i=1}^{n} \Psi_{ji}(d\delta q_i - \delta dq_i) \qquad (j = 1, \ldots, n) \tag{5.49}$$

where $F_j \neq 0$, in general.

There are two principal choices which we can make concerning transpositional relations, namely, (1) $d\delta q_i - \delta dq_i = 0$ and $d\delta\theta_j - \delta d\theta_j \neq 0$, or (2) $d\delta\theta_j - \delta d\theta_j = 0$ and $d\delta q_i - \delta dq_i \neq 0$. For a system with independent qs, we usually choose $d\delta q_i - \delta dq_i = 0$ since this is consistent with a continuous varied path when integral methods are used.

Further transpositional relations can be obtained by first recalling from (4.3) that

$$dq_i = \sum_{j=1}^{n} \Phi_{ij}(q, t)d\theta_j + \Phi_{it}(q, t)dt \qquad (i = 1, \ldots, n) \tag{5.50}$$

and thus

$$\delta q_i = \sum_{j=1}^{n} \Phi_{ij}(q, t)\delta\theta_j \qquad (i = 1, \ldots, n) \tag{5.51}$$

Multiply (5.49) by Φ_{rj} and sum over j. Thus, we obtain

$$d\delta q_r - \delta dq_r = -\sum_{j=1}^{n} \Phi_{rj}F_j + \sum_{j=1}^{n} \Phi_{rj}(d\delta\theta_j - \delta d\theta_j) \qquad (r = 1, \ldots, n) \tag{5.52}$$

where we recall that

$$\sum_{j=1}^{n} \Phi_{rj}\Psi_{ji} = \delta_{ri} \tag{5.53}$$

and δ_{ri} is the Kronecker delta. In detail, we find that

$$d\delta q_r - \delta dq_r = -\sum_{i=1}^{n}\sum_{j=1}^{n}\sum_{k=1}^{n} \Phi_{rj} \left(\frac{\partial \Psi_{ji}}{\partial q_k} - \frac{\partial \Psi_{jk}}{\partial q_i} \right) dq_k\delta q_i$$

$$- \sum_{i=1}^{n}\sum_{j=1}^{n} \Phi_{rj} \left(\frac{\partial \Psi_{ji}}{\partial t} - \frac{\partial \Psi_{jt}}{\partial q_i} \right) dt\delta q_i$$

$$+ \sum_{j=1}^{n} \Phi_{rj}(d\delta\theta_j - \delta d\theta_j) \qquad (r = 1, \ldots, n) \tag{5.54}$$

Using (5.48), (5.50), and (5.51), we can write F_j in terms of the Hamel coefficients. We see that

$$F_j = \sum_{i=1}^{n}\sum_{k=1}^{n}\sum_{r=1}^{n} \left(\frac{\partial \Psi_{ji}}{\partial q_k} - \frac{\partial \Psi_{jk}}{\partial q_i} \right) \Phi_{ir}\delta\theta_r \left(\sum_{l=1}^{n} \Phi_{kl}d\theta_l + \Phi_{kt}dt \right)$$

$$+ \sum_{i=1}^{n}\sum_{r=1}^{n} \left(\frac{\partial \Psi_{ji}}{\partial t} - \frac{\partial \Psi_{jt}}{\partial q_i} \right) \Phi_{ir}dt\delta\theta_r \qquad (j = 1, \ldots, n) \tag{5.55}$$

or

$$F_j = \sum_{l=1}^{n}\sum_{r=1}^{n} \gamma_{rl}^{j}d\theta_l\delta\theta_r + \sum_{r=1}^{n} \gamma_r^{j}dt\delta\theta_r \qquad (j = 1, \ldots, n) \tag{5.56}$$

where

$$\gamma_{rl}^j = \sum_{i=1}^n \sum_{k=1}^n \left(\frac{\partial \Psi_{ji}}{\partial q_k} - \frac{\partial \Psi_{jk}}{\partial q_i} \right) \Phi_{kl} \Phi_{ir} \tag{5.57}$$

$$\gamma_r^j = \sum_{i=1}^n \sum_{k=1}^n \left(\frac{\partial \Psi_{ji}}{\partial q_k} - \frac{\partial \Psi_{jk}}{\partial q_i} \right) \Phi_{kt} \Phi_{ir} + \sum_{i=1}^n \left(\frac{\partial \Psi_{ji}}{\partial t} - \frac{\partial \Psi_{jt}}{\partial q_i} \right) \Phi_{ir} \tag{5.58}$$

Then we can express (5.52) in the form

$$d\delta q_s - \delta d q_s = -\sum_{j=1}^n \sum_{l=1}^n \sum_{r=1}^n \Phi_{sj} \gamma_{rl}^j d\theta_l \delta\theta_r - \sum_{j=1}^n \sum_{r=1}^n \Phi_{sj} \gamma_r^j dt \delta\theta_r$$

$$+ \sum_{j=1}^n \Phi_{sj} (d\delta\theta_j - \delta d\theta_j) \qquad (s = 1, \ldots, n) \tag{5.59}$$

The basic transpositional equations such as (5.47), (5.54), and (5.59) are written for an *unconstrained system*. The application of constraints, however, will be shown to be quite simple.

Nonholonomic constraints

Consider a system with n generalized coordinates and m nonholonomic constraints which are linear in the $\dot{q}$s. We can represent these constraints by setting to zero the last m equations of (5.42) or (5.43). Thus, we have

$$d\theta_j = \sum_{i=1}^n \Psi_{ji}(q, t) dq_i + \Psi_{jt}(q, t) dt \qquad (j = 1, \ldots, n - m) \tag{5.60}$$

$$d\theta_j = \sum_{i=1}^n \Psi_{ji}(q, t) dq_i + \Psi_{jt}(q, t) dt = 0 \qquad (j = n - m + 1, \ldots, n) \tag{5.61}$$

or, in terms of quasi-velocities,

$$u_j = \sum_{i=1}^n \Psi_{ji}(q, t)\dot{q}_i + \Psi_{jt}(q, t) \qquad (j = 1, \ldots, n - m) \tag{5.62}$$

$$u_j = \sum_{i=1}^n \Psi_{ji}(q, t)\dot{q}_i + \Psi_{jt}(q, t) = 0 \qquad (j = n - m + 1, \ldots, n) \tag{5.63}$$

The m constraints are given by (5.63). Note that the first $(n - m)$ us given by (5.62) are *independent*. Thus, as velocity variables, we can use the $(n - m)$ us rather than n constrained $\dot{q}$s.

In a similar fashion, using (5.44), we can write the variational equations for the constrained system.

$$\delta\theta_j = \sum_{i=1}^n \Psi_{ji}(q, t)\delta q_i \qquad (j = 1, \ldots, n - m) \tag{5.64}$$

$$\delta\theta_j = \sum_{i=1}^n \Psi_{ji}(q, t)\delta q_i = 0 \qquad (j = n - m + 1, \ldots, n) \tag{5.65}$$

The m virtual or instantaneous constraint equations are given by (5.65).
   With the application of the constraints, we see that

$$d\delta\theta_j = 0, \qquad \delta d\theta_j = 0 \qquad (j = n - m + 1, \ldots, n) \tag{5.66}$$

so (5.49) becomes

$$d\delta\theta_j - \delta d\theta_j = F_j + \sum_{i=1}^{n} \Psi_{ji}(d\delta q_i - \delta dq_i) \qquad (j = 1, \ldots, n - m) \tag{5.67}$$

and (5.52) reduces to

$$d\delta q_r - \delta dq_r = -\sum_{i=1}^{n} \Phi_{ri} F_i + \sum_{j=1}^{n-m} \Phi_{rj}(d\delta\theta_j - \delta d\theta_j) \qquad (r = 1, \ldots, n) \tag{5.68}$$

The theorem of Frobenius

If the expressions for F_j given in (5.48) are all equal to zero for an unconstrained system, then the exactness conditions apply for the n differential forms of (5.42). These exactness conditions are *sufficient* for the integrability of the differential forms, implying that the θs are true generalized coordinates rather than quasi-coordinates.

   A question arises concerning the necessary and sufficient conditions for the integrability of the differential forms if there are constraints. This is answered by the *Theorem of Frobenius: A system of differential equations such as (5.42) is completely integrable to yield* $\theta_j = \theta_j(q, t)$ *for* $j = 1, \ldots, n$ *if and only if* $F_j = 0$ *for all* dq_i *and* δq_i *satisfying the constraints.* Recall that F_j is given by (5.48).

   As an example, consider a constraint which is represented by the differential form

$$d\theta = a(x, y, z)dx + b(x, y, z)dy + c(x, y, z)dz = 0 \tag{5.69}$$

Also,

$$\delta\theta = a(x, y, z)\delta x + b(x, y, z)\delta y + c(x, y, z)\delta z = 0 \tag{5.70}$$

This constraint is integrable, that is, *holonomic* if $F_j = 0$ or, in detail, if

$$\left(\frac{\partial a}{\partial y} - \frac{\partial b}{\partial x}\right)(dy\delta x - dx\delta y) + \left(\frac{\partial b}{\partial z} - \frac{\partial c}{\partial y}\right)(dz\delta y - dy\delta z)$$

$$+ \left(\frac{\partial c}{\partial x} - \frac{\partial a}{\partial z}\right)(dx\delta z - dz\delta x) = 0 \tag{5.71}$$

where, from (5.69) and (5.70),

$$dz = -\frac{1}{c}(adx + bdy), \qquad \delta z = -\frac{1}{c}(a\delta x + b\delta y) \tag{5.72}$$

Upon making these substitutions and simplifying, we obtain

$$\left[\frac{\partial a}{\partial y} - \frac{\partial b}{\partial x} + \frac{a}{c}\left(\frac{\partial b}{\partial z} - \frac{\partial c}{\partial y}\right) + \frac{b}{c}\left(\frac{\partial c}{\partial x} - \frac{\partial a}{\partial z}\right)\right](dy\delta x - dx\delta y) = 0 \tag{5.73}$$

where $dx, \delta x, dy, \delta y$ are independent. Hence, we find that

$$a\left(\frac{\partial b}{\partial z} - \frac{\partial c}{\partial y}\right) + b\left(\frac{\partial c}{\partial x} - \frac{\partial a}{\partial z}\right) + c\left(\frac{\partial a}{\partial y} - \frac{\partial b}{\partial x}\right) = 0 \tag{5.74}$$

This is the necessary and sufficient condition for the integrability of (5.69).

Geometrical considerations

First, let us consider an *unconstrained* system described in terms of n quasi-velocities which are given by (5.43). We wish to compare points on an actual path in n-dimensional q-space with the corresponding points on a varied path. This can be visualized by considering a small quadrilateral $ABCD$, as shown in Fig. 5.3a. The infinitesimal vector $d\mathbf{q}$ occurs on the actual path during a small time interval dt. The variation $\delta\mathbf{q}$ is an infinitesimal vector drawn from a point on the actual path to a corresponding point (at the same time) on the varied path. Thus, the operator d refers to differences that occur along a solution path with the passage of time; whereas, δ refers to differences in going from the actual path to a varied path with time held constant.

For this *unconstrained* case, the small quadrilateral $ABCD$ is closed. We see that the vector displacements ABC and ADC are equal. Thus

$$d\mathbf{q} + \delta\mathbf{q} + d\delta\mathbf{q} = \delta\mathbf{q} + d\mathbf{q} + \delta d\mathbf{q} \tag{5.75}$$

and

$$d\delta\mathbf{q} = \delta d\mathbf{q} \tag{5.76}$$

In terms of components,

$$d\delta q_i - \delta d q_i = 0 \qquad (i = 1, \dots, n) \tag{5.77}$$

This result applies to a *holonomic* system for which all $F_j = 0$. On the other hand, even for an unconstrained or holonomic system, if (5.77) applies, then (5.49) shows that for each $F_j \neq 0$ due to the presence of quasi-velocities, there is a corresponding transpositional term

$$d\delta\theta_j - \delta d\theta_j \neq 0 \tag{5.78}$$

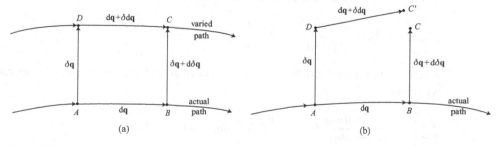

Figure 5.3.

Now let us consider the more general case in which there are m *nonholonomic constraints* that are applied to the system by setting the last m differential forms equal to zero. Thus, we assume that

$$d\theta_j = 0, \qquad \delta\theta_j = 0 \qquad (j = n - m + 1, \ldots, n) \tag{5.79}$$

as in (5.61) and (5.65). Since the virtual constraint applies at B and the actual constraint applies at D in Fig. 5.3b, we see that

$$d\delta\theta_j = 0, \qquad \delta d\theta_j = 0 \qquad (j = n - m + 1, \ldots, n) \tag{5.80}$$

and therefore

$$d\delta\theta_j - \delta d\theta_j = 0 \qquad (j = n - m + 1, \ldots, n) \tag{5.81}$$

Assuming that $F_j \neq 0$ for $j = n - m + 1, \ldots, n$, we see from (5.49) that at least m of the $d\delta q_i - \delta dq_i$ are nonzero, implying that

$$d\delta\mathbf{q} - \delta d\mathbf{q} \neq 0 \tag{5.82}$$

Thus, the quadrilateral in Fig. 5.3b does not close, and the vector in n-space directed from C' to C is equal to $d\delta\mathbf{q} - \delta d\mathbf{q}$. This lack of closure indicates that, even at the differential level, there is path dependence due to the nonintegrability of the nonholonomic constraints. For example, if there is one nonholonomic constraint, it is kinematically possible to obtain closure for all but one of the component qs. However, the remaining $d\delta q_i - \delta dq_i$ must be nonzero, resulting in nonclosure in n-space. Similarly, if there are m nonholonomic constraints, then at least m of the $d\delta q_i - \delta dq_i$ must be nonzero.

For a nominal point A on the actual path, it is important to understand how the $\Psi_{ji}(q, t)$ and $\Psi_{jt}(q, t)$ coefficients are evaluated when applying the constraints at points B and D. At A we have the constraint equations

$$\sum_{i=1}^{n} \Psi_{ji} dq_i + \Psi_{jt} dt = 0 \qquad (j = n - m + 1, \ldots, n) \tag{5.83}$$

$$\sum_{i=1}^{n} \Psi_{ji} \delta q_i = 0 \qquad (j = n - m + 1, \ldots, n) \tag{5.84}$$

where the Ψ_{ji} and Ψ_{jt} coefficients are evaluated at A. At B we have

$$(\Psi_{ji})_B = \Psi_{ji} + \sum_{k=1}^{n} \frac{\partial \Psi_{ji}}{\partial q_k} dq_k + \frac{\partial \Psi_{ji}}{\partial t} dt \tag{5.85}$$

where the terms on the right are again evaluated at A. Thus, for the virtual constraint at B, we obtain

$$\sum_{i=1}^{n} \left[\Psi_{ji} + \sum_{k=1}^{n} \frac{\partial \Psi_{ji}}{\partial q_k} dq_k + \frac{\partial \Psi_{ji}}{\partial t} dt \right] (\delta q_i + d\delta q_i) = 0 \tag{5.86}$$

and, keeping terms to second order, there remains

$$\sum_{i=1}^{n} \Psi_{ji}\delta q_i + \sum_{i=1}^{n} \Psi_{ji}d\delta q_i + \sum_{i=1}^{n}\sum_{k=1}^{n} \frac{\partial \Psi_{ji}}{\partial q_k}dq_k\delta q_i + \sum_{i=1}^{n} \frac{\partial \Psi_{ji}}{\partial t}dt\delta q_i = 0$$

$$(j = n - m + 1, \ldots, n) \qquad (5.87)$$

Similarly, at D the coefficients are

$$(\Psi_{jk})_D = \Psi_{jk} + \sum_{i=1}^{n} \frac{\partial \Psi_{jk}}{\partial q_i}\delta q_i \qquad (5.88)$$

$$(\Psi_{jt})_D = \Psi_{jt} + \sum_{i=1}^{n} \frac{\partial \Psi_{jt}}{\partial q_i}\delta q_i \qquad (5.89)$$

Hence, the actual constraint equation at D requires that

$$\sum_{k=1}^{n} \left[\Psi_{jk} + \sum_{i=1}^{n} \frac{\partial \Psi_{jk}}{\partial q_i}\delta q_i \right](dq_k + \delta dq_k) + \left[\Psi_{jt} + \sum_{i=1}^{n} \frac{\partial \Psi_{jt}}{\partial q_i}\delta q_i \right]dt = 0 \qquad (5.90)$$

and, omitting third-order terms, we obtain

$$\sum_{k=1}^{n} \Psi_{jk}dq_k + \sum_{k=1}^{n} \Psi_{jk}\delta dq_k + \sum_{i=1}^{n}\sum_{k=1}^{n} \frac{\partial \Psi_{jk}}{\partial q_i}dq_k\delta q_i + \Psi_{jt}dt + \sum_{i=1}^{n} \frac{\partial \Psi_{jt}}{\partial q_i}dt\delta q_i = 0$$

$$(j = n - m + 1, \ldots, n) \quad (5.91)$$

Now subtract (5.91) from (5.87) and recall the constraint equations at A, namely, (5.83) and (5.84). The result is

$$\sum_{i=1}^{n} \Psi_{ji}(d\delta q_i - \delta dq_i) + \sum_{i=1}^{n}\sum_{k=1}^{n} \left(\frac{\partial \Psi_{ji}}{\partial q_k} - \frac{\partial \Psi_{jk}}{\partial q_i} \right)dq_k\delta q_i$$

$$+ \sum_{i=1}^{n} \left(\frac{\partial \Psi_{ji}}{\partial t} - \frac{\partial \Psi_{jt}}{\partial q_i} \right)dt\delta q_i = 0 \qquad (j = n - m + 1, \ldots, n) \qquad (5.92)$$

This is identical to the basic equation (5.47) for the case of m nonholonomic constraints with

$$d\delta\theta_j - \delta d\theta_j = 0 \qquad (j = n - m + 1, \ldots, n) \qquad (5.93)$$

We have presented a derivation which emphasizes the assumptions concerning the values of the coefficients at points B and D which are slightly displaced from the nominal reference point A. Note, however, that ultimately all calculations involve values at A; and this applies as well to calculations of F_j.

As a result of the lack of closure of the differential quadrilaterals if there are nonholonomic constraints, it is not possible to find a continuous varied path which simultaneously satisfies the actual constraints and the virtual or instantaneous constraints. One must choose one or the other. If one chooses the stationarity principle in which the varied paths are required to satisfy the actual constraints but not the virtual constraints, then incorrect equations of motion result, as we found earlier. On the other hand, if the varied paths satisfy the virtual constraints but not the actual constraints, the variational process produces correct equations of motion, as we found in the discussion of Hamilton's principle.

5.3   The Boltzmann–Hamel equation, transpositional form

Derivation

The general form of the Boltzmann–Hamel equation, as given in (4.85), is

$$\frac{d}{dt}\left(\frac{\partial T}{\partial u_r}\right) - \frac{\partial T}{\partial \theta_r} + \sum_{j=1}^{n}\sum_{l=1}^{n-m}\frac{\partial T}{\partial u_j}\gamma_{rl}^{j}u_l + \sum_{j=1}^{n}\frac{\partial T}{\partial u_j}\gamma_r^{j} = Q_r \qquad (r = 1, \ldots, n - m) \tag{5.94}$$

where $T(q, u, t)$ is the *unconstrained* kinetic energy and Q_r is the generalized applied force associated with u_r. The $(n - m)$ us are independent and $u_r = \dot{\theta}_r$ where θ_r is a quasi-coordinate.

Multiply (5.94) by $\delta\theta_r dt$ and sum over r. We obtain

$$\sum_{r=1}^{n-m}\left[\frac{d}{dt}\left(\frac{\partial T}{\partial u_r}\right) - \frac{\partial T}{\partial \theta_r} - Q_r\right]\delta\theta_r dt$$

$$+ \sum_{j=1}^{n}\sum_{r=1}^{n-m}\left[\sum_{l=1}^{n-m}\frac{\partial T}{\partial u_j}\gamma_{rl}^{j}d\theta_l\delta\theta_r + \frac{\partial T}{\partial u_j}\gamma_r^{j}dt\delta\theta_r\right] = 0 \tag{5.95}$$

Now recall that with the application of the m constraints,

$$d\theta_l = 0, \qquad \delta\theta_l = 0 \qquad (l = n - m + 1, \ldots, n) \tag{5.96}$$

and thus, from (5.56),

$$F_j = \sum_{l=1}^{n-m}\sum_{r=1}^{n-m}\gamma_{rl}^{j}d\theta_l\delta\theta_r + \sum_{r=1}^{n-m}\gamma_r^{j}dt\delta\theta_r \qquad (j = 1, \ldots, n) \tag{5.97}$$

Hence, we find that

$$\sum_{r=1}^{n-m}\left[\frac{d}{dt}\left(\frac{\partial T}{\partial u_r}\right) - \frac{\partial T}{\partial \theta_r} - Q_r\right]\delta\theta_r dt + \sum_{j=1}^{n}\frac{\partial T}{\partial u_j}F_j = 0 \tag{5.98}$$

Let us assume that

$$d\delta\theta_j - \delta d\theta_j = 0 \qquad (j = 1, \ldots, n - m) \tag{5.99}$$

so that this transpositional expression is now equal to zero for all j. Then, from (5.49), we obtain

$$F_j = -\sum_{i=1}^{n}\Psi_{ji}(d\delta q_i - \delta dq_i) \tag{5.100}$$

Finally, dividing by dt, we can write (5.98) in the form

$$\sum_{r=1}^{n-m}\left[\frac{d}{dt}\left(\frac{\partial T}{\partial u_r}\right)-\frac{\partial T}{\partial\theta_r}-Q_r\right]\delta\theta_r-\sum_{i=1}^{n}\sum_{j=1}^{n}\frac{\partial T}{\partial u_j}\Psi_{ji}\left[\frac{d}{dt}(\delta q_i)-\delta\dot{q}_i\right]=0 \qquad (5.101)$$

This is the *transpositional form* of the *Boltzmann–Hamel equation*. The $(n-m)$ equations of motion are obtained by writing the transpositional expressions in terms of the $\delta\theta$s and then setting the coefficient of each $\delta\theta_r$ equal to zero. This approach results in the same equations of motion as (5.94) but is somewhat simpler to apply.

Example 5.2 A dumbbell consists of two particles, each of mass m, connected by a massless rod of length l. There is a knife-edge constraint at particle 1 (Fig. 5.4). We wish to find the equations of motion as the system moves in the horizontal xy-plane.

Let us use (5.101). The generalized coordinates are (x, y, ϕ) and the us are

$$u_1 = v = \dot{x}\cos\phi + \dot{y}\sin\phi$$
$$u_2 = \dot{\phi} \qquad (5.102)$$
$$u_3 = w = -\dot{x}\sin\phi + \dot{y}\cos\phi = 0$$

where the last equation is the nonholonomic constraint equation. The coefficient matrix is

$$\Psi = \begin{bmatrix} \cos\phi & \sin\phi & 0 \\ 0 & 0 & 1 \\ -\sin\phi & \cos\phi & 0 \end{bmatrix} \qquad (5.103)$$

As quasi-coordinates, let us choose s and η, where $\dot{s} = v$ and $\dot{\eta} = w$. The angle ϕ is a true coordinate.

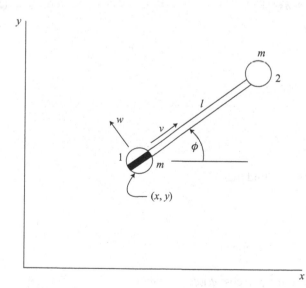

Figure 5.4.

In evaluating the transpositional terms we note that, for the unconstrained system,

$$\dot{x} = v\cos\phi - w\sin\phi \tag{5.104}$$
$$\dot{y} = v\sin\phi + w\cos\phi \tag{5.105}$$

so

$$\delta x = \cos\phi\, \delta s - \sin\phi\, \delta\eta \tag{5.106}$$
$$\delta y = \sin\phi\, \delta s + \cos\phi\, \delta\eta \tag{5.107}$$

From (5.99), we see that

$$\frac{d}{dt}(\delta s) = \delta v, \qquad \frac{d}{dt}(\delta\phi) = \delta\dot{\phi}, \qquad \frac{d}{dt}(\delta\eta) = \delta w \tag{5.108}$$

Then

$$\frac{d}{dt}(\delta x) = \cos\phi\, \delta v - \dot{\phi}\sin\phi\, \delta s - \sin\phi\, \delta w - \dot{\phi}\cos\phi\, \delta\eta \tag{5.109}$$
$$\delta\dot{x} = \cos\phi\, \delta v - v\sin\phi\, \delta\phi - \sin\phi\, \delta w - w\cos\phi\, \delta\phi \tag{5.110}$$

Now apply the actual and virtual constraint equations resulting in

$$w = 0, \qquad \delta w = 0, \qquad \delta\eta = 0 \tag{5.111}$$

and we obtain

$$\frac{d}{dt}(\delta x) - \delta\dot{x} = \sin\phi(v\delta\phi - \dot{\phi}\delta s) \tag{5.112}$$

In a similar manner, we find that

$$\frac{d}{dt}(\delta y) = \sin\phi\, \delta v + \dot{\phi}\cos\phi\, \delta s \tag{5.113}$$
$$\delta\dot{y} = \sin\phi\, \delta v + v\cos\phi\, \delta\phi \tag{5.114}$$

and thus

$$\frac{d}{dt}(\delta y) - \delta\dot{y} = \cos\phi(\dot{\phi}\delta s - v\delta\phi) \tag{5.115}$$

Notice from (5.108) that

$$\frac{d}{dt}(\delta\phi) - \delta\dot{\phi} = 0 \tag{5.116}$$

Then, with the aid of (5.103), we find that

$$\sum_{i=1}^{n} \Psi_{ji}\left[\frac{d}{dt}(\delta q_i) - \delta\dot{q}_i\right] = 0 \tag{5.117}$$

for $j = 1$ and $j = 2$. For $j = 3$, we obtain

$$\sum_{i=1}^{n} \Psi_{ji}\left[\frac{d}{dt}(\delta q_i) - \delta\dot{q}_i\right] = (\sin^2\phi + \cos^2\phi)(\dot{\phi}\delta s - v\delta\phi)$$
$$= \dot{\phi}\delta s - v\delta\phi \tag{5.118}$$

The unconstrained kinetic energy is

$$T = m(v^2 + w^2) + \frac{1}{2}ml^2\dot{\phi}^2 + ml\dot{\phi}w \tag{5.119}$$

Furthermore, all the Qs are equal to zero. Now we can substitute into the general equation (5.101), resulting in

$$2m\dot{v}\delta s + ml^2\ddot{\phi}\delta\phi - ml\dot{\phi}(\dot{\phi}\delta s - v\delta\phi) = 0 \tag{5.120}$$

The v equation of motion, obtained by setting the coefficient of δs equal to zero, is

$$2m\dot{v} - ml\dot{\phi}^2 = 0 \tag{5.121}$$

Similarly, the coefficient of $\delta\phi$ is

$$ml^2\ddot{\phi} + mlv\dot{\phi} = 0 \tag{5.122}$$

yielding the second equation of motion. A comparison shows that this approach is somewhat less complicated than the usual Boltzmann–Hamel method.

5.4   The central equation

Derivation

Consider a system of N particles. Let us begin with d'Alembert's principle in the form

$$\sum_{i=1}^{N}(\mathbf{F}_i - m_i\ddot{\mathbf{r}}_i) \cdot \delta\mathbf{r}_i = 0 \tag{5.123}$$

where $\mathbf{r}_i$ is the position vector of the ith particle and $\mathbf{F}_i$ is the applied force acting on it. The virtual displacements $\delta\mathbf{r}_i$ satisfy the instantaneous (virtual) constraints.

We can write the kinematic identity

$$\frac{d}{dt}(\mathbf{v}_i \cdot \delta\mathbf{r}_i) = \ddot{\mathbf{r}}_i \cdot \delta\mathbf{r}_i + \mathbf{v}_i \cdot \frac{d}{dt}(\delta\mathbf{r}_i)$$

$$= \ddot{\mathbf{r}}_i \cdot \delta\mathbf{r}_i + \delta\left(\frac{\mathbf{v}_i \cdot \mathbf{v}_i}{2}\right) + \mathbf{v}_i \cdot \left[\frac{d}{dt}(\delta\mathbf{r}_i) - \delta\mathbf{v}_i\right] \tag{5.124}$$

where the particle velocity $\mathbf{v}_i = \dot{\mathbf{r}}_i$. Now let

$$\delta P = \sum_{i=1}^{N} m_i\mathbf{v}_i \cdot \delta\mathbf{r}_i \tag{5.125}$$

$$\delta T = \sum_{i=1}^{N} m_i\delta\left(\frac{\mathbf{v}_i \cdot \mathbf{v}_i}{2}\right) \tag{5.126}$$

$$\delta D = \sum_{i=1}^{N} m_i\mathbf{v}_i \cdot \left[\frac{d}{dt}(\delta\mathbf{r}_i) - \delta\mathbf{v}_i\right] \tag{5.127}$$

$$\delta W = \sum_{i=1}^{N} \mathbf{F}_i \cdot \delta\mathbf{r}_i \tag{5.128}$$

From (5.123)–(5.128), we obtain the *central equation* of Heun and Hamel:

$$\frac{d}{dt}(\delta P) = \delta T + \delta W + \delta D \tag{5.129}$$

Explicit form

The central equation has the same general validity as d'Alembert's principle from which it was derived. The principal assumption is that the δrs satisfy the virtual constraints. There is a need, however, to put the central equation in a more convenient, usable form.

First, let us transform to generalized coordinates. In terms of velocity coefficients, we have

$$\mathbf{v}_i = \sum_{k=1}^{n} \boldsymbol{\gamma}_{ik}(q, t)\dot{q}_k + \boldsymbol{\gamma}_{it}(q, t) \qquad (i = 1, \ldots, N) \tag{5.130}$$

$$\delta \mathbf{r}_i = \sum_{k=1}^{n} \boldsymbol{\gamma}_{ik}(q, t)\delta q_k \qquad (i = 1, \ldots, N) \tag{5.131}$$

Thus, we find that

$$\frac{d}{dt}(\delta \mathbf{r}_i) = \sum_{k=1}^{n} \boldsymbol{\gamma}_{ik}\frac{d}{dt}(\delta q_k) + \sum_{k=1}^{n}\sum_{l=1}^{n} \frac{\partial \boldsymbol{\gamma}_{il}}{\partial q_k}\dot{q}_k\delta q_l + \sum_{l=1}^{n} \frac{\partial \boldsymbol{\gamma}_{il}}{\partial t}\delta q_l \tag{5.132}$$

and

$$\delta \mathbf{v}_i = \sum_{k=1}^{n} \boldsymbol{\gamma}_{ik}\delta \dot{q}_k + \sum_{k=1}^{n}\sum_{l=1}^{n} \frac{\partial \boldsymbol{\gamma}_{ik}}{\partial q_l}\dot{q}_k\delta q_l + \sum_{l=1}^{n} \frac{\partial \boldsymbol{\gamma}_{it}}{\partial q_l}\delta q_l \tag{5.133}$$

But

$$\boldsymbol{\gamma}_{ik} = \frac{\partial \mathbf{v}_i}{\partial \dot{q}_k} = \frac{\partial \mathbf{r}_i}{\partial q_k}, \qquad \boldsymbol{\gamma}_{it} = \frac{\partial \mathbf{r}_i}{\partial t} \tag{5.134}$$

so we obtain

$$\frac{\partial \boldsymbol{\gamma}_{il}}{\partial q_k} = \frac{\partial^2 \mathbf{r}_i}{\partial q_k \partial q_l} = \frac{\partial \boldsymbol{\gamma}_{ik}}{\partial q_l}, \qquad \frac{\partial \boldsymbol{\gamma}_{il}}{\partial t} = \frac{\partial^2 \mathbf{r}_i}{\partial t \partial q_l} = \frac{\partial \boldsymbol{\gamma}_{it}}{\partial q_l} \tag{5.135}$$

for $i = 1, \ldots, N$ and $k, l = 1, \ldots, n$. Thus, from (5.132)–(5.135), we find that

$$\frac{d}{dt}(\delta \mathbf{r}_i) - \delta \mathbf{v}_i = \sum_{k=1}^{n} \boldsymbol{\gamma}_{ik}\left[\frac{d}{dt}(\delta q_k) - \delta \dot{q}_k\right] \tag{5.136}$$

The kinetic energy is

$$T = \frac{1}{2}\sum_{i=1}^{N} m_i \mathbf{v}_i \cdot \mathbf{v}_i \tag{5.137}$$

where the ith particle velocity $\mathbf{v}_i$ is given by (5.130). Thus, for the *unconstrained* kinetic energy $T(q, \dot{q}, t)$, we obtain

$$\frac{\partial T}{\partial \dot{q}_k} = \sum_{i=1}^{N} m_i \mathbf{v}_i \cdot \frac{\partial \mathbf{v}_i}{\partial \dot{q}_k} = \sum_{i=1}^{N} m_i \mathbf{v}_i \cdot \boldsymbol{\gamma}_{ik} \tag{5.138}$$

Then, from (5.127), (5.136), and (5.138), we find that

$$\delta D = \sum_{k=1}^{n} \frac{\partial T}{\partial \dot{q}_k} \left[\frac{d}{dt}(\delta q_k) - \delta \dot{q}_k \right] \tag{5.139}$$

Let us consider the general case where the *unconstrained* kinetic energy is written in terms of quasi-velocities, that is, let $T = T^*(q, u, t)$. Then

$$\delta T^* = \sum_{i=1}^{n} \frac{\partial T^*}{\partial q_i} \delta q_i + \sum_{j=1}^{n} \frac{\partial T^*}{\partial u_j} \delta u_j \tag{5.140}$$

where

$$\delta q_i = \sum_{j=1}^{n} \Phi_{ij} \delta \theta_j \tag{5.141}$$

$$\delta u_j = \frac{d}{dt}(\delta \theta_j) - \left[\frac{d}{dt}(\delta \theta_j) - \delta u_j \right] \tag{5.142}$$

Thus,

$$\delta T^* = \sum_{j=1}^{n} \left[\frac{\partial T^*}{\partial \theta_j} \delta \theta_j + \frac{\partial T^*}{\partial u_j} \frac{d}{dt}(\delta \theta_j) \right] - \sum_{j=1}^{n} \frac{\partial T^*}{\partial u_j} \left[\frac{d}{dt}(\delta \theta_j) - \delta u_j \right] \tag{5.143}$$

where we use the notation

$$\frac{\partial T^*}{\partial \theta_j} = \sum_{i=1}^{n} \frac{\partial T^*}{\partial q_i} \Phi_{ij} \qquad (j = 1, \ldots, n) \tag{5.144}$$

The virtual work of the applied forces is

$$\delta W = \sum_{r=1}^{n} Q_r \delta \theta_r \tag{5.145}$$

where Q_r is the applied generalized force associated with u_r or $\delta \theta_r$.

Now let us integrate the central equation with respect to time over the fixed interval t_1 to t_2. The system follows the actual path in q-space; and the δqs, which equal zero at the end-points, satisfy the virtual constraints.

$$\int_{t_1}^{t_2} \left[\frac{d}{dt}(\delta P) - \delta T^* - \delta W - \delta D \right] dt = 0 \tag{5.146}$$

First consider

$$\int_{t_1}^{t_2} \frac{d}{dt}(\delta P) dt = [\delta P]_{t_1}^{t_2} = \left[\sum_{i=1}^{N} m_i \mathbf{v}_i \cdot \delta \mathbf{r}_i \right]_{t_1}^{t_2} - 0 \tag{5.147}$$

where we note that $\delta \mathbf{r}_i = 0$ at the end-points. Next, consider the integral of δT^*. We note that, upon integrating the second term of (5.143) by parts, we obtain

$$\int_{t_1}^{t_2} \sum_{j=1}^n \frac{\partial T^*}{\partial u_j} \frac{d}{dt}(\delta\theta_j)\, dt = \left[\sum_{j=1}^n \frac{\partial T^*}{\partial u_j} \delta\theta_j\right]_{t_1}^{t_2} - \int_{t_1}^{t_2} \sum_{r=1}^n \frac{d}{dt}\left(\frac{\partial T^*}{\partial u_r}\right) \delta\theta_r\, dt \tag{5.148}$$

where the $\delta\theta$s equal zero at the end-points. Thus, we find that

$$\int_{t_1}^{t_2} \delta T^*\, dt = -\int_{t_1}^{t_2} \sum_{r=1}^n \left[\frac{d}{dt}\left(\frac{\partial T^*}{\partial u_r}\right) - \frac{\partial T^*}{\partial\theta_r}\right] \delta\theta_r\, dt - \int_{t_1}^{t_2} \sum_{j=1}^n \frac{\partial T^*}{\partial u_j}\left[\frac{d}{dt}(\delta\theta_j) - \delta u_j\right] dt \tag{5.149}$$

Now we can write (5.146) in the form

$$\int_{t_1}^{t_2} \left\{ \sum_{r=1}^n \left[\frac{d}{dt}\left(\frac{\partial T^*}{\partial u_r}\right) - \frac{\partial T^*}{\partial\theta_r} - Q_r\right] \delta\theta_r + \sum_{j=1}^n \frac{\partial T^*}{\partial u_j}\left[\frac{d}{dt}(\delta\theta_j) - \delta u_j\right] \right.$$
$$\left. - \sum_{i=1}^n \frac{\partial T}{\partial \dot{q}_i}\left[\frac{d}{dt}(\delta q_i) - \delta\dot{q}_i\right] \right\} dt = 0 \tag{5.150}$$

The end times t_1 and t_2 are arbitrary, so the integrand must equal zero continuously, resulting in the general equation

$$\sum_{r=1}^n \left[\frac{d}{dt}\left(\frac{\partial T^*}{\partial u_r}\right) - \frac{\partial T^*}{\partial\theta_r} - Q_r\right] \delta\theta_r + \sum_{j=1}^n \frac{\partial T^*}{\partial u_j}\left[\frac{d}{dt}(\delta\theta_j) - \delta u_j\right]$$
$$- \sum_{i=1}^n \frac{\partial T}{\partial \dot{q}_i}\left[\frac{d}{dt}(\delta q_i) - \delta\dot{q}_i\right] = 0 \tag{5.151}$$

Up to this point, no assumptions have been made concerning the transpositional terms in (5.151). But now let us assume that

$$\frac{d}{dt}(\delta\theta_j) - \delta u_j = 0 \quad (j = 1, \ldots, n) \tag{5.152}$$

Furthermore, let us apply the m virtual constraints by letting $\delta\theta_r = 0$ for $r = n - m + 1, \ldots, n$. The remaining $(n - m)$ $\delta\theta$s are *independent*. Then we obtain the *explicit form of the central equation*:

$$\sum_{r=1}^{n-m} \left[\frac{d}{dt}\left(\frac{\partial T^*}{\partial u_r}\right) - \frac{\partial T^*}{\partial\theta_r} - Q_r\right] \delta\theta_r - \sum_{i=1}^n \frac{\partial T}{\partial \dot{q}_i}\left[\frac{d}{dt}(\delta q_i) - \delta\dot{q}_i\right] = 0 \tag{5.153}$$

Notice that this result is similar to (5.101), the transpositional form of the Boltzmann–Hamel equation.

The $(n - m)$ differential equations of motion are obtained by writing the last term of (5.153) as a function of the $\delta\theta$s, and then setting the coefficient of each $\delta\theta_r$ in (5.153) equal to zero. As mentioned earlier, the kinetic energy $T^*(q, u, t)$ may be written for the *constrained* system, that is, the order of differentiation and the application of constraints makes no difference in this case. But the kinetic energy $T(q, \dot{q}, t)$ must be written for the *unconstrained* system.

The Boltzmann–Hamel equation can be derived from the general result given in (5.151). Let us begin by first noting that

$$\frac{\partial T}{\partial \dot{q}_i} = \sum_{j=1}^{n} \frac{\partial T^*}{\partial u_j} \Psi_{ji} \qquad (i = 1, \ldots, n) \tag{5.154}$$

and then recalling from (5.47) that

$$\frac{d}{dt}(\delta \theta_j) - \delta u_j - \sum_{i=1}^{n} \Psi_{ji} \left[\frac{d}{dt}(\delta q_i) - \delta \dot{q}_i \right] = \sum_{i=1}^{n} \sum_{k=1}^{n} \left(\frac{\partial \Psi_{ji}}{\partial q_k} - \frac{\partial \Psi_{jk}}{\partial q_i} \right) \dot{q}_k \delta q_i$$

$$+ \sum_{i=1}^{n} \left(\frac{\partial \Psi_{ji}}{\partial t} - \frac{\partial \Psi_{jt}}{\partial q_i} \right) \delta q_i \qquad (j = 1, \ldots, n) \tag{5.155}$$

Moreover,

$$\dot{q}_k = \sum_{l=1}^{n} \Phi_{kl} u_l + \Phi_{kt} \qquad (k = 1, \ldots, n) \tag{5.156}$$

$$\delta q_i = \sum_{r=1}^{n} \Phi_{ir} \delta \theta_r \qquad (i = 1, \ldots, n) \tag{5.157}$$

Then (5.151) takes the form

$$\sum_{r=1}^{n} \left[\frac{d}{dt}\left(\frac{\partial T^*}{\partial u_r} \right) - \frac{\partial T^*}{\partial \theta_r} - Q_r + \sum_{i=1}^{n} \sum_{j=1}^{n} \sum_{k=1}^{n} \frac{\partial T^*}{\partial u_j} \left(\frac{\partial \Psi_{ji}}{\partial q_k} - \frac{\partial \Psi_{jk}}{\partial q_i} \right) \right.$$

$$\left. \times \left(\sum_{l=1}^{n} \Phi_{kl} u_l + \Phi_{kt} \right) \Phi_{ir} + \sum_{i=1}^{n} \sum_{j=1}^{n} \frac{\partial T^*}{\partial u_j} \left(\frac{\partial \Psi_{ji}}{\partial t} - \frac{\partial \Psi_{jt}}{\partial q_i} \right) \Phi_{ir} \right] \delta \theta_r = 0 \tag{5.158}$$

Now apply the instantaneous constraints by setting $\delta \theta_r = 0$ for $r = n - m + 1, \ldots, n$. The remaining $(n - m)$ $\delta \theta$s are *independent* so each coefficient must equal zero. Thus we obtain

$$\frac{d}{dt}\left(\frac{\partial T^*}{\partial u_r} \right) - \frac{\partial T^*}{\partial \theta_r} + \sum_{i=1}^{n} \sum_{j=1}^{n} \sum_{k=1}^{n} \sum_{l=1}^{n} \frac{\partial T^*}{\partial u_j} \left(\frac{\partial \Psi_{ji}}{\partial q_k} - \frac{\partial \Psi_{jk}}{\partial q_i} \right) \Phi_{kl} \Phi_{ir} u_l$$

$$+ \sum_{i=1}^{n} \sum_{j=1}^{n} \sum_{k=1}^{n} \frac{\partial T^*}{\partial u_j} \left(\frac{\partial \Psi_{ji}}{\partial q_k} - \frac{\partial \Psi_{jk}}{\partial q_i} \right) \Phi_{kt} \Phi_{ir}$$

$$+ \sum_{i=1}^{n} \sum_{j=1}^{n} \frac{\partial T^*}{\partial u_j} \left(\frac{\partial \Psi_{ji}}{\partial t} - \frac{\partial \Psi_{jt}}{\partial q_i} \right) \Phi_{ir} = Q_r \qquad (r = 1, \ldots, n - m) \tag{5.159}$$

This is a detailed form of the Boltzmann–Hamel equation, as in (4.82). Here the derivation involves an integral method using the central equation. We note from (5.155) that, in spite of transpositional terms in the integral expressed in (5.150), the varied path is actually determined by δqs or $\delta \theta$s which satisfy the virtual constraints.

Example 5.3   A thin uniform disk of mass m and radius r rolls without slipping on the horizontal xy-plane (Fig. 5.5). Let us find the differential equations of motion, using the explicit central equation, namely,

$$\sum_{r=1}^{n-m} \left[\frac{d}{dt}\left(\frac{\partial T^*}{\partial u_r}\right) - \frac{\partial T^*}{\partial \theta_r} - Q_r \right] \delta \theta_r - \sum_{i=1}^{n} \frac{\partial T}{\partial \dot{q}_i}\left[\frac{d}{dt}(\delta q_i) - \delta \dot{q}_i \right] = 0 \tag{5.160}$$

As generalized coordinates, let us choose $(\phi, \theta, \psi, x, y)$ where (ϕ, θ, ψ) are Euler angles and where (x, y) is the location of the contact point C. The independent us are $(\omega_d, \dot{\theta}, \Omega)$. There are two additional us that are set equal to zero to represent the nonholonomic constraints. Thus, we have

$$u_1 = \omega_d = \dot{\phi} \sin \theta \tag{5.161}$$

$$u_2 = \dot{\theta} \tag{5.162}$$

$$u_3 = \Omega = \dot{\phi} \cos \theta + \dot{\psi} \tag{5.163}$$

$$u_4 = r\dot{\psi} \cos \phi + \dot{x} = 0 \tag{5.164}$$

$$u_5 = r\dot{\psi} \sin \phi + \dot{y} = 0 \tag{5.165}$$

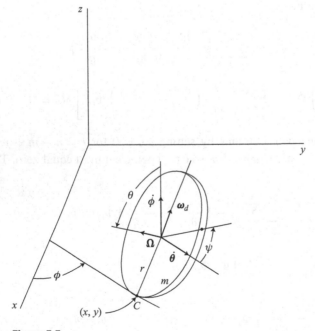

Figure 5.5.

The two constraint equations state that there is no slipping at the contact point C. With this assumption, the contact point moves with a speed $r\dot\psi$ on the xy-plane. The rotations $\dot\phi$ and $\dot\theta$ occur about C and do not cause it to move.

The *constrained* kinetic energy $T^*(q, u)$ is equal to the sum of the translational and rotational portions. We obtain

$$
T^* = \frac{1}{2}mr^2(\dot\theta^2 + \Omega^2) + \frac{1}{2}\left(\frac{mr^2}{2}\right)\Omega^2 + \frac{1}{2}\left(\frac{mr^2}{4}\right)(\omega_d^2 + \dot\theta^2)
$$

$$
= \frac{1}{8}mr^2u_1^2 + \frac{5}{8}mr^2u_2^2 + \frac{3}{4}mr^2u_3^2 \tag{5.166}
$$

where we note that the moments of inertia about the center are

$$
I_a = \frac{1}{2}mr^2, \qquad I_t = \frac{1}{4}mr^2 \tag{5.167}
$$

The *unconstrained* kinetic energy $T(q, \dot q)$ is more complicated. Again we sum the translational and rotational parts, obtaining

$$
T = \frac{1}{2}m[(\dot x - r\dot\phi\cos\phi\cos\theta + r\dot\theta\sin\phi\sin\theta)^2 + (\dot y - r\dot\phi\sin\phi\cos\theta - r\dot\theta\cos\phi\sin\theta)^2
$$

$$
+ r^2\dot\theta^2\cos^2\theta] + \frac{1}{8}mr^2(\dot\phi^2\sin^2\theta + \dot\theta^2) + \frac{1}{4}mr^2(\dot\phi\cos\theta + \dot\psi)^2
$$

$$
= \frac{1}{8}mr^2\dot\phi^2(1 + 5\cos^2\theta) + \frac{5}{8}mr^2\dot\theta^2 + \frac{1}{4}mr^2\dot\psi^2 + \frac{1}{2}m(\dot x^2 + \dot y^2)
$$

$$
+ \frac{1}{2}mr^2\dot\phi\dot\psi\cos\theta - mr\dot\phi\dot x\cos\phi\cos\theta + mr\dot\theta\dot x\sin\phi\sin\theta
$$

$$
- mr\dot\phi\dot y\sin\phi\cos\theta - mr\dot\theta\dot y\cos\phi\sin\theta \tag{5.168}
$$

We wish to express the last term of (5.160) in terms of us and $\delta\theta$s. If (5.161)–(5.165) are solved for the $\dot q$s, the result is

$$
\dot\phi = u_1\csc\theta \tag{5.169}
$$

$$
\dot\theta = u_2 \tag{5.170}
$$

$$
\dot\psi = -u_1\cot\theta + u_3 \tag{5.171}
$$

$$
\dot x = ru_1\cos\phi\cot\theta - ru_3\cos\phi + u_4 \tag{5.172}
$$

$$
\dot y = ru_1\sin\phi\cot\theta - ru_3\sin\phi + u_5 \tag{5.173}
$$

where $u_4 = u_5 = 0$. Similarly,

$$
\delta\phi = \csc\theta\,\delta\theta_1 \tag{5.174}
$$

$$
\delta\theta = \delta\theta_2 \tag{5.175}
$$

$$
\delta\psi = -\cot\theta\,\delta\theta_1 + \delta\theta_3 \tag{5.176}
$$

$$
\delta x = r\cos\phi\cot\theta\,\delta\theta_1 - r\cos\phi\,\delta\theta_3 \tag{5.177}
$$

$$
\delta y = r\sin\phi\cot\theta\,\delta\theta_1 - r\sin\phi\,\delta\theta_3 \tag{5.178}
$$

In addition we assume that, for all j,

$$\frac{d}{dt}(\delta\theta_j) - \delta u_j = 0 \tag{5.179}$$

Then we find that

$$\frac{d}{dt}(\delta\phi) - \delta\dot{\phi} = \csc\theta\cot\theta(u_1\delta\theta_2 - u_2\delta\theta_1) \tag{5.180}$$

$$\frac{\partial T}{\partial\dot{\phi}} = \frac{1}{4}mr^2 u_1\sin\theta + \frac{3}{2}mr^2 u_3\cos\theta \tag{5.181}$$

and therefore

$$\frac{\partial T}{\partial\dot{\phi}}\left[\frac{d}{dt}(\delta\phi) - \delta\dot{\phi}\right] = \frac{1}{4}mr^2\cot\theta(u_1 + 6u_3\cot\theta)(u_1\delta\theta_2 - u_2\delta\theta_1) \tag{5.182}$$

Since (5.162) is integrable and (5.179) applies, we obtain

$$\frac{d}{dt}(\delta\theta) - \delta\dot{\theta} = 0 \tag{5.183}$$

Similarly, we find that

$$\frac{\partial T}{\partial\dot{\psi}}\left[\frac{d}{dt}(\delta\psi) - \delta\dot{\psi}\right] = -\frac{1}{2}mr^2 u_3\csc^2\theta(u_1\delta\theta_2 - u_2\delta\theta_1) \tag{5.184}$$

and

$$\frac{\partial T}{\partial\dot{x}}\left[\frac{d}{dt}(\delta x) - \delta\dot{x}\right] = mr^2(u_2\sin\phi\cos\phi\csc\theta - u_3\cos^2\phi\csc^2\theta)(u_1\delta\theta_2 - u_2\delta\theta_1)$$
$$+ mr^2(u_2\sin^2\phi - u_3\sin\phi\cos\phi\csc\theta)(u_1\delta\theta_3 - u_3\delta\theta_1) \tag{5.185}$$

Furthermore,

$$\frac{\partial T}{\partial\dot{y}}\left[\frac{d}{dt}(\delta y) - \delta\dot{y}\right] = -mr^2(u_2\sin\phi\cos\phi\csc\theta + u_3\sin^2\phi\csc^2\theta)(u_1\delta\theta_2 - u_2\delta\theta_1)$$
$$+ mr^2(u_2\cos^2\phi + u_3\sin\phi\cos\phi\csc\theta)(u_1\delta\theta_3 - u_3\delta\theta_1) \tag{5.186}$$

Adding (5.182)–(5.186), we obtain

$$\sum_{i=1}^{5}\frac{\partial T}{\partial\dot{q}_i}\left[\frac{d}{dt}(\delta q_i) - \delta\dot{q}_i\right] = mr^2\left(\frac{1}{4}u_1\cot\theta - \frac{3}{2}u_3\right)(u_1\delta\theta_2 - u_2\delta\theta_1)$$
$$+ mr^2 u_2(u_1\delta\theta_3 - u_3\delta\theta_1)$$
$$= mr^2\left(-\frac{1}{4}u_1 u_2\cot\theta + \frac{1}{2}u_2 u_3\right)\delta\theta_1$$
$$+ mr^2\left(\frac{1}{4}u_1^2\cot\theta - \frac{3}{2}u_1 u_3\right)\delta\theta_2$$
$$+ mr^2 u_1 u_2\delta\theta_3 \tag{5.187}$$

Referring to (5.166), we have

$$\sum_{r=1}^{n} \left[\frac{d}{dt} \left(\frac{\partial T^*}{\partial u_r} \right) - \frac{\partial T^*}{\partial \theta_r} - Q_r \right] \delta\theta_r$$

$$= \frac{1}{4}mr^2\dot{u}_1\delta\theta_1 + \left(\frac{5}{4}mr^2\dot{u}_2 + mgr\cos\theta \right)\delta\theta_2 + \frac{3}{2}mr^2\dot{u}_3\delta\theta_3 \qquad (5.188)$$

where

$$Q_1 = 0, \qquad Q_2 = -mgr\cos\theta, \qquad Q_3 = 0 \qquad (5.189)$$

Finally, using the explicit form of the central equation, we obtain the equations of motion. The u_1 equation, found by setting the coefficient of $\delta\theta_1$ equal to zero, is

$$\frac{1}{4}mr^2\dot{u}_1 + \frac{1}{4}mr^2u_1u_2\cot\theta - \frac{1}{2}mr^2u_2u_3 = 0 \qquad (5.190)$$

Similarly, the u_2 equation is

$$\frac{5}{4}mr^2\dot{u}_2 - \frac{1}{4}mr^2u_1^2\cot\theta + \frac{3}{2}mr^2u_1u_3 = -mgr\cos\theta \qquad (5.191)$$

The u_3 equation is

$$\frac{3}{2}mr^2\dot{u}_3 - mr^2u_1u_2 = 0 \qquad (5.192)$$

Using the original notation for angular velocity components, we can write the equations of motion as follows:

$$\frac{1}{4}mr^2\dot{\omega}_d + \frac{1}{4}mr^2\omega_d\dot{\theta}\cot\theta - \frac{1}{2}mr^2\dot{\theta}\Omega = 0 \qquad (5.193)$$

$$\frac{5}{4}mr^2\ddot{\theta} - \frac{1}{4}mr^2\omega_d^2\cot\theta + \frac{3}{2}mr^2\omega_d\Omega = -mgr\cos\theta \qquad (5.194)$$

$$\frac{3}{2}mr^2\dot{\Omega} - mr^2\omega_d\dot{\theta} = 0 \qquad (5.195)$$

These equations are identical with (4.259)–(4.261) obtained earlier.

5.5   Suslov's principle

Dependent and independent coordinates

Thus far, we have assumed that the linear nonholonomic constraint equations have the form

$$\sum_{i=1}^{n} a_{ji}(q,t)\dot{q}_i + a_{jt}(q,t) = 0 \qquad (j = 1,\ldots,m) \qquad (5.196)$$

This form has the characteristic that all the qs are treated equally. Now let us change the viewpoint and arbitrarily designate m of the qs as *dependent*; whereas, the remaining

$(n - m)$ qs are termed *independent*. Equation (5.196) can be solved for the dependent $\dot{q}_D$s in terms of the independent $\dot{q}_I$s. We obtain m nonholonomic constraint equations written in the form

$$\dot{q}_D = \sum_{I=1}^{n-m} B_{DI}(q, t)\dot{q}_I + B_{Dt}(q, t) \qquad (D = n - m + 1, \ldots, n) \qquad (5.197)$$

or, more briefly,

$$\dot{q}_D = \phi_D(q, \dot{q}_I, t) \qquad (D = n - m + 1, \ldots, n) \qquad (5.198)$$

Suslov's principle

Let us begin with the general nonholonomic form of Hamilton's principle, namely,

$$\int_{t_1}^{t_2} (\delta T + \delta W)dt = 0 \qquad (5.199)$$

where each varied path satisfies the virtual constraints, and the δqs are equal to zero at the end-points in extended q-space (Fig. 5.1).

Now introduce the *constrained* kinetic energy $T^0(q, \dot{q}_I, t)$ which is obtained from the unconstrained kinetic energy $T(q, \dot{q}, t)$ by substituting for the $\dot{q}_D$s using the constraint equations (5.197). For any realizable motion, the energies T^0 and T are equal. Their variations δT^0 and δT are not equal, however. To see how this comes about, consider

$$\delta T = \sum_{i=1}^{n} \frac{\partial T}{\partial q_i}\delta q_i + \sum_{i=1}^{n} \frac{\partial T}{\partial \dot{q}_i}\delta\dot{q}_i \qquad (5.200)$$

For a given varied path, we can take

$$\frac{d}{dt}(\delta q_i) - \delta\dot{q}_i = 0 \qquad (i = 1, \ldots, n) \qquad (5.201)$$

and therefore we obtain that, in (5.199),

$$\delta T = \sum_{i=1}^{n} \frac{\partial T}{\partial q_i}\delta q_i + \sum_{i=1}^{n} \frac{\partial T}{\partial \dot{q}_i}\frac{d}{dt}(\delta q_i) \qquad (5.202)$$

For the constrained system, however, we know that at least m of the transpositional expressions of (5.201) must be nonzero. Let us assume that

$$\frac{d}{dt}(\delta q_I) - \delta\dot{q}_I = 0 \qquad (I = 1, \ldots, n - m) \qquad (5.203)$$

$$\frac{d}{dt}(\delta q_D) - \delta\dot{q}_D \neq 0 \qquad (D = n - m + 1, \ldots, n) \qquad (5.204)$$

where

$$\delta\dot{q}_D = \delta\phi_D = \sum_{i=1}^{n} \frac{\partial\phi_D}{\partial q_i}\delta q_i + \sum_{I=1}^{n-m} \frac{\partial\phi_D}{\partial\dot{q}_I}\delta\dot{q}_I \qquad (5.205)$$

Then we see that

$$\delta T^0 = \sum_{i=1}^{n} \frac{\partial T}{\partial q_i} \delta q_i + \sum_{I=1}^{n-m} \frac{\partial T}{\partial \dot{q}_I} \frac{d}{dt}(\delta q_I) + \sum_{D=n-m+1}^{n} \frac{\partial T}{\partial \dot{q}_D} \delta \dot{q}_D \qquad (5.206)$$

A comparison of (5.202) and (5.206) results in

$$\delta T = \delta T^0 + \sum_{D=n-m+1}^{n} \frac{\partial T}{\partial \dot{q}_D} \left[\frac{d}{dt}(\delta q_D) - \delta \dot{q}_D \right] \qquad (5.207)$$

Now substitute the expression for δT from (5.207) into (5.199) which is Hamilton's principle. The result is

$$\int_{t_1}^{t_2} \left\{ \delta T^0 + \delta W + \sum_{D=n-m+1}^{n} \frac{\partial T}{\partial \dot{q}_D} \left[\frac{d}{dt}(\delta q_D) - \delta \dot{q}_D \right] \right\} dt = 0 \qquad (5.208)$$

This is *Suslov's principle*.

The virtual work is

$$\delta W = \sum_{i=1}^{n} Q_i \delta q_i = \sum_{I=1}^{n-m} Q_I \delta q_I + \sum_{D=n-m+1}^{n} Q_D \delta q_D \qquad (5.209)$$

where the Qs are generalized applied forces for the case of unconstrained δqs. We actually have constrained δqs, however, with

$$\delta q_D = \sum_{I=1}^{n-m} B_{DI} \delta q_I \qquad (5.210)$$

Hence, the virtual work becomes

$$\delta W = \sum_{I=1}^{n-m} Q_I \delta q_I + \sum_{I=1}^{n-m} \sum_{D=n-m+1}^{n} Q_D B_{DI} \delta q_I \qquad (5.211)$$

or

$$\delta W = \sum_{I=1}^{n-m} Q_I^0 \delta q_I \qquad (5.212)$$

where the generalized applied force associated with q_I in the *constrained* system is

$$Q_I^0 = Q_I + \sum_{D=n-m+1}^{n} Q_D B_{DI} \qquad (I = 1, \ldots, n-m) \qquad (5.213)$$

Equations of motion

To obtain the differential equations of motion from Suslov's principle, first let us write

$$\delta T^0 = \sum_{I=1}^{n-m} \frac{\partial T^0}{\partial q_I} \delta q_I + \sum_{D=n-m+1}^{n} \frac{\partial T^0}{\partial q_D} \delta q_D + \sum_{I=1}^{n-m} \frac{\partial T^0}{\partial \dot{q}_I} \frac{d}{dt}(\delta q_I) \qquad (5.214)$$

where we have used (5.203). Now integrate δT^0 with respect to time, using integration by parts on the last term and noting that δq_I vanishes at t_1 and t_2. Upon multiplying by -1 and recalling (5.210), the integral of (5.208) can be written in the form

$$\int_{t_1}^{t_2} \left\{ \sum_{I=1}^{n-m} \left[\frac{d}{dt}\left(\frac{\partial T^0}{\partial \dot{q}_I}\right) - \left(\frac{\partial T^0}{\partial q_I} + \sum_{D=n-m+1}^{n} \frac{\partial T^0}{\partial q_D} B_{DI}\right) - Q_I^0 \right] \delta q_I \right.$$
$$\left. - \sum_{D=n-m+1}^{n} \frac{\partial T}{\partial \dot{q}_D} \left[\frac{d}{dt}(\delta q_D) - \delta \dot{q}_D \right] \right\} dt = 0 \qquad (5.215)$$

The transpositional term can be written as a homogeneous linear expression in the δq_Is which are independent. Hence, the integrand must vanish for all t, resulting in the *explicit form of Suslov's principle*:

$$\sum_{I=1}^{n-m} \left[\frac{d}{dt}\left(\frac{\partial T^0}{\partial \dot{q}_I}\right) - \left(\frac{\partial T^0}{\partial q_I} + \sum_{D=n-m+1}^{n} \frac{\partial T^0}{\partial q_D} B_{DI}\right) - Q_I^0 \right] \delta q_I$$
$$- \sum_{D=n-m+1}^{n} \frac{\partial T}{\partial \dot{q}_D} \left[\frac{d}{dt}(\delta q_D) - \delta \dot{q}_D \right] = 0 \qquad (5.216)$$

The equations of motion are obtained by expressing the last term in terms of δq_Is and then equating to zero the coefficient of each δq_I. This results in $(n-m)$ second-order differential equations. There are also m kinematical constraint equations, giving a total of n equations to solve for the n qs.

Comparing this result with the explicit central equation, (5.153), we see that the Suslov equation has the advantage that the last summation is over the m dependent $\dot{q}$s only rather than over the complete set of n. On the other hand, Suslov's equation has the possible disadvantage that it does not allow quasi-velocities to be used.

Example 5.4   As an example of the application of Suslov's equation (5.216), consider the motion of a dumbbell with a knife-edge constraint sliding on the horizontal xy-plane (Fig. 5.6). As generalized coordinates, let us choose (ϕ, x, y), where (ϕ, x) are considered independent and y is dependent.

The nonholonomic constraint equation is

$$\dot{q}_D = \dot{y} = \dot{x} \tan \phi \qquad (5.217)$$

and the corresponding virtual constraint is

$$\delta q_D = \delta y = \delta x \tan \phi \qquad (5.218)$$

The unconstrained kinetic energy, from (1.127), is

$$T = m(\dot{x} + \dot{y}^2) + \frac{1}{2}ml^2\dot{\phi}^2 + ml\dot{\phi}(-\dot{x}\sin\phi + \dot{y}\cos\phi) \qquad (5.219)$$

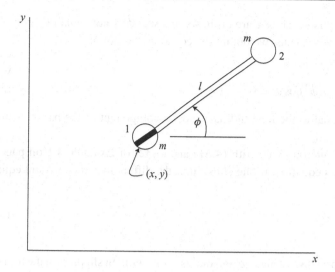

Figure 5.6.

The constrained kinetic energy, upon substitution from (5.217), is

$$T^0 = m(1 + \tan^2 \phi)\dot{x}^2 + \frac{1}{2}ml^2\dot{\phi}^2 = m\dot{x}^2 \sec^2 \phi + \frac{1}{2}ml^2\dot{\phi}^2 \tag{5.220}$$

We see that

$$\frac{d}{dt}(\delta y) - \delta\dot{y} = \sec^2 \phi \,(\dot{\phi}\delta x - \dot{x}\delta\phi) \tag{5.221}$$

where we note from (5.203) that

$$\frac{d}{dt}(\delta x) - \delta\dot{x} = 0 \tag{5.222}$$

Then, using (5.217), we obtain

$$\frac{\partial T}{\partial \dot{y}}\left[\frac{d}{dt}(\delta y) - \delta\dot{y}\right] = (2m\dot{x}\tan\phi + ml\dot{\phi}\cos\phi)\sec^2\phi\,(\dot{\phi}\delta x - \dot{x}\delta\phi) \tag{5.223}$$

Differentiation of the constrained kinetic energy yields

$$\frac{\partial T^0}{\partial \phi} = 2m\dot{x}^2 \sec^2 \phi \tan\phi, \qquad \frac{\partial T^0}{\partial x} = \frac{\partial T^0}{\partial y} = 0 \tag{5.224}$$

$$\frac{\partial T^0}{\partial \dot{\phi}} = ml^2\dot{\phi}, \qquad \frac{\partial T^0}{\partial \dot{x}} = 2m\dot{x}\sec^2\phi \tag{5.225}$$

Also,

$$Q_\phi^0 = 0, \qquad Q_x^0 = 0 \tag{5.226}$$

Finally, a substitution in Suslov's equation, (5.216), results in

$$(ml^2\ddot{\phi} + ml\dot{\phi}\dot{x}\sec\phi)\delta\phi + (2m\ddot{x} + 2m\dot{\phi}\dot{x}\tan\phi - ml\dot{\phi}^2\cos\phi)\sec^2\phi\,\delta x = 0 \tag{5.227}$$

The virtual displacements $\delta\phi$ and δx are arbitrary and $\sec^2\phi$ is not equal to zero, so each coefficient must equal zero. Thus, we obtain the equations of motion

$$ml^2\ddot{\phi} + ml\dot{\phi}\dot{x}\sec\phi = 0 \tag{5.228}$$

$$2m\ddot{x} + 2m\dot{\phi}\dot{x}\tan\phi - ml\dot{\phi}^2\cos\phi = 0 \tag{5.229}$$

Note that $\dot{x}\sec\phi$ is equal to the longitudinal velocity component of the particles, that is, the velocity along the rod.

These equations of motion agree with (4.31) and (4.32) of Example 4.1 on page 220, obtained using Maggi's equation, if one substitutes the differentiated constraint equation, namely

$$\ddot{y} = \ddot{x}\tan\phi + \dot{\phi}\dot{x}\sec^2\phi \tag{5.230}$$

Example 5.5 A uniform disk of mass m and radius r rolls without slipping on the horizontal xy-plane (Fig. 5.7). We wish to obtain the equations of motion using the Suslov equation, (5.216).

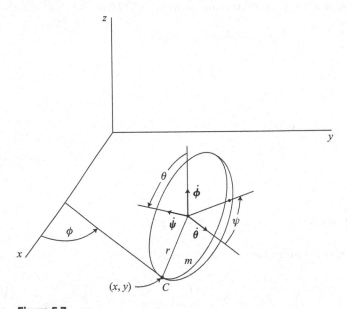

Figure 5.7.

The configuration of the system is given by the location of the contact point C and the Euler angles. Thus, we can choose $(\phi, \theta, \psi, x, y)$ as generalized coordinates, with the first three as independent q_Is and the last two as dependent q_Ds. The constraint equations of the form $\dot{q}_D = \phi_D(q, \dot{q}_I)$ are

$$\dot{x} = -r\dot{\psi}\cos\phi \tag{5.231}$$

$$\dot{y} = -r\dot{\psi}\sin\phi \tag{5.232}$$

and the corresponding virtual displacements are

$$\delta x = -r \cos \phi \, \delta \psi \tag{5.233}$$

$$\delta y = -r \sin \phi \, \delta \psi \tag{5.234}$$

The unconstrained kinetic energy is equal to that due to the translation of the center of mass, plus that due to rotation about the center of mass. Thus, we obtain

$$
\begin{aligned}
T = \; &\frac{1}{2}m[(\dot{x}\cos\phi + \dot{y}\sin\phi - r\dot\phi\cos\theta)^2 \\
&+ (-\dot{x}\sin\phi + \dot{y}\cos\phi - r\dot\theta\sin\theta)^2 + r^2\dot\theta^2\cos^2\theta] \\
&+ \frac{1}{8}mr^2(\dot\phi^2\sin^2\theta + \dot\theta^2) + \frac{1}{4}mr^2(\dot\phi\cos\theta + \dot\psi)^2 \\
= \; &\frac{1}{8}mr^2\dot\phi^2(1 + 5\cos^2\theta) + \frac{5}{8}mr^2\dot\theta^2 + \frac{1}{4}mr^2\dot\psi^2 + \frac{1}{2}m(\dot{x}^2 + \dot{y}^2) \\
&+ \frac{1}{2}mr^2\dot\phi\dot\psi\cos\theta - mr\dot\phi\cos\theta(\dot{x}\cos\phi + \dot{y}\sin\phi) \\
&+ mr\dot\theta\sin\theta(\dot{x}\sin\phi - \dot{y}\cos\phi)
\end{aligned}
\tag{5.235}
$$

The constrained kinetic energy, obtained by substituting the constraint equations for $\dot{x}$ and $\dot{y}$ into (5.235), is

$$T^0 = \frac{1}{8}mr^2\dot\phi^2(1 + 5\cos^2\theta) + \frac{5}{8}mr^2\dot\theta^2 + \frac{3}{4}mr^2\dot\psi^2 + \frac{3}{2}mr^2\dot\phi\dot\psi\cos\theta \tag{5.236}$$

This assumes that there is no slipping at the contact point C.

The generalized applied forces for the constrained system are

$$Q_1^0 = 0, \qquad Q_2^0 = -mgr\cos\theta, \qquad Q_3^0 = 0 \tag{5.237}$$

In accordance with (5.203), we have

$$\frac{d}{dt}(\delta\phi) - \delta\dot\phi = 0, \qquad \frac{d}{dt}(\delta\theta) - \delta\dot\theta = 0, \qquad \frac{d}{dt}(\delta\psi) - \delta\dot\psi = 0 \tag{5.238}$$

Then, differentiation of (5.231) and (5.233) results in

$$\frac{d}{dt}(\delta x) - \delta\dot{x} = r\sin\phi(\dot\phi\delta\psi - \dot\psi\delta\phi) \tag{5.239}$$

Similarly,

$$\frac{d}{dt}(\delta y) - \delta\dot{y} = -r\cos\phi(\dot\phi\delta\psi - \dot\psi\delta\phi) \tag{5.240}$$

Furthermore,

$$
\begin{aligned}
\frac{\partial T}{\partial \dot{x}} &= m\dot{x} - mr\dot\phi\cos\phi\cos\theta + mr\dot\theta\sin\phi\sin\theta \\
&= -mr\dot\phi\cos\phi\cos\theta + mr\dot\theta\sin\phi\sin\theta - mr\dot\psi\cos\phi
\end{aligned}
\tag{5.241}
$$

and

$$\frac{\partial T}{\partial \dot{y}} = -mr\dot{\phi} \sin \phi \cos \theta - mr\dot{\theta} \cos \phi \sin \theta - mr\dot{\psi} \sin \phi \tag{5.242}$$

Hence, we find that

$$\frac{\partial T}{\partial \dot{x}} \left[\frac{d}{dt}(\delta x) - \delta \dot{x} \right] + \frac{\partial T}{\partial \dot{y}} \left[\frac{d}{dt}(\delta y) - \delta \dot{y} \right] = mr^2 \dot{\theta} \sin \theta (\dot{\phi} \delta \psi - \dot{\psi} \delta \phi) \tag{5.243}$$

To obtain the ϕ equation of motion, we first evaluate

$$\frac{d}{dt}\left(\frac{\partial T^0}{\partial \dot{\phi}}\right) - \frac{\partial T^0}{\partial \phi} = \frac{1}{4}mr^2 \ddot{\phi}(1 + 5\cos^2 \theta) + \frac{3}{2}mr^2 \ddot{\psi} \cos \theta$$
$$- \frac{5}{2}mr^2 \dot{\phi}\dot{\theta} \sin \theta \cos \theta - \frac{3}{2}mr^2 \dot{\theta}\dot{\psi} \sin \theta \tag{5.244}$$

Then, using Suslov's equation, the coefficient of $\delta \phi$ is the ϕ equation:

$$\frac{1}{4}mr^2 \ddot{\phi}(1 + 5\cos^2 \theta) + \frac{3}{2}mr^2 \ddot{\psi} \cos \theta - \frac{5}{2}mr^2 \dot{\phi}\dot{\theta} \sin \theta \cos \theta$$
$$- \frac{1}{2}mr^2 \dot{\theta}\dot{\psi} \sin \theta = 0 \tag{5.245}$$

Next, we obtain

$$\frac{d}{dt}\left(\frac{\partial T^0}{\partial \dot{\theta}}\right) - \frac{\partial T^0}{\partial \theta} = \frac{5}{4}mr^2 \ddot{\theta} + \frac{5}{4}mr^2 \dot{\phi}^2 \sin \theta \cos \theta + \frac{3}{2}mr^2 \dot{\phi}\dot{\psi} \sin \theta \tag{5.246}$$

and the θ equation is

$$\frac{5}{4}mr^2 \ddot{\theta} + \frac{5}{4}mr^2 \dot{\phi}^2 \sin \theta \cos \theta + \frac{3}{2}mr^2 \dot{\phi}\dot{\psi} \sin \theta = -mgr \cos \theta \tag{5.247}$$

Finally,

$$\frac{d}{dt}\left(\frac{\partial T^0}{\partial \dot{\psi}}\right) - \frac{\partial T^0}{\partial \psi} = \frac{3}{2}mr^2 \ddot{\psi} + \frac{3}{2}mr^2 \ddot{\phi} \cos \theta - \frac{3}{2}mr^2 \dot{\phi}\dot{\theta} \sin \theta \tag{5.248}$$

and the ψ equation is

$$\frac{3}{2}mr^2 \ddot{\psi} + \frac{3}{2}mr^2 \ddot{\phi} \cos \theta - \frac{5}{2}mr^2 \dot{\phi}\dot{\theta} \sin \theta = 0 \tag{5.249}$$

These equations of motion are equivalent to those obtained in Example 4.2 on page 222 using Maggi's equation, but the Suslov approach is simpler and more direct.

5.6   Summary of integral methods

A review of integral methods in the analysis of nonholonomic systems must begin with Hamilton's principle, namely,

$$\int_{t_1}^{t_2} (\delta T + \delta W) dt = 0 \tag{5.250}$$

Here the kinetic energy $T(q, \dot{q}, t)$ is unconstrained and

$$\delta T = \sum_{i=1}^{n} \frac{\partial T}{\partial q_i} \delta q_i + \sum_{i=1}^{n} \frac{\partial T}{\partial \dot{q}_i} \frac{d}{dt}(\delta q_i) \tag{5.251}$$

since we assume that

$$\frac{d}{dt}(\delta q_i) - \delta \dot{q}_i = 0 \tag{5.252}$$

Furthermore, the varied path determined by the δqs satisfies the virtual constraints of (5.4) but does not satisfy, in general, the actual constraints of (5.11).

Hamilton's principle leads to equations of motion of the form given by Lagrange's principle or Maggi's equation.

The other integral methods can be derived directly from Hamilton's principle. For example, if the kinetic energy in terms of quasi-velocities is $T^*(q, u, t)$ and we use the substitution

$$\delta T = \delta T^* + \sum_{i=1}^{n} \frac{\partial T}{\partial \dot{q}_i} \left[\frac{d}{dt}(\delta q_i) - \delta \dot{q}_i \right] \tag{5.253}$$

in (5.250), the result after integration by parts reduces to the explicit form of the central equation, (5.153).

On the other hand, if we designate the qs as either independent or dependent, and write the *constrained* kinetic energy $T^0(q, \dot{q}_I, t)$, then the substitution

$$\delta T = \delta T^0 + \sum_{D=n-m+1}^{n} \frac{\partial T}{\partial \dot{q}_D} \left[\frac{d}{dt}(\delta q_D) - \delta \dot{q}_D \right] \tag{5.254}$$

in (5.250) results in the explicit form of Suslov's principle, (5.216). Here we assume that

$$\frac{d}{dt}(\delta q_I) - \delta \dot{q}_I = 0 \qquad (I = 1, \ldots, n - m) \tag{5.255}$$

All these methods utilize varied paths, determined by the δqs, which satisfy the virtual constraints but not the actual constraints. Furthermore, for a given system and varied path, the integrals used in the various methods will all be equal to δP at any given time t, where $t_1 < t < t_2$. Thus, differing notation and variables are being used to express the same basic integral of Hamilton's principle. The differing approaches, however, can result in one method being easier than another, depending upon the particular system being analyzed.

5.7   Bibliography

Greenwood, D. T. *Classical Dynamics*. Meneola, NY: Dover Publications, 1997.

Lanczos, C. *The Variational Principles of Mechanics*. Toronto: University of Toronto Press, 1949.

Neimark, Ju.I. and Fufaev, N. A. *Dynamics of Nonholonomic Systems*. Translations of Mathematical Monographs, Vol. 33. Providence, RI: American Mathematical Society, 1972.

Papastavridis, J. G. *Analytical Mechanics*. Oxford: Oxford University Press, 2002.

Pars, L. A. *A Treatise on Analytical Dynamics*. London: William Heinemann, 1965.

5.8   Problems

5.1. Consider the nonholonomic form of Hamilton's principle, as given by (5.27). Show that the integral

$$\int_{t_1}^{t} (\delta T + \delta W) dt$$

where $t_1 < t < t_2$, is equal to δP of the central equation (5.129).

5.2. (a) Show that, for a system of N particles,

$$\delta P = \sum_{j=1}^{n} \frac{\partial T^*}{\partial u_j} \delta \theta_j$$

where $T^*(q, u, t)$ is the unconstrained kinetic energy written in terms of quasi-velocities. (b) Starting with the central equation in the form of (5.129), and without using the fixed end-points and integration by parts of the integral method, derive (5.151) and the explicit form of the central equation.

5.3. A nonholonomic system has m constraints that are expressed in the dependent form

$$\dot{q}_k = \sum_{i=1}^{n-m} B_{ki}(q) \dot{q}_i \qquad (k = n - m + 1, \ldots, n)$$

Assuming that the independent qs satisfy

$$\frac{d}{dt}(\delta q_i) - \delta \dot{q}_i = 0 \qquad (i = 1, \ldots, n - m)$$

show that, for the dependent qs,

$$\frac{d}{dt}(\delta q_k) - \delta \dot{q}_k = \sum_{i=1}^{n-m} \sum_{j=1}^{n-m} \beta_{ij}^k \dot{q}_j \delta q_i \qquad (k = n - m + 1, \ldots, n)$$

where

$$\beta_{ij}^k = \frac{\partial B_{ki}}{\partial q_j} - \frac{\partial B_{kj}}{\partial q_i} + \sum_{l=n-m+1}^{n} \left(\frac{\partial B_{ki}}{\partial q_l} B_{lj} - \frac{\partial B_{kj}}{\partial q_l} B_{li} \right)$$

A substitution of this result into the Suslov equation (5.216) results in *Voronets'* *equation:*

$$\frac{d}{dt}\left(\frac{\partial T^0}{\partial \dot{q}_i}\right) - \left(\frac{\partial T^0}{\partial q_i} + \sum_{k=n-m+1}^{n} \frac{\partial T^0}{\partial q_k} B_{ki}\right) - \sum_{j=1}^{n-m} \sum_{k=n-m+1}^{n} \frac{\partial T}{\partial \dot{q}_k} \beta_{ij}^k \dot{q}_j = Q_i^0$$

where

$$Q_i^0 = Q_i + \sum_{k=n-m+1}^{n} Q_k B_{ki} \qquad (i = 1, \ldots, n - m)$$

5.4. Consider the rotational motion of a rigid body using $(\omega_x, \omega_y, \omega_z)$ as quasi-velocities, where xyz is a body-fixed principal axis system at the center of mass. (a) Use classical Euler angles as qs and show that

$$\frac{d}{dt}(\delta\theta_x) - \delta\omega_x = \omega_z\delta\theta_y - \omega_y\delta\theta_z$$

and similarly for the y and z components. (b) Use the central equation in the general form of (5.151) to derive the Euler equations of rotational motion. Assume that $d/dt\,(\delta q_i) - \delta\dot{q}_i = 0$ for each of the Euler angles.

5.5. A pair of wheels, each considered to be a thin uniform disk of mass m and radius r, are connected by a massless axle of length L. They roll without slipping on the xy-plane which is inclined at an angle α from the horizontal. The independent us are $(v, \dot{\phi})$ where v is the speed of the center of mass, and $\dot{\phi}$ is the angular velocity of the axle. The qs are $(x, y, \phi, \psi_1, \psi_2)$. (a) Use the explicit form of the central equation to obtain the differential equations of motion. (b) Assume the initial conditions are $v(0) = v_0, \phi(0) = 0, \dot{\phi}(0) = \omega_0$. Solve for $\dot{x}$ and $\dot{y}$ as functions of time and find their average values.

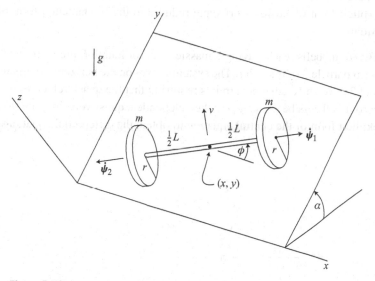

Figure P 5.5.

5.6. A dumbbell consists of two particles, each of mass m, connected by a massless rod of length l (Fig. P 5.6). Particle 1 has a knife-edge constraint oriented perpendicular to the rod. A constant force $\mathbf{F} = F\mathbf{i}$ is applied at particle 2. (a) Use (x, y, ϕ) as generalized coordinates and obtain the differential equations of motion from the explicit form of Suslov's principle. Let x be the dependent coordinate. (b) Show that the speed v of particle 1 is constant. (c) Show that the angle ϕ obeys a differential equation of the same form as for a simple pendulum, even though particle 1 is moving.

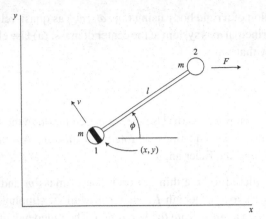

Figure P 5.6.

5.7. Consider a holonomic system with n qs and m constraints of the form

$$q_D = \Phi_D(q_I, t)$$

where $I = 1, \ldots, n - m$ and $D = n - m + 1, \ldots, n$. Show that, for this system, the explicit form of Suslov's principle reduces to the fundamental form of Lagrange's equation.

5.8. Two dumbbells, each having a massless rod of length l, are connected by a joint to form particle 3 of mass $2m$. The system moves on the horizontal xy-plane. Particles 1 and 2 have knife-edge constraints requiring that the quasi-velocities u_3 and u_4 equal zero. Let the qs be (x, y, ψ, ϕ); the independent quasi-velocities are (u_1, u_2). Use the explicit form of the central equation to obtain the differential equations of motion.

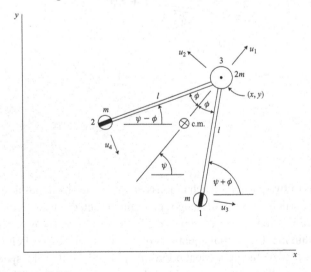

Figure P 5.8.

5.9. Consider again the system of Problem 5.8. Choose (x, y) as dependent qs and (ψ, ϕ) as independent qs. (a) Find the unconstrained kinetic energy $T(q, \dot{q})$ and the constrained kinetic energy $T^0(q, \dot{q}_I)$. (b) Use the explicit form of Suslov's principle to obtain the differential equations of motion.

5.10. A disk of mass $2m$, radius r, and central moment of inertia I has a knife-edge constraint at its center C. A particle of mass m is connected by a massless rod of length l to a point P on the circumference by means of a joint. The system moves on the horizontal xy-plane. Using (θ, ϕ, x, y) as qs and $(v, \dot{\theta}, \dot{\phi})$ as us, obtain the differential equations of motion by means of the explicit central equation.

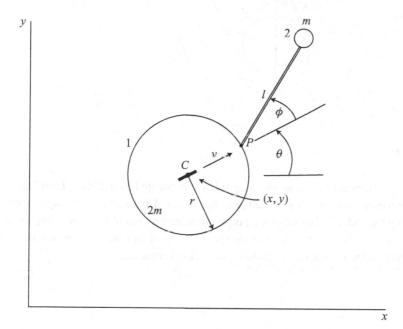

Figure P 5.10.

5.11. A thin uniform disk of mass m and radius r is constrained to remain vertical as it rolls on the horizontal xy-plane (Fig. P 5.11). A series of knife edges are embedded in the xy-plane and are aligned parallel to the x-axis. As a result, the contact point P on the disk cannot slip in the y direction, but can slide without friction in the x direction. Let θ be the angle between the plane of the disk and the vertical yz-plane. Choose (θ, ψ, x, y) as qs and $(\dot{\theta}, \dot{\psi}, \dot{x})$ as independent $\dot{q}$s, where $\dot{\psi}$ is the rotation rate about the axis of symmetry. (a) Use the explicit form of Suslov's principle to obtain the differential equations of motion. (b) Assume the initial conditions $\theta(0) = 0$, $\dot{\theta}(0) = \dot{\theta}_0$, $\dot{\psi}(0) = \dot{\psi}_0$, $\dot{x}(0) = \dot{x}_0$. Find the minimum and maximum values of $\dot{\psi}$ and of the speed v of the center C.

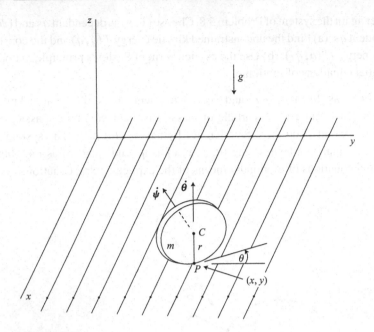

Figure P 5.11.

5.12. Two rods, each of mass m and length l, are connected by a joint at B and move in the horizontal xy-plane. A knife-edge constraint at A permits no velocity component w along the rod at A but allows a perpendicular component v. Choose $(\theta_1, \theta_2, x, y)$ as qs and $(\dot{\theta}_1, \dot{\theta}_2, v, w)$ as us. Obtain the differential equations of motion by using the transpositional form of the Boltzmann–Hamel equation.

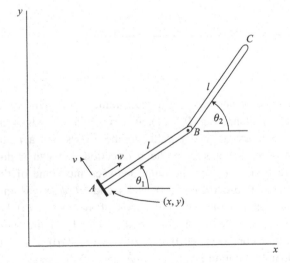

Figure P 5.12.

6 Introduction to numerical methods

Digital computers are used extensively in the analysis of dynamical systems. The equations describing the motion of a system are partly dynamic and partly kinematic in nature. In either case, they take the form of ordinary differential equations which are nonlinear, in general. These differential equations are usually written with time as the independent variable, and time is assumed to vary in a continuous manner from some initial value, frequently zero, to a final value.

In any representation using digital computers, the time can assume only a finite number of discrete values. The differential equations of the system are replaced by difference equations. The solutions of these difference equations are not the same, in general, as the solutions of the corresponding differential equations at the given discrete times, thereby introducing computational errors. More importantly, if parameters such as step size are not properly chosen, a given numerical procedure may produce an apparent unstable response for a system that is actually stable.

In this chapter, we shall begin with a brief discussion of the interpolation and extrapolation of digital data. Then we will proceed with a discussion of various algorithms for the numerical integration of ordinary differential equations. The errors arising from these numerical procedures will be discussed, and questions of numerical stability will be considered.

Another important consideration in the numerical analysis of dynamical systems lies in the proper representation of geometrical constraints. Even if the physical system is stable, there may be numerical instabilities resulting from the method of applying constraints. We shall discuss methods of representing constraints and will analyze their stability.

Finally, a topic of great interest is that of error detection and correction. One approach is to use *integrals of the motion*, that is, functions whose values remain constant during the motion. Examples are the energy or angular momentum functions associated with certain systems. Any deviations from the expected values of these functions serve as indicators of errors and are the starting point for possible corrections. These possibilities will be discussed.

6.1 Interpolation

Polynomial approximations

Consider a function of time $f(t)$ whose values are given at the $(n + 1)$ distinct points $t_0, t_1, \ldots, t_n$. It is always possible to find an interpolating polynomial of degree n which passes exactly through these $(n + 1)$ points. Let us choose

$$P_n(t) = a_0 + a_1 t + \cdots + a_n t^n \tag{6.1}$$

as the approximating polynomial to $y = f(t)$ and let

$$y_i = f(t_i) \tag{6.2}$$

We can determine the $(n + 1)$ coefficients $a_o, a_1, \ldots, a_n$ from the $(n + 1)$ equations

$$y_i = \sum_{k=0}^{n} a_k t_i^k \qquad (i = 0, 1, \ldots, n) \tag{6.3}$$

which are linear in the as. For distinct times $t_0, t_1, \ldots, t_n$, a solution for the as is always possible because the determinant of their coefficients is always nonzero. Thus, we can obtain the interpolating polynomial $P_n(t)$.

Lagrange's interpolation formula

An alternative form of the same interpolating polynomial can be obtained by first using the notation

$$\Pi_i(t) = (t - t_0)(t - t_1) \cdots (t - t_{i-1})(t - t_{i+1}) \cdots (t - t_n) \tag{6.4}$$

We note that the factor $(t - t_i)$ is omitted. Now define a *sampling polynomial*

$$\delta_i(t) = \frac{\Pi_i(t)}{\Pi_i(t_i)} \qquad (i = 0, 1, \ldots, n) \tag{6.5}$$

which is a polynomial of degree n in t. It has the property that it is equal to zero at all the sampling times $t_0, t_1, \ldots$ except at t_i where its value is one. It is apparent, then, that a polynomial of degree n which passes through the $(n + 1)$ discrete points is

$$P_n(t) = \sum_{i=0}^{n} y_i \delta_i(t) = \sum_{i=0}^{n} y_i \frac{\Pi_i(t)}{\Pi_i(t_i)} \tag{6.6}$$

This is the *Lagrange interpolation formula*.

An estimate of the error in $P_n(t)$ can be expressed as a polynomial of degree $(n + 1)$ in t. Moreover, we know that the error is zero at the $(n + 1)$ points $t_0, t_1, \ldots, t_n$. So let us assume that

$$y(t) \approx P_n(t) + C(t - t_0)(t - t_1) \cdots (t - t_n) \tag{6.7}$$

where C is a constant to be determined. Now let us differentiate this equation $(n + 1)$ times with respect to t. The terms arising from $P_n(t)$ will disappear completely, and we need to consider only the term $C t^{n+1}$ since the others will vanish upon differentiation. With the aid of the mean value theorem, we can write

$$y^{(n+1)}(\xi) = (n + 1)! \, C \qquad (t_0 < \xi < t_n) \tag{6.8}$$

where $y^{(n+1)}$ is the $(n+1)$th time derivative of $y(t)$. Thus, from (6.7) and (6.8), we obtain

$$y(t) = P_n(t) + \frac{(t - t_0)(t - t_1) \cdots (t - t_n)}{(n + 1)!} y^{(n+1)}(\xi) \tag{6.9}$$

The second term on the right represents the error $R_n(t)$, that is,

$$R_n(t) = \frac{(t - t_0)(t - t_1) \cdots (t - t_n)}{(n + 1)!} y^{(n+1)}(\xi) \tag{6.10}$$

It should be noted that the error in the polynomial approximation does not necessarily decrease with increasing n. This is particularly true when the sample times $t_0, \ldots, t_n$ are equally spaced and the evaluation occurs at a time t near the limits of the overall interval $[t_0, t_n]$. The problem arises for functions such as $\tan t$, $\ln t$, and $1/(1 + t)^2$ whose Taylor series each have a finite radius of convergence. Although there is an exact match of $P_n(t)$ and $y(t)$ at the $(n + 1)$ discrete points, there is no restriction on the derivatives at these points. Thus, for polynomials of relatively high degree, it is possible for large errors to occur between sample times. This is known as the *Runge phenomenon*. On the other hand, functions such as $\sin t$ and e^t, whose Taylor series representations have no convergence problems, will not be subject to these large errors.

Divided difference

Consider the polynomial $P_n(t)$ passing through the $(n + 1)$ points (y_i, t_i) where $i = 0, 1, \ldots, n$. Define the *divided difference*

$$[t_a, t_b] = [t_b, t_a] = \frac{y_b - y_a}{t_b - t_a} \tag{6.11}$$

Similarly, a second-order divided difference is

$$[t_a, t_b, t_c] = \frac{[t_b, t_c] - [t_a, t_b]}{t_c - t_a} \tag{6.12}$$

Again, the order of the arguments makes no difference. More generally, for the $(n + 1)$ times $t_0, \ldots, t_n$, the nth-order divided difference, written in terms of $(n - 1)$th-order divided differences, is

$$[t_0, t_1, \ldots, t_n] = \frac{[t_1, t_2, \ldots, t_n] - [t_0, t_1, \ldots, t_{n-1}]}{t_n - t_0} \tag{6.13}$$

Divided differences can be used in expressing the interpolating polynomial $P_n(t)$. Consider *Newton's interpolation formula*:

$$P_n(t) = y_0 + (t - t_0)\{[t_0, t_1] + (t - t_1)[\,[t_0, t_1, t_2] + (t - t_2)([t_0, t_1, t_2, t_3] + \cdots)]\} \tag{6.14}$$

where the expression is continued until $[t_0, t_1, \ldots, t_n]$ appears.

The coefficients to be used in Newton's interpolation formula can be computed with the aid of a *divided difference table*. For example, consider the case of four sample times

t_0, t_1, t_2, and t_3. The corresponding divided difference table is

$$
\begin{array}{c|ccccc}
t_0 & y_0 \\
 & & [t_0, t_1] \\
t_1 & y_1 & & [t_0, t_1, t_2] \\
 & & [t_1, t_2] & & [t_0, t_1, t_2, t_3] \\
t_2 & y_2 & & [t_1, t_2, t_3] \\
 & & [t_2, t_3] \\
t_3 & y_3
\end{array}
$$

As a numerical example, consider the function $y = \log_{10} t$ and assume unit time intervals.

t	$\log_{10} t$	$[\cdot, \cdot]$	$[\cdot, \cdot, \cdot]$	$[\cdot, \cdot, \cdot, \cdot]$
1	0.00000			
		0.30103		
2	0.30103		−0.06247	
		0.17609		0.01230
3	0.47712		−0.02558	
		0.12494		
4	0.60206			

The cubic curve passing through the four given points is found by using Newton's interpolation formula with the numerical coefficients obtained from the first entry of each column. Thus, we obtain

$$P_3(t) = 0 + (t - 1)\{0.30103 + (t - 2)[-0.06247 + (t - 3)(0.01230)]\}$$

For the case $t = 2.5$, for example, the interpolation result is

$$P_3(2.5) = 1.5\{0.30103 + 0.5[-0.06247 - 0.5(0.01230)]\} = 0.40008$$

The actual value is $\log_{10} 2.5 = 0.39794$.

An alternate form of Newton's interpolation formula is

$$
\begin{aligned}
P_n(t) = y_0 &+ (t - t_0)[t_0, t_1] + (t - t_0)(t - t_1)[t_0, t_1, t_2] \\
&+ \cdots + (t - t_0) \cdots (t - t_{n-1})[t_0, \ldots, t_n]
\end{aligned}
\tag{6.15}
$$

This leads to an error or remainder term of the form

$$R_n(t) = (t - t_0) \cdots (t - t_n)[t_0, t_1, \ldots, t_n, t] \tag{6.16}$$

which is obtained by considering t to be an additional data point and writing the next term, recalling that there would be zero error at time t if the additional term $R_n(t)$ is included. Comparing this result with (6.10), and assuming that t is an interior point, we find that

$$[t_0, t_1, \ldots, t_n, t] = \frac{y^{(n+1)}(\xi)}{(n + 1)!} \qquad (t_0 < \xi < t_n) \tag{6.17}$$

Now eliminate t and obtain

$$[t_0, \ldots, t_n] = \frac{y^{(n)}(\xi)}{n!} \qquad (t_0 < \xi < t_n) \tag{6.18}$$

For the case $n = 0$, (6.17) reduces to

$$[t_0, t] = \frac{y(t) - y_0}{t - t_0} = \dot{y}(\xi) \qquad (t_0 < \xi < t) \tag{6.19}$$

which is the *mean value theorem*.

Forward and backward differences

Let us consider differences associated with a uniform time interval $\Delta t = h$. The forward difference operator Δ which operates on a function of time $f(t)$ is given by

$$\Delta f(t) = f(t + h) - f(t) \tag{6.20}$$

or

$$\Delta f_n = f_{n+1} - f_n \tag{6.21}$$

The forward difference operator has the property of *linearity*, that is,

$$\Delta[af(t) + bg(t)] = a\Delta f + b\Delta g \tag{6.22}$$

where a and b are constants. Furthermore, for a product of two functions of time,

$$\Delta[f(t)g(t)] = f(t + h)g(t + h) - f(t)g(t) \tag{6.23}$$

or

$$\begin{aligned}
\Delta(f_n g_n) &= f_{n+1}g_{n+1} - f_n g_n \\
&= f_{n+1}g_{n+1} - f_{n+1}g_n + f_{n+1}g_n - f_n g_n \\
&= f_{n+1}\Delta g_n + g_n \Delta f_n \\
&= g_{n+1}\Delta f_n + f_n \Delta g_n
\end{aligned} \tag{6.24}$$

where the last equality is obtained by symmetry. For a quotient, we have

$$\Delta\left(\frac{f_n}{g_n}\right) = \frac{g_n \Delta f_n - f_n \Delta g_n}{g_n g_{n+1}} \tag{6.25}$$

The second forward difference is

$$\begin{aligned}
\Delta^2 f_n = \Delta(\Delta f_n) &= (f_{n+2} - f_{n+1}) - (f_{n+1} - f_n) \\
&= f_{n+2} - 2f_{n+1} + f_n
\end{aligned} \tag{6.26}$$

In a similar manner, we find that

$$\Delta^3 f_n = \Delta(\Delta^2 f_n) = f_{n+3} - 3f_{n+2} + 3f_{n+1} - f_n \tag{6.27}$$

$$\Delta^4 f_n = f_{n+4} - 4f_{n+3} + 6f_{n+2} - 4f_{n+1} + f_n \tag{6.28}$$

and so on.

If $f(t)$ is a polynomial in t, then each forward difference operation reduces its degree by one. For example, suppose that

$$f(t) = a_N t^N + a_{N-1} t^{N-1} + \cdots + a_0 \tag{6.29}$$

Then

$$\begin{aligned}
\Delta f &= a_N [(t+h)^N - t^N] + \cdots \\
&= a_N N h t^{N-1} + \cdots \tag{6.30}
\end{aligned}$$
$$\Delta^2 f = a_N N(N-1) h^2 t^{N-2} + \cdots \tag{6.31}$$
$$\Delta^k f = a_N N(N-1) \cdots (N-k+1) h^k t^{N-k} + \cdots \tag{6.32}$$

Hence, we find that

$$\Delta^N f = a_N N! h^N \tag{6.33}$$

and

$$\Delta^{N+1} f = 0 \tag{6.34}$$

Forward differences are related to the corresponding derivatives with respect to time at the middle of the sampling range in t. For example,

$$\frac{\Delta f_n}{h} = \frac{f_{n+1} - f_n}{h} \approx \dot{y}_{n+\frac{1}{2}} \tag{6.35}$$
$$\frac{\Delta^2 f_n}{h^2} = \frac{f_{n+2} - 2f_{n+1} + f_n}{h^2} \approx \ddot{y}_{n+1} \tag{6.36}$$

which can be regarded as central difference approximations to the time derivatives at the given points. More generally, we find that

$$\frac{\Delta^k f_n}{h^k} \approx \frac{d^k f}{dt^k} \left(t_n + \frac{1}{2} kh \right) \tag{6.37}$$

and in addition,

$$\frac{\Delta^k f_n}{h_k} = f^{(k)}(\xi) \qquad (t_n < \xi < t_n + kh) \tag{6.38}$$

As an example, suppose that

$$f(t) = a_3 t^3 + a_2 t^2 + a_1 t + a_0 \tag{6.39}$$

Then,

$$\begin{aligned}
\Delta f(t) &= f(t+h) - f(t) \\
&= a_3 (3ht^2 + 3h^2 t + h^3) + a_2 (2ht + h^2) + a_1 h \tag{6.40}
\end{aligned}$$

On the other hand,

$$\dot{f}(t) = 3a_3 t^2 + 2a_2 t + a_1 \tag{6.41}$$

and

$$hf\left(t + \frac{1}{2}h\right) = 3a_3h\left(t^2 + ht + \frac{1}{4}h^2\right) + 2a_2h\left(t + \frac{1}{2}h\right) + a_1h \tag{6.42}$$

Upon comparing the right-hand sides of (6.40) and (6.42), we see that they differ only in terms of order h^3.

From (6.21) and (6.26), we find that forward differences involve present and future data, but one may also define a difference operator which uses present and past data. This is the *backward difference* operator ∇. Its defining equation is

$$\nabla f(t) = f(t) - f(t - h) \tag{6.43}$$

or

$$\nabla f_n = f_n - f_{n-1} \tag{6.44}$$

In addition, we find that

$$\nabla^2 f_n = f_n - 2f_{n-1} + f_{n-2} \tag{6.45}$$
$$\nabla^3 f_n = f_n - 3f_{n-1} + 3f_{n-2} - f_{n-3} \tag{6.46}$$
$$\nabla^4 f_n = f_n - 4f_{n-1} + 6f_{n-2} - 4f_{n-3} + f_{n-4} \tag{6.47}$$

and

$$\frac{\nabla^k f_n}{h^k} \approx \frac{d^k f}{dt^k}\left(t_n - \frac{1}{2}kh\right) \tag{6.48}$$

In applying the interpolating polynomial $P_n(t)$ we have assumed that the time t lies in the interval $[t_0, t_n]$. On the other hand, we can use the same polynomial $P_n(t)$ in the process of *extrapolation* to estimate values of $f(t)$ for times outside the given interval. If t is slightly greater than t_n, for example, then (6.17) becomes

$$[t_0, t_1, \ldots, t_n, t] = \frac{y^{(n+1)}(\xi)}{(n+1)!} \qquad (t_0 < \xi < t) \tag{6.49}$$

More explicitly, if we assume a uniform time interval h, and set $t = t_{n+1}$, then we find from (6.16), that the error is equal to the remainder term

$$R_{n+1} = (t_{n+1} - t_0)\cdots(t_{n+1} - t_n)\frac{y^{(n+1)}(\xi)}{(n+1)!}$$
$$= h^{n+1}y^{(n+1)}(\xi) \qquad (t_0 < \xi < t_{n+1}) \tag{6.50}$$

This is the error in extrapolating to the $(n+1)$th step, using the data of the previous n steps, including the initial value y_0.

6.2   Numerical integration

In the study of dynamical systems we consider, in general, ordinary differential equations with time as the independent variable. If one uses Lagrangian methods, the resulting second-order equations of motion are nonlinear, in general, but are always linear in the $\ddot{q}$s. Usually each second-order differential equation is converted to two first-order equations before numerical integration takes place, resulting in the solution for the dependent variables as functions of time. Numerical integration is accomplished by first converting the differential equations to *difference equations*. These equations are then solved at discrete instants of time. The solutions of the difference equations ideally should be the same as the solutions of the differential equations evaluated at the discrete times. Now let us consider some of the numerical procedures or *algorithms* used in the integration of ordinary differential equations. The errors associated with these methods will be evaluated.

Euler's method

Let us begin with a single first-order differential equation

$$\dot{y} = f(y, t) \tag{6.51}$$

Choose a step size $\Delta t = h$ and assume that the initial condition $y(t_0) = y_0$ is given. The numerical solution for $y(t)$ at $t = t_1, t_2, t_3, \ldots$ is obtained by repeating the following sequence of calculations:

(1) $\dot{y}_n = f(y_n, t_n)$ (6.52)

(2) $y_{n+1} = y_n + h\dot{y}_n$ (6.53)

for $n = 0, 1, 2, \ldots$. Note that the result of the first calculation is used in the second, and the result of the second calculation is used in the next repetition of step (1).

Next, let us consider the second-order differential equation

$$\ddot{y} = f(y, \dot{y}, t) \tag{6.54}$$

This equation is replaced by the two first-order equations

$$\dot{y} = v \tag{6.55}$$

$$\dot{v} = f(y, v, t) \tag{6.56}$$

with the given initial conditions y_0 and v_0. The Euler integration method for this case consists of the following procedure:

(1) $\dot{v}_n = f(y_n, v_n, t_n)$ (6.57)

(2) $y_{n+1} = y_n + hv_n$ (6.58)

(3) $v_{n+1} = v_n + h\dot{v}_n$ (6.59)

which is repeated at each time step.

More generally, if a system is described by a set of N coupled first-order equations in N state variables, then equations similar to (6.52) and (6.53) are used in sequence for each of the state variables at each time step. These computed variables are then used as initial values for the next step.

Truncation errors

Numerical integration involves the use of known data at time t_n, and possibly at previous times, along with the given differential equations, to estimate the values of the variables at time $t_{n+1} = t_n + h$. This is an extrapolation process. If one can represent the true solution of a differential equation in the neighborhood of time t_n by a Taylor series about t_n, then the algorithm is a representation of the *truncated series*, that is, its first few terms. The error due to using a limited number of terms is called the *truncation error*, and is expressed by giving the first omitted term or first error term. The truncation error is a function of step size h, and it approaches zero as h goes to zero.

To illustrate the nature of a truncation error, let us consider a single first-order differential equation such as (6.51), and assume that it is integrated numerically by the Euler method. We need to distinguish between the true or exact value y_{n+1} at $t = t_{n+1}$ and the corresponding computed value y_{n+1}^*. Suppose, for example, that the initial condition is $y(t_0) = y_0$ and the corresponding exact solution of the differential equation is $y(t)$, as shown in Fig. 6.1. Other slightly different initial conditions would result in corresponding roughly parallel solution curves, as shown in the figure. The computed solution $y^*(t)$, however, suffers a truncation error at each step which, in effect, transfers it to an adjacent solution curve. Thus, as shown in Fig. 6.1 in exaggerated form, the truncation error can continue to accumulate. The truncation error for a single step is called the *local truncation error*. On the other hand, the accumulated error over a given time interval is known as the *global truncation error*.

Consider the use of Euler's method in the integration of the single first-order equation

$$\dot{y} = f(y, t) \tag{6.60}$$

Assuming that there is no error in the value y_n at time t_n, the computed value at time t_{n+1} is

$$y_{n+1}^* = y_n + h\dot{y}_n \tag{6.61}$$

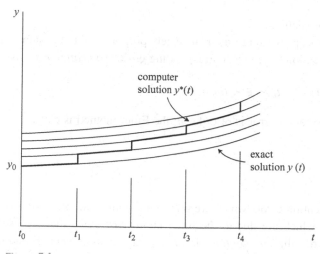

Figure 6.1.

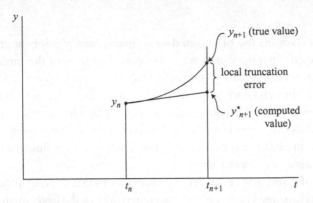

Figure 6.2.

where

$$\dot{y}_n = f(y_n, t_n) \tag{6.62}$$

as shown in Fig. 6.2. To obtain the truncation error, consider a Taylor series about the time t_n,

$$y_{n+1} = y_n + h\dot{y}_n + \frac{h^2}{2!}\ddot{y}_n + \frac{h^3}{3!}\dddot{y}_n + \cdots \tag{6.63}$$

The *local truncation error* can be approximated by the first error term, that is

$$E_{n+1} = y_{n+1} - y^*_{n+1} = \frac{h^2}{2!}\ddot{y}_n \tag{6.64}$$

More accurately, using the mean value theorem, the local truncation error is

$$E_{n+1} = \frac{h^2}{2!}\ddot{y}(\xi) \qquad (t_n < \xi < t_{n+1}) \tag{6.65}$$

for the Euler integration method.

Since the number of steps per unit time is inversely proportional to the step size h, and assuming additive truncation errors, the corresponding *global truncation error* is

$$\frac{E_{n+1}}{h} = \frac{h}{2}\ddot{y}(\xi) = O(h) \qquad (t_n < \xi < t_{n+1}) \tag{6.66}$$

Because the global truncation error is of order h, the Euler method is called a *first-order method*.

Roundoff errors

In addition to the truncation errors which are due to the numerical integration algorithm, there are other errors which are due to the finite number of digits that are used in the computations. These are called *roundoff errors*. A typical situation where roundoff errors can become important occurs when the calculations involve the small difference of large numbers, in a relative sense. By subtracting the nearly equal numbers, one can lose many significant digits. Furthermore, roundoff errors become relatively more important if the truncation errors are drastically reduced by using a very small step size.

To illustrate the nature of roundoff errors, suppose we are given $y = e^t$ and we desire to calculate $y_{n+1} - y_n$ for the case $t_n = 10$ and $h = 0.001$. If one uses a calculator which employs ten decimal digits, the result is

$$e^{10.001} - e^{10} = 22048.50328 - 22026.46579$$
$$= 22.03749$$

We notice that, although the numbers to be subtracted are given to ten significant digits, the result has only seven significant digits, and there is some doubt concerning the accuracy of the last digit. In order to avoid errors due to the small difference of large numbers, one should attempt to recast the problem in a form in which the difference is obtained directly, possibly through a change of variables. For example, in the above problem, we notice that

$$y_{n+1} - y_n = (e^h - 1)e^{t_n} \tag{6.67}$$

and we can use a truncated series for the exponential function to obtain

$$e^h - 1 = h + \frac{1}{2}h^2 + \frac{1}{6}h^3 + \cdots$$
$$= 10^{-3} + \frac{1}{2} \times 10^{-6} + \frac{1}{6} \times 10^{-9} = 1.000500167 \times 10^{-3}$$

Then

$$y_{n+1} - y_n = 1.000500167 \times 22.02646579 = 22.03748270$$

This result is accurate to the full ten digits. The improved accuracy is due to the replacement of a difference by a product involving a convergent series.

Another example in which roundoff errors can become significant occurs in dynamical problems using Euler angles as generalized coordinates, where these coordinate values are near a singular point. For type I Euler angles, this singular orientation occurs for the attitude angle θ near $\pm \pi/2$. Suppose, for example, that these Euler angles are used in writing the dynamical equations of a rigid body. A basic physical variable in the rotational dynamics is the angular velocity component ω_x, where

$$\omega_x = \dot{\phi} - \dot{\psi} \sin \theta \tag{6.68}$$

For the values of θ near $\pm \pi/2$, the variables $\dot{\phi}$ and $\dot{\psi}$ may be very large and almost equal, with ω_x a relatively small quantity. Thus, we have the small difference of large quantities entering the equations of motion, and resulting in significant roundoff errors. At the same time it also may be necessary to use a small step size h in order to keep the truncation errors under control. Perhaps the best strategy under these conditions is to avoid the use of Euler angles altogether in the dynamical equations. Another approach such as Euler's rotational equations could be used for the dynamics, with the orientation being specified in the kinematical equations by using four Euler parameters, which have no singularity problems.

Trapezoidal method

Consider the first-order equation

$$\dot{y} = f(y, t) \tag{6.69}$$

The *trapezoidal integration algorithm* is

$$y_{n+1} = y_n + \frac{1}{2}h(\dot{y}_{n+1} + \dot{y}_n) \tag{6.70}$$

where

$$\dot{y}_{n+1} = f(y_{n+1}, t_{n+1}) \tag{6.71}$$

This is an *implicit method* because y_{n+1} appears on both sides of the equation. Usually this implies that an iterative method of solution must be used. For example, one might start with a first estimate of y_{n+1} and use (6.71) to obtain a first estimate of $\dot{y}_{n+1}$ that is substituted into (6.70) to obtain a second estimate of y_{n+1}. This procedure is repeated until successive estimates of y_{n+1} are sufficiently close, assuming that the iterative process converges. These values are then used as initial data for the next step.

The values of y_{n+1} and $\dot{y}_{n+1}$ obtained by this iterative process are not, in general, equal to the true values. Let us use the notation y_{n+1}^* and $\dot{y}_{n+1}^*$ for these computed values, where, in accordance with the trapezoidal algorithm,

$$y_{n+1}^* = y_n + \frac{1}{2}h(\dot{y}_{n+1}^* + \dot{y}_n) \tag{6.72}$$

Here, y_n and $\dot{y}_n$ are assumed to be known. Now let us use a Taylor series to obtain

$$\dot{y}_{n+1}^* \approx \dot{y}_{n+1} = \dot{y}_n + h\ddot{y}_n + \frac{h^2}{2}\dddot{y}_n + \cdots \tag{6.73}$$

with the result that

$$y_{n+1}^* = y_n + h\dot{y}_n + \frac{h^2}{2}\ddot{y}_n + \frac{h^3}{4}\dddot{y}_n + \cdots \tag{6.74}$$

The Taylor series for the true value y_{n+1} is

$$y_{n+1} = y_n + h\dot{y}_n + \frac{h^2}{2}\ddot{y}_n + \frac{h^3}{6}\dddot{y}_n + \cdots \tag{6.75}$$

Hence, the local truncation error is

$$E_{n+1} = y_{n+1} - y_{n+1}^* = -\frac{h^3}{12}\dddot{y}_n \tag{6.76}$$

which is of the order h^3. The global truncation error is $O(h^2)$, so the trapezoidal integration algorithm is a second-order method.

Modified Euler method

Now let us combine the two methods we have introduced up to this point. Let us consider a first-order system

$$\dot{y} = f(y, t) \tag{6.77}$$

with $y(t_0) = y_0$. The *modified Euler method*, or *Heun's method*, is a *two-pass method*, that is, the dynamical equation (6.77) is evaluated twice for each step. It consists of the Euler algorithm as a *predictor* followed by the trapezoidal algorithm as a *corrector* for each step. The method is *self-starting* because it requires no data for times previous to the nominal time t_n. It is considered to be an *explicit* method because there are no iterations, even though the trapezoidal algorithm is used once per step, and we have classed the iterated trapezoidal algorithm as implicit.

For given y_n and t_n, the modified Euler method uses the following procedure

(1) $\quad \dot{y}_n = f(y_n, t_n)$ $\hfill (6.78)$

(2) $\quad \tilde{y}_{n+1} = y_n + h\dot{y}_n$ $\hfill (6.79)$

(3) $\quad \dot{\tilde{y}}_{n+1} = f(\tilde{y}_{n+1}, t_{n+1})$ $\hfill (6.80)$

(4) $\quad y_{n+1} = y_n + \dfrac{1}{2}h(\dot{\tilde{y}}_{n+1} + \dot{y}_n)$ $\hfill (6.81)$

We see that, in (6.79), the Euler method is used to obtain a *first estimate* or *predicted value* of y_{n+1}, indicated by a tilde ($\sim$). This $\tilde{y}_{n+1}$ is used in (6.80) to obtain an estimated value of $\dot{y}_{n+1}$, which is then used in the trapezoidal formula of (6.81) to obtain the computed value y_{n+1} to be used in the following time step. This is the *corrected value*.

Now let us find an expression for the truncation error of the modified Euler method. We begin by recalling from (6.64) that the local truncation error due to the use of the Euler method in calculating $\tilde{y}_{n+1}$ is

$$y_{n+1} - \tilde{y}_{n+1} = \frac{1}{2}h^2\ddot{y}_n \tag{6.82}$$

This error will result in a corresponding error in $\dot{\tilde{y}}_{n+1}$, as compared with the true value $\dot{y}_{n+1}$. Assuming small errors, we find that

$$\dot{\tilde{y}}_{n+1} = \dot{y}_{n+1} + (\tilde{y}_{n+1} - y_{n+1})f'_{n+1} \tag{6.83}$$

where we use the notation $f'_{n+1} = (\partial f / \partial y)_{n+1}$. Then, using (6.81)–(6.83), we obtain the computed value

$$y^*_{n+1} = y_n + \frac{1}{2}h(\dot{y}_{n+1} + \dot{y}_n) - \frac{1}{4}h^3 f'_{n+1}\ddot{y}_n \tag{6.84}$$

By using Taylor expansions for y_{n+1} and $\dot{y}_{n+1}$, we find that

$$y_n + \frac{1}{2}h(\dot{y}_{n+1} + \dot{y}_n) = y_{n+1} + \frac{1}{12}h^3\dddot{y}_n \tag{6.85}$$

Finally, we obtain the local truncation error

$$E_{n+1} = y_{n+1} - y_{n+1}^* = -\frac{1}{12}h^3(\dddot{y}_n - 3f_n'\ddot{y}_n) \tag{6.86}$$

where we have substituted f_n' for f_{n+1}' since they differ by $O(h)$.

We notice that, although the order of the local truncation error of the Euler predictor is $O(h^2)$ whereas that of the trapezoidal corrector is $O(h^3)$, this apparent mismatch does not result in a significant loss of accuracy. As an example, consider the *linear first-order* system described by

$$\dot{y} = \lambda y \tag{6.87}$$

where λ is a constant. Then $\ddot{y} = \lambda^2 y$, $\dddot{y} = \lambda^3 y$, etc., and $f' = \partial f/\partial y = \lambda$. The local truncation error of the modified Euler method is

$$E_{n+1} = -\frac{1}{12}h^3(\dddot{y}_n - 3f_n'\ddot{y}_n) = \frac{1}{6}h^3\lambda^3 y_n \tag{6.88}$$

If we compare this error with that given by (6.76), we see that this error is twice as large as the error for the fully iterated and converged trapezoidal method. The *global truncation error* of the modified Euler method is $\frac{1}{6}h^2\lambda^3 y_n$ for this system. Because it is of order h^2, it is classified as a second-order method.

Runge–Kutta methods

The Runge–Kutta integration algorithms are characterized by being explicit, self-starting, multiple-pass methods. Consider first the *second-order Runge–Kutta (RK-2) method*. When applied to the general first-order system of (6.77), the RK-2 procedure is as follows:

$$(1) \quad \dot{y}_n = f(y_n, t_n) \tag{6.89}$$

$$(2) \quad \tilde{y}_{n+\frac{1}{2}} = y_n + \frac{1}{2}h\dot{y}_n \tag{6.90}$$

$$(3) \quad \dot{\tilde{y}}_{n+\frac{1}{2}} = f\left(\tilde{y}_{n+\frac{1}{2}}, t_{n+\frac{1}{2}}\right) \tag{6.91}$$

$$(4) \quad y_{n+1} = y_n + h\dot{\tilde{y}}_{n+\frac{1}{2}} \tag{6.92}$$

Notice that step (2) uses the Euler method to estimate the value of y at $t_{n+\frac{1}{2}}$. This estimate $\tilde{y}_{n+\frac{1}{2}}$ is substituted into the differential equation to obtain an estimate $\dot{\tilde{y}}_{n+\frac{1}{2}}$ of the midpoint velocity. In step (4), this estimate is used as an approximation of the average velocity in the calculation of y_{n+1}.

The local truncation error can be shown to be

$$E_{n+1} = \frac{1}{24}h^3(\dddot{y}_n + 3f_n'\ddot{y}_n) \tag{6.93}$$

For the linear first-order system given by (6.87) the local truncation error for RK-2 is

$$E_{n+1} = \frac{1}{6}h^3\lambda^3 y_n \tag{6.94}$$

This is equal to the truncation error for the modified Euler method.

Now let us consider the widely-used *fourth-order Runge–Kutta (RK-4) method*. Again assume a general first-order system. The procedure for the RK-4 method is as follows:

(1) $\dot{y}_n = f(y_n, t_n)$ (6.95)

(2) $\tilde{y}_{n+\frac{1}{2}} = y_n + \frac{1}{2}h\dot{y}_n$ (6.96)

(3) $\dot{\tilde{y}}_{n+\frac{1}{2}} = f\left(\tilde{y}_{n+\frac{1}{2}}, t_{n+\frac{1}{2}}\right)$ (6.97)

(4) $\tilde{\tilde{y}}_{n+\frac{1}{2}} = y_n + \frac{1}{2}h\dot{\tilde{y}}_{n+\frac{1}{2}}$ (6.98)

(5) $\dot{\tilde{\tilde{y}}}_{n+\frac{1}{2}} = f\left(\tilde{\tilde{y}}_{n+\frac{1}{2}}, t_{n+\frac{1}{2}}\right)$ (6.99)

(6) $\tilde{y}_{n+1} = y_n + h\dot{\tilde{\tilde{y}}}_{n+\frac{1}{2}}$ (6.100)

(7) $\dot{\tilde{y}}_{n+1} = f(\tilde{y}_{n+1}, t_{n+1})$ (6.101)

(8) $y_{n+1} = y_n + \frac{1}{6}h\left(\dot{y}_n + 2\dot{\tilde{y}}_{n+\frac{1}{2}} + 2\dot{\tilde{\tilde{y}}}_{n+\frac{1}{2}} + \dot{\tilde{y}}_{n+1}\right)$ (6.102)

This is an explicit, self-starting, four-pass method with a global truncation error of $O(h^4)$. The first three steps of the procedure are identical with those of the RK-2 method. Then two more passes are added in order to improve the accuracy without requiring more initial data at each time step.

A general analysis of the truncation error of the RK-4 method is rather complicated, but for the case of a linear system described by $\dot{y} = \lambda y$ of (6.87), it can be shown that the local truncation error is

$$E_{n+1} = y_{n+1} - y_{n+1}^* = \frac{(\lambda h)^5}{120}y_n$$ (6.103)

Adams–Bashforth predictors

Predictor algorithms are explicit algorithms which can be used individually or can be coupled with correctors to form an explicit predictor–corrector combination. In general, Adams–Bashforth (AB) predictors are not self-starting because they use past data which is not available at the initial time. Hence, some self-starting method such as RK-4 is often used for the first few time steps before switching to the chosen predictor algorithm.

In order to obtain the various Adams–Bashforth predictors, one can first recall the definition of a backward difference, as given by (6.44),

$$\nabla y_n = y_n - y_{n-1}$$ (6.104)

Higher-order backward differences are defined by a repeated application of this operator, with the results given in (6.45)–(6.47). It can be shown that

$$y_{n+1} = y_n + h\left(\dot{y}_n + \frac{1}{2}\nabla\dot{y}_n + \frac{5}{12}\nabla^2\dot{y}_n + \frac{3}{8}\nabla^3\dot{y}_n + \frac{251}{720}\nabla^4\dot{y}_n + \cdots\right)$$ (6.105)

where

$$\nabla \dot{y}_n = \dot{y}_n - \dot{y}_{n-1} \tag{6.106}$$

$$\nabla^2 \dot{y}_n = \dot{y}_n - 2\dot{y}_{n-1} + \dot{y}_{n-2} \tag{6.107}$$

$$\nabla^3 \dot{y}_n = \dot{y}_n - 3\dot{y}_{n-1} + 3\dot{y}_{n-2} - \dot{y}_{n-3} \tag{6.108}$$

$$\nabla^4 \dot{y}_n = \dot{y}_n - 4\dot{y}_{n-1} + 6\dot{y}_{n-2} - 4\dot{y}_{n-3} + \dot{y}_{n-4} \tag{6.109}$$

The various Adams–Bashforth predictors are obtained by truncating the series of (6.105) at corresponding points. For example, the AB-1 algorithm is identical to the Euler method and uses only the first two terms. Thus, we take

$$y_{n+1} = y_n + h\dot{y}_n \tag{6.110}$$

and obtain the first-order term, or local truncation error, from the next term in (6.105), namely,

$$E_{n+1} = \frac{h}{2}\nabla \dot{y}_n = \frac{h}{2}(\dot{y}_n - \dot{y}_{n-1}) \tag{6.111}$$

Now expand $\dot{y}_{n-1}$ in a Taylor series about t_n and obtain

$$\dot{y}_{n-1} = \dot{y}_n - h\ddot{y}_n + \cdots \tag{6.112}$$

Then, from (6.111) and (6.112), we obtain the first error term for the AB-1 algorithm.

$$E_{n+1} = \frac{1}{2}h^2 \ddot{y}_n \tag{6.113}$$

In a similar manner, we can obtain other Adams–Bashforth predictor algorithms and the corresponding first error terms. They can be summarized as follows:

AB-2

$$y_{n+1} = y_n + \frac{3}{2}h\dot{y}_n - \frac{1}{2}h\dot{y}_{n-1} \tag{6.114}$$

$$E_{n+1} = \frac{5}{12}h^3 \dddot{y}_n \tag{6.115}$$

AB-3

$$y_{n+1} = y_n + \frac{h}{12}(23\dot{y}_n - 16\dot{y}_{n-1} + 5\dot{y}_{n-2}) \tag{6.116}$$

$$E_{n+1} = \frac{3}{8}h^4 y_n^{(4)} \tag{6.117}$$

AB-4

$$y_{n+1} = y_n + \frac{h}{24}(55\dot{y}_n - 59\dot{y}_{n-1} + 37\dot{y}_{n-2} - 9\dot{y}_{n-3}) \tag{6.118}$$

$$E_{n+1} = \frac{251}{720}h^5 y_n^{(5)} \tag{6.119}$$

The global truncation errors are equal to E_{n+1}/h in each case. Notice that data from more steps in the past are required by the higher-order methods.

Adams–Moulton correctors

The Adams–Moulton (AM) algorithms differ from the predictor algorithms by being *implicit*, that is, variables at time t_{n+1} appear on both sides of the integration algorithm. The derivation of AM algorithms begins with the backward difference expression

$$y_{n+1} = y_n + h\left(\dot{y}_{n+1} - \frac{1}{2}\nabla\dot{y}_{n+1} - \frac{1}{12}\nabla^2\dot{y}_{n+1} - \frac{1}{24}\nabla^3\dot{y}_{n+1} - \frac{19}{720}\nabla^4\dot{y}_{n+1} + \cdots\right)$$

(6.120)

where

$$\dot{y}_{n+1} = f(y_{n+1}, t_{n+1})$$

(6.121)

and

$$\nabla\dot{y}_{n+1} = \dot{y}_{n+1} - \dot{y}_n$$

(6.122)

$$\nabla^2\dot{y}_{n+1} = \dot{y}_{n+1} - 2\dot{y}_n + \dot{y}_{n-1}$$

(6.123)

$$\nabla^3\dot{y}_{n+1} = \dot{y}_{n+1} - 3\dot{y}_n + 3\dot{y}_{n-1} - \dot{y}_{n-2}$$

(6.124)

$$\nabla^4\dot{y}_{n+1} = \dot{y}_{n+1} - 4\dot{y}_n + 6\dot{y}_{n-1} - 4\dot{y}_{n-2} + \dot{y}_{n-3}$$

(6.125)

Using procedures similar to those used in deriving the AB algorithms, the following Adams–Moulton corrector algorithms, and the corresponding first error terms, can be obtained.

AM-2 (Trapezoidal)

$$y_{n+1} = y_n + \frac{h}{2}(\dot{y}_{n+1} + \dot{y}_n)$$

(6.126)

$$E_{n+1} = -\frac{h^3}{12}\dddot{y}_n$$

(6.127)

AM-3

$$y_{n+1} = y_n + \frac{h}{12}(5\dot{y}_{n+1} + 8\dot{y}_n - \dot{y}_{n-1})$$

(6.128)

$$E_{n+1} = -\frac{h^4}{24}y_n^{(4)}$$

(6.129)

AM-4

$$y_{n+1} = y_n + \frac{h}{24}(9\dot{y}_{n+1} + 19\dot{y}_n - 5\dot{y}_{n-1} + \dot{y}_{n-2})$$

(6.130)

$$E_{n+1} = -\frac{19}{720}h^5 y_n^{(5)}$$

(6.131)

The first error terms of the AM algorithms are observed to be considerably smaller than the errors for the corresponding AB algorithms. One reason for the greater accuracy of the AM algorithms is that they are *implicit*, and the values of y_{n+1} are those which are achieved after the convergence of the iteration process.

Truncation errors of predictor–corrector methods

We shall consider Adams–Bashforth predictors followed by Adams–Moulton correctors, but with only a single pass of the corrector algorithm per step, that is, without iteration. This leads to truncation errors of the same order in h as the corrector, but generally with a larger coefficient.

Let us begin with AB-1, AM-2 or the *modified Euler* method. This algorithm has been analyzed previously, and the local truncation error, or first error term, is

$$E_{n+1} = -\frac{h^3}{12}(\dddot{y}_n - 3f'_n \ddot{y}_n) \tag{6.132}$$

as in (6.86).

Next consider the AB-2, AM-3 predictor–corrector algorithm. The AB-2 predictor results in an estimated value of y_{n+1} which is

$$\tilde{y}_{n+1} = y_n + \frac{3}{2}h\dot{y}_n - \frac{h}{2}\dot{y}_{n-1} \tag{6.133}$$

The first error term of this expression is

$$y_{n+1} - \tilde{y}_{n+1} = E_{n+1} = \frac{5}{12}h^3 \dddot{y}_n \tag{6.134}$$

Now apply perturbation theory at time t_{n+1} to the system differential equation

$$\dot{y} = f(y, t) \tag{6.135}$$

with the result that

$$
\begin{aligned}
\tilde{\dot{y}}_{n+1} &= \dot{y}_{n+1} + (f'_{n+1})(\tilde{y}_{n+1} - y_{n+1}) \\
&= \dot{y}_{n+1} - \frac{5}{12}h^3 f'_n \dddot{y}_n
\end{aligned} \tag{6.136}
$$

The AM-3 algorithm gives a computed value

$$
\begin{aligned}
y^*_{n+1} &= y_n + \frac{h}{12}(5\tilde{\dot{y}}_{n+1} + 8\dot{y}_n - \dot{y}_{n-1}) \\
&= y_n + \frac{h}{12}(5\dot{y}_{n+1} + 8\dot{y}_n - \dot{y}_{n-1}) - \frac{25}{144}h^4 f'_n \dddot{y}_n
\end{aligned} \tag{6.137}
$$

But from (6.128) and (6.129),

$$y_{n+1} = y_n + \frac{h}{12}(5\dot{y}_{n+1} + 8\dot{y}_n - \dot{y}_{n-1}) - \frac{h^4}{24}y_n^{(4)} \tag{6.138}$$

Thus, we find that the local truncation error for the AB-2, AM-3 predictor–corrector algorithm is

$$E_{n+1} = y_{n+1} - y^*_{n+1} = -\frac{h^4}{144}\left(6y_n^{(4)} - 25f'_n \dddot{y}_n\right) \tag{6.139}$$

For the particular case of the linear system $\dot{y} = \lambda y$, where λ is a constant, the local truncation error is

$$E_{n+1} = \frac{19}{144}(\lambda h)^4 y_n \tag{6.140}$$

If one analyzes the AB-3, AM-3 combination, the truncation error is the same as for the corrector AM-3 alone, namely,

$$E_{n+1} = -\frac{h^4}{24}y_n^{(4)} \tag{6.141}$$

Here, however, no iteration is involved.

For the case of a linear system, this results in

$$E_{n+1} = -\frac{1}{24}(\lambda h)^4 y_n \tag{6.142}$$

Thus, we find that raising the order of the predictor from AB-2 to AB-3 does not change the order of the truncation error, but it may reduce the magnitude of the numerical coefficient.

Finally, for the AB-3, AM-4 predictor–corrector combination, it can be shown that the local truncation error is

$$E_{n+1} = -\frac{h^5}{2880}\left(70y_n^{(5)} - 405 f_n' y_n^{(4)}\right) \tag{6.143}$$

Special integration algorithms

If the differential equations are of a particular form, it may be possible to use an integration algorithm which is especially suitable. For example, suppose that the acceleration is not a function of velocity, that is, it has the form

$$\ddot{y} = f(y, t) \tag{6.144}$$

Let us look for an integration algorithm which applies directly to this second-order differential equation rather than to the equivalent set of two first-order equations. Let us begin with the approximations

$$\dot{y}_{n+\frac{1}{2}} \approx \frac{y_{n+1} - y_n}{h}, \qquad \dot{y}_{n-\frac{1}{2}} \approx \frac{y_n - y_{n-1}}{h} \tag{6.145}$$

Then we obtain the basic *central difference equation*

$$\ddot{y}_n = \frac{\dot{y}_{n+\frac{1}{2}} - \dot{y}_{n-\frac{1}{2}}}{h} = \frac{1}{h^2}(y_{n+1} - 2y_n + y_{n-1}) \tag{6.146}$$

and the following *central difference algorithm*:

$$y_{n+1} = 2y_n - y_{n-1} + h^2 \ddot{y}_n \tag{6.147}$$

A substitution of the Taylor series for y_{n-1} results in the computed value

$$y_{n+1}^* = y_n + h\dot{y}_n + \frac{h^2}{2}\ddot{y}_n + \frac{h^3}{6}\dddot{y}_n - \frac{h^4}{24}y_n^{(4)} \tag{6.148}$$

A comparison of this computed value with the Taylor series for y_{n+1} yields the local truncation error

$$E_{n+1} = y_{n+1} - y_{n+1}^* = \frac{h^4}{12} y_n^{(4)} \tag{6.149}$$

The global truncation error is $O(h^3)$.

Next, consider the *midpoint method* as it applies to the system of (6.144). Let us use the implicit algorithm

$$\dot{y}_{n+1} = \dot{y}_n + hf\left(\frac{y_{n+1} + y_n}{2}, t_{n+\frac{1}{2}}\right) \tag{6.150}$$

$$y_{n+1} = y_n + \frac{h}{2}(\dot{y}_{n+1} + \dot{y}_n) \tag{6.151}$$

We see that the acceleration $\ddot{y}_{n+\frac{1}{2}}$ is obtained by using the midpoint approximation

$$y_{n+\frac{1}{2}} = \frac{1}{2}(y_{n+1} + y_n) \tag{6.152}$$

in (6.144).

Let us use the notation

$$k(y, t) \equiv -\frac{\partial f}{\partial y} \tag{6.153}$$

and assume that k is slowly varying over any time step h so $k_n \approx k_{n+1}$. Then

$$\ddot{y}_{n+\frac{1}{2}} = \ddot{y}_n + \left(\frac{\partial f}{\partial y}\right)\left(\frac{y_{n+1} - y_n}{2}\right) = \ddot{y}_n - \frac{k}{2}(y_{n+1} - y_n) \tag{6.154}$$

Thus, (6.150) becomes

$$\dot{y}_{n+1} = \dot{y}_n + h\ddot{y}_n - \frac{hk}{2}(y_{n+1} - y_n) \tag{6.155}$$

A substitution of this expression for $\dot{y}_{n+1}$ into (6.151) yields the computed value

$$y_{n+1}^* = y_n + h\dot{y}_n + \frac{h^2}{2}\ddot{y}_n - \frac{h^3 k}{8}(\dot{y}_{n+1} + \dot{y}_n)$$

$$= y_n + h\dot{y}_n + \frac{h^2}{2}\ddot{y}_n - \frac{h^3 k}{4}\dot{y}_n + \cdots \tag{6.156}$$

where we note that $\dot{y}_{n+1} \approx \dot{y}_n$. A comparison of this result with the Taylor series

$$y_{n+1} = y_n + h\dot{y}_n + \frac{h^2}{2}\ddot{y}_n + \frac{h^3}{6}\dddot{y}_n + \cdots \tag{6.157}$$

indicates that the local truncation error is

$$E_{n+1} = y_{n+1} - y_{n+1}^* = \frac{h^3}{12}(2\dddot{y}_n + 3k\dot{y}_n) \tag{6.158}$$

The midpoint method, also known as the *midpoint rule*, has the advantages of being numerically stable and, for certain formulations, can also preserve angular momentum.

6.3   Numerical stability

One of the recurring problems in the numerical analysis of physical systems is that the numerical results may show an instability which is not present in the physical system. The stability of a numerical method depends on the step size h and on the stability characteristics of the physical system. Other factors may influence the stability, including the numerical methods used in representing constraints. In this section, we will consider the numerical stability characteristics of various integration algorithms.

Euler integration

Consider first the relatively simple case of a single first-order differential equation

$$\dot{y} = f(y, t) \tag{6.159}$$

The Euler algorithm is

$$y_{n+1} = y_n + h\dot{y}_n \tag{6.160}$$

The numerical stability is analyzed by the perturbation equation

$$\delta y_{n+1} = \delta y_n + h f'_n \delta y_n \tag{6.161}$$

where we use the notation

$$f'_n \equiv \frac{\partial f}{\partial y}(y_n, t_n) \tag{6.162}$$

We consider (6.161) as a *difference equation* in δy and assume that f'_n varies slowly enough that it can be assumed to be constant. Let

$$\frac{\partial f}{\partial y} = \lambda \tag{6.163}$$

where λ is a real constant. Then (6.161) becomes

$$\delta y_{n+1} - (1 + \lambda h)\delta y_n = 0 \tag{6.164}$$

Now assume a solution of the form

$$\delta y_n = A\rho^n \tag{6.165}$$

where A is a constant and ρ is the ratio $\delta y_{n+1}/\delta y_n$. From (6.164) and (6.165) we obtain

$$(\rho - 1 - \lambda h)A\rho^n = 0 \tag{6.166}$$

We seek nonzero values of ρ in order to avoid trivial zero solutions for δy. Thus, we obtain

$$\rho = 1 + \lambda h \tag{6.167}$$

The *necessary condition for numerical stability* is that $|\rho| \leq 1$. The step size h is assumed to be positive, and we see that $-2 \leq \lambda h \leq 0$ for numerical stability.

Now let us consider the more general case of a system of m first-order differential equations which are to be integrated using the Euler method. We have

$$\dot{y}_i = f_i(y, t) \qquad (i = 1, \ldots, m) \tag{6.168}$$

The corresponding perturbation equations are

$$\delta \dot{y}_i = \sum_{k=1}^{m} \frac{\partial f_i}{\partial y_k} \delta y_k = \sum_{k=1}^{m} g_{ik} \delta y_k \qquad (i = 1, \ldots, m) \tag{6.169}$$

where $g_{ik} \equiv \partial f_i / \partial y_k$ is assumed to be varying slowly enough to be considered constant in the perturbation equations and the corresponding difference equations. Let us assume solutions for δy of the form $e^{\lambda t}$. Then, for a single mode involving one value of λ,

$$\frac{d}{dt}(\delta y_i) = \delta \dot{y}_i = \lambda \delta y_i \qquad (i = 1, \ldots, m) \tag{6.170}$$

From (6.169) and (6.170), upon setting the determinant of the coefficients of the δys equal to zero, we obtain the characteristic equation

$$\det |(g_{ik} - \lambda \delta_{ik})| = 0 \tag{6.171}$$

where δ_{ik} is the Kronecker delta. There are m eigenvalues (λs) which are real or occur in complex conjugate pairs. The perturbation equations of the *physical system* are stable if all the λs lie in the left half of the complex plane, that is, if they have negative real parts.

The Euler algorithm, when applied to the perturbation equations (6.170), results in difference equations of the form

$$\delta y_{i(n+1)} = \delta y_{in} + \lambda h \delta y_{in} = (1 + \lambda h) \delta y_{in} \qquad (i = 1, \ldots, m) \tag{6.172}$$

where the first subscript indicates the state variable and the second subscript refers to the time. This is consistent with (6.164) obtained earlier.

Let us assume solutions to (6.172) of the form

$$\delta y_{in} = A_i \rho^n \tag{6.173}$$

Then, we obtain

$$A_i \rho^{n+1} = (1 + \lambda h) A_i \rho^n \tag{6.174}$$

For each eigenvalue λ_j, we have

$$\rho_j = 1 + \lambda_j h \qquad (j = 1, \ldots, m) \tag{6.175}$$

The necessary condition for stability is

$$|\rho_j| \leq 1 \qquad (j = 1, \ldots, m) \tag{6.176}$$

This result can be expressed conveniently as a plot of the stable region on the complex plane, as shown in Fig. 6.3. If λ is real, then λh must lie in the range $[-2, 0]$ for numerical stability. On the other hand, if λ is imaginary, as it would be for an undamped second-order system, then the Euler integration method is at least somewhat unstable, no matter how small the step size.

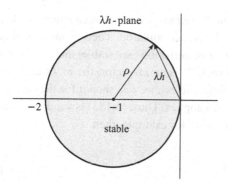

Figure 6.3.

Trapezoidal method

Let us consider the first-order system

$$\dot{y} = f(y, t) \tag{6.177}$$

The trapezoidal algorithm is

$$y_{n+1} = y_n + \frac{h}{2}(\dot{y}_{n+1} + \dot{y}_n) \tag{6.178}$$

Since $\dot{y}_{n+1}$ is a function of y_{n+1}, this is an *implicit* method and uses iteration.

The perturbation equation corresponding to (6.177) is

$$\delta\dot{y} = \frac{\partial f}{\partial y}\delta y = \lambda\delta y \tag{6.179}$$

where we assume that λ can be considered to be constant. The trapezoidal algorithm applied to the perturbation equation results in

$$\left(1 - \frac{\lambda h}{2}\right)\delta y_{n+1} - \left(1 + \frac{\lambda h}{2}\right)\delta y_n = 0 \tag{6.180}$$

Assume solutions of the form

$$\delta y_n = A\rho^n \tag{6.181}$$

and substitute into (6.180). After dividing by $A\rho^n$, the characteristic equation is

$$\left(1 - \frac{\lambda h}{2}\right)\rho - \left(1 + \frac{\lambda h}{2}\right) = 0 \tag{6.182}$$

or

$$\rho = \frac{1 + \dfrac{\lambda h}{2}}{1 - \dfrac{\lambda h}{2}} \tag{6.183}$$

Notice that λ is the real eigenvalue for the linearized perturbation equation, while ρ is the corresponding stability parameter for the difference equation. There is numerical stability if $|\rho| \leq 1$, and this will occur if the physical perturbations are stable, that is, if $\lambda \leq 0$.

For the more general case where y in (6.177) is an m-vector, the m λs and the corresponding m ρs are complex in general. Nevertheless, as was shown for the case of Euler integration, the equation relating λ and ρ still applies. Thus, (6.183) is valid for each eigenvalue λ and the corresponding ρ. If $\lambda = a + ib$, for example, then

$$\rho = \frac{1 + \dfrac{ah}{2} + i\dfrac{bh}{2}}{1 - \dfrac{ah}{2} - i\dfrac{bh}{2}} \tag{6.184}$$

We see that $|\rho| < 1$ for $a < 0$. In other words, if the physical system is stable and all its roots are in the left half-plane, then the numerical representation using the trapezoidal method is also stable for any step size h.

Runge–Kutta methods

Consider a first-order system whose perturbation equation has the form

$$\delta \dot{y} = \lambda \delta y \tag{6.185}$$

where λ is a constant. Referring to the RK-2 algorithm given by (6.89)–(6.92), we find that

$$\delta \tilde{y}_{n+\frac{1}{2}} = \delta y_n + \frac{1}{2}\lambda h \delta y_n \tag{6.186}$$

$$\delta \tilde{\dot{y}}_{n+\frac{1}{2}} = \lambda \delta y_n + \frac{1}{2}\lambda^2 h \delta y_n \tag{6.187}$$

$$\delta y_{n+1} = \delta y_n + \lambda h \delta y_n + \frac{1}{2}(\lambda h)^2 \delta y_n$$

$$= \left[1 + \lambda h + \frac{1}{2}(\lambda h)^2\right] \delta y_n \tag{6.188}$$

Now assume a solution of the form

$$\delta y_n = A\rho^n \tag{6.189}$$

Upon substituting into (6.188) and dividing by $A\rho^n$, we obtain

$$\rho = 1 + \lambda h + \frac{1}{2}(\lambda h)^2 \tag{6.190}$$

For this system with λ real, the value of ρ is never negative, and $\rho = 1$ for $\lambda h = 0, -2$. Thus, there is numerical stability for $\rho \leq 1$ or for $-2 \leq \lambda h \leq 0$.

Now consider the more general case in which δy is an m-vector, as in (6.170). Equation (6.190) still applies in this case, but the λs and the corresponding ρs may be complex. In order to obtain the stability in the the complex λh-plane, we take $|\rho| = 1$ and let

$$\rho = e^{i\phi} \tag{6.191}$$

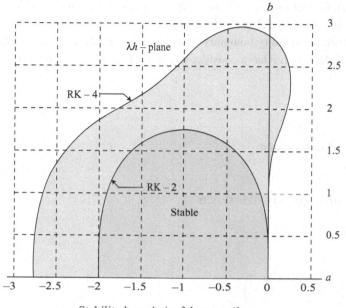

Stability boundaries $\lambda h = a + ib$

Figure 6.4.

Then we solve for λh as ϕ is varied. For example, suppose we assume

$$\lambda h = a + ib \tag{6.192}$$

and substitute into (6.190). By equating real and imaginary parts separately, we obtain the following equations for a and b:

$$a = -1 \pm \left(\cos\phi - \frac{1}{2} + \sqrt{\frac{5}{4} - \cos\phi} \right)^{\frac{1}{2}} \tag{6.193}$$

$$b = \frac{\sin\phi}{a+1} \tag{6.194}$$

The results are plotted in Fig. 6.4. For convenience, only the upper half of the complex plane is shown. Note that the stability boundary for RK-2 is symmetric about the real axis and also about the line $a = -1$. The maximum value of b is $\sqrt{3}$.

The stability boundary for the RK-4 algorithm is found by using a similar procedure. In this case, we find that, in general,

$$\rho = 1 + \lambda h + \frac{1}{2}(\lambda h)^2 + \frac{1}{6}(\lambda h)^3 + \frac{1}{24}(\lambda h)^4 \tag{6.195}$$

and, for a point on the stability boundary,

$$\rho = e^{i\phi} = \cos\phi + i\sin\phi \tag{6.196}$$

It can be shown that, for real values of λh, there is numerical stability for $-2.7853 \le \lambda h \le 0$. Furthermore, for an imaginary value of λh, as might occur in the analysis of an undamped second-order system, the stability boundary occurs at $\lambda h = \pm i 2\sqrt{2}$. From Fig. 6.4 it can be seen that the RK-4 algorithm has a considerably larger region of numerical stability than RK-2.

Midpoint method

Consider the second-order system

$$\ddot{y} = f(y, t) \tag{6.197}$$

The implicit algorithm applied to this system is

$$y_{n+1} = y_n + \frac{h}{2}(\dot{y}_{n+1} + \dot{y}_n) \tag{6.198}$$

where

$$\dot{y}_{n+1} = \dot{y}_n + hf\left(\frac{y_{n+1} + y_n}{2}, t_{n+\frac{1}{2}}\right) \tag{6.199}$$

We see that y_{n+1} appears on both sides of the equation, implying an iterated solution.

The numerical stability is analyzed by considering the perturbation equations

$$\delta y_{n+1} = \delta y_n + \frac{h}{2}(\delta \dot{y}_{n+1} + \delta \dot{y}_n) \tag{6.200}$$

$$\delta \dot{y}_{n+1} = \delta \dot{y}_n + h\left(\frac{\partial f}{\partial y}\right)_{n+\frac{1}{2}}\left(\frac{1}{2}\delta y_{n+1} + \frac{1}{2}\delta y_n\right) \tag{6.201}$$

Let us use the notation

$$\frac{\partial f}{\partial y} \equiv -k \tag{6.202}$$

and assume that k varies slowly enough to be considered constant in the stability analysis. Then (6.201) becomes

$$\delta \dot{y}_{n+1} = \delta \dot{y}_n - \frac{hk}{2}(\delta y_{n+1} + \delta y_n) \tag{6.203}$$

Now consider y and $\dot{y}$ as separate variables in the difference equations, and assume the solutions

$$\delta y_n = A\rho^n \tag{6.204}$$

$$\delta \dot{y}_n = B\rho^n \tag{6.205}$$

A substitution into (6.200) and (6.203) yields

$$(\rho - 1)A\rho^n - \frac{h}{2}(\rho + 1)B\rho^n = 0 \tag{6.206}$$

$$(\rho + 1)\frac{hk}{2}A\rho^n + (\rho - 1)B\rho^n = 0 \tag{6.207}$$

Divide each equation by ρ^n and set the determinant of the coefficients of A and B equal to zero. The resulting characteristic equation is

$$\left(1 + \frac{h^2 k}{4}\right)\rho^2 + \left(\frac{h^2 k}{2} - 2\right)\rho + 1 + \frac{h^2 k}{4} = 0 \tag{6.208}$$

Assuming that k is *positive*, we obtain

$$\rho = \left(1 + \frac{h^2 k}{4}\right)^{-1}\left(1 - \frac{h^2 k}{4} \pm i h \sqrt{k}\right) \tag{6.209}$$

Thus,

$$|\rho|^2 = \left(1 + \frac{h^2 k}{4}\right)^{-2}\left[\left(1 - \frac{h^2 k}{4}\right)^2 + h^2 k\right] = 1 \tag{6.210}$$

We see that this iterative midpoint method is numerically stable for all h and positive k.

In addition to the numerical stability, one should also consider the convergence properties of the iteration procedure. This procedure consists of evaluating $\dot{y}_{n+1}$ and y_{n+1} alternately, using (6.198) and (6.199), meanwhile keeping constant y_n and $\dot{y}_n$. We wish to determine whether perturbations of y_{n+1} will grow or diminish in amplitude in successive iterations. From (6.200) and (6.201) we obtain

$$\delta y_{n+1}^{(j+1)} = h\delta\dot{y}_n - \frac{h^2 k}{4}\delta y_{n+1}^{(j)} + \left(1 - \frac{h^2 k}{4}\right)\delta y_n \tag{6.211}$$

where $\delta\dot{y}_n$ and δy_n are set equal to zero. The superscript (j) or $(j+1)$ refers to the iteration number. We see that

$$\left|\delta y_{n+1}^{(j+1)}\right| < \left|\delta y_{n+1}^{(j)}\right|$$

if $0 < h\sqrt{k} < 2$. Previously, we assumed that k is positive. If k is negative, the midpoint method is numerically unstable for all positive values of step size h.

We conclude, then, that the convergence of the iteration procedure is critical in determining the range of h for overall numerical stability of the midpoint method.

Extraneous roots

When one uses higher-order algorithms such as Adams–Bashforth or Adams–Moulton, the characteristic equation associated with the difference equations is of higher degree than the characteristic equation obtained from the linearized differential equations describing perturbations of the system. Thus, in addition to the roots approximating the true values, there are additional *extraneous* roots introduced in the numerical solution. These extraneous roots arise because the algorithms include values of the variables previous to time t_n.

As a simple example, consider the AB-2 algorithm, namely,

$$y_{n+1} = y_n + \frac{3}{2}h\dot{y}_n - \frac{1}{2}h\dot{y}_{n-1} \tag{6.212}$$

Suppose we apply this algorithm to the first-order equation

$$\dot{y} = \lambda y \tag{6.213}$$

where λ is a constant. Assuming solutions of the form

$$y_n = A\rho^n \tag{6.214}$$

the characteristic equation associated with (6.212) is

$$\rho^2 - \left(1 + \frac{3}{2}\lambda h\right)\rho + \frac{1}{2}\lambda h = 0 \tag{6.215}$$

Assuming that λh is real, the AB-2 algorithm is numerically stable for $-1 \le \lambda h \le 0$.

As a specific example, let $\lambda h = -0.1$. The corresponding characteristic equation is

$$\rho^2 - 0.85\rho - 0.05 = 0 \tag{6.216}$$

and the roots are $\rho_{1,2} = 0.9052, -0.0552$. If the numerical procedure were exact, then we should have $\rho = e^{\lambda h} = 0.9048$. Thus, we see that the root -0.0552 is extraneous.

As another example, let $\lambda h = -1$ which is on the boundary of the stable region. The characteristic roots in this case are $\rho = 0.5, -1$. The root -1 is on the stability boundary and is *extraneous*. Thus, it is the extraneous root which causes numerical instability for more negative values of λh.

In general, the number of extraneous roots is determined by the number of time increments in the past at which data are used in the algorithm. For example, the AB-3 algorithm involves $\dot{y}_{n-2}$. This results in two extraneous roots for each variable, assuming first-order differential equations.

For multiple-pass algorithms such as the various predictor–corrector methods, the algorithm as a whole determines the number of extraneous roots. For example, the AB-3, AM-4 combination involves $\dot{y}_{n-2}$ and will introduce two extraneous roots per variable.

6.4   Frequency response methods

Transfer functions

Consider a linear time-invariant system which has a sinusoidal input. The steady-state output will be sinusoidal with the same frequency as the input but, in general, with a different amplitude and phase. If one uses complex notation, the amplitude and phase of the output relative to the input are expressed by the *transfer function*

$$G(i\omega) = Me^{i\phi} \tag{6.217}$$

where $M(\omega)$ is the relative amplitude and $\phi(\omega)$ is the relative phase.

The system under consideration may be originally linear or may be a *linearized* system which is represented by a set of perturbation equations. The data resulting from a numerical integration of the linear equations can be considered to be samples taken from a continuous output, albeit an output slightly in error compared to the true output. The transfer function relating this computed output to the same sinusoidal input is of the form

$$G^*(i\omega) = M^* e^{i\phi^*} \tag{6.218}$$

where M^* and ϕ^* are functions of ω. Our problem is to find the relationship of $G^*(i\omega)$ to $G(i\omega)$ for some combination of linear equations and integration algorithm.

As a simple example, consider a *pure integrator* with a unit sinusoidal input. Its differential equation can be written in the complex form

$$\dot{y} = e^{i\omega t} \tag{6.219}$$

Let us use the Euler integration algorithm and the discrete time

$$t_n = nh \tag{6.220}$$

We obtain the difference equation

$$y_{n+1} = y_n + h e^{i\omega nh} \tag{6.221}$$

The steady-state solution of this equation has the form

$$y_n = G^*(i\omega) e^{i\omega nh} \tag{6.222}$$

where $G^*(i\omega)$ is complex, in general. This solution is sinusoidal and of the same frequency as the input, but $G^*(i\omega)$ allows for a different amplitude and phase at the output y_n.

If (6.222) is substituted into (6.221), we obtain

$$\left(e^{i\omega h(n+1)} - e^{i\omega nh}\right) G^*(i\omega) = h e^{i\omega nh}$$

or

$$G^*(i\omega) = \frac{h}{e^{i\omega h} - 1} \tag{6.223}$$

Compare this result with that for the integrator of (6.219) where

$$G(i\omega) = \frac{1}{i\omega} \tag{6.224}$$

We see that one can obtain $G^*(i\omega)$ from $G(i\omega)$ by replacing $i\omega$ with $(e^{i\omega h} - 1)/h$. This substitution is valid for any transfer function $G(i\omega)$ if Euler integration is used.

In general, if one knows the transfer operator $G(s)$, where s is the Laplacian variable, then one obtains $G(i\omega)$ by the substitution $s = i\omega$. $G^*(i\omega)$ is obtained by letting $s = (e^{i\omega h} - 1)/h$ for the case of Euler integration.

As a second example, suppose we again consider an integrator with a sinusoidal input, as given by (6.219), and use the RK-2 algorithm. The resulting difference equation is

$$y_{n+1} = y_n + h e^{i\omega h(n+\frac{1}{2})} \tag{6.225}$$

Assume steady-state solutions of the form given by (6.222). We obtain

$$(e^{i\omega h(n+1)} - e^{i\omega hn}) G^* = h e^{i\omega h(n+\frac{1}{2})} \tag{6.226}$$

which leads to

$$G^*(i\omega) = \frac{h}{e^{i\omega h/2} - e^{-i\omega h/2}} = \frac{h}{i2\sin(\omega h/2)} \qquad (6.227)$$

We note that as the step size h approaches zero, $G^*(i\omega)$ approaches $G(i\omega)$, as given by (6.224).

It is important to realize that, for multiple-pass algorithms such as RK-2, one can no longer obtain $G^*(i\omega)$ by making the appropriate substituting for s in $G(s)$. The form of $G^*(i\omega)$ is complicated by the presence of intermediate variables in making extra passes.

Example 6.1   Consider a second-order system described by the differential equation

$$\ddot{y} + 2\zeta\omega_n\dot{y} + \omega_n^2 y = f(t) \qquad (6.228)$$

where ω_n is the undamped natural frequency and ζ is the damping ratio. Assuming that all initial conditions are zero, the Laplace transform of this equation is

$$\left(s^2 + 2\zeta\omega_n s + \omega_n^2\right)Y(s) = F(s) \qquad (6.229)$$

This leads to

$$G(s) = \frac{Y(s)}{F(s)} = \frac{1}{s^2 + 2\zeta\omega_n s + \omega_n^2} \qquad (6.230)$$

For a sinusoidal input,

$$f(t) = \cos\omega t \qquad (6.231)$$

the steady-state response is found by letting $s = i\omega$ in (6.230) and obtaining the transfer function

$$G(i\omega) = \frac{1}{\omega_n^2 - \omega^2 + i2\zeta\omega_n\omega} = \frac{\left(\dfrac{1}{\omega_n^2}\right)}{1 - \left(\dfrac{\omega}{\omega_n}\right)^2 + i2\zeta\left(\dfrac{\omega}{\omega_n}\right)} \qquad (6.232)$$

Now consider a specific numerical example with $\omega_n = 1$, $\omega = 0.95$, $\zeta = 0.5$, and obtain

$$G = \frac{1}{0.0975 + i0.95} = 1.047131\angle - 1.468523$$

where the latter form gives the amplitude and the phase angle in radians relative to the sinusoidal input.

Suppose we solve this problem using Euler integration with a step size $h = 0.01$. If we replace $i\omega$ by $(e^{i\omega h} - 1)/h$ in (6.232), we obtain

$$G^*(i\omega) = \frac{h^2}{(e^{i\omega h} - 1)^2 + 2\zeta\omega_n h(e^{i\omega h} - 1) + \omega_n^2 h^2} \qquad (6.233)$$

This results in the numerical value

$$G^* = \frac{1}{0.093035 + i0.941412} = 1.057085\angle - 1.472291$$

We see that the steady-state amplitude of the numerical solution is about one percent too large. The numerical phase lag is more accurate but is also slightly large.

Example 6.2   Now consider a first-order system which is analyzed using the RK-2 algorithm. We desire the steady-state response to a unit sinusoidal input. The differential equation is

$$\dot{y} = \lambda y + \cos \omega t \tag{6.234}$$

In complex form we have

$$\dot{y} = \lambda y + e^{i\omega t} \tag{6.235}$$

The RK-2 algorithm results in

$$\dot{y}_n = \lambda y_n + e^{i\omega h n} \tag{6.236}$$

$$\tilde{y}_{n+\frac{1}{2}} = y_n + \frac{1}{2}h\dot{y}_n \tag{6.237}$$

$$\tilde{\dot{y}}_{n+\frac{1}{2}} = \lambda \tilde{y}_{n+\frac{1}{2}} + e^{i\omega h(n+\frac{1}{2})} \tag{6.238}$$

$$y_{n+1} = y_n + h\tilde{\dot{y}}_{n+\frac{1}{2}}$$

$$= \left(1 + \lambda h + \frac{(\lambda h)^2}{2}\right) y_n + \frac{\lambda h^2}{2}e^{i\omega h n} + he^{i\omega h(n+\frac{1}{2})} \tag{6.239}$$

The steady-state numerical solution has the form

$$y_n = G^*(i\omega)e^{i\omega h n} \tag{6.240}$$

A substitution of this solution into (6.239) yields the transfer function

$$G^*(i\omega) = \frac{h\left(1 + \frac{\lambda h}{2}e^{-i\omega h/2}\right)}{e^{i\omega h/2} - \left(1 + \lambda h + \frac{(\lambda h)^2}{2}\right)e^{-i\omega h/2}} \tag{6.241}$$

or

$$G^*(i\omega) = \frac{h + \frac{\lambda h^2}{2}\cos\frac{\omega h}{2} - i\frac{\lambda h^2}{2}\sin\frac{\omega h}{2}}{-\lambda h\left(1 + \frac{\lambda h}{2}\right)\cos\frac{\omega h}{2} + i\left(2 + \lambda h + \frac{(\lambda h)^2}{2}\right)\sin\frac{\omega h}{2}} \tag{6.242}$$

The corresponding transfer function for the actual system with continuous time is

$$G(i\omega) = \frac{1}{-\lambda + i\omega} \tag{6.243}$$

Note that G^* approaches G as the step size h approaches zero.

Let us consider a numerical example with $\lambda = -1$, $\omega = 1$ and $h = 0.01$. With these values, the RK-2 algorithm yields

$$G^* = 0.707104\angle - 0.785390$$

The result for continuous time is

$$G = 0.707107 \angle -0.785398$$

We see that the sinusoidal response using RK-2 is slightly small both in amplitude and in phase lag compared to the true values.

Free sinusoidal response

The effects of truncation errors are most readily apparent in the numerical analysis of systems having neutral stability. For these systems, any change in damping due to the integration algorithm appears as a sinusoidal oscillation which is growing or shrinking in amplitude.

As an illustrative example, consider a linear second-order system described by

$$\ddot{y} + \omega_0^2 y = 0 \tag{6.244}$$

where ω_0 is the natural frequency in radians per second. If we assume a solution of the form e^{st}, the corresponding characteristic equation is

$$s^2 + \omega_0^2 = 0 \tag{6.245}$$

which has the roots

$$s_{1,2} = \pm i\omega_0 \tag{6.246}$$

Case 1

Let us analyze the free unforced response of this system by using the Euler integration algorithm. First, we write (6.244) in the form of two first-order equations, namely,

$$\dot{y} = v \tag{6.247}$$
$$\dot{v} = -\omega_0^2 y \tag{6.248}$$

The corresponding difference equations are

$$y_{n+1} = y_n + h v_n \tag{6.249}$$
$$v_{n+1} = v_n - h\omega_0^2 y_n \tag{6.250}$$

Assume the solutions

$$y_n = A\rho^n, \qquad v_n = B\rho^n \tag{6.251}$$

and substitute into the difference equations. We obtain

$$(\rho - 1)\rho^n A - h\rho^n B = 0 \tag{6.252}$$
$$h\omega_0^2 \rho^n A + (\rho - 1)\rho^n B = 0 \tag{6.253}$$

After dividing these equations by ρ^n, the characteristic equation that results is obtained from the determinant of the coefficients, namely,

$$(\rho - 1)^2 + h^2 \omega_0^2 = 0 \tag{6.254}$$

with the roots

$$\rho_{1,2} = 1 \pm i\omega_0 h \tag{6.255}$$

Choose the positive sign in accordance with the usual convention that ρ rotates in the counterclockwise sense in the complex plane. Use the notation

$$\rho = Me^{i\phi} = M\angle\phi \tag{6.256}$$

where

$$M = \sqrt{1 + \omega_0^2 h^2}, \qquad \phi = \tan^{-1}(\omega_0 h) \tag{6.257}$$

Since $M > 1$ for all nonzero values of the step size h, we see that the numerical solution is unstable for this case of Euler integration.

In general, the numerical solution can be considered to be composed of samples taken at every time step from a solution of the form e^{st} where we now allow the response to be damped. Thus, the corresponding characteristic equation has the form

$$s^2 + 2\zeta\omega_n s + \omega_n^2 = 0 \tag{6.258}$$

with the roots

$$s_{1,2} = -\zeta\omega_n + i\omega_n\sqrt{1 - \zeta^2} \tag{6.259}$$

where ω_n is the undamped natural frequency and ζ is the damping ratio of the numerical solution. Now ρ and s are related by the equation

$$\rho = e^{sh} \tag{6.260}$$

Equivalently, we can write

$$M = e^{-\zeta\omega_n h} \tag{6.261}$$
$$\phi = \omega_n h\sqrt{1 - \zeta^2} \tag{6.262}$$

where we assume that $0 \le \zeta \le 1$. If ρ is known, that is, if M and ϕ are given, then ω_n and ζ can be obtained from the following equations:

$$\omega_n = \frac{1}{h}\sqrt{\ln^2 M + \phi^2} \tag{6.263}$$

$$\zeta = \frac{-\ln M}{\omega_n h} = \frac{-\ln M}{\sqrt{\ln^2 M + \phi^2}} \tag{6.264}$$

Case 2

Now let us analyze the same problem using the RK-2 algorithm. The resulting difference equations are

$$y_{n+1} = \left(1 - \frac{\omega_0^2 h^2}{2}\right) y_n + h v_n \tag{6.265}$$

$$v_{n+1} = \left(1 - \frac{\omega_0^2 h^2}{2}\right) v_n - \omega_0^2 h y_n \tag{6.266}$$

Again we assume the solution forms of (6.251) and obtain

$$\left[\rho - \left(1 - \frac{\omega_0^2 h^2}{2}\right)\right] \rho^n A - h \rho^n B = 0 \tag{6.267}$$

$$\omega_0^2 h \rho^n A + \left[\rho - \left(1 - \frac{\omega_0^2 h^2}{2}\right)\right] \rho^n B = 0 \tag{6.268}$$

The resulting characteristic equation is

$$\left[\rho - \left(1 - \frac{\omega_0^2 h^2}{2}\right)\right]^2 + \omega_0^2 h^2 = 0 \tag{6.269}$$

with the roots

$$\rho_{1,2} = 1 - \frac{\omega_0^2 h^2}{2} \pm i \omega_0 h \tag{6.270}$$

Thus, we obtain

$$M = |\rho| = \sqrt{1 + \frac{1}{4} \omega_0^4 h^4} \tag{6.271}$$

$$\phi = \tan^{-1}\left(\frac{\omega_0 h}{1 - \frac{1}{2} \omega_0^2 h^2}\right) \tag{6.272}$$

The values of ω_n and ζ for the numerical solution are obtained from (6.263) and (6.264). Since $M > 1$, the numerical solution is slightly unstable for the usual small values of $\omega_0 h$.

Case 3

Next, let us use the AB-3 algorithm. In this case, the difference equations are

$$y_{n+1} = y_n + \frac{h}{12}(23 v_n - 16 v_{n-1} + 5 v_{n-2}) \tag{6.273}$$

$$v_{n+1} = v_n - \frac{\omega_0^2 h}{12}(23 y_n - 16 y_{n-1} + 5 y_{n-2}) \tag{6.274}$$

If we assume the solutions $y_n = A\rho^n$ and $v_n = B\rho^n$, we find that the characteristic equation is

$$\rho^4(\rho - 1)^2 + \left[\frac{\omega_0 h}{12}(23\rho^2 - 16\rho + 5)\right]^2 = 0 \qquad (6.275)$$

This equation has six roots even though the characteristic equation of the actual system has only two roots. Thus, there are four extraneous roots. These extra roots arise from the fact that the AB-3 algorithm requires data from two previous steps for each of the two variables. In the usual case, the four extraneous roots will be stable, and the remaining two roots will approximate e^{sh} for $s = \pm i\omega_0$.

Case 4

Finally, let us consider the RK-4 algorithm being applied to this problem. The resulting difference equations are

$$y_{n+1} = \left(1 - \frac{\omega_0^2 h^2}{2} + \frac{\omega_0^4 h^4}{24}\right) y_n + \left(1 - \frac{\omega_0^2 h^2}{6}\right) h v_n \qquad (6.276)$$

$$v_{n+1} = \left(-\omega_0^2 h + \frac{\omega_0^4 h^3}{6}\right) y_n + \left(1 - \frac{\omega_0^2 h^2}{2} + \frac{\omega_0^4 h^4}{24}\right) v_n \qquad (6.277)$$

The corresponding characteristic equation is

$$\left(\rho - 1 + \frac{\omega_0^2 h^2}{2} - \frac{\omega_0^4 h^4}{24}\right)^2 + \left(\omega_0 h - \frac{\omega_0^3 h^3}{6}\right)^2 = 0 \qquad (6.278)$$

and the resulting roots are

$$\rho_{1,2} = 1 - \frac{\omega_0^2 h^2}{2} + \frac{\omega_0^4 h^4}{24} \pm i\left(\omega_0 h - \frac{\omega_0^3 h^3}{6}\right) \qquad (6.279)$$

In polar form, we obtain

$$M = \sqrt{1 - \frac{\omega_0^6 h^6}{72}\left(1 - \frac{\omega_0^2 h^2}{8}\right)} \qquad (6.280)$$

$$\phi = \tan^{-1}\left(\frac{24\omega_0 h - 4\omega_0^3 h^3}{24 - 12\omega_0^2 h^2 + \omega_0^4 h^4}\right) \qquad (6.281)$$

Numerical comparisons

In order to obtain a better grasp of the relative accuracies of the various integration algorithms, consider a specific numerical example in which $\omega_0 = 1$ rad/s and $h = 0.1$ s. In this case, the theoretical value of ρ is

$$\rho = e^{i0.1} = 0.995004165 + i0.099833417 \qquad (6.282)$$

Table 6.1. A comparison of errors

Method	$O(h)$	$M - 1$	$\phi - 0.1$	$M(2\pi)$
Euler	1	0.004988	−0.0003313	1.3684
RK-2	2	0.00001250	0.0001662	1.0007844
AB-3	3	−0.00003727	0.000003959	0.997661
RK-4	4	−0.00000000694	−0.00000008304	0.999999566

or

$$M = 1.0, \qquad \phi = 0.1 \text{ rad} \tag{6.283}$$

The errors in amplitude and phase are shown in Table 6.1. It is interesting to note that the methods with odd orders of accuracy in h tend to be more accurate in phase, while the methods of even order in h are relatively more accurate in amplitude. On an absolute basis, the RK-4 method is quite accurate in both amplitude and phase.

The last column represents the amplitude of the computed solution after one complete cycle of the oscillation, consisting of approximately 63 time steps. We see that the Euler solution has increased in amplitude by about 37%, while the amplitude of the RK-4 solution has an error of less than one part in a million in the analysis of this neutrally stable system.

6.5   Kinematic constraints

An important consideration in the numerical analysis of dynamical systems lies in the proper representation of kinematical constraints. Even if the physical system is stable, there may be numerical instabilities resulting from the method of applying constraints. In this section, we shall present methods of representing constraints and will analyze their stability.

Baumgarte's method for holonomic constraints

Let us consider first a dynamical system which is subject to m holonomic constraints of the form

$$\phi_j(q, t) = 0 \qquad (j = 1, \ldots, m) \tag{6.284}$$

Suppose there are n second-order dynamical equations written in the fundamental Lagrangian form

$$\frac{d}{dt}\left(\frac{\partial T}{\partial \dot{q}_i}\right) - \frac{\partial T}{\partial q_i} = \sum_{j=1}^{m} \lambda_j a_{ji} + Q_i(q, \dot{q}, t) \qquad (i = 1, \ldots, n) \tag{6.285}$$

where the λs are Lagrange multipliers, the Qs are generalized applied forces, and where

$$a_{ji}(q,t) = \frac{\partial \phi_j}{\partial q_i} \tag{6.286}$$

At this point there are n dynamical equations which are linear in the n $\ddot{q}$s and the m λs. We need m additional equations to solve for the variables. These additional equations can be obtained by differentiating the constraint equations twice with respect to time. First,

$$\dot{\phi}_j = \sum_{i=1}^{n} a_{ji}(q,t)\dot{q}_i + a_{jt}(q,t) \tag{6.287}$$

where

$$a_{jt}(q,t) = \frac{\partial \phi_j}{\partial t} \tag{6.288}$$

Then $\ddot{\phi}_j$ has the form

$$\ddot{\phi}_j = \sum_{i=1}^{n} a_{ji}(q,t)\ddot{q}_i + G_j(q,\dot{q},t) \qquad (j = 1, \ldots, m) \tag{6.289}$$

If these expressions for $\ddot{\phi}_j$ are set equal to zero then, with the aid of (6.285), one can solve for the $\ddot{q}_i(q,\dot{q},t)$ and $\lambda_j(q,\dot{q},t)$. The n $\ddot{q}_i$ expressions can be integrated numerically for given initial conditions, thereby obtaining $\dot{q}_i$ and q_i as functions of time.

The problem with this approach is that the resulting numerical solutions will be unstable even though the physical system may actually be stable. This instability arises from the fact that, in effect, each constraint equation in the differentiated form

$$\ddot{\phi}_j = 0 \qquad (j = 1, \ldots, m) \tag{6.290}$$

is also being integrated twice with respect to time. This equation has a repeated zero characteristic root and is therefore unstable.

In order to stabilize the numerical representation of holonomic constraints, Baumgarte proposed that (6.290) be changed to

$$\ddot{\phi}_j + \alpha \dot{\phi}_j + \beta \phi_j = 0 \qquad (j = 1, \ldots, m) \tag{6.291}$$

where α and β are suitably chosen constants whose values may depend on the step size h. This allows the roots of the corresponding characteristic equation to be shifted to the left half-plane where the constraint response can be heavily damped.

In detail, the Baumgarte procedure for a system having holonomic constraints is as follows:

1 Write the differential equations using Lagrange multipliers

$$\frac{d}{dt}\left(\frac{\partial T}{\partial \dot{q}_i}\right) - \frac{\partial T}{\partial q_i} = \sum_{j=1}^{m} \lambda_j a_{ji}(q,t) + Q_i(q,\dot{q},t) \qquad (i = 1, \ldots, n) \tag{6.292}$$

2 Solve these n equations for the n $\ddot{q}$s

$$\ddot{q}_i = \ddot{q}_i(q,\dot{q},\lambda,t) \qquad (i = 1, \ldots, n) \tag{6.293}$$

3 Define the stabilizing function U_j

$$U_j \equiv -\alpha\dot{\phi}_j(q, \dot{q}, t) - \beta\phi_j(q, t) \qquad (j = 1, \ldots, m) \tag{6.294}$$

where $\dot{\phi}_j$ is given by (6.287).

4 Stabilize each constraint by setting

$$\ddot{\phi}_j = U_j(q, \dot{q}, t) \qquad (j = 1, \ldots, m) \tag{6.295}$$

or, using (6.289),

$$\sum_{i=1}^{n} a_{ji}(q, t)\ddot{q}_i + G_j(q, \dot{q}, t) = U_j(q, \dot{q}, t) \qquad (j = 1, \ldots, m) \tag{6.296}$$

5 Substitute the $\ddot{q}_i$ expressions from (6.293) into (6.296) and solve these m equations for the m λs

$$\lambda_j = \lambda_j(q, \dot{q}, t) \qquad (j = 1, \ldots, m) \tag{6.297}$$

6 Substitute from (6.297) back into (6.293), obtaining dynamical equations of the form

$$\ddot{q}_i = \ddot{q}_i(q, \dot{q}, t) \qquad (i = 1, \ldots, n) \tag{6.298}$$

These n second-order equations are integrated numerically, usually by converting them to $2n$ first-order equations.

Satisfactory constraint damping can be achieved by setting

$$\alpha = \frac{2}{h}, \qquad \beta = \frac{1}{h^2} \tag{6.299}$$

This corresponds to critical damping of the constraint error ϕ_j which may be nonzero. The constraint errors are driven primarily by truncation errors associated with the numerical integration algorithm. Hence, they will not disappear entirely, even with the stabilization procedure.

The numerical stability of a particular constraint function $\phi(q)$ can be analyzed by considering the integration algorithm to be applied directly to the Baumgarte equation

$$\ddot{\phi} + \alpha\dot{\phi} + \beta\phi = 0 \tag{6.300}$$

This assumes that the physical system is stable and the step size h is small enough that the higher-order terms may be neglected in any Taylor expansion.

As a simple example, the Euler integration algorithm results in

$$\phi_{n+1} = \phi_n + h\dot{\phi}_n \tag{6.301}$$

$$\dot{\phi}_{n+1} = \dot{\phi}_n + h\ddot{\phi}_n = (1 - \alpha h)\dot{\phi}_n - \beta h\phi_n \tag{6.302}$$

These difference equations have solutions with the forms

$$\phi_n = A\rho^n \tag{6.303}$$

$$\dot{\phi}_n = B\rho^n \tag{6.304}$$

Upon the substitution of these solutions into (6.301) and (6.302), we obtain

$$(\rho^{n+1} - \rho^n)A - h\rho^n B = 0 \tag{6.305}$$
$$\beta h \rho^n A + (\rho^{n+1} - \rho^n + \alpha h \rho^n)B = 0 \tag{6.306}$$

After dividing each equation by ρ^n and setting the determinant of the coefficients equal to zero, we obtain the characteristic equation

$$\rho^2 + (\alpha h - 2)\rho + (1 - \alpha h + \beta h^2) = 0 \tag{6.307}$$

If we take α proportional to $1/h$ and β proportional to $1/h^2$, the roots of the characteristic equation are independent of the step size. This means that the numerical stability of the solution for the constraint function $\phi(q)$ is also independent of the step size. As an example, suppose we choose $\alpha = 2/h$ and $\beta = 1/h^2$. Then the characteristic equation (6.307) has a double zero root, indicating a "deadbeat" response. This means that, in theory, the constraint error ϕ should go to zero in one step, provided that the initial conditions (at $n = 0$) are in accordance with the A/B ratio of the corresponding eigenvector. For arbitrary initial conditions and the given values of α and β, the error will go to zero in two steps, in theory. In the actual case, however, the constraint error does not go to zero because of new disturbances at each step due to truncation and roundoff errors.

Example 6.3 A particle of mass m moves on a frictionless rigid wire in the form of a horizontal circle of radius R (Fig. 6.5). The equation of constraint in terms of Cartesian coordinates is

$$\phi(x, y) = \sqrt{x^2 + y^2} - R = 0 \tag{6.308}$$

Thus, we obtain

$$a_{1x} = \frac{\partial \phi}{\partial x} = \frac{x}{\sqrt{x^2 + y^2}}, \qquad a_{1y} = \frac{\partial \phi}{\partial y} = \frac{y}{\sqrt{x^2 + y^2}} \tag{6.309}$$

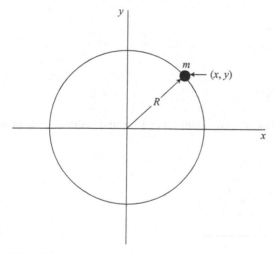

Figure 6.5.

The kinetic energy is

$$T = \frac{1}{2}m(\dot{x}^2 + \dot{y}^2) \tag{6.310}$$

so (6.292) results in the following dynamical equations:

$$m\ddot{x} = \frac{\lambda x}{\sqrt{x^2 + y^2}} \tag{6.311}$$

$$m\ddot{y} = \frac{\lambda y}{\sqrt{x^2 + y^2}} \tag{6.312}$$

Next, differentiate the constraint function with respect to time,

$$\dot{\phi} = \frac{1}{\sqrt{x^2 + y^2}}(x\dot{x} + y\dot{y}) \tag{6.313}$$

A second differentiation with respect to time results in

$$\ddot{\phi} = \frac{1}{\sqrt{x^2 + y^2}}\left[x\ddot{x} + y\ddot{y} + \dot{x}^2 + \dot{y}^2 - \frac{1}{x^2 + y^2}(x\dot{x} + y\dot{y})^2\right] = U \tag{6.314}$$

in accordance with (6.295). In this instance, from (6.294) we have

$$U = -\frac{\alpha}{\sqrt{x^2 + y^2}}(x\dot{x} + y\dot{y}) - \beta(\sqrt{x^2 + y^2} - R) \tag{6.315}$$

Now substitute the expressions for $\ddot{x}$ and $\ddot{y}$ from (6.311) and (6.312) into (6.314) and solve for λ/m. We obtain

$$\frac{\lambda}{m} = \frac{-(x\dot{y} - y\dot{x})^2}{(x^2 + y^2)^{3/2}} + U \tag{6.316}$$

Finally, substitute this result back into (6.311) and (6.312) to obtain the following dynamical equations:

$$\ddot{x} = \frac{-x(x\dot{y} - y\dot{x})^2}{(x^2 + y^2)^2} + \frac{x}{\sqrt{x^2 + y^2}}U \tag{6.317}$$

$$\ddot{y} = \frac{-y(x\dot{y} - y\dot{x})^2}{(x^2 + y^2)^2} + \frac{y}{\sqrt{x^2 + y^2}}U \tag{6.318}$$

These equations can be replaced by an equivalent set of four first-order equations, namely,

$$\dot{x} = v_x \tag{6.319}$$

$$\dot{y} = v_y \tag{6.320}$$

$$\dot{v}_x = -\frac{x}{(x^2 + y^2)^2}(xv_y - yv_x)^2 + \frac{x}{\sqrt{x^2 + y^2}}U \tag{6.321}$$

$$\dot{v}_y = -\frac{y}{(x^2 + y^2)^2}(xv_y - yv_x)^2 + \frac{y}{\sqrt{x^2 + y^2}}U \tag{6.322}$$

where, from (6.315),

$$U = -\frac{\alpha(x v_x + y v_y)}{\sqrt{x^2 + y^2}} - \beta(\sqrt{x^2 + y^2} - R) \tag{6.323}$$

These equations are integrated numerically.

An explanation of the dynamical equations (6.317) and (6.318) can be obtained by transforming to polar coordinates using

$$r = \sqrt{x^2 + y^2} \tag{6.324}$$

$$\theta = \tan^{-1} \frac{y}{x} \tag{6.325}$$

Then we find that

$$\omega = \dot{\theta} = \frac{x \dot{y} - y \dot{x}}{x^2 + y^2} \tag{6.326}$$

For the case of no constraint stabilization ($U = 0$), the magnitudes of $\ddot{x}$ and $\ddot{y}$ are equal to the corresponding components of the centripetal acceleration $r \omega^2$. Also, it can be shown from the dynamical equations that

$$\frac{d}{dt}(r^2 \omega) = 0 \tag{6.327}$$

Thus, the angular momentum about the origin is conserved.

Baumgarte's method for nonholonomic constraints

For systems with nonholonomic constraints of the form

$$g_j = \sum_{i=1}^{n} a_{ji}(q, t) \dot{q}_i + a_{jt}(q, t) = 0 \qquad (j = 1, \ldots, m) \tag{6.328}$$

one can use Lagrange's equations in the form of (6.285). More generally, if the nonholonomic constraints have the form

$$g_j(q, \dot{q}, t) = 0 \qquad (j = 1, \ldots, m) \tag{6.329}$$

where g_j is not necessarily linear in the $\dot{q}$s, we can write Lagrange's equation as follows:

$$\frac{d}{dt} \left(\frac{\partial T}{\partial \dot{q}_i} \right) - \frac{\partial T}{\partial q_i} = \sum_{j=1}^{m} \lambda_j \frac{\partial g_j}{\partial \dot{q}_i} + Q_i(q, \dot{q}, t) \qquad (i = 1, \ldots, n) \tag{6.330}$$

Note that (6.330) reduces to (6.285) for the usual case in which the nonholonomic constraint equations have the linear form of (6.328).

The Baumgarte method for nonholonomic constraints is similar to the holonomic case, except that the nonholonomic constraint function $g_j(q, \dot{q}, t)$ replaces $\dot{\phi}_j(q, \dot{q}, t)$. In detail, the procedure is as follows:

1 Write the differential equations of motion using (6.330). Solve for the n $\ddot{q}$s

$$\ddot{q}_i = \ddot{q}_i(q, \dot{q}, \lambda, t) \qquad (i = 1, \ldots, n) \tag{6.331}$$

2 Define the stabilizing function U_j

$$U_j = -\alpha g_j(q, \dot{q}, t) - \beta \int_0^t g_j dt \qquad (j = 1, \ldots, m) \tag{6.332}$$

As before, α and β are constants which depend on the step size.

3 Differentiate each constraint function $g_j(q, \dot{q}, t)$ once with respect to time and set the resulting function equal to the corresponding U_j

$$\dot{g}_j(q, \dot{q}, \ddot{q}, t) = U_j \qquad (j = 1, \ldots, m) \tag{6.333}$$

4 Substitute the $\ddot{q}_i$ expressions from (6.331) into (6.333) and solve for the m λs

$$\lambda_j = \lambda_j(q, \dot{q}, U, t) \qquad (j = 1, \ldots, m) \tag{6.334}$$

5 Substitute these λ expressions back into (6.331) and obtain the final dynamical equations

$$\ddot{q}_i = \ddot{q}_i(q, \dot{q}, U, t) \qquad (i = 1, \ldots, n) \tag{6.335}$$

The Us are given by (6.332). Usually the dynamical equations are written as $2n$ first-order equations.

In general, α should be chosen to be inversely proportional to h, and β should be inversely proportional to h^2. Acceptable accuracy can be obtained with β set equal to zero. For improved accuracy, however, one should choose β to be nonzero at the cost of an additional numerical integration for each constraint.

Example 6.4 Let us apply the Baumgarte method to a nonholonomic system whose dynamics have been analyzed previously. In these instances, however, the question of numerical stability was not considered.

The system consists of two particles, each of mass m, which are connected by a massless rod of length l, and which move on the horizontal xy-plane. There is a knife-edge constraint at particle 1, as shown in Fig. 6.6, resulting in a nonholonomic constraint of the form

$$g(q, \dot{q}) = -\dot{x} \sin \phi + \dot{y} \cos \phi = 0 \tag{6.336}$$

This equation states that the velocity component of particle 1 in a direction normal to the rod is always zero.

Let us use Lagrange's equation in the form

$$\frac{d}{dt}\left(\frac{\partial T}{\partial \dot{q}_i}\right) - \frac{\partial T}{\partial q_i} = \sum_{j=1}^m \lambda_j \frac{\partial g_j}{\partial \dot{q}_i} \qquad (i = 1, \ldots, n) \tag{6.337}$$

The generalized coordinates are (x, y, ϕ) and the kinetic energy, assuming no constraints, is

$$T = m(\dot{x}^2 + \dot{y}^2) + \frac{1}{2}ml^2\dot{\phi}^2 + ml\dot{\phi}(-\dot{x} \sin \phi + \dot{y} \cos \phi) \tag{6.338}$$

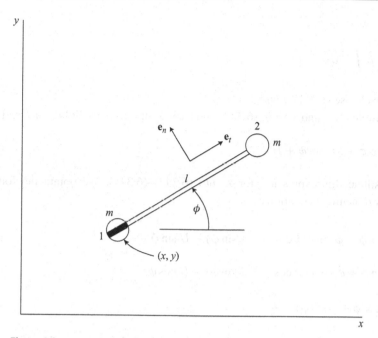

Figure 6.6.

Then, using (6.337), the three dynamical equations are

$$2m\ddot{x} - ml\ddot{\phi}\sin\phi - ml\dot{\phi}^2\cos\phi = -\lambda\sin\phi \tag{6.339}$$

$$2m\ddot{y} + ml\ddot{\phi}\cos\phi - ml\dot{\phi}^2\sin\phi = \lambda\cos\phi \tag{6.340}$$

$$ml^2\ddot{\phi} - ml\ddot{x}\sin\phi + ml\ddot{y}\cos\phi = 0 \tag{6.341}$$

These equations can be solved for the $\ddot{q}$s; the result is

$$\ddot{x} = \frac{1}{2}l\dot{\phi}^2\cos\phi - \frac{\lambda}{m}\sin\phi \tag{6.342}$$

$$\ddot{y} = \frac{1}{2}l\dot{\phi}^2\sin\phi + \frac{\lambda}{m}\cos\phi \tag{6.343}$$

$$\ddot{\phi} = -\frac{\lambda}{ml} \tag{6.344}$$

In order to stabilize the nonholonomic constraint, we differentiate the constraint function once with respect to time, and then we set

$$\dot{g} + \alpha g + \beta \int_0^t g\,dt = 0 \tag{6.345}$$

Thus, we obtain

$$\dot{g} = -\ddot{x}\sin\phi + \ddot{y}\cos\phi - \dot{\phi}(\dot{x}\cos\phi + \dot{y}\sin\phi) = U \tag{6.346}$$

where

$$U = -\alpha g - \beta \int_0^t g \, dt \tag{6.347}$$

Typically, we choose $\alpha = 1/h$ and $\beta = 1/h^2$.

Now substitute for $\ddot{x}$ and $\ddot{y}$ from (6.342) and (6.343) into (6.346). Solve for λ and obtain

$$\lambda = m(\dot{x}\dot{\phi}\cos\phi + \dot{y}\dot{\phi}\sin\phi + U) \tag{6.348}$$

Finally, substitute this expression for λ into (6.342)–(6.344). We obtain the following second-order dynamical equations:

$$\ddot{x} = \frac{1}{2}l\dot{\phi}^2\cos\phi - \dot{\phi}\sin\phi(\dot{x}\cos\phi + \dot{y}\sin\phi) - U\sin\phi \tag{6.349}$$

$$\ddot{y} = \frac{1}{2}l\dot{\phi}^2\sin\phi + \dot{\phi}\cos\phi(\dot{x}\cos\phi + \dot{y}\sin\phi) + U\cos\phi \tag{6.350}$$

$$\ddot{\phi} = -\frac{\dot{\phi}}{l}(\dot{x}\cos\phi + \dot{y}\sin\phi) - \frac{U}{l} \tag{6.351}$$

These equations are integrated numerically. For the unstabilized case ($U = 0$), it can be shown that (6.349)–(6.351) are equivalent to (4.31)–(4.33) which were obtained earlier by using Maggi's equations. One can conclude that the differentiated constraint equation (4.33) would result in numerical instability unless there is stabilization by some method such as that of Baumgarte.

One-step method for holonomic constraints

Although the Baumgarte method works quite well in stabilizing kinematic constraints, one can improve the accuracy of constraint representations by using a one-step method. The improvement is particularly noticeable when there is more than one constraint. When the Baumgarte method is used, for example, there is the possibility that the terms added to stabilize a given constraint may actually increase the error of a second constraint. Thus, there is an interaction of corrective efforts.

The goal of the one-step method is to completely eliminate the constraint errors at the end of each time step, insofar as the constraint linearizations are valid. In practice, this means that if relatively small truncation and roundoff errors occur with each step, then the constraint accuracy will be excellent. Another consideration is that the corrections should be made in a direction in configuration space such that the dynamical response of the system will not be altered appreciably.

The procedure for the one-step method begins in a manner similar to the Baumgarte method. The equations of motion are written with the aid of Lagrange multipliers, and then are solved for the accelerations $\ddot{q}_i(q, \dot{q}, \lambda, t)$, where $i = 1, \ldots, n$. The m holonomic constraint functions are differentiated twice with respect to time, with the result

$$\ddot{\phi}_j(q, \dot{q}, \ddot{q}, t) = 0 \qquad (j = 1, \ldots, m) \tag{6.352}$$

Then the n $\ddot{q}$ expressions obtained from the dynamical equations are substituted into (6.352). These m equations are solved for the m Lagrange multipliers $\lambda_j(q, \dot{q}, t)$ which are then substituted back into the $\ddot{q}$ expressions. At this point, we have n dynamical equations without λs, namely,

$$\ddot{q}_i = \ddot{q}_i(q, \dot{q}, t) \qquad (i = 1, \ldots, n) \tag{6.353}$$

These equations are identical with (6.335) of the Baumgarte method, except that all the Us have been set equal to zero. The n equations are integrated numerically and, at each time step, a one-step error correction takes place.

In order to explain the geometric aspects of the one-step error correction method, let us consider first the case of a single holonomic constraint

$$\phi(q, t) = 0 \tag{6.354}$$

This constraint can be represented as a surface in configuration space (q-space). The configuration point P, whose position is given by the n-vector $\mathbf{q}$, must move on the constraint surface if the constraint equation (6.354) is satisfied exactly. But, in general, the value of the constraint function ϕ will not be exactly zero at the end of each step. It needs to be corrected to zero. To accomplish this, first find the gradient vector

$$\nabla\phi = \left[\frac{\partial\phi}{\partial q_1}, \ldots, \frac{\partial\phi}{\partial q_n} \right]^{\mathrm{T}} \tag{6.355}$$

where the axes are orthogonal in n-space. In an effort to minimize the effect of the constraint correction on other aspects of the solution such as periods or damping, let us arbitrarily choose a correction vector $\Delta\mathbf{q}$ of minimum length. Assuming a small constraint error, this implies that the correction should be made in a direction normal to the constraint surface in n-space, that is, in the direction of the gradient $\nabla\phi$. So let us choose

$$\Delta q_i = C\frac{\partial\phi}{\partial q_i} \qquad (i = 1, \ldots, n) \tag{6.356}$$

where C is a constant whose value is found from the condition that the correction should exactly cancel the constraint errors. Thus, we require that

$$\nabla\phi \cdot \Delta\mathbf{q} = C\sum_{i=1}^{n} \left(\frac{\partial\phi}{\partial q_i} \right)^2 = -\phi \tag{6.357}$$

and we obtain

$$C = \frac{-\phi}{\displaystyle\sum_{i=1}^{n} \left(\frac{\partial\phi}{\partial q_i} \right)^2} \tag{6.358}$$

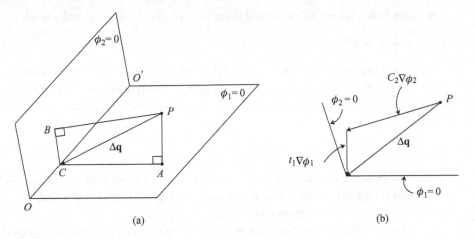

Figure 6.7.

Then we use (6.356) to find the correction components

$$\Delta q_i = \frac{-\left(\dfrac{\partial \phi}{\partial q_i}\right)\phi}{\displaystyle\sum_{i=1}^{n}\left(\dfrac{\partial \phi}{\partial q_i}\right)^2} \qquad (i = 1, \ldots, n) \tag{6.359}$$

These corrections are applied at the end of each time step. If a multiple-pass algorithm is used, it is still preferable to make the correction once per step rather than once per pass.

Now let us generalize the results to the case of m holonomic constraints of the form

$$\phi_j(q, t) = 0 \qquad (j = 1, \ldots, m) \tag{6.360}$$

The solution vector $\mathbf{q}(t)$ must move in the $(n - m)$-dimensional space formed by the common intersection of the m constraint surfaces. This is visualized most easily for the case of two constraints in a three-dimensional space, as shown in Fig. 6.7. Here the operating point P does not lie in either constraint surface, although it should lie on their line of intersection OO' if the constraints are to be satisfied exactly.

The correction vector $\Delta \mathbf{q} = PC$ is chosen to be orthogonal to OO' and is expressed as a linear combination of the gradient vectors $\nabla \phi_1$ and $\nabla \phi_2$, having the directions PA and PB, respectively. Thus,

$$\Delta \mathbf{q} = C_1 \nabla \phi_1 + C_2 \nabla \phi_2 \tag{6.361}$$

as shown in Fig. 6.7b. Here we have assumed that the constraint errors ϕ_1 and ϕ_2 are small at P, and the constraint surfaces in this vicinity can be approximated by planes. The coefficients C_1 and C_2 are found from the condition that the constraint errors are removed completely, that is,

$$\nabla \phi_1 \cdot \Delta \mathbf{q} = C_1 (\nabla \phi_1)^2 + C_2 \nabla \phi_1 \cdot \nabla \phi_2 = -\phi_1 \tag{6.362}$$

$$\nabla \phi_2 \cdot \Delta \mathbf{q} = C_1 \nabla \phi_1 \cdot \nabla \phi_2 + C_2 (\nabla \phi_2)^2 = -\phi_2 \tag{6.363}$$

After solving for C_1 and C_2, the actual Δq corrections are found from (6.361).

$$\Delta q_1 = C_1 \frac{\partial \phi_1}{\partial q_1} + C_2 \frac{\partial \phi_2}{\partial q_1} \tag{6.364}$$

$$\Delta q_2 = C_1 \frac{\partial \phi_1}{\partial q_2} + C_2 \frac{\partial \phi_2}{\partial q_2} \tag{6.365}$$

$$\Delta q_3 = C_1 \frac{\partial \phi_1}{\partial q_3} + C_2 \frac{\partial \phi_2}{\partial q_3} \tag{6.366}$$

Now let us generalize by assuming there are n qs and m holonomic constraints. The correction vector $\Delta \mathbf{q}$ is written as a linear combination of the constraint gradient vectors. It has the form

$$\Delta \mathbf{q} = \sum_{k=1}^{m} C_k \nabla \phi_k \tag{6.367}$$

In terms of scalar components, we have

$$\Delta q_i = \sum_{k=1}^{m} C_k \frac{\partial \phi_k}{\partial q_i} \qquad (i = 1, \ldots, n) \tag{6.368}$$

The Δqs are chosen such that

$$\nabla \phi_j \cdot \Delta \mathbf{q} = -\phi_j \qquad (j = 1, \ldots, m) \tag{6.369}$$

or, in detail,

$$\sum_{i=1}^{n} \sum_{k=1}^{m} C_k \frac{\partial \phi_j}{\partial q_i} \frac{\partial \phi_k}{\partial q_i} = -\phi_j \qquad (j = 1, \ldots, m) \tag{6.370}$$

These m equations are solved for the m Cs that are then substituted into (6.368) to obtain the individual Δq_i corrections that are added to the corresponding qs.

The procedure will usually result in very small constraint errors at the end of each time step, provided that the linearization assumptions are valid. A problem remains, however. It arises because although the ϕ functions are all approximately equal to zero, the corresponding constraint error rates ($\dot\phi$s) are not necessarily zero, and may actually increase in magnitude as the integration proceeds. To avoid this possible difficulty, let us use a similar one-step procedure to correct the *velocities* in a manner such that all the $\dot\phi$s go to zero. In other words, the correction $\Delta \dot{\mathbf{q}}$ in *velocity space* is formed of a linear combination of the gradient vectors of the individual $\dot\phi$ functions.

Let us begin by noting that the gradient in velocity space of $\dot\phi_j(q, \dot q, t)$ is

$$\nabla \dot\phi_j = \left[\frac{\partial \dot\phi_j}{\partial \dot q_1}, \ldots, \frac{\partial \dot\phi_j}{\partial \dot q_n} \right]^{\mathrm{T}} \tag{6.371}$$

Choose the velocity vector correction $\Delta \dot{\mathbf{q}}$ of the form

$$\Delta \dot{\mathbf{q}} = \sum_{k=1}^{m} K_k \nabla \dot\phi_k \tag{6.372}$$

where the values of the Ks are such that

$$\nabla \phi_j \cdot \Delta \dot{\mathbf{q}} = -\dot{\phi}_j \qquad (j = 1, \ldots, m) \tag{6.373}$$

We note that $\dot{\phi}_j(q, \dot{q}, t)$ is linear in the $\dot{q}$s, so

$$\frac{\partial \dot{\phi}_j}{\partial \dot{q}_i} = \frac{\partial \phi_j}{\partial q_i} \tag{6.374}$$

Hence, we can write (6.372) and (6.373) more explicitly as

$$\Delta \dot{q}_i = \sum_{k=1}^{m} K_k \frac{\partial \phi_k}{\partial q_i} \qquad (i = 1, \ldots, n) \tag{6.375}$$

where the Ks are obtained by solving

$$\sum_{i=1}^{n} \sum_{k=1}^{m} K_k \frac{\partial \phi_j}{\partial q_i} \frac{\partial \phi_k}{\partial q_i} = -\dot{\phi}_j \qquad (j = 1, \ldots, m) \tag{6.376}$$

Notice that the coefficients of the Ks in (6.376) are identical with the coefficients of the Cs found earlier in (6.370). Thus, the solution for the Ks is relatively easy. Knowing the values of the Ks, one obtains the velocity corrections from (6.375).

To summarize, the one-step method proceeds by first obtaining the dynamical equations in the form of (6.353), that is, after eliminating the Lagrange multipliers and solving for the individual accelerations ($\ddot{q}$s) as functions of $(q, \dot{q}, t)$. These equations are integrated numerically. At the end of each time step, the one-step corrections are made, first the Δq_is using (6.368) and then the $\Delta \dot{q}_i$s using (6.375). If the initial errors in the ϕ_js or $\dot{\phi}_j$s are unusually large due to the use of a large step-size, the one-step corrections can be repeated to yield negligibly small final constraint errors.

Example 6.5   A particle mass $m = 1$ moves in a uniform gravitational field. It is constrained to follow the elliptical line of intersection of a cone and a plane, as shown in Fig. 6.8. The equation of the cone is

$$\phi_1 = x^2 + y^2 - \frac{1}{4}z^2 = 0 \tag{6.377}$$

and that of the plane is

$$\phi_2 = x + y + z - 1 = 0 \tag{6.378}$$

The equations of motion can be obtained from Lagrange's equation in the form

$$\frac{d}{dt}\left(\frac{\partial L}{\partial \dot{q}_i}\right) - \frac{\partial L}{\partial q_i} = \sum_{j=1}^{2} \lambda_j a_{ji} \qquad (i = 1, 2, 3) \tag{6.379}$$

where

$$L = T - V = \frac{1}{2}(\dot{x}^2 + \dot{y}^2 + \dot{z}^2) - gz \tag{6.380}$$

$$a_{ji} = \frac{\partial \phi_j}{\partial q_i} \tag{6.381}$$

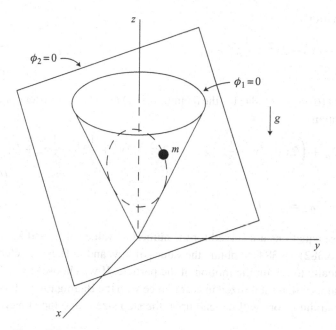

Figure 6.8.

The λs are Lagrange multipliers and g is the acceleration of gravity. The resulting equations of motion are

$$\ddot{x} = 2x\lambda_1 + \lambda_2 \tag{6.382}$$

$$\ddot{y} = 2y\lambda_1 + \lambda_2 \tag{6.383}$$

$$\ddot{z} = -g - \frac{1}{2}z\lambda_1 + \lambda_2 \tag{6.384}$$

Now differentiate the constraint equations twice with respect to time. We obtain

$$\dot{\phi}_1 = 2x\dot{x} + 2y\dot{y} - \frac{1}{2}z\dot{z} \tag{6.385}$$

$$\dot{\phi}_2 = \dot{x} + \dot{y} + \dot{z} \tag{6.386}$$

$$\ddot{\phi}_1 = 2x\ddot{x} + 2y\ddot{y} - \frac{1}{2}z\ddot{z} + 2\dot{x}^2 + 2\dot{y}^2 - \frac{1}{2}\dot{z}^2 \tag{6.387}$$

$$\ddot{\phi}_2 = \ddot{x} + \ddot{y} + \ddot{z} \tag{6.388}$$

Consider first Baumgarte's method and set

$$\ddot{\phi}_j = U_j \qquad (j = 1, 2) \tag{6.389}$$

where we take

$$U_j = -\alpha\dot{\phi}_j - \beta\phi_j \qquad (j = 1, 2) \tag{6.390}$$

This results in the equations

$$2x\ddot{x} + 2y\ddot{y} - \frac{1}{2}z\ddot{z} = -2\dot{x}^2 - 2\dot{y}^2 + \frac{1}{2}\dot{z}^2 + U_1 \tag{6.391}$$

$$\ddot{x} + \ddot{y} + \ddot{z} = U_2 \tag{6.392}$$

Now substitute from (6.382)–(6.384) for the $\ddot{q}$s into (6.391) and (6.392). After collecting terms in the λs, we obtain

$$\left(4x^2 + 4y^2 + \frac{1}{4}z^2\right)\lambda_1 + \left(2x + 2y - \frac{1}{2}z\right)\lambda_2 = -2\dot{x}^2 - 2\dot{y}^2 + \frac{1}{2}\dot{z}^2 - \frac{1}{2}gz + U_1 \tag{6.393}$$

$$\left(2x + 2y - \frac{1}{2}z\right)\lambda_1 + 3\lambda_2 = g + U_2 \tag{6.394}$$

At each time step these equations are solved for the numerical values of λ_1 and λ_2 that are then substituted into (6.382)–(6.384) to obtain the values of $\ddot{x}$, $\ddot{y}$, and $\ddot{z}$. These accelerations are integrated numerically to obtain the motion of the particle. If we choose $\alpha = 2/h$ and $\beta = 1/h^2$, the constraints will be stabilized in accordance with the Baumgarte method. The magnitude of the constraint errors will depend upon the step size h and the choice of the integration algorithm.

As a comparison, let us now use the *one-step method* to enforce the constraints. The analysis proceeds as before except that $U_1 = U_2 = 0$ in obtaining the acceleration components which are integrated numerically for one step. The one-step corrections are found using

$$\Delta\mathbf{q} = C_1\nabla\phi_1 + C_2\nabla\phi_2 \tag{6.395}$$

where

$$\Delta\mathbf{q} = \Delta x\mathbf{i} + \Delta y\mathbf{j} + \Delta z\mathbf{k} \tag{6.396}$$

$$\nabla\phi_1 = 2x\mathbf{i} + 2y\mathbf{j} - \frac{1}{2}z\mathbf{k} \tag{6.397}$$

$$\nabla\phi_2 = \mathbf{i} + \mathbf{j} + \mathbf{k} \tag{6.398}$$

and $(\mathbf{i}, \mathbf{j}, \mathbf{k})$ are the Cartesian unit vectors. The Cs are found from the equations

$$\nabla\phi_1 \cdot \Delta\mathbf{q} = C_1(\nabla\phi_1)^2 + C_2\nabla\phi_1 \cdot \nabla\phi_2 = -\phi_1 \tag{6.399}$$

$$\nabla\phi_2 \cdot \Delta\mathbf{q} = C_1\nabla\phi_1 \cdot \nabla\phi_2 + C_2(\nabla\phi_2)^2 = -\phi_2 \tag{6.400}$$

where

$$(\nabla\phi_1)^2 = 4x^2 + 4y^2 + \frac{1}{4}z^2 \tag{6.401}$$

$$\nabla\phi_1 \cdot \nabla\phi_2 = 2x + 2y - \frac{1}{2}z \tag{6.402}$$

$$(\nabla\phi_2)^2 = 3 \tag{6.403}$$

Notice that these coefficients are identical to the coefficients of the λs found earlier in (6.393) and (6.394). This occurs because both the correction vector $\Delta\mathbf{q}$ and the total constraint force vector are expressed in terms of the same nonorthogonal components whose directions lie along the normals to the constraint surfaces.

The Δqs, found by using (6.395), are

$$\Delta x = 2x C_1 + C_2 \tag{6.404}$$

$$\Delta y = 2y C_1 + C_2 \tag{6.405}$$

$$\Delta z = -\frac{1}{2} z C_1 + C_2 \tag{6.406}$$

The *velocity corrections* are obtained by solving first for the Ks, using the equations

$$\left(4x^2 + 4y^2 + \frac{1}{4}z^2\right) K_1 + \left(2x + 2y - \frac{1}{2}z\right) K_2 = -\dot{\phi}_1 \tag{6.407}$$

$$\left(2x + 2y - \frac{1}{2}z\right) K_1 + 3K_2 = -\dot{\phi}_2 \tag{6.408}$$

Then the Ks are substituted into (6.372). The results are

$$\Delta \dot{x} = 2x K_1 + K_2 \tag{6.409}$$

$$\Delta \dot{y} = 2y K_1 + K_2 \tag{6.410}$$

$$\Delta \dot{z} = -\frac{1}{2} z K_1 + K_2 \tag{6.411}$$

A comparison of the numerical results obtained by using the Baumgarte method or the one-step method of representing holonomic constraints shows that the one-step method is much more accurate, although either method may be acceptable for a relatively small step size. On the other hand, the two methods show about the same accuracy with respect to the time response, as measured by the period. Time errors are determined primarily by truncation errors associated with the integration algorithm.

One-step method for nonholonomic constraints

The procedure for applying the one-step method to systems with nonholonomic constraints begins in the same way as with the Baumgarte method, except that the Us are set equal to zero. Thus, assuming m nonholonomic constraints of the form

$$g_j(q, \dot{q}, t) = 0 \qquad (j = 1, \ldots, m) \tag{6.412}$$

one can write the dynamical equations as in (6.330), namely,

$$\frac{d}{dt}\left(\frac{\partial T}{\partial \dot{q}_i}\right) - \frac{\partial T}{\partial q_i} = \sum_{j=1}^{m} \lambda_j \frac{\partial g_j}{\partial \dot{q}_i} + Q_i(q, \dot{q}, t) \qquad (i = 1, \ldots, n) \tag{6.413}$$

Next, differentiate the constraint functions with respect to time, obtaining

$$\dot{g}_j(q, \dot{q}, \ddot{q}, t) = 0 \qquad (j = 1, \ldots, m) \tag{6.414}$$

One can solve the n equations of (6.413) for the n $\ddot{q}$s as functions of $(q, \dot{q}, \lambda, t)$ and these expressions are then substituted into (6.414), obtaining m equations which are linear in the

λs. After solving these equations for $\lambda_j(q, \dot{q}, t)$, $(j = 1, \ldots, m)$, we substitute back into the $\ddot{q}$ expressions to obtain the final dynamical equations of the form

$$\ddot{q}_i = \ddot{q}_i(q, \dot{q}, t) \qquad (i = 1, \ldots, n) \tag{6.415}$$

At this point, we should explain that another method such as Maggi's equation plus (6.414) might be used in obtaining (6.415).

The n second-order dynamical equations can be converted to $2n$ first-order equations which are integrated numerically. At the end of each step, we wish to apply a one-step correction to the solution. The fact that the system is nonholonomic results in an important difference in the correction strategy, as compared to the holonomic case. Due to the nonintegrable nature of the constraints, there is no direct way of detecting errors in the configuration, that is, the qs. On the other hand, for any given configuration, velocity errors result in a separation of the solution point from the constraint surface in velocity space. These errors can be detected and corrections made to the $\dot{q}$s. Another reason for this approach is the fact that, at least for holonomic systems, errors in the $\dot{\phi}$s tend to be much larger than configuration errors as measured by the ϕs. In other words, the errors in the $\dot{q}$s tend to be larger than errors in the qs due to the smoothing effect of the integration process. Our approach, then, will be to make one-step corrections in a manner similar to the velocity corrections for holonomic constraints.

First, we require that the velocity correction vector be a linear combination of the individual constraint gradient vectors in velocity space.

$$\Delta \dot{q}_i = \sum_{k=1}^{m} K_k \frac{\partial g_k}{\partial \dot{q}_i} \qquad (i = 1, \ldots, n) \tag{6.416}$$

Next, the corrections must exactly cancel the constraint errors, that is,

$$\sum_{i=1}^{n} \frac{\partial g_j}{\partial \dot{q}_i} \Delta \dot{q}_i = -g_j \qquad (j = 1, \ldots, m) \tag{6.417}$$

or

$$\sum_{i=1}^{n} \sum_{k=1}^{m} K_k \frac{\partial g_j}{\partial \dot{q}_i} \frac{\partial g_k}{\partial \dot{q}_i} = -g_j \qquad (j = 1, \ldots, m) \tag{6.418}$$

These m equations are solved for the m Ks and then the velocity corrections are made using (6.416).

Example 6.6 Let us consider once again a nonholonomic system consisting of a dumbbell sliding on a horizontal plane and constrained by a knife edge at one of the particles (Fig. 6.6). The nonholonomic constraint equation is

$$g_1(q, \dot{q}) = -\dot{x} \sin \phi + \dot{y} \cos \phi = 0 \tag{6.419}$$

which states that the transverse velocity at particle 1 is equal to zero. Following the procedure of Example 6.4 on page 370 for the case in which $U = 0$, we obtain the dynamical equations:

$$\ddot{x} = \frac{1}{2}l\dot{\phi}^2 \cos\phi - \dot{\phi}\sin\phi(\dot{x}\cos\phi + \dot{y}\sin\phi) \tag{6.420}$$

$$\ddot{y} = \frac{1}{2}l\dot{\phi}^2 \sin\phi + \dot{\phi}\cos\phi(\dot{x}\cos\phi + \dot{y}\sin\phi) \tag{6.421}$$

$$\ddot{\phi} = -\frac{\dot{\phi}}{l}(\dot{x}\cos\phi + \dot{y}\sin\phi) \tag{6.422}$$

The equations are integrated numerically. At the end of each time step, the nonholonomic constraint function g_1 is evaluated. From (6.418), we find that

$$K(\sin^2\phi + \cos^2\phi) = -g_1 \tag{6.423}$$

or

$$K = -g_1 \tag{6.424}$$

Then, from (6.416), the one-step velocity corrections are

$$\Delta\dot{x} = -g_1\sin\phi \tag{6.425}$$
$$\Delta\dot{y} = g_1\cos\phi \tag{6.426}$$
$$\Delta\dot{\phi} = 0 \tag{6.427}$$

Thus, at the end of each time step in the numerical integration, the value of the nonholonomic constraint function is approximately zero.

A comparison of constraint enforcement methods

If the given holonomic and nonholonomic examples are analyzed numerically for typical cases, some observations and conclusions can be made. For the holonomic system of Example 6.5 on page 376, the results are much more accurate for the one-step method than for the Baumgarte method. This is particularly true of the velocities and amounts to about two additional decimal digits of accurate data for the same step size, using the RK-2 algorithm. The configuration accuracy is also much superior for the one-step method, although both methods show satisfactory accuracy for a relatively small step size. A comparison of computing times shows that the one-step method requires about one-third more time for the same step size.

If one compares the Baumgarte and one-step methods, as they are applied to nonholonomic systems such as Example 6.6 on page 380, the one-step method again provides superior accuracy. Both methods, however, exhibit some loss of accuracy for nonholonomic constraints. This arises because there is no direct way of detecting configuration errors for nonholonomic systems as the computation proceeds.

A comparison of the solution accuracies of the two methods with respect to time shows roughly the same accuracy. This occurs because time accuracy is not closely related to configuration accuracy. Rather, the time aspect of the computations, as indicated by the period of the motion, for example, is strongly influenced by the choice of the integration algorithm and the step size. The truncation errors are the most important factor. For conservative systems such as we have been studying, the overall accuracy can be improved by making further corrections based on energy errors. These errors tend to appear as velocity errors whose magnitudes increase with time. This affects the periods and other important characteristics of the motion. The use of energy methods in improving computational accuracy will be considered in Section 6.6.

Euler parameter constraints

The four Euler parameters $(\epsilon_x, \epsilon_y, \epsilon_z, \eta)$ were introduced in Section 3.1 as generalized coordinates to represent the orientation of a rigid body relative to some frame, usually an inertial frame. Since a rigid body has only three rotational degrees of freedom, there must be an equation of constraint relating the Euler parameters. It is the holonomic constraint

$$\phi(\epsilon_x, \epsilon_y, \epsilon_z, \eta) = \epsilon_x^2 + \epsilon_y^2 + \epsilon_z^2 + \eta^2 - 1 = 0 \tag{6.428}$$

In accordance with (3.80), the Euler parameters are generated by the numerical integration of four first-order kinematical equations, namely,

$$\dot{\epsilon}_x = \frac{1}{2}(\omega_z \epsilon_y - \omega_y \epsilon_z + \omega_x \eta) \tag{6.429}$$

$$\dot{\epsilon}_y = \frac{1}{2}(\omega_x \epsilon_z - \omega_z \epsilon_x + \omega_y \eta) \tag{6.430}$$

$$\dot{\epsilon}_z = \frac{1}{2}(\omega_y \epsilon_x - \omega_x \epsilon_y + \omega_z \eta) \tag{6.431}$$

$$\dot{\eta} = -\frac{1}{2}(\omega_x \epsilon_x + \omega_y \epsilon_y + \omega_z \epsilon_z) \tag{6.432}$$

The body-axis angular velocity components $(\omega_x, \omega_y, \omega_z)$ are obtained by the numerical integration of the rotational equations of motion.

Let us use the one-step method to correct any constraint errors. First, calculate the gradient vector in q-space at the operating point.

$$\nabla \phi = [2\epsilon_x \ \ 2\epsilon_y \ \ 2\epsilon_z \ \ 2\eta]^T \tag{6.433}$$

Next, we require the correction vector $\Delta \mathbf{q}$ to exactly compensate for the constraint error at the end of each time step, that is,

$$\nabla \phi \cdot \Delta \mathbf{q} = -\phi \tag{6.434}$$

where

$$\Delta \mathbf{q} = C \nabla \phi \tag{6.435}$$

Thus, we take the correction vector in the direction of the gradient $\nabla \phi$. From (6.434) and (6.435), the coefficient C is

$$C = \frac{-\phi}{\nabla \phi \cdot \nabla \phi} = \frac{-\phi}{4(\epsilon_x^2 + \epsilon_y^2 + \epsilon_z^2 + \eta^2)} \approx -\frac{\phi}{4} \tag{6.436}$$

In detail, the corrections to the generalized coordinates are

$$\Delta\epsilon_x = -\frac{1}{2}\epsilon_x\phi, \qquad \Delta\epsilon_y = -\frac{1}{2}\epsilon_y\phi, \qquad \Delta\epsilon_z = -\frac{1}{2}\epsilon_z\phi, \qquad \Delta\eta = -\frac{1}{2}\eta\phi \qquad (6.437)$$

Note that this correction $\Delta\mathbf{q}$ does not change the direction of the configuration vector $\mathbf{q}$ in 4-space.

The correction of the Euler parameter values does not cause a risk of numerical instability because, in the usual case of rigid body motion, the Euler parameter values do not enter the dynamical equations. Thus, there is no need to set $\dot{\phi}$ or $\ddot{\phi}$ equal to zero, thereby avoiding tendencies toward numerical instability. Instead, the dynamical equations are written in terms of quasi-velocities (ωs) and the Euler parameter calculations for the orientation are performed separately.

6.6   Energy and momentum methods

In assessing the accuracy of numerically computed solutions of the equations of motion of a dynamical system, it is helpful to have relatively simple check solutions available. A common approach is to use integrals of the motion, that is, functions whose values remain constant during the course of the solution, as checks on accuracy. For example, the energy integral $E = T_2 - T_0 + V$ may be used for conservative systems. Other systems may have one or more components of momentum or angular momentum conserved. Integrals of the motion are particularly effective in detecting programming errors. Of course, these methods are ineffective against errors in the values of the physical parameters.

Energy corrections

Let us consider a conservative system whose energy integral has the form

$$E(q, \dot{q}) = T_2 - T_0 + V = E_0 \qquad (6.438)$$

At each time step in the numerical solution, we wish to correct for an energy error ΔE given by

$$\Delta E = E - E_0 \qquad (6.439)$$

where E_0 is evaluated from the initial conditions.

First let us note that, for a given configuration and energy constant E_0, the equation

$$E(q, \dot{q}) - E_0 = 0 \qquad (6.440)$$

represents a surface in n-dimensional velocity space (Fig. 6.9). Since the kinetic energy function $T(q, \dot{q})$ is positive definite and quadratic in the $\dot{q}$s, the form of the energy surface is ellipsoidal for the particular case of 3-space.

The velocity correction $\Delta\dot{q}$ will be taken in the direction normal to the energy surface, that is, in the direction of the gradient ∇E or ∇T. In accordance with the one-step method,

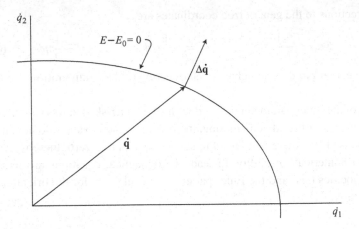

Figure 6.9.

let us assume that

$$\Delta\dot{\mathbf{q}} = K\nabla T \tag{6.441}$$

and choose the value of K such that

$$\nabla T \cdot \Delta\dot{\mathbf{q}} = -\Delta E \tag{6.442}$$

which results in the exact cancellation of the energy error. Thus, we obtain

$$K = \frac{-\Delta E}{(\nabla T)^2} = -\Delta E \left[\sum_{j=1}^{n} \left(\frac{\partial T}{\partial\dot{q}_j} \right)^2 \right]^{-1} \tag{6.443}$$

The individual velocity corrections are

$$\Delta\dot{q}_i = K\frac{\partial T}{\partial\dot{q}_i} \qquad (i = 1, \ldots, n) \tag{6.444}$$

If the conservative system is subject to kinematic constraints, the corresponding constraint surfaces in velocity space are typically planes passing through the origin. The velocity vector $\dot{\mathbf{q}}$ is constrained to lie in the common intersection of these planes. Any velocity correction is taken in a direction normal to $\dot{\mathbf{q}}$. But from Fig. 6.9 we see that the velocity correction $\Delta\dot{\mathbf{q}}$ for an energy error is approximately the same direction as $\dot{\mathbf{q}}$. (It is exactly the same direction if the energy surface is spherical.) Thus, the velocity corrections for constraint errors and energy errors are approximately normal to each other and do not interact very strongly.

From (6.443) and (6.444), we see that $\Delta\dot{\mathbf{q}}$ can be quite large if the magnitude of $\dot{\mathbf{q}}$ or the generalized momentum vector is very small. For most dynamical systems, this will not occur. But if the kinetic energy T_2 does attain a small value compared to its maximum during the motion of the conservative system, then significant errors in the corrections $\Delta\dot{q}_i$ can occur because the linear theory assumes small perturbations. In this case, it is preferable to make the corrections in configuration space rather than velocity space, so long as T_2 and $\dot{\mathbf{q}}$ remain small.

Proceeding, then, along these lines, we can write

$$E = T_2 + V' = E_0 \tag{6.445}$$

where

$$V'(q) = V - T_0 \tag{6.446}$$

Then, using a similar analysis, and taking the correction $\Delta\mathbf{q}$ in the direction of the gradient of V' in configuration space, we obtain

$$\Delta q_i = -\Delta E \left[\sum_{j=1}^{n} \left(\frac{\partial V'}{\partial q_j} \right)^2 \right]^{-1} \frac{\partial V'}{\partial q_i} \qquad (i = 1, \dots, n) \tag{6.447}$$

When T_2 is no longer small, it is preferable to return to corrections in velocity space since velocity errors tend to be larger.

If the kinetic energy is expressed in terms of quasi-velocities (us) rather than true velocities ($\dot{q}$s), the analysis proceeds similarly to that resulting in (6.443) and (6.444). Thus, substituting u_j for $\dot{q}_j$, we obtain the corrections

$$\Delta u_i = -\Delta E \left[\sum_{j=1}^{n-m} \left(\frac{\partial T}{\partial u_j} \right)^2 \right]^{-1} \frac{\partial T}{\partial u_i} \qquad (i = 1, \dots, n-m) \tag{6.448}$$

where there are m constraints and $(n - m)$ independent us.

Example 6.7 Let us consider the correction of energy errors in the free rotational motion of a rigid body in space. The rotational kinetic energy is

$$T = \frac{1}{2} I_{xx}\omega_x^2 + \frac{1}{2} I_{yy}\omega_y^2 + \frac{1}{2} I_{zz}\omega_z^2 \tag{6.449}$$

where the principal moments of inertia are taken about the center of mass. The angular velocity components ($\omega_x, \omega_y, \omega_z$) are quasi-velocities, so (6.448) applies. The corrections are

$$\Delta\omega_i = \frac{-\Delta E\, I_{ii}\omega_i}{I_{xx}\omega_x^2 + I_{yy}\omega_y^2 + I_{zz}\omega_z^2} \qquad (i = x, y, z) \tag{6.450}$$

Notice that the correction vector $\Delta\omega$ has the same direction as the angular momentum.

$$\mathbf{H} = I_{xx}\omega_x\mathbf{i} + I_{yy}\omega_y\mathbf{j} + I_{zz}\omega_z\mathbf{k} \tag{6.451}$$

where $(\mathbf{i}, \mathbf{j}, \mathbf{k})$ are body-fixed Cartesian unit vectors. This $\Delta\omega$ is of minimum magnitude for the given energy correction.

Work and energy

We have been concerned with numerical checks and corrections as they are applied to conservative systems. Now let us extend the theory to nonconservative systems. We shall consider two basic methods.

First method

Let us apply the principle of work and kinetic energy to a system of N rigid bodies. The velocity of the reference point P_i of the ith body is $\mathbf{v}_i$ and its angular velocity is $\boldsymbol{\omega}_i$. The total effect of all the forces, including constraint forces, which act on the ith body are equivalent to a force $\mathbf{F}_i$ acting at the reference point plus a couple of moment $\mathbf{M}_i$. Then, since

$$\dot{T} = \dot{W} \tag{6.452}$$

where, from (4.357),

$$\dot{W} = \sum_{i=1}^{N} (\mathbf{F}_i \cdot \mathbf{v}_i + \mathbf{M}_i \cdot \boldsymbol{\omega}_i) \tag{6.453}$$

we can integrate with respect to time to obtain

$$\Delta T = \int_{t_1}^{t_2} \sum_{i=1}^{N} (\mathbf{F}_i \cdot \mathbf{v}_i + \mathbf{M}_i \cdot \boldsymbol{\omega}_i) dt \tag{6.454}$$

At any time t_2, this value of ΔT can be used as a check by comparing it with a value of ΔT obtained from the solution of the differential equations. To be effective, the separate integration given by (6.454) must be at least as accurate as the ΔT obtained from the numerical solution of the equations of motion.

If a portion of $\mathbf{F}_i$ and of $\mathbf{M}_i$ arise from a potential energy function $V(q, t)$, but $\mathbf{F}'_i$ and $\mathbf{M}'_i$ are the portions that are not due to potential energy, then in accordance with (4.364) we can write

$$\Delta(T + V) = \int_{t_1}^{t_2} \left[\sum_{i=1}^{N} (\mathbf{F}'_i \cdot \mathbf{v}_i + \mathbf{M}'_i \cdot \boldsymbol{\omega}_i) + \frac{\partial V}{\partial t} \right] dt \tag{6.455}$$

This calculation can also be used as a check on the numerical solution of differential equations.

The above approach is relatively direct and effective when it is applicable. It turns out, however, that $\mathbf{F}'_i$ and $\mathbf{M}'_i$ frequently include unknown constraint forces.

Second method

Another approach is to make energy checks using the energy integral $E = T_2 - T_0 + V$ rather than the total energy $T + V$. Consider a nonholonomic system consisting of N rigid bodies, with a kinetic energy $T(q, u, t)$ and potential energy $V(q, t)$. Suppose there are n qs and m constraints. The n $\dot{q}$s and $(n - m)$ us are related by

$$\dot{q}_i = \sum_{j=1}^{n-m} \Phi_{ij}(q, t) u_j + \Phi_{it}(q, t) \quad (i = 1, \ldots, n) \tag{6.456}$$

The velocity of the *center of mass* of the ith body is

$$\mathbf{v}_i = \sum_{j=1}^{n-m} \gamma_{ij}(q,t)u_j + \gamma_{it}(q,t) \qquad (i = 1, \ldots, N) \tag{6.457}$$

and the angular velocity is

$$\boldsymbol{\omega}_i = \sum_{j=1}^{n-m} \beta_{ij}(q,t)u_j + \beta_{it}(q,t) \qquad (i = 1, \ldots, N) \tag{6.458}$$

Then, in accordance with (4.380), we find that the energy rate is

$$\dot{E} = \dot{T}_2 - \dot{T}_0 + \dot{V}$$
$$= \sum_{j=1}^{n-m} Q'_j u_j + \frac{\partial V}{\partial t} + \sum_{k=1}^{n} \frac{\partial V}{\partial q_k} \Phi_{kt} - \sum_{i=1}^{N} \mathbf{p}_i \cdot \dot{\gamma}_{it} - \sum_{i=1}^{N} \mathbf{H}_{ci} \cdot \dot{\beta}_{it} \tag{6.459}$$

where $\mathbf{p}_i$ is the linear momentum of the ith body and $\mathbf{H}_{ci}$ is the angular momentum about its center of mass. Integration of $\dot{E}$ with respect to time results in

$$\Delta E = \int_{t_1}^{t_2} \left[\sum_{j=1}^{n-m} Q'_j u_j + \frac{\partial V}{\partial t} + \sum_{k=1}^{n} \frac{\partial V}{\partial q_k} \Phi_{kt} - \sum_{i=1}^{N} \mathbf{p}_i \cdot \dot{\gamma}_{it} - \sum_{i=1}^{N} \mathbf{H}_{ci} \cdot \dot{\beta}_{it} \right] dt \tag{6.460}$$

This result can be used as a numerical check on the accuracy of ΔE found from the numerical integration of the equations of motion. For the usual case of ideal constraints and independent us, the Q'_j generalized forces will not involve constraint forces, and thus will be known functions or equal to zero. Note that $\dot{\gamma}_{it}$ and $\dot{\beta}_{it}$ represent linear or angular accelerations when all the us are set equal to zero, that is, they generally appear only if there are moving constraints. Thus, in spite of the apparent complexity, (6.460) is often relatively easy to evaluate.

Example 6.8 Let us consider work and energy relationships for a rheonomic system consisting of a top whose point is constrained to move uniformly around a horizontal circle of radius r (Fig. 6.10). The configuration is given by type II Euler angles. Let us choose $(\dot{\phi}, \dot{\theta}, \Omega)$ as us, where Ω is the angular velocity of the top about its symmetry axis. The angular velocity of the point P along its path is ω_0. The $\mathbf{i}\,\mathbf{j}\,\mathbf{k}$ unit vectors are as shown, with $\mathbf{i}$ remaining horizontal and $\mathbf{j}\,\mathbf{k}$ defining a vertical plane. Since the motion of the point P is known, the easiest approach is to consider that point fixed, and then apply a horizontal inertial force at the center of mass due to the actual acceleration of P. With this assumption, the velocity of the center of mass is linear and homogeneous in the us, so

$$\gamma_{1t} = 0 \tag{6.461}$$

The angular velocity of the top is

$$\boldsymbol{\omega}_1 = \dot{\theta}\mathbf{i} + \dot{\phi}\sin\theta\,\mathbf{j} + \Omega\mathbf{k} \tag{6.462}$$

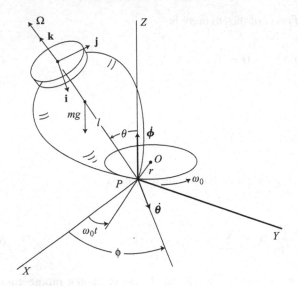

Figure 6.10.

Note that

$$\dot{\phi} = u_1, \qquad \dot{\theta} = u_2, \qquad \dot{\psi} = -u_1 \cos\theta + u_3 \tag{6.463}$$

We see from (6.462) and (6.463) that

$$\beta_{11} = \sin\theta\, \mathbf{j}, \qquad \beta_{12} = \mathbf{i}, \qquad \beta_{13} = \mathbf{k}, \qquad \beta_{1t} = 0 \tag{6.464}$$

and also that

$$\Phi_{1t} = \Phi_{2t} = \Phi_{3t} = 0 \tag{6.465}$$

The potential energy is

$$V = mgl\cos\theta \tag{6.466}$$

From the general energy rate equation (6.459), we see that the last four terms vanish, leaving

$$\dot{E} = \dot{T}_2 - \dot{T}_0 + \dot{V} = \sum_{j=1}^{3} Q'_j u_j \tag{6.467}$$

where the Q'_j generalized forces are inertial.

The acceleration of P in the actual motion is $r\omega_0^2$ and is directed from P toward O. The corresponding inertial force applied at the center of mass results in

$$
\begin{aligned}
Q'_1 &= mlr\omega_0^2 \sin\theta \cos(\phi - \omega_0 t) \\
Q'_2 &= mlr\omega_0^2 \cos\theta \sin(\phi - \omega_0 t) \\
Q'_3 &= 0
\end{aligned}
\tag{6.468}
$$

where the Q'_j generalized forces are moments about the Z, $\dot{\theta}$, and $\mathbf{k}$ axes, respectively. Thus, from (6.467) and (6.468), the energy rate is

$$\dot{E} = \dot{T}_2 - \dot{T}_0 + \dot{V}$$
$$= mlr\omega_0^2[\dot{\phi}\sin\theta\cos(\phi - \omega_0 t) + \dot{\theta}\cos\theta\sin(\phi - \omega_0 t)] \tag{6.469}$$

The kinetic energy, assuming a fixed point P is

$$T = T_2 = \frac{1}{2}I_t(\dot{\theta}^2 + \dot{\phi}^2\sin^2\theta) + \frac{1}{2}I_a\Omega^2 \tag{6.470}$$

where the Is are taken about P and the total spin Ω is constant.

Finally, integrating over an arbitrary time interval, we obtain

$$\Delta E = \Delta(T_2 + V)$$
$$= mlr\omega_0^2 \int_{t_1}^{t_2} [\dot{\phi}\sin\theta\cos(\phi - \omega_0 t) + \dot{\theta}\cos\theta\sin(\phi - \omega_0 t)]dt \tag{6.471}$$

The same result is obtained for $\dot{E}$ if we follow the straightforward procedure of considering the actual motion of P in finding the velocity of the center of mass and the total kinetic energy. The calculations are more complicated, however. The Q' terms equal zero, and we obtain

$$\dot{E} = -\mathbf{p}_1 \cdot \dot{\boldsymbol{\gamma}}_{1t} \tag{6.472}$$

where $\mathbf{p}_1$ and $\dot{\boldsymbol{\gamma}}_{1t}$ are nonzero. The result given in (6.471) can be used to check the accuracy of the numerical solution of the differential equations of motion (which have not been used in obtaining the expression for $\dot{E}$). The integration of $\dot{E}$ is accomplished concurrently with the integration of the equations of motion, and E is considered as an extra variable. This check will detect most programming errors.

Conservation of momentum

One can use momentum-like integrals of the motion to improve the accuracy of numerical solutions of the equations of motion for dynamical systems. A common approach, when it applies, is to use the Routhian method, as discussed in Chapter 2. This method applies to systems described by the standard holonomic form of Lagrange's equation, and which have one or more *ignorable coordinates*, that is, coordinates which do not appear in the Lagrangian function although the corresponding $\dot{q}$s do appear. For example, if there are k ignorable coordinates, then there are k integrals of the motion having the form

$$\frac{\partial L}{\partial \dot{q}_i} = \beta_i \qquad (i = 1, \ldots, k) \tag{6.473}$$

where the βs are constants and $\partial L/\partial \dot{q}_i$ is a generalized momentum. The Routhian method proceeds by defining the Routhian function

$$R(q, \dot{q}, t) = L - \sum_{i=1}^{k} \beta_i \dot{q}_i \tag{6.474}$$

where the k ignorable $\dot{q}$s have been eliminated by solving for them from the k equations of (6.473) and then substituting into (6.474). The $(n - k)$ differential equations of motion are obtained from

$$\frac{d}{dt}\left(\frac{\partial R}{\partial \dot{q}_i}\right) - \frac{\partial R}{\partial q_i} = 0 \qquad (i = k + 1, \ldots, n) \tag{6.475}$$

More generally, the differential equations of motion for the nonignored coordinates can include general applied forces. In this case, we obtain

$$\frac{d}{dt}\left(\frac{\partial R}{\partial \dot{q}_i}\right) - \frac{\partial R}{\partial q_i} = Q_i' \qquad (i = k + 1, \ldots, n) \tag{6.476}$$

We see from (6.475) that the effective number of degrees of freedom is reduced to $(n - k)$, thereby simplifying the analysis, particularly if one is not interested in solving for the time history of the ignored coordinates. It should be noted from (6.474) that one cannot simply use (6.473) to eliminate the ignorable $\dot{q}$s from the Lagrangian function and then continue to use the standard form of Lagrange's equations. This results in incorrect equations of motion. The reason is that the standard Lagrange's equations require that $L(q, \dot{q}, t)$ contain a full set of n independent $\dot{q}$s if the holonomic system has n kinematic degrees of freedom. So-called momentum or energy constraints do not reduce the number of kinematic degrees of freedom.

In addition to the Routhian method, there are other approaches which use conservation of momentum to reduce the number of degrees of freedom in the dynamical analysis. For example, isolated systems such as dynamical systems in space will have conservation of both linear and angular momentum. Since each is a vector quantity in 3-space, if one considers the Cartesian components of each vector, there are immediately available six integrals of the motion. The conservation of linear momentum implies that the center of mass of the system translates with constant velocity. Therefore, one can find an inertial frame in which the center of mass is fixed. By using this reference frame and choosing the generalized coordinates accordingly, one can reduce the number of degrees of freedom in the dynamical analysis by three. A similar reduction in the rotational degrees of freedom is not necessarily available because any rotating frame is noninertial. Nevertheless, special situations such as planar rotational motion can be used to simplify the analysis.

Angular momentum corrections

The Routhian method of reducing the effective degrees of freedom does not apply if the kinetic energy is expressed in terms of quasi-velocities. But quasi-velocities are commonly used in the rotational analysis of rigid bodies; that is, the angular velocity of a rigid body is expressed in terms of its body-axis components, and these components are quasi-velocities. If one considers the free rotational motion of a rigid body, there is conservation of angular momentum, and this can be used to correct errors and improve the accuracy of numerical solutions for its motion.

The simplest approach is to correct for *amplitude errors only* rather than correcting the three angular momentum components separately. The form of the square of the angular momentum magnitude is quadratic in the us. In that respect, it resembles a kinetic energy function, and it can be treated similarly.

As an example, consider a rigid body that is rotating in free space. Let us choose the square of the angular momentum magnitude as an integral of the motion, that is,

$$P(\omega) = I_{xx}^2\omega_x^2 + I_{yy}^2\omega_y^2 + I_{zz}^2\omega_z^2 = H_0^2 \tag{6.477}$$

We assume principal axes at the center of mass, and H_0 is the constant magnitude of the angular momentum.

The function $P(\omega)$ represents an ellipsoid in three-dimensional velocity space (ω-space) and is called the *momentum ellipsoid*. As the rotational motion proceeds the vector ω, drawn from the origin of the body-fixed frame, moves such that its point always lies on the momentum ellipsoid. Due to numerical errors, however, the computed value of ω may lie on an ellipsoid with a slightly different value of H^2, corresponding to an error in the angular momentum integral

$$\Delta P = H^2 - H_0^2 \tag{6.478}$$

The correction $\Delta\omega$ is made in a direction normal to the momentum ellipsoid at the operating point, that is, in the direction of the gradient. This minimizes the magnitude of the correction $\Delta\omega$ resulting in the required correction ΔH in angular momentum magnitude. Thus, we take

$$\Delta\omega = K\nabla P \tag{6.479}$$

where the gradient of P in ω-space is

$$\nabla P = \frac{\partial P}{\partial\omega_x}\mathbf{i} + \frac{\partial P}{\partial\omega_y}\mathbf{j} + \frac{\partial P}{\partial\omega_z}\mathbf{k}$$
$$= 2I_{xx}^2\omega_x\mathbf{i} + 2I_{yy}^2\omega_y\mathbf{j} + 2I_{zz}^2\omega_z\mathbf{k} \tag{6.480}$$

The constant K is chosen to provide an exact cancellation of the error if the linearization is valid. Thus, we obtain

$$\nabla P \cdot \Delta\omega = K(\nabla P)^2 = -\Delta P \tag{6.481}$$

which results in

$$K = \frac{-\Delta P}{4\left(I_{xx}^4\omega_x^2 + I_{yy}^4\omega_y^2 + I_{zz}^4\omega_z^2\right)} \tag{6.482}$$

Then, using (6.479), the individual corrections are

$$\Delta\omega_x = 2KI_{xx}^2\omega_x$$
$$\Delta\omega_y = 2KI_{yy}^2\omega_y \tag{6.483}$$
$$\Delta\omega_z = 2KI_{zz}^2\omega_z$$

It is interesting to observe that, for a rigid body rotating in free space, there is also conservation of kinetic energy. This can be expressed as

$$2T = I_{xx}\omega_x^2 + I_{yy}\omega_y^2 + I_{zz}\omega_z^2 = 2E_0 \tag{6.484}$$

which is the equation of an ellipsoid in body-fixed ω-space. It is called the *energy ellipsoid*. Since both the angular momentum and the kinetic energy are conserved, the angular velocity must satisfy (6.477) and (6.484). In other words, the point of the ω-vector must move on the curve defined by the intersection of the momentum and energy ellipsoids. This closed curve in body-fixed ω-space is called a *polhode*. Thus, for general initial conditions, the path of the ω-vector relative to body axes is periodic. On the other hand, the path of ω in inertial space is not periodic in the general case of rigid body motion.

Example 6.9 Consider the planar motion of a solid cylinder of mass m and radius r as it moves in the xy-plane (Fig. 6.11). An inextensible string is wound around a cylinder and the straight portion OP has a variable length l. We wish to obtain the differential equations of motion and establish correction procedures based on the conservation of energy and angular momentum.

First, notice that the angular velocity of the cylinder is

$$\omega = \dot{\theta} + \frac{l}{r} \tag{6.485}$$

where $\dot{\theta}$ is the angular velocity of the unit vectors $\mathbf{e}_l$, $\mathbf{e}_\theta$, and the second term on the right is the angular velocity of the cylinder relative to the unit vectors. The velocity $\mathbf{v}$ of the center of the cylinder is equal to the velocity of the tangent point P, fixed in the string, plus the velocity of C relative to P. Thus, we obtain

$$\mathbf{v} = r\omega\mathbf{e}_l + l\dot{\theta}\mathbf{e}_\theta$$
$$= (\dot{l} + r\dot{\theta})\mathbf{e}_l + l\dot{\theta}\mathbf{e}_\theta \tag{6.486}$$

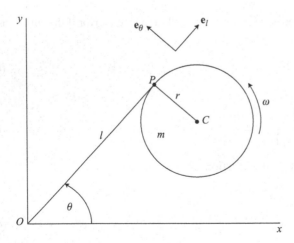

Figure 6.11.

Since the moment of inertia of the cylinder about its center is $I = \frac{1}{2}mr^2$, we find that the kinetic energy is

$$T = \frac{1}{2}m[(\dot{l} + r\dot{\theta})^2 + l^2\dot{\theta}^2] + \frac{1}{4}mr^2\left(\dot{\theta} + \frac{\dot{l}}{r}\right)^2$$

$$= \frac{3}{4}m\dot{l}^2 + \frac{1}{2}m\left(l^2 + \frac{3}{2}r^2\right)\dot{\theta}^2 + \frac{3}{2}mr\dot{l}\dot{\theta} \tag{6.487}$$

There is no potential energy, so Lagrange's equation has the form

$$\frac{d}{dt}\left(\frac{\partial T}{\partial \dot{q}_i}\right) - \frac{\partial T}{\partial q_i} = 0 \tag{6.488}$$

This leads to the following l equation:

$$\frac{3}{2}m\ddot{l} + \frac{3}{2}mr\ddot{\theta} - ml\dot{\theta}^2 = 0 \tag{6.489}$$

The θ equation is

$$m\left(l^2 + \frac{3}{2}r^2\right)\ddot{\theta} + \frac{3}{2}mr\ddot{l} + 2ml\dot{l}\dot{\theta} = 0 \tag{6.490}$$

These two equations can be solved for the individual accelerations, resulting in

$$\ddot{l} = \frac{1}{l}\left[\frac{2}{3}\left(l^2 + \frac{3}{2}r^2\right)\dot{\theta}^2 + 2rl\dot{\theta}\right] \tag{6.491}$$

$$\ddot{\theta} = -\frac{1}{l}(2\dot{l}\dot{\theta} + r\dot{\theta}^2) \tag{6.492}$$

These equations are then integrated numerically to obtain l and θ as functions of time.

The accuracy of the computations can be improved by using conservation of energy and of angular momentum. Conservation of energy is expressed by the equation

$$E = \frac{3}{4}m\dot{l}^2 + \frac{1}{2}m\left(l^2 + \frac{3}{2}r^2\right)\dot{\theta}^2 + \frac{3}{2}mr\dot{l}\dot{\theta} = E_0 \tag{6.493}$$

where E_0 is a constant evaluated from initial conditions. Similarly, the conservation of angular momentum about O is expressed as

$$P = m\left(l^2 + \frac{3}{2}r^2\right)\dot{\theta} + \frac{3}{2}mr\dot{l} = P_0 \tag{6.494}$$

Because the motion is planar, this is a scalar equation. For this reason, we can correct magnitude errors directly, rather than using the square of the magnitude, as we did in the case of three-dimensional rotations. Note that P is also the generalized momentum p_θ.

The errors in energy and angular momentum are

$$\Delta E = E - E_0 \tag{6.495}$$

$$\Delta P = P - P_0 \tag{6.496}$$

where E and P are evaluated numerically at each time step of the computation. Let us choose the velocity corrections in accordance with

$$\Delta \dot{q} = K_1 \nabla E + K_2 \nabla P \tag{6.497}$$

where the gradients are taken in velocity space. The Ks are chosen such that

$$\nabla E \cdot \Delta \dot{q} = -\Delta E \tag{6.498}$$

$$\nabla P \cdot \Delta \dot{q} = -\Delta P \tag{6.499}$$

Thus, we obtain

$$K_1 (\nabla E)^2 + K_2 \nabla E \cdot \nabla P = -\Delta E \tag{6.500}$$

$$K_1 \nabla E \cdot \nabla P + K_2 (\nabla P)^2 = -\Delta P \tag{6.501}$$

The solution for the Ks yields

$$K_1 = \frac{1}{\Delta} [(\nabla E \cdot \nabla P) \Delta P - (\nabla P)^2 \Delta E] \tag{6.502}$$

$$K_2 = \frac{1}{\Delta} [(\nabla E \cdot \nabla P) \Delta E - (\nabla E)^2 \Delta P] \tag{6.503}$$

where the determinant of the coefficients is

$$\Delta = (\nabla E)^2 (\nabla P)^2 - (\nabla E \cdot \nabla P)^2 \tag{6.504}$$

Let us choose a velocity space with components $\dot{l}$ and $r\dot{\theta}$ in order to have dimensional homogeneity. Then, we find that

$$\frac{\partial E}{\partial \dot{l}} = \frac{3}{2} m (\dot{l} + r\dot{\theta}) \tag{6.505}$$

$$\frac{\partial E}{\partial (r\dot{\theta})} = \frac{3}{2} m \left[\dot{l} + \left(1 + \frac{2l^2}{3r^2} \right) (r\dot{\theta}) \right] \tag{6.506}$$

$$(\nabla E)^2 = \frac{9}{2} m^2 \dot{l}^2 + \frac{9}{2} m^2 \dot{l} r\dot{\theta} \left(2 + \frac{2l^2}{3r^2} \right) + \frac{9}{4} m^2 (r\dot{\theta})^2 \left(2 + \frac{4l^2}{3r^2} + \frac{4l^4}{9r^4} \right) \tag{6.507}$$

Similarly,

$$\frac{\partial P}{\partial \dot{l}} = \frac{3}{2} mr \tag{6.508}$$

$$\frac{\partial P}{\partial (r\dot{\theta})} = \frac{3}{2} mr \left(1 + \frac{2l^2}{3r^2} \right) \tag{6.509}$$

$$(\nabla P)^2 = \frac{9}{4} m^2 r^2 \left(2 + \frac{4l^2}{3r^2} + \frac{4l^4}{9r^4} \right) \tag{6.510}$$

Also,

$$\nabla E \cdot \nabla P = \frac{9}{4}m^2 r \left[\left(2 + \frac{2l^2}{3r^2} \right) l + \left(2 + \frac{4l^2}{3r^2} + \frac{4l^4}{9r^4} \right) r\dot\theta \right] \tag{6.511}$$

Then, substituting into (6.504), we obtain

$$\Delta = \left(\frac{3m^2 l^2 l}{2r} \right)^2 \tag{6.512}$$

The corrections obtained by using (6.497) are

$$\Delta l = \frac{3}{2}mK_1(l + r\dot\theta) + \frac{3}{2}mr K_2 \tag{6.513}$$

$$\Delta(r\dot\theta) = \frac{3}{2}mK_1 \left[l + \left(1 + \frac{2l^2}{3r^2} \right) r\dot\theta \right] + \frac{3}{2}mr K_2 \left(1 + \frac{2l^2}{3r^2} \right) \tag{6.514}$$

In this example, we chose the straightforward procedure of obtaining second-order equations (6.491) and (6.492) which are integrated numerically. The accuracy of the integrations can be improved by using conservation of energy and angular momentum. But now let us consider a *second approach*.

We note that θ is an ignorable coordinate, and therefore the corresponding generalized momentum is constant in value.

$$\frac{\partial T}{\partial \dot\theta} = m \left(l^2 + \frac{3}{2}r^2 \right) \dot\theta + \frac{3}{2}mrl = \beta_\theta \tag{6.515}$$

where β_θ is a constant. This equation can be solved for $\dot\theta$, giving the result

$$\dot\theta = \frac{\beta_\theta - \frac{3}{2}mrl}{m \left(l^2 + \frac{3}{2}r^2 \right)} \tag{6.516}$$

Upon substituting this expression for $\dot\theta$ into (6.491), we obtain a single differential equation of motion, namely,

$$\ddot{l} = \frac{4\beta_\theta^2 - 9m^2 r^2 l^2}{6m^2 l \left(l^2 + \frac{3}{2}r^2 \right)} \tag{6.517}$$

Numerical integration of (6.517) results in values of $\dot l$ and l as functions of time. For a known $\dot l$, the value of $\dot\theta$ is obtained from (6.516), and this can be integrated numerically to obtain θ as a function of time.

Equation (6.517) can also be obtained by using the Routhian method. The conservation of angular momentum is inherent in the formulation, but the conservation of energy can be used for error correction. By using (6.487) and (6.516), the total energy of the system can be written in the form

$$E = \frac{3m^2 l^2 \dot{l}^2 + 2\beta_\theta^2}{4m\left(l^2 + \frac{3}{2}r^2\right)} \tag{6.518}$$

For an energy error ΔE, the velocity correction is found from

$$\nabla E \cdot \Delta \dot{q} = -\Delta E \tag{6.519}$$

which results in the correction

$$\Delta \dot{l} = -\frac{\Delta E}{\left(\frac{\partial E}{\partial \dot{l}}\right)} = -\frac{2\left(l^2 + \frac{3}{2}r^2\right)}{3m l^2 \dot{l}} \Delta E \tag{6.520}$$

This correction should not be applied if $\dot{l}$ is very small compared to its maximum value.

6.7 Bibliography

Baumgarte, J. W. A new method of stabilization for holonomic constraints. *Journal of Applied Mechanics*, 1983, Vol. 50, pp. 869–70.

Burden, R. L. and Faires, J. D. *Numerical Analysis*, 3rd edn. Boston: PWS Publishers, 1985.

Dahlquist, G. and Bjorck, A. *Numerical Methods*. Englewood Cliffs, NJ: Prentice Hall, 1974.

Hamming, R. W. *Numerical Methods for Scientists and Engineers*, 2nd edn. New York: Dover Publications, 1973.

6.8 Problems

6.1. The differential equation $\dot{y} + y = 0$ is to be integrated numerically using the modified Euler method. Use a step size $h = 0.1$ and let $y_n = 1$. (a) Solve for the actual numerical error after one step. Compare this result with the value of E_{n+1} obtained from (6.86). (b) Using the mean value theorem to obtain E_{n+1}, solve for the value of ξ, where $0 < \xi < 0.1$ and $t_n = 0$.

6.2. Given a second-order differential equation of the form $\ddot{y} = f(y, t)$. Suppose it is integrated using the algorithm

$$y_{n+1} = 2y_n - y_{n-1} + \frac{h^2}{12}(\ddot{y}_{n+1} + 10\ddot{y}_n + \ddot{y}_{n-1})$$

(a) Find the first error term. (b) Consider the differential equation $\ddot{y} + \omega_0^2 y = 0$, where ω_0 is a positive constant. The given algorithm is used to integrate this equation. Analyze the numerical stability by obtaining the corresponding characteristic equation. Determine the range of step size h for numerical instability.

6.3. An extrapolation formula

$$v_{n+\frac{1}{2}} = \frac{1}{h}(ay_n + by_{n-1} + cy_{n-2})$$

is used to estimate the velocity $v \equiv \dot{y}$ at time $t_{n+\frac{1}{2}}$. The step size h is uniform and the values of y_n, y_{n-1} and y_{n-2} are assumed to be known. (a) Solve for the coefficients a, b, and c. (b) Find the first error term.

6.4. Given the system

$$\dot{x} = -x + y$$
$$\dot{y} = -3y$$

Suppose that Euler integration is used to obtain a solution. (a) Obtain the characteristic equation for analyzing numerical stability. (b) Solve for the roots and give the range of step size h for stability.

6.5. Two thin uniform rods, each of mass m and length l, are connected by a pivot at A. There is a fixed pivot at O, and end B can slide freely on a rigid horizontal wire through O. (a) Derive the differential equation of motion. (b) Let $l = 1$ m, $m = 1$ kg, $g = 9.8$ m/s^2 and assume the initial conditions $\theta(0) = 0$, $\omega(0) = 0$ where $\omega \equiv \dot{\theta}$. Tabulate numerical values of θ, ω, and $\dot{\omega}$ as functions of time for $0 \leq t \leq 1.5$ s. (c) Find: (1) ω_{max} and the corresponding time t and angle θ; (2) the time when $\theta = \pi/2$; (3) θ_{max} and its time, assuming that B can pass through O.

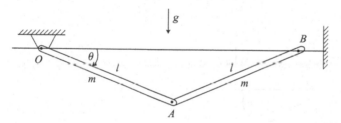

Figure P 6.5.

6.6. A particle of unit mass is constrained to follow the elliptical line of intersection of a cone and a plane, as shown in Fig. 6.8 and discussed in Example 6.5 on page 376. The equation of the cone represents the holonomic constraint

$$\phi_1 = x^2 + y^2 - \frac{1}{4}z^2 = 0$$

whereas the plane represents the second constraint

$$\phi_2 = x + y + z - 1 = 0$$

Using units of meters and seconds, let $g = 9.8$ m/s^2 and choose the initial conditions $x(0) = y(0) = -\frac{1}{2}(1 + \sqrt{2})$, $z(0) = 2 + \sqrt{2}$, $\dot{x}(0) = 1$, $\dot{y}(0) = -1$, $\dot{z}(0) = 0$. (a) Starting with the differential equations in Lagrange multiplier form, use numerical integration and the Baumgarte stabilization method with $\alpha = 2/h$ and

$\beta = 1/h^2$ to obtain the motion of the system for $0 \le t \le 2$. Tabulate the values of $x, y, z, v_x, v_y, v_z, \phi_1, \phi_2$ and find the times of z_{min} and z_{max}. (b) Repeat the calculations, but use the one-step method of stabilization. (c) Repeat step (b) but add a one-step energy correction.

6.7. Consider a dumbbell with a knife-edge constraint sliding on the horizontal xy-plane, as discussed in Examples 6.4 and 6.6, on pages 370 and 380 respectively, and as shown in Fig. 6.6. Choose (x, y, ϕ) as qs and let $v_x \equiv \dot{x}, v_y \equiv \dot{y}, \omega \equiv \dot{\phi}$. (a) Derive the six first-order differential equations of motion. (b) Assume $m = 1, l = 2$, and the initial conditions are $x(0) = 0, y(0) = 0, \phi(0) = 0, v_x(0) = 0, v_y(0) = 0, \omega(0) = 1$. The nonholonomic constraint equation is $g_1 = -\dot{x}\sin\phi + \dot{y}\cos\phi = 0$. Use the one-step constraint stabilization method. Find the numerical values of $x, y, \phi, v_x, v_y, \omega$, and g_1, at the times $t = 0.1, t = 1.0$, and $t = 10$.

6.8. A pendulum of particle mass m and length l is forced to rotate about a vertical axis at a constant rate Ω. The pendulum angle θ is measured downward from horizontal. (a) Derive the differential equation for θ. (b) Assume that $g/l = 4 \text{ s}^{-2}, \Omega = \sqrt{10}$ rad/s and the initial conditions are $\theta(0) = 0, \omega(0) = 0$ where $\omega \equiv \dot{\theta}$. Integrate the differential equation numerically using RK-4 with $h = 0.001$ s. Tabulate values of θ, ω, and $\dot{\omega}$ over the time interval $0 \le t \le 1.5$ s. (c) Find θ_{max} and ω_{max}, giving the time when each occurs.

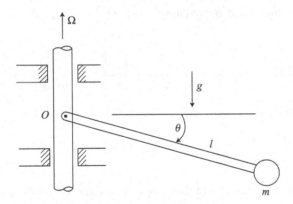

Figure P 6.8.

6.9. A rigid body has an inertia matrix

$$I = \begin{bmatrix} 1000 & 0 & 0 \\ 0 & 2500 & 0 \\ 0 & 0 & 3000 \end{bmatrix} \text{ kg·m}^2$$

with respect to its center of mass. Euler parameters are used to specify the orientation. A constant body-axis moment $M_z = 15\,000$ N·m is applied to the body. The initial conditions are $\omega_x(0) = 15$ rad/s, $\omega_y(0) = 0, \omega_z(0) = 0, \epsilon_x(0) = 0, \epsilon_y(0) = 0$,

$\epsilon_z(0) = 0$, $\eta(0) = 1$. (a) Use Euler's dynamical equations and the $(\dot{\epsilon}, \dot{\eta})$ kinematical equations to obtain numerical values of $\omega_x, \omega_y, \omega_z, \epsilon_x, \epsilon_y, \epsilon_z, \eta$ by integrating over the time interval $0 \le t \le 1$ s. For times within this interval, calculate the work W done by M_z and also the angular deviation α of the x-axis from its original orientation, where $\alpha = \cos^{-1}(1 - 2\epsilon_y^2 - 2\epsilon_z^2)$. (b) Find W_{max} and the time at which it occurs. (c) Determine α_{max} and the time of occurrence. (d) Find the time at which $\omega_x, \omega_y, \omega_z$ simultaneously return to their original values.

6.10. A thin uniform disk of mass m and radius r rolls without slipping on a horizontal surface. The generalized coordinates are $(\psi, \theta, \phi, X, Y)$ where (ψ, θ, ϕ) are type I Euler angles, and (X, Y) is the location of the contact point. The contact point has a velocity $r\dot{\phi}$ at an angle ψ with the positive X-axis, that is, $\dot{X} = r\dot{\phi}\cos\psi$ and $\dot{Y} = -r\dot{\phi}\sin\psi$. Choose $(\dot{\psi}, \dot{\theta}, \Omega)$ as us where $\Omega = \dot{\phi} - \dot{\psi}\sin\theta$ is the angular velocity of the disk about its symmetry axis. Assume that $g/r = 20$ s^{-2}, where $g = 9.81$ m/s^2 and $r = 0.4905$ m. (a) Derive the three differential equations of motion, that is, equations for $\dot{u}$s in terms of qs and us. (b) Assume the initial conditions $\psi(0) = 0$, $\theta(0) = \pi/6$, $\phi(0) = 0$, $X(0) = 0$, $Y(0) = 0$, $\dot{\psi}(0) = 4$ rad/s, $\dot{\theta}(0) = 0$, $\Omega(0) = 2$ rad/s. Use numerical integration to obtain the values of the qs and us over the interval $0 \le t \le 3.5$ s. Find θ_{min} and the following θ_{max} with their corresponding times, as well as the times when $\theta = 0$. Show that the motion of θ is periodic, but the path of the contact point on the XY-plane is not periodic.

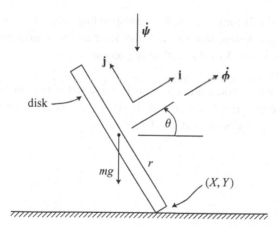

Figure P 6.10.

6.11. A dumbbell consists of two particles, each of mass m, connected by a massless rod of length l (Fig. P 6.11). Given that $m = 1$ kg, $l = 1$ m, and $g = 9.8$ m/s^2. Choose (x, θ) as generalized coordinates and assume the initial conditions $x(0) = \frac{1}{2}$ m, $\dot{x}(0) = 0$, $\theta(0) = 0$, $\dot{\theta}(0) = \frac{1}{2}$ rad/s. As the dumbbell falls due to gravity and slides without friction, particle A hits the wall inelastically. Then it slides down the wall until it finally leaves and moves to the right. Use the notation $v \equiv \dot{x}$, $\omega \equiv \dot{\theta}$. Consider three phases of the motion.

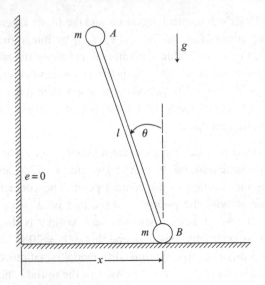

Figure P 6.11.

Phase 1 – before impact   (a) Derive the four first-order differential equations of motion. (b) Integrate these equations numerically and tabulate values of x, θ, v, ω, at each 0.1 s from zero to just before the time of impact.

Phase 2 – impact   (c) Find the time of impact and the corresponding values of x and θ. (d) What are the values of $\dot{x}$ and $\dot{\theta}$ just before impact. (e) Solve for $\dot{x}$ and $\dot{\theta}$ immediately after impact and also the horizontal impulse $\hat{\lambda}$ of the wall acting on particle A.

Phase 3 – after impact   (f) Given the constraint equation $\dot{x} - l\dot{\theta}\cos\theta = 0$, obtain expressions for $\ddot{x}$ and $\ddot{\theta}$ which are to be integrated numerically. (g) Find the time at which particle A leaves the wall and give the corresponding values of x, θ, v, and ω.

Appendix

A.1 Answers to problems

Chapter 1

1.1. (a) $\dot{x} = m\sqrt{\dfrac{2gr}{m_0(m + m_0)}}, \qquad \dot{\theta} = -\sqrt{\dfrac{2g(m + m_0)}{m_0 r}}$

   (b) $x_{\max} = \dfrac{2mr}{m + m_0}$

1.2. $\theta_{\max} = 120°$

1.3. $\mathbf{v}_1 = -\dfrac{1}{5}\, v_0 \mathbf{i}, \qquad \mathbf{v}_2, \mathbf{v}_3 = \dfrac{v_0}{5}\,(3\mathbf{i} \pm \sqrt{3}\mathbf{j})$

1.4. $v = \dfrac{1}{3}\, v_0$, all particles

1.5. $\rho = \dfrac{v_0^2 \cos^2 \theta_0}{g \cos^3 \theta}$

1.6. 1.79 cm higher on west bank

1.7. $m\ddot{l} + kl - ml\dot{\theta}^2 - mg\cos\theta = kL$
   $ml\ddot{\theta} + 2m\dot{l}\dot{\theta} + mg\sin\theta = 0$

1.8. (a) $v = v_0\left(1 - e^{-\frac{c}{m}t}\right), \qquad$ (b) $W = mv_0^2\left(1 - e^{-\frac{c}{m}t}\right),$
   (c) $W = mv_0^2 = 2T,$

1.9. $\hat{F}_i = c\Delta l$

1.10. (a) $v_{10} = 0.6302\, v_0, \qquad$ (b) $v_9 = 0.03317 v_0$

1.11. (a) $\dot{x} = -\dfrac{1}{3}\, v_0, \qquad \dot{\theta} = -\dfrac{2\sqrt{2}\, v_0}{3l}$

   (b) $\hat{Q}_y = \dfrac{4}{3} m v_0$

1.12. (a) $\dot{x} = \dfrac{v_0}{\sqrt{2}}$, $\qquad \dot{y} = \dfrac{v_0}{3\sqrt{2}}$, $\qquad \omega = \dfrac{4v_0}{3l}$

(b) $\hat{F} = \dfrac{2}{3}mv_0$

1.13. (a) $\omega = \dfrac{v_0}{2\sqrt{2}l}$, $\qquad$ (b) $\mathbf{v}_A = \dfrac{1}{4}v_0\mathbf{i}$

1.14. (a) $l\ddot{\phi} + r\omega_0^2 \sin\phi = 0$, $\qquad$ (b) $\phi_{\max} = 60°$

1.15. $\Delta\omega = \dfrac{2r\hat{F}}{m(r^2 + l^2 - 2rl\cos\theta)}$, $\qquad \hat{R}_h = \dfrac{(r^2 - l^2)\hat{F}}{r^2 + l^2 - 2rl\cos\theta}$

$\hat{R}_v = \dfrac{2rl\sin\theta\,\hat{F}}{r^2 + l^2 - 2rl\cos\theta}$

1.16. (a) $\dot{\theta} = \dfrac{2\hat{F}}{(5 - \sqrt{3}\mu)ml}$, $\qquad \dot{x} = \dfrac{2(1 - \sqrt{3}\mu)\hat{F}}{3(5 - \sqrt{3}\mu)m}$,

(b) $\mu_{\min} = \dfrac{1}{\sqrt{3}}$

1.17. (a) $2mr\ddot{\theta} = mg(1 - \mu)\sin\theta - mg(1 + \mu)\cos\theta + 2\mu mr\dot{\theta}^2$

(b) $P = \dfrac{1}{2}mg[(1 + \mu)\sin\theta + (1 - \mu)\cos\theta]$

1.18. $v_0 = \dfrac{ch}{m}\left(1 - \dfrac{\pi}{4}\right)$

1.19. (a) $\ddot{x} = \dfrac{l\dot{\theta}^2\cos\theta}{1 + \cos^2\theta}$, $\qquad \ddot{\theta} = \dfrac{\dot{\theta}^2\sin\theta\cos\theta}{1 + \cos^2\theta}$, $\qquad P = \dfrac{ml\dot{\theta}^2}{1 + \cos^2\theta}$

(b) $\dot{x} = \dfrac{v_0\sin\theta}{\sqrt{2(1 + \cos^2\theta)}}$, $\qquad \dot{\theta} = \sqrt{\dfrac{2}{1 + \cos^2\theta}}\dfrac{v_0}{l}$

1.20. (a) $m_1\ddot{x}_1 = k(x_2 - x_1)\left(1 - \dfrac{L}{l}\right) + F$

$m_1\ddot{y}_1 = k(y_2 - y_1)\left(1 - \dfrac{L}{l}\right)$

$m_2\ddot{x}_2 = k(x_1 - x_2)\left(1 - \dfrac{L}{l}\right)$

$m_2\ddot{y}_2 = k(y_1 - y_2)\left(1 - \dfrac{L}{l}\right)$

where $l = \sqrt{(x_2 - x_1)^2 + (y_2 - y_1)^2}$

(b) $m_1\ddot{x}_1 = k(x_2 - x_1)\left(1 - \dfrac{L}{l}\right) + F$

$$m_1 \ddot{y}_1 = k(x_2 - y_1)\left(1 - \frac{L}{l}\right)$$

$$m_2 \ddot{x}_2 = -\frac{1}{2}k(2x_2 - x_1 - y_1)\left(1 - \frac{L}{l}\right)$$

1.21. (a) $\theta_{max} = \tan^{-1} 1.1732 = 49.56°$

(b) $\theta = 34.86°$

1.22. (a) Upward $\ddot{y} = \dfrac{\dot{y}^2 - gy}{L - y}$

Downward $\ddot{y} = \left(\dfrac{2L - 3y}{3L - 3y}\right)g$

1.23. (a) $r\ddot{\theta} - r\omega_0^2 \sin\theta = 0$, (b) $\dot{\theta} = \omega_0$,

(c) $F_r = \dfrac{1}{2}mr\omega_0^2$

Chapter 2

2.1. (a) $m\ddot{x} = 2x\lambda$, $m\ddot{y} + mg = \dfrac{8}{3}y\lambda$

(b) $x_{max} = 0.9063L$, $P = 1.077mg$

2.2. (a) $m(2\ddot{x} - l\ddot{\phi}\sin\phi - l\dot{\phi}^2\cos\phi) = \mp \mu mg\sin\phi$

$m(2\ddot{y} + l\ddot{\phi}\cos\phi - l\dot{\phi}^2\sin\phi) = \pm \mu mg\cos\phi$

$ml(l\ddot{\phi} - \ddot{x}\sin\phi + \ddot{y}\cos\phi) = 0$ or $\dfrac{ml^2}{2}\ddot{\phi} = \mp\dfrac{1}{2}\mu mgl$

(b) $|(\dot{x}\cos\phi + \dot{y}\sin\phi)\dot{\phi}| > \mu g$

2.3. (a) $m\ddot{x} = 2x\lambda$, $m\ddot{y} = -\lambda - mg$, $\ddot{x} = -\dfrac{2x}{4x^2 + 1}(2\dot{x}^2 + g)$

(b) $\dot{x} = -\sqrt{\dfrac{2g\left(x_0^2 - x^2\right)}{4x^2 + 1}}$

2.4. (a) $\dot{\phi} = \sqrt{1 + \dfrac{2r}{L}\cos\phi + \dfrac{r^2}{L^2}}\,\omega$, (b) $\dot{\theta} = \sqrt{1 + \dfrac{r^2}{l^2}}\,\omega$

2.5. (a) $m(L\cos\phi - R\theta)\ddot{\theta} - mR\dot{\theta}^2 - 2mL\dot{\theta}\dot{\phi}\sin\phi = 0$

$mL^2\ddot{\phi} + mL\dot{\theta}^2(L\cos\phi - R\theta)\sin\phi - mgL\cos\phi = 0$

(b) $F = m[L\dot{\phi}^2 + (L\cos\phi - R\theta)\dot{\theta}^2\cos\phi + g\sin\phi]$

2.6. (a) $mR^2\ddot{\theta} + mR^2(\dot{\phi} + \Omega)^2\sin\theta\cos\theta + cR^2\dot{\theta} = 0$

$mR^2\ddot{\phi}\cos^2\theta - 2mR^2(\dot{\phi} + \Omega)\dot{\theta}\sin\theta\cos\theta + cR^2\dot{\phi}\cos^2\theta = 0$

2.7. $m\left[2\ddot{x} - \dfrac{l}{\sqrt{2}}\ddot{\phi}(\sin\phi + \cos\phi) + \dfrac{l\dot{\phi}^2}{\sqrt{2}}(\sin\phi - \cos\phi)\right] = -\lambda\sin\phi$

$m\left[2\ddot{y} + \dfrac{l}{\sqrt{2}}\ddot{\phi}(\cos\phi - \sin\phi) - \dfrac{l\dot{\phi}^2}{\sqrt{2}}(\sin\phi + \cos\phi)\right] = \lambda\cos\phi$

$ml\left[l\ddot{\phi} - \dfrac{\ddot{x}}{\sqrt{2}}(\sin\phi + \cos\phi) + \dfrac{\ddot{y}}{\sqrt{2}}(\cos\phi - \sin\phi)\right] = 0$

2.8. $m[2\ddot{x} - (l_o + s)\ddot{\phi}\sin\phi + \ddot{s}\cos\phi - (l_0 + s)\dot{\phi}^2\cos\phi - 2\dot{\phi}\dot{s}\sin\phi] = -\lambda\sin\phi$

$m[2\ddot{y} + (l_0 + s)\ddot{\phi}\cos\phi + \ddot{s}\sin\phi - (l_0 + s)\dot{\phi}^2\sin\phi + 2\dot{\phi}\dot{s}\cos\phi] = \lambda\cos\phi$

$m[(l_0 + s)^2\ddot{\phi} - (l_0 + s)\ddot{x}\sin\phi + (l_0 + s)\ddot{y}\cos\phi + 2(l_0 + s)\dot{\phi}\dot{s}] = 0$

$m[\ddot{s} + \ddot{x}\cos\phi + \ddot{y}\sin\phi - (l_0 + s)\dot{\phi}^2] + ks = 0$

2.9. (a) $\dot{x} = \dfrac{2}{3}v_0,$ (b) $\hat{P} = \dfrac{1}{3}\sqrt{2}\,mv_0$

2.10. (a) $2ml^2\ddot{\theta}\left(1 - \dfrac{2}{3}\sin^2\theta\right) - \dfrac{4}{3}ml^2\dot{\theta}^2\sin\theta\cos\theta - \dfrac{ml^2\omega_0^2\sin\theta\cos\theta}{3\left(1 - \dfrac{2}{3}\cos^2\theta\right)^2} = 0$

(b) $\dot{\theta} = \dfrac{1}{2}\sqrt{3}\,\omega_0$

2.11. $mr\ddot{\theta} - mr\Omega^2\sin\theta\cos\theta + \mu\sqrt{N_r^2 + N_\phi^2}\,\mathrm{sgn}(\dot{\theta}) = mg\sin\theta$

where $N_r = -m(r\dot{\theta}^2 + r\Omega^2\sin^2\theta) + mg\cos\theta$

$N_\phi = 2mr\Omega\dot{\theta}\cos\theta$

2.12. (a) $ml^2\ddot{\theta} - ml^2(\dot{\phi} + \omega)^2\sin\theta\cos\theta - mRl\omega^2\cos\theta\cos\phi + mgl\sin\theta = 0$

$ml^2\ddot{\phi}\sin^2\theta + 2ml^2\dot{\theta}(\dot{\phi} + \omega)\sin\theta\cos\theta + mRl\omega^2\sin\theta\sin\phi = 0$

(b) $E = \dfrac{1}{2}ml^2(\dot{\theta}^2 + \dot{\phi}^2\sin^2\theta) - \dfrac{1}{2}m\omega^2(R^2 + l^2\sin^2\theta + 2Rl\sin\theta\cos\phi)$

$- mgl\cos\theta$

2.13. (a) $\dot{x} = \dot{x}_0,$ $\dot{y} = v_0$ (b) $\Delta T_2 = \dfrac{1}{2}mv_0^2,$

$\Delta T = \dfrac{1}{2}mv_0^2 + m\Omega x_0 v_0,$ (c) $\hat{C} = mv_0\mathbf{j}$

2.14. (a) $m\ddot{r} - mr\dot{\theta}^2 = \lambda,$ $mr^2\ddot{\theta} + 2mr\dot{r}\dot{\theta} = -k\lambda$

(b) $\dot{r} = \dfrac{kv_0}{\sqrt{r^2 + k^2}},$ $N = \dfrac{(r^2 + 2k^2)mv_0^2}{(r^2 + k^2)^{3/2}}$

(c) $\ddot{r} = -\dfrac{r\dot{r}^2}{r^2 + k^2} - \dfrac{\mu\dot{r}^2(r^2 + 2k^2)}{k(r^2 + k^2)}$

2.15. (a) $\ddot{x} = \dfrac{1}{l^2}[g(\mu(x^2 - y^2) + (1 - \mu^2)xy) - (x - \mu y)(\dot{x}^2 + \dot{y}^2)] - \mu g$

$\ddot{y} = \dfrac{1}{l^2}[g(y + \mu x)^2 - (y + \mu x)(\dot{x}^2 + \dot{y}^2)] - g$

(b) $\ddot{\theta} = \dfrac{g}{l}[(1 - \mu^2)\sin\theta - 2\mu\cos\theta] + \mu\dot{\theta}^2$

2.16. (a) $2m\ddot{r}_1 - 2mr_1\dot{\theta}^2 = r_1\lambda,$ $2m\ddot{r}_2 - 2mr_2\dot{\theta}^2 = r_2\lambda,$ $2ml^2\ddot{\theta} = 0$

$\ddot{r}_1 + \dfrac{r_1}{l^2 - r_1^2}\dot{r}_1^2 = 0,$ $\ddot{r}_2 + \dfrac{r_2}{l^2 - r_2^2}\dot{r}_2^2 = 0,$ $\ddot{\theta} = 0$

(b) $r_1 = l\sin\left(\dfrac{\sqrt{2}\,v_0 t}{l} + \dfrac{\pi}{4}\right),$ $r_2 = l\cos\left(\dfrac{\sqrt{2}\,v_0 t}{l} + \dfrac{\pi}{4}\right),$ $\theta = \omega_0 t$

2.17. (a) At $t = 0$, $\hat{C}_r = mv_0$. At $t = R/v_0$, $\hat{C}_r = -mv_0$

(b) $v = \dfrac{1}{2}R\omega_0$, (c) $W = \dfrac{1}{2}mv_0^2$ (first impulse)

$W = -\dfrac{1}{2}mv_0^2$ (second impulse), $W = -\dfrac{3}{8}mR^2\omega_0^2$

2.18. (a) $\sqrt{3}\,\sin\theta_1 - \sin\theta_2 = 0$

(b) $T = \dfrac{(3 - \sin^4\theta_2)}{2(3 - \sin^2\theta_2)}ml^2\dot{\theta}_2^2,$ $V = mgl(\cos\theta_2 + \sqrt{3 - \sin^2\theta_2})$

(c) $v_B = 1.0824\sqrt{gl}$ downward,   (d) $\ddot{y}_A = 2(\sqrt{2} - 1)g$

2.19. (a) $\hat{N} = 4mv_0,$ $\dot{\theta}(0+) = \dot{\phi}(0+) = -\dfrac{v_0}{l},$ $\dot{x}(0+) = 0,$ $\dot{y}(0+) = \dfrac{1}{3}v_0,$

θ, ϕ, x, y unchanged

(b) $2ml^2\ddot{\theta}\left(1 - \dfrac{2}{3}\sin^2\theta\right) - \dfrac{4}{3}ml^2(\dot{\theta}^2 + \dot{\phi}^2)\sin\theta\cos\theta = 0$

$2ml^2\ddot{\phi}\left(1 - \dfrac{2}{3}\cos^2\theta\right) + \dfrac{8}{3}ml^2\dot{\theta}\dot{\phi}\sin\theta\cos\theta = 0$

(c) $\theta_{\min} = 0,$ $\dot{\phi} = -\dfrac{2v_0}{l}$

2.20. (a) $\ddot{x} = 0,$ $\ddot{y} = 0,$ $\ddot{\theta} = 0,$ $\ddot{\phi} = 0$

(b) $F_{AB} = F_{CD} = \dfrac{1}{2}ml(\dot{\theta}_0 - \dot{\phi}_0)^2,$ $F_{AD} = F_{BC} = \dfrac{1}{2}ml(\dot{\theta}_0 + \dot{\phi}_0)^2$

(c) Period $= \pi/\dot{\phi}_0$

2.21. (a) $\dfrac{ml^2}{2}\ddot{\theta}(1 + \cos^2\theta) + \dfrac{1}{2}ml^2(\dot{\phi}^2 - \dot{\theta}^2)\sin\theta\cos\theta + mgl\cos\theta = 0$

$\dfrac{ml^2}{2}\ddot{\phi}\cos^2\theta - ml^2\dot{\theta}\dot{\phi}\sin\theta\cos\theta = 0$

(b) $\dot{\theta} = -\sqrt{\dfrac{1}{8}\dot{\phi}_0^2 + \sqrt{2}\dfrac{g}{l}}$

(c) $|\dot{\phi}_0|_{\min} = 2.3784\sqrt{\dfrac{g}{l}}$

2.22. (a) $(2\sin^2\theta + 1)\ddot{\theta} + 2\dot{\theta}^2\sin\theta\cos\theta + \dfrac{g}{2r}\cos\theta = 0$

(b) $\theta = 28.26°$,   (c) $v_2 = 0.4608\sqrt{gr}$

Chapter 3

3.1. (a) $\boldsymbol{\omega} = \dfrac{\sqrt{3}}{20}\mathbf{J} + \dfrac{1}{20}\mathbf{K} = \dfrac{\sqrt{3}}{20}\mathbf{j} - \dfrac{1}{20}\mathbf{k}$ (b) $C = \begin{bmatrix} 1 & 0 & 0 \\ 0 & \dfrac{1}{2} & \dfrac{\sqrt{3}}{2} \\ 0 & -\dfrac{\sqrt{3}}{2} & \dfrac{1}{2} \end{bmatrix}$

$\dot{C} = \begin{bmatrix} 0 & \dfrac{1}{20} & -\dfrac{\sqrt{3}}{20} \\ \dfrac{1}{20} & 0 & 0 \\ \dfrac{\sqrt{3}}{20} & 0 & 0 \end{bmatrix}$

3.2. $[(I_{xx}\sin^2\psi + I_{yy}\cos^2\psi)\sin^2\theta + I_{zz}\cos^2\theta]\ddot{\phi} + (I_{xx} - I_{yy})\ddot{\theta}\sin\theta\sin\psi\cos\psi$

$\quad + I_{zz}\ddot{\psi}\cos\theta + [2(I_{xx}\sin^2\psi + I_{yy}\cos^2\psi)\sin\theta\cos\theta - 2I_{zz}\sin\theta\cos\theta]\dot{\phi}\dot{\theta}$

$\quad + 2(I_{xx} - I_{yy})\dot{\phi}\dot{\psi}\sin^2\theta\sin\psi\cos\psi + (I_{xx} - I_{yy})\dot{\theta}^2\cos\theta\sin\psi\cos\psi$

$\quad + [(I_{xx} - I_{yy})\sin\theta(\cos^2\psi - \sin^2\psi) - I_{zz}\sin\theta]\dot{\theta}\dot{\psi} = Q_\phi$

$(I_{xx}\cos^2\psi + I_{yy}\sin^2\psi)\ddot{\theta} + (I_{xx} - I_{yy})\ddot{\phi}\sin\theta\sin\psi\cos\psi$

$\quad - (I_{xx}\sin^2\psi + I_{yy}\cos^2\psi - I_{zz})\dot{\phi}^2\sin\theta\cos\theta$

$\quad + [(I_{xx} - I_{yy})(\cos^2\psi - \sin^2\psi) + I_{zz}]\dot{\phi}\dot{\psi}\sin\theta$

$\quad - 2(I_{xx} - I_{yy})\dot{\theta}\dot{\psi}\sin\psi\cos\psi = Q_\theta$

$I_{zz}\ddot{\psi} + I_{zz}\ddot{\phi}\cos\theta - (I_{xx} - I_{yy})\dot{\phi}^2\sin^2\theta\sin\psi\cos\psi$

$\quad - [(I_{xx} - I_{yy})(\cos^2\psi - \sin^2\psi) + I_{zz}]\dot{\phi}\dot{\theta}\sin\theta$

$\quad + (I_{xx} - I_{yy})\dot{\theta}^2\sin\psi\cos\psi = Q_\psi$

3.3. (a) $\mathbf{v}_1 = \dfrac{v_0}{\sqrt{2}}\left(\mathbf{i} + \dfrac{1-e}{2}\mathbf{j}\right)$, $\mathbf{v}_2 = \left(\dfrac{1+e}{2\sqrt{2}}\right)v_0\mathbf{j}$

(b) $\mathbf{v}_1 = \dfrac{v_0}{\sqrt{2}}\left[\left(1 - \dfrac{1+e}{2}\mu\right)\mathbf{i} + \dfrac{1-e}{2}\mathbf{j}\right]$

$$\mathbf{v}_2 = \frac{v_0}{2\sqrt{2}}(1+e)(\mu\mathbf{i}+\mathbf{j})$$

$$\boldsymbol{\omega}_1 = \boldsymbol{\omega}_2 = \left(\frac{1+e}{2\sqrt{2}\,I}\right)\mu r m v_0 \mathbf{k}$$

3.4. (a) $\omega_x = \frac{5M_y}{4I_t\Omega}\left(1 - \cos\frac{4}{5}\Omega t\right), \qquad \omega_y = \frac{5M_y}{4I_t\Omega}\sin\frac{4}{5}\Omega t$

(b) $t = 10\pi/\Omega$

3.5. (a) $\dot{\psi} = -H\left(\frac{\sin^2\phi}{I_{yy}} + \frac{\cos^2\phi}{I_{zz}}\right)$

$$\dot{\theta} = H\left(\frac{1}{I_{zz}} - \frac{1}{I_{yy}}\right)\cos\theta\sin\phi\cos\phi$$

$$\dot{\phi} = H\left(\frac{1}{I_{xx}} - \frac{\sin^2\phi}{I_{yy}} - \frac{\cos^2\phi}{I_{zz}}\right)\sin\theta$$

(b) $\dot{\psi} = -\frac{H}{I_t}, \qquad \dot{\theta} = 0, \qquad \dot{\phi} = H\left(\frac{1}{I_a} - \frac{1}{I_t}\right)\sin\theta$

3.6. (a) $\left(\frac{3}{2}m + m_0\right)\ddot{x}_1 + \frac{1}{4}m\ddot{x}_2 = F, \qquad \left(\frac{3}{2}m + m_0\right)\ddot{x}_2 + \frac{1}{4}m\ddot{x}_1 = 0$

(b) $x_2 = -\frac{1}{14},$ (c) $P_1 = -\frac{11}{195}F$ (upper face), $\qquad P_2 = \frac{3}{195}F$ (lower face)

3.7. (a) $\omega_x = \frac{123}{190}\omega_0, \qquad \omega_y = \frac{12}{95}\omega_0, \qquad \omega_z = \frac{21}{190}\omega_0$

(b) $T = 0.9711\,ma^2\omega_0^2$

3.8. $\dot{\psi} = \pm\sqrt{\dfrac{mgr\tan\theta}{\dfrac{R}{r}I_a + I_t\sin\theta + mRr}}$

3.9. (a) $m\ddot{x} + mr\ddot{\theta} - mx\dot{\theta}^2 - mg\,\sin\theta = 0$

$$m\left(r^2 + x^2 + \frac{l^2}{12}\right)\ddot{\theta} + mr\ddot{x} + 2mx\dot{x}\dot{\theta} - mg\,(r\sin\theta + x\cos\theta) = 0$$

3.10. (a) $I_t\ddot{\theta} - I_a\Omega\dot{\phi}\sin\theta - I_t\dot{\phi}^2\sin\theta\cos\theta + mgl\sin\theta = 0$

$$I_t\ddot{\phi}\sin\theta + 2I_t\dot{\theta}\dot{\phi}\cos\theta + I_a\Omega\dot{\theta} = 0$$

(b) $\theta_{\min} = 34.63°$

3.11. (a) $(I_a\sin^2\theta + I_t\cos^2\theta)\ddot{\psi} - I_a\ddot{\phi}\sin\theta = Q_\psi = 0$

$$I_a(\ddot{\phi} - \ddot{\psi}\sin\theta) = I_a\dot{\Omega} = Q_\phi$$

(b) Q_ϕ is internal.

3.12. $\left[2mR^2\sin^2\theta + \frac{1}{2}ml^2(\sin^2\theta\sin^2\psi + \cos^2\theta)\right]\ddot{\phi}$

$\qquad + \frac{1}{2}ml^2\ddot{\theta}\sin\theta\sin\psi\cos\psi + \frac{1}{2}ml^2\ddot{\psi}\cos\theta$

$\qquad + (4mR^2 - ml^2\cos^2\psi)\dot{\phi}\dot{\theta}\sin\theta\cos\theta + \frac{1}{2}ml^2\dot{\theta}^2\cos\theta\sin\psi\cos\psi$

$\qquad + ml^2\dot{\phi}\dot{\psi}\sin^2\theta\sin\psi\cos\psi - ml^2\dot{\theta}\dot{\psi}\sin\theta\sin^2\psi = 0$

$\left(2mR^2 + \frac{1}{2}ml^2\cos^2\psi\right)\ddot{\theta} + \frac{1}{2}ml^2\ddot{\phi}\sin\theta\sin\psi\cos\psi$

$\qquad - \left(2mR^2 - \frac{1}{2}ml^2\cos^2\psi\right)\dot{\phi}^2\sin\theta\cos\theta + ml^2\dot{\phi}\dot{\psi}\sin\theta\cos^2\psi$

$\qquad - ml^2\dot{\theta}\dot{\psi}\sin\psi\cos\psi + 2mgR\sin\theta = 0$

$\frac{1}{2}ml^2\left[\ddot{\psi} + \ddot{\phi}\cos\theta - \dot{\phi}^2\sin^2\theta\sin\psi\cos\psi + \dot{\theta}^2\sin\psi\cos\psi\right.$

$\qquad \left. - 2\dot{\phi}\dot{\theta}\sin\theta\cos^2\psi\right] = 0$

3.13. $\dot{x} = \dot{x}_0, \qquad \dot{y} = \dfrac{12a^2\dot{y}_0 + al^2\omega_0}{12a^2 + l^2}, \qquad \omega = \dfrac{12a\dot{y}_0 + l^2\omega_0}{12a^2 + l^2}$

3.14. (a) $\ddot{\theta} = \dfrac{12}{ml^2(1+3\sin^2\theta)}\left(\dfrac{12\beta^2\cos\theta}{ml^2\sin^3\theta} - \dfrac{ml^2}{4}\dot{\theta}^2\sin\theta\cos\theta + \dfrac{mgl}{2}\sin\theta\right)$

$\qquad$ where $\beta = \dfrac{ml^2}{12}\dot{\phi}(0)\sin^2\theta(0) = \dfrac{ml^2}{24}\omega_0$

$\qquad$ (b) $N = -\dfrac{ml\omega_0^2}{10\sqrt{2}} + \dfrac{2}{5}mg$

3.15. (a) $(I_t + ml^2\cos^2\theta)\ddot{\theta} - ml^2\dot{\theta}^2\sin\theta\cos\theta + I_t\dot{\psi}^2\sin\theta\cos\theta + I_a\Omega\dot{\psi}\cos\theta$

$\qquad = -mgl\cos\theta$

$\qquad I_t\ddot{\psi}\cos\theta - 2I_t\dot{\theta}\dot{\psi}\sin\theta - I_a\Omega\dot{\theta} = 0$

$\qquad$ (b) $\theta_{\min} = 30°,\qquad$ (c) $\dot{\psi} = -2\sqrt{\dfrac{2g}{l}}$

3.16. (a) $T = \dfrac{1}{2}(3m + m_0)v^2 + \dfrac{1}{2}\left(\dfrac{mr^2}{2} + \dfrac{3}{4}mL^2 + \dfrac{m_0L^2}{12}\right)\dot{\phi}^2$

$\qquad \mathbf{m} = \begin{bmatrix} (3m + m_0) & 0 \\ 0 & \left(\dfrac{mr^2}{2} + \dfrac{3}{4}mL^2 + \dfrac{m_0L^2}{12}\right) \end{bmatrix}$

$\qquad$ (b) $v = \dfrac{\hat{F}}{3m + m_0}, \qquad \phi = \dfrac{\hat{F}c}{\dfrac{mr^2}{2} + \dfrac{3}{4}mL^2 + \dfrac{m_0L^2}{12}}$

3.17. $\dot{x} = \dfrac{(l\omega_0 - 6v_0 \cos\theta)\sin\theta}{2(1 + 3\cos^2\theta)}$, $\dot{\theta} = \dfrac{l\omega_0 - 6v_0 \cos\theta}{l(1 + 3\cos^2\theta)}$

$\hat{C}_y = \dfrac{m\left(v_0 + \frac{1}{2}l\omega_0 \cos\theta\right)}{1 + 3\cos^2\theta}$

3.18. (a) $u_1 = -\dfrac{2\sqrt{2}\hat{F}}{5m}$, $u_2 = \dfrac{18\hat{F}}{5ml}$

(b) $\hat{C} = \dfrac{2\sqrt{2}}{5}\hat{F}$

3.19. (a) $\omega = \sqrt{\dfrac{g}{R}\left(2 - \dfrac{l}{R}\right)}$,   (b) $\omega = -\sqrt{\dfrac{g}{2R - l}}$

3.20. (a) $\theta_{\max} = 65.07°$,   (b) $(\omega_0)_{\min} = 2\sqrt{\dfrac{g}{r}}$

3.21. $\omega = \dfrac{6v_0}{7b}\mathbf{i} + \dfrac{6v_0}{7a}\mathbf{j}$, $\mathbf{v}_c = -\dfrac{6}{7}v_0\mathbf{k}$

3.22. (a) $\mathbf{v}_1 = v_{1t}\mathbf{e}_t + v_{1l}\mathbf{e}_l$, $\mathbf{v}_2 = v_{2t}\mathbf{e}_t + v_{2l}\mathbf{e}_l$

where $v_{1l} = v_{2l} = \dfrac{-\left(\frac{L}{2} + r\right)^2 v_0}{\frac{4}{5}r^2 + 3\left(\frac{L}{2} + r\right)^2}$

$v_{1t} = \dfrac{\frac{4}{5}r^2 + 4\left(\frac{L}{2} + r\right)^2}{\frac{4}{5}r^2 + 3\left(\frac{L}{2} + r\right)^2}v_0$, $v_{2t} = \dfrac{\frac{4}{5}r^2 v_0}{\frac{4}{5}r^2 + 3\left(\frac{L}{2} + r\right)^2}$

(b) $\hat{M}_1 = \dfrac{\frac{4}{5}r^2\left(\frac{L}{2} + r\right)mv_0}{\frac{4}{5}r^2 + 3\left(\frac{L}{2} + r\right)^2}$,   (c) $\hat{P} = \dfrac{\frac{4}{5}r^2 mv_0}{\frac{4}{5}r^2 + 3\left(\frac{L}{2} + r\right)^2}$

3.23. $\dot{\theta}_1 = \dfrac{v_1}{2\sqrt{2}l} - \dfrac{3v_2}{8l}$, $\dot{\theta}_2 = -\dfrac{v_1}{2l} + \dfrac{5v_2}{4\sqrt{2}l}$

$\dot{\theta}_3 = -\dfrac{v_1}{2\sqrt{2}l} - \dfrac{5v_2}{8l}$

3.24. (a) $\dfrac{ml^2}{3}\ddot{\theta}\sin^2\alpha - \dfrac{\mu ml^2}{3}\dot{\theta}^2\sin\alpha\cos\alpha$

$= \dfrac{1}{2}mgl(\sin\theta\sin\alpha - \mu\cos\theta\cos\alpha)$

(b) $\theta = 48.19°$

3.25. (a) $\mathbf{v}_B = \dfrac{\hat{F}}{m}(-\sin\alpha\,\mathbf{i} + 4\cos\alpha\,\mathbf{j})$, $\overline{m} = \dfrac{m}{1 + 3\cos^2\alpha}$

3.26. $\overline{m}_1 = \dfrac{5}{4}m$, $\overline{m}_2 = \dfrac{5}{16}m$

3.27. $\dfrac{4}{3} ml^2\ddot{\theta}_1 - \dfrac{3}{2} ml\ddot{x}\sin\theta_1 + \dfrac{3}{2} ml\ddot{y}\cos\theta_1 + \dfrac{1}{2} ml^2\ddot{\theta}_2\cos(\theta_2 - \theta_1)$

$$-\dfrac{1}{2} ml^2\dot{\theta}_2^2\sin(\theta_2 - \theta_1) = 0$$

$\dfrac{1}{3} ml^2\ddot{\theta}_2 - \dfrac{1}{2} ml\ddot{x}\sin\theta_2 + \dfrac{1}{2} ml\ddot{y}\cos\theta_2 + \dfrac{1}{2} ml^2\ddot{\theta}_1\cos(\theta_2 - \theta_1)$

$$+\dfrac{1}{2} ml^2\dot{\theta}_1^2\sin(\theta_2 - \theta_1) = 0$$

$2m\ddot{x}\sin\theta_1 - 2\,m\ddot{y}\cos\theta_1 - \dfrac{3}{2} ml\ddot{\theta}_1 - \dfrac{1}{2} ml\ddot{\theta}_2\cos(\theta_2 - \theta_1)$

$$+\dfrac{1}{2} ml\dot{\theta}_2^2\sin(\theta_2 - \theta_1) = 0$$

Chapter 4

4.1. (a) $T = \dfrac{1}{2}m(\dot{X}^2 + \dot{Y}^2 + l^2\dot{\phi}^2\sin^2\theta + l^2\dot{\theta}^2 + 2l\dot{\phi}\dot{X}\cos\phi\sin\theta$

$$+ 2l\dot{\phi}\dot{Y}\sin\phi\sin\theta + 2l\dot{\theta}\dot{X}\sin\phi\cos\theta - 2l\dot{\theta}\dot{Y}\cos\phi\cos\theta)$$

$$+ \dfrac{1}{2}I_{xx}(\dot{\phi}\sin\theta\sin\psi + \dot{\theta}\cos\psi)^2 + \dfrac{1}{2}I_{yy}(\dot{\phi}\sin\theta\cos\psi - \dot{\theta}\sin\psi)^2$$

$$+ \dfrac{1}{2}I_{zz}(\dot{\psi} + \dot{\phi}\cos\theta)^2, \qquad V = mgl\cos\theta$$

(b) $u_4 = -r\dot{\theta}\sin\phi + r\dot{\psi}\cos\phi\sin\theta + \dot{X} = 0$

$u_5 = r\dot{\theta}\cos\phi + r\dot{\psi}\sin\phi\sin\theta + \dot{Y} = 0$

(c) $Q_1 = 0, \qquad Q_2 = mgl\sin\theta, \qquad Q_3 = Q_4 = Q_5 = 0$

$$\Phi_{ij} = \begin{bmatrix} 1 & 0 & 0 \\ 0 & 1 & 0 \\ 0 & 0 & 1 \\ 0 & r\sin\phi & -r\cos\phi\sin\theta \\ 0 & -r\cos\phi & -r\sin\phi\sin\theta \end{bmatrix}$$

4.2. (a) $2m\dot{v}_r - 2mr\Omega^2\cos(\phi - \theta) - ml(\dot{\phi} + \Omega)^2 = 0$

$$ml^2\ddot{\phi} + mlv_r\dot{\phi} + 2ml\Omega v_r + mlr\Omega^2\sin(\phi - \theta) = 0$$

(b) $\dot{E} = \dot{T}_2 - \dot{T}_0 = 0$

4.3. $I\dot{\omega}_1 + I\left(\dfrac{r}{R - r}\right)\omega_2\omega_3 = 0$

$(I + mr^2)\dot{\omega}_2 - I\left(\dfrac{r}{R - r}\right)\omega_1\omega_3 = 0$

$(I + mr^2)\dot{\omega}_3 = mgr\sin\theta$

4.4. (a) $(I + mr^2)\dot{\omega}_x - mr\dot{v}\cos\alpha = mgr\sin\alpha$

$(I + mr^2)\dot{\omega}_y = 0, \qquad I\dot{\omega}_z = 0$

$(m + m_0)\dot{v} - mr\dot{\omega}_x\cos\alpha = 0$

(b) $\dot{x} = r\omega_y, \qquad \dot{y} = -r\omega_x$

4.5. (a) $I\dot{\omega}_r + \left(\dfrac{r}{R-r}\right)I\omega_\phi\omega_z = 0$

$(I + mr^2)\dot{\omega}_\phi - \left(\dfrac{r}{R-r}\right)I\omega_r\omega_z = -mgr$

$(I + mr^2)\dot{\omega}_z = 0$

(b) $\dot{z} = r\omega_\phi, \qquad \omega_\phi = A\cos\sqrt{\dfrac{I}{I+mr^2}}\dot{\phi}t + B\sin\sqrt{\dfrac{I}{I+mr^2}}\dot{\phi}t$

4.6. $(I_a + mr^2\sin^2\theta)\dot{\omega}_1 + mr(l + r\cos\theta)\dot{\omega}_2\sin\theta$

$\qquad + mr(l + r\cos\theta)\omega_2\omega_3\cos\theta + mr^2\omega_1\omega_3\sin\theta\cos\theta = 0$

$\qquad [I_p + mr(2l + r\cos\theta)\cos\theta]\dot{\omega}_2 + mr(l + r\cos\theta)\dot{\omega}_1\sin\theta$

$\qquad + [I_a + mr(l + r\cos\theta)\cos\theta]\omega_1\omega_3$

$\qquad + [I_p + mr(2l + r\cos\theta)\cos\theta]\omega_2\omega_3\cot\theta = 0$

$\qquad (I_p + mr^2 + 2mrl\cos\theta)\dot{\omega}_3 - (I_a + mr^2 + mrl\cos\theta)\omega_1\omega_2$

$\qquad -[(I_p + mr^2 + 2mrl\cos\theta)\cot\theta + mrl\sin\theta]\omega_2^2 - mrl\omega_3^2\sin\theta$

$\qquad\qquad = mgl\sin\theta$

4.7. (a) $(I + mr^2)\dot{\omega}_x - I\Omega\omega_y = -mgr, \qquad I\dot{\omega}_y + I\Omega\omega_x = 0,$

$(I + mr^2)\dot{\omega}_z + mr\Omega^2 x = 0$

(b) $\dot{z} = r\omega_x = \dfrac{-mgr^2}{\Omega\sqrt{I(I+mr^2)}}\sin\sqrt{\dfrac{I}{I+mr^2}}\Omega t$

(c) $v_0 > \dfrac{1}{2}r\Omega$

4.8. $\left[\dfrac{5}{4}mr^2 + m_0(r + l\sin\psi)^2\right]\ddot{\theta} - m_0(r + l\sin\psi)l\cos\psi(\dot{\omega}_d$

$\qquad + \dot{\theta}\omega_d\cot\theta - 2\dot{\theta}\Omega) - \left[\dfrac{1}{4}mr^2 + m_0(r + l\sin\psi)l\sin\psi\right]\omega_d^2\cot\theta$

$\qquad + \left[\dfrac{3}{2}mr^2 + m_0(r + l\sin\psi)(r + 2l\sin\psi)\right]\omega_d\Omega$

$\qquad\qquad = -[mr + m_0(r + l\sin\psi)]g\cos\theta$

$$\left(\frac{1}{4}mr^2 + m_0l^2\cos^2\psi\right)\dot{\omega}_d - m_0(r + l\sin\psi)l\ddot{\theta}\cos\psi$$

$$+ \frac{1}{4}mr^2(\dot{\theta}\omega_d\cot\theta - 2\dot{\theta}\Omega) + m_0l\cos\psi\left[l\omega_d^2\cot\theta\sin\psi\right.$$

$$\left.+ l\dot{\theta}\omega_d\cot\theta\cos\psi - 2l\dot{\theta}\Omega\cos\psi - (r + 2l\sin\psi)\omega_d\Omega\right] = m_0gl\cos\theta\cos\psi$$

$$\left[\frac{3}{2}mr^2 + m_0(r^2 + l^2 + 2rl\sin\psi)\right]\dot{\Omega}$$

$$- \left[mr^2 + m_0(r + l\sin\psi)^2 - m_0l^2\cos^2\psi\right]\dot{\theta}\omega_d$$

$$+ m_0l\cos\psi\left[(r + l\sin\psi)(\omega_d^2 - \dot{\theta}^2) + r(\Omega^2 - \omega_d\Omega\cot\theta)\right] = -m_0gl\sin\theta\cos\psi$$

4.9. $(m_0 + 3m)\dot{v} + m_0l\dot{\theta}^2 = 0$

$$\left(I_c + m_0l^2 + 3mb^2 + \frac{1}{2}mr^2\right)\ddot{\theta} - m_0lv\dot{\theta} = 0$$

4.10. $I\dot{\omega}_1 + Ib\omega_2\omega_3\cos\alpha = 0$

$$(I + mr^2)(\dot{\omega}_2 - b\omega_3^2\sin\alpha) - Ib\omega_1\omega_3\cos\alpha = -mgr\cos\alpha$$

$$(I + mr^2)(\dot{\omega}_3 + b\omega_2\omega_3\sin\alpha) = 0$$

4.11. $(5 + \cos^2\phi)m\dot{v} + mL\ddot{\theta}\sin\phi\cos\phi + mL\dot{\theta}^2\left(\frac{L}{l}\cos^3\phi - \cos^2\phi - 3\right)$

$$+ \frac{mv^2}{l}\sin^2\phi\cos\phi + mv\dot{\theta}\sin\phi\cos\phi\left(1 - \frac{2L}{l}\cos\phi\right) = F$$

$$mL^2\ddot{\theta}(3 + \sin^2\phi) + mL\dot{v}\sin\phi\cos\phi + mLv\dot{\theta}\left(3 + \sin^2\phi - \frac{2L}{l}\sin^2\phi\cos\phi\right)$$

$$+ mL^2\dot{\theta}^2\sin\phi\cos\phi\left(\frac{L}{l}\cos\phi - 1\right) + \frac{mL}{l}v^2\sin^3\phi = 0$$

4.12. (a) $m\left(\dot{v}_\theta + \frac{r}{R}v_\theta\omega_\theta\right) = 0, \qquad I\left(\dot{\omega}_r - \frac{1}{R}v_\theta\omega_\theta\right) = 0$

$$(I + mr^2)\dot{\omega}_\theta - \frac{mr}{R}v_\theta^2 + \frac{I}{R}v_\theta\omega_r = 0, \qquad I\dot{\omega}_z = 0$$

(b) $\mathbf{H}_0 = (I\omega_r - mrv_\theta)\mathbf{e}_r + (I + mr^2)\omega_\theta\mathbf{e}_\theta + (I\omega_z + mRv_\theta)\mathbf{k}$

(c) Final values: $\theta = 113.58°, \qquad v_\theta = 0, \qquad \omega_r = 1, \qquad \omega_\theta = 0.6547$

4.13. (a) $2m\left(1 + \frac{l^2}{4R^2}\right)\dot{v} - ml\left(1 - \frac{l^2}{4R^2}\right)\omega_3^2 = 2mg\sin\theta\sin\phi$

$$ml^2\dot{\omega}_3 + \left(1 - \frac{l^2}{4R^2}\right)mlv\omega_3 = mgl\sin\theta\cos\phi$$

(b) $\dot{\alpha} = \frac{1}{R\sin\theta}\left(v\cos\phi - \frac{1}{2}l\omega_3\sin\phi\right)$

$$\dot{\theta} = -\frac{1}{R}\left(v\sin\phi + \frac{1}{2}l\omega_3\cos\phi\right)$$

$$\dot{\phi} = \left(1 + \frac{l}{2R}\cot\theta\sin\phi\right)\omega_3 - \frac{v}{R}\cot\theta\cos\phi$$

4.14. (a) $I_t\ddot{\phi}\sin\theta + 2I_t\dot{\phi}\dot{\theta}\cos\theta - I_a\Omega\dot{\theta} = mlr\omega_0^2\cos(\phi - \omega_0 t)$

$I_t\ddot{\theta} - I_t\dot{\phi}^2\sin\theta\cos\theta + I_a\Omega\dot{\phi}\sin\theta$
$$= mgl\sin\theta + mlr\omega_0^2\cos\theta\sin(\phi - \omega_0 t)$$
$I_a\dot{\Omega} = 0$

(b) $\dot{E} = mlr\omega_0^2[\dot{\phi}\sin\theta\cos(\phi - \omega_0 t) + \dot{\theta}\cos\theta\sin(\phi - \omega_0 t)]$

(c) $\dot{\phi}_{1,2} = \dfrac{I_a\Omega\tan\theta_0}{2(I_t\sin\theta_0 + mlr)}\left[1 \pm \sqrt{1 - \dfrac{4mgl(I_t\sin\theta_0 + mlr)}{I_a^2\Omega^2\tan\theta_0}}\right]$

4.15. (a) $m\dot{v} - ml\ddot{\theta}\sin\theta - ml\dot{\theta}^2\cos\theta - ml\dot{\phi}^2\cos\theta = 0$

$(I_{xx}\sin^2\theta + I_{zz}\cos^2\theta)\ddot{\phi} + 2(I_{xx} - I_{zz})\dot{\phi}\dot{\theta}\sin\theta\cos\theta + mlv\dot{\phi}\cos\theta = 0$

$I_{yy}\ddot{\theta} - (I_{xx} - I_{zz})\dot{\phi}^2\sin\theta\cos\theta - ml\dot{v}\sin\theta = -mgl\cos\theta$

(b) $\dot{X} = -v\sin\phi, \qquad \dot{Y} = v\cos\phi$

4.16. (a) $(I + mr^2)\dot{\omega}_1 + (I + mr^2)\left(\dfrac{r}{R + r}\right)\omega_1\omega_2\cot\theta$

$$+ I\left(\frac{r}{R + r}\right)\omega_2\omega_3 = 0$$

$(I + mr^2)\dot{\omega}_2 - (I + mr^2)\left(\dfrac{r}{R + r}\right)\omega_1^2\cot\theta - I\left(\dfrac{r}{R + r}\right)\omega_1\omega_3 = mgr\sin\theta$

$I\dot{\omega}_3 = 0$

(b) In terms of $\dot{\phi}, \dot{\theta}, \ddot{\phi}, \ddot{\theta}$, we obtain

$(I + mr^2)\left(\dfrac{R + r}{r}\right)\ddot{\phi}\sin\theta + 2(I + mr^2)\left(\dfrac{R + r}{r}\right)\dot{\phi}\dot{\theta}\cos\theta - I\Omega\dot{\theta} = 0$

$(I + mr^2)\left(\dfrac{R + r}{r}\right)\ddot{\theta} - (I + mr^2)\left(\dfrac{R + r}{r}\right)\dot{\phi}^2\sin\theta\cos\theta + I\Omega\dot{\phi}\sin\theta$
$$= mgr\sin\theta$$

which are the top equations with correspondences

$$I_t = (I + mr^2)\left(\frac{R + r}{r}\right), \qquad I_a = I, \qquad l = r$$

4.17. $m\dot{v} = F_x\cos\beta\cos\phi, \qquad mv\dot{\phi}\cos\beta = -F_x\sin\phi$

4.18. (a) $(I + mr^2)\dot{\omega}_x - mr^2\Omega\omega_y + mr\Omega^2 y = 0$

$(I + mr^2)\dot{\omega}_y + mr^2\Omega\omega_x - mr\Omega^2 x = 0$

$I\dot{\omega}_z = 0$

(b) $x(t) = \dfrac{1}{\Omega^*}(\dot{x}_0 \sin \Omega^* t + \dot{y}_0 \cos \Omega^* t) + x_0 - \dfrac{\dot{y}_0}{\Omega^*}$

$y(t) = \dfrac{1}{\Omega^*}(\dot{y}_0 \sin \Omega^* t - \dot{x}_0 \cos \Omega^* t) + y_0 + \dfrac{\dot{x}_0}{\Omega^*}$

where $\Omega^* = \dfrac{I\Omega}{I + mr^2}$

(c) $\mathbf{H}_{0h} = [(I + mr^2)\omega_x - mr\Omega x]\mathbf{i} + [(I + mr^2)\omega_y - mr\Omega y]\mathbf{j} = \text{const}$

4.19. (a) $(I + mr^2)\dot{\omega}_x = \dfrac{mgr\pi h}{L} \sin \dfrac{2\pi y}{L}$

$(I + mr^2)\dot{\omega}_y - \dfrac{2\pi^2 mr^3 h}{L^2}\omega_x\omega_z \cos \dfrac{2\pi y}{L} = 0$

$I\dot{\omega}_z = 0$

(b) $\dot{y} = -r\omega_x \left(1 + \dfrac{2\pi^2 hr}{L^2} \cos \dfrac{2\pi y}{L}\right)$

4.20. (a) $\dfrac{1}{4} mr^2\ddot{\theta}(1 + 4\cos^2 \theta) - mr^2\dot{\theta}^2 \sin\theta \cos\theta - \dfrac{1}{4} mr^2\omega_d^2 \cot\theta$

$+ \dfrac{1}{2} mr^2\Omega\omega_d = -mgr\cos\theta$

$\dfrac{1}{4} mr^2\dot{\omega}_d - \dfrac{1}{2} mr^2\Omega\dot{\theta} + \dfrac{1}{4} mr^2\dot{\theta}\omega_d \cot\theta = 0, \qquad \dot{\Omega} = 0$

(b) $\omega_d = \sqrt{\dfrac{2g}{r}} (\sqrt{\theta^2 + 2\theta} - \theta), \qquad \dot{\phi} = \dfrac{\omega_d}{\theta}$

Chapter 5

5.5. (a) $3m\dot{v} = -mg \sin\alpha \cos\phi, \qquad \ddot{\phi} = 0$

(b) $\dot{x} = -v_0 \sin\omega_0 t + \dfrac{g}{3\omega_0} \sin\alpha \sin^2 \omega_0 t$

$\dot{y} = v_0 \cos\omega_0 t - \dfrac{g}{6\omega_0} \sin\alpha \sin 2\omega_0 t$

$(\dot{x})_{av} = \dfrac{g}{6\omega_0} \sin\alpha, \qquad (\dot{y})_{av} = 0$

5.6. (a) $2m\ddot{y} + ml\ddot{\phi} \cos\phi + 2m\dot{y}\dot{\phi} \tan\phi = -F \sin\phi \cos\phi$

$ml^2\ddot{\phi} \cos\phi + ml\ddot{y} + ml\dot{y}\dot{\phi} \tan\phi = -Fl \sin\phi \cos\phi$

(c) $ml^2\ddot{\phi} + Fl \sin\phi = 0$

5.8. $2m\dot{u}_1(1 + \cos^2 \phi) + \dfrac{2mu_1^2}{l} \sin^2 \phi \cos\phi - \dfrac{2mu_2^2}{l} \cos\phi = 0$

$2m\dot{u}_2(1 + \sin^2 \phi) + \dfrac{2mu_1 u_2}{l} \cos^3 \phi = 0$

5.9. (a) $T = 2m(\dot{x}^2 + \dot{y}^2) + ml^2(\dot{\psi}^2 + \dot{\phi}^2)$

$$+ 2ml\dot{x}(\dot{\psi}\sin\psi\cos\phi + \dot{\phi}\cos\psi\sin\phi)$$

$$+ 2ml\dot{y}(-\dot{\psi}\cos\psi\cos\phi + \dot{\phi}\sin\psi\sin\phi)$$

$$T^0 = ml^2\dot{\psi}^2(2\tan^2\phi + 1) + ml^2\dot{\phi}^2(2\cot^2\phi + 1)$$

(b) $2ml^2\left[(2\tan^2\phi + 1)\ddot{\psi} + \left(\dfrac{2\tan^2\phi - 1}{\sin\phi\cos\phi}\right)\dot{\psi}\dot{\phi}\right] = 0$

$$\dfrac{2ml^2}{\sin^2\phi}\left[(\cos^2\phi + 1)\ddot{\phi} + \dfrac{\dot{\psi}^2\sin\phi}{\cos\phi} - 2\dot{\phi}^2\cot\phi\right] = 0$$

5.10. $3m\dot{v} - ml\ddot{\theta}\sin\phi - ml\ddot{\phi}\sin\phi - ml(\dot{\theta} + \dot{\phi})^2\cos\phi - mr\dot{\theta}^2 = 0$

$[I + m(r^2 + l^2 + 2rl\cos\phi)]\ddot{\theta} + ml(l + r\cos\phi)\ddot{\phi}$

$- mrl(2\dot{\theta} + \dot{\phi})\dot{\phi}\sin\phi - ml\dot{v}\sin\phi + m(r + l\cos\phi)v\dot{\theta} = 0$

$ml^2\ddot{\phi} + ml(l + r\cos\phi)\ddot{\theta} - ml\dot{v}\sin\phi + mlv\dot{\theta}\cos\phi + mrl\dot{\theta}^2\sin\phi = 0$

5.11. (a) $\dfrac{1}{4}mr^2\ddot{\theta} = 0,\qquad \dfrac{1}{2}mr^2\ddot{\psi}(1 + 2\cos^2\theta) - mr^2\dot{\theta}\dot{\psi}\sin\theta\cos\theta = 0$

$m\ddot{x} = 0$

(b) $\dot{\psi}_{min} = \dot{\psi}_0,\qquad \dot{\psi}_{max} = \sqrt{3}\dot{\psi}_0$

$v_{min} = \dot{x}_0,\qquad v_{max} = \sqrt{\dot{x}_0^2 + r^2\dot{\psi}_0^2}$

5.12. $\dfrac{4}{3}ml^2\ddot{\theta}_1 + \dfrac{1}{2}ml^2\ddot{\theta}_2\cos(\theta_2 - \theta_1) + \dfrac{3}{2}ml\dot{v} - \dfrac{1}{2}ml^2\dot{\theta}_2^2\sin(\theta_2 - \theta_1) = 0$

$\dfrac{1}{3}ml^2\ddot{\theta}_2 + \dfrac{1}{2}ml^2\ddot{\theta}_1\cos(\theta_2 - \theta_1) + \dfrac{1}{2}ml\dot{v}\cos(\theta_2 - \theta_1)$

$+ \dfrac{1}{2}ml\dot{\theta}_1(v + l\dot{\theta}_1)\sin(\theta_2 - \theta_1) = 0$

$2m\dot{v} + \dfrac{3}{2}ml\ddot{\theta}_1 + \dfrac{1}{2}ml\ddot{\theta}_2\cos(\theta_2 - \theta_1) - \dfrac{1}{2}ml^2\dot{\theta}_2^2\sin(\theta_2 - \theta_1) = 0$

Chapter 6

6.1. (a) $y_{n+1} - y^*_{n+1} = -0.0001626,\qquad E_{n+1} = -0.0001667$

(b) $\xi = 0.024813$

6.2. (a) $-\dfrac{h^6}{240}y_n^{(6)}$

(b) $\rho^2 + b\rho + 1 = 0$ where $b = 2\left(\dfrac{5h^2\omega_0^2 - 12}{h^2\omega_0^2 + 12}\right)$

Unstable for $h > \dfrac{\sqrt{6}}{\omega_0}$

6.3. (a) $a = 2$, $\qquad b = -3$, $\qquad c = 1$

(b) $\dfrac{23}{24} h^2 \dddot{y}_n$

6.4. (a) $\rho^2 + (4h - 2)\rho + (2h^2 - 4h + 1) = 0$

(b) $\rho_{1,2} = 1 - 2h \pm \sqrt{2}h$, stable for $0 \le h \le 0.5858$

6.5. (a) $\dfrac{2}{3} ml^2 \ddot{\theta}(1 + 3\sin^2 \theta) + 2ml^2 \dot{\theta}^2 \sin \theta \cos \theta - mgl \cos \theta = 0$

(b)

t	θ	ω	$\dot{\omega}$
0.5	1.131533	2.774195	−0.762194
1.0	2.518895	2.913179	0.059869
1.5	2.914747	−2.396043	−9.159207

(c) $\omega_{max} = 2.913254$ at $t = 0.319054$, $\qquad \theta = 0.615481$

$\theta = \dfrac{\pi}{2}$ at $t = 0.660765$

$\omega_{max} = 2.913254$ at $t = 1.002475$, $\qquad \theta = 2.526105$

$\theta_{max} = 3.141593$ at $t = 1.321531$

6.6. Correct results:

$t = 0.5 \quad x = -0.345172 \quad v_x = \quad 2.262982$
$\qquad\qquad y = -1.251733 \quad v_y = \quad 1.142012$
$\qquad\qquad z = \quad 2.596904 \quad v_z = -3.404994$

$t = 1.0 \quad x = -0.066909 \quad v_x = -5.738694$
$\qquad\qquad y = \quad 0.351428 \quad v_y = \quad 1.211603$
$\qquad\qquad z = \quad 0.715481 \quad v_z = \quad 4.527092$

$t = 1.5 \quad x = -1.331422 \quad v_x = -0.600386$
$\qquad\qquad y = -0.548644 \quad v_y = -2.141998$
$\qquad\qquad z = \quad 2.880066 \quad v_z = \quad 2.742384$

$z_{min} = 0.585786$ at $t = 0.953950$

$z_{max} = 3.414214$ at $t = 0$ and $t = 1.907900$

6.7. (a) $\dot{x} = v_x$, $\qquad \dot{y} = v_y$, $\qquad \dot{\phi} = \omega$

$$\dot{v}_x = \frac{1}{2} l\omega^2 \cos \phi - \omega \sin \phi (v_x \cos \phi + v_y \sin \phi)$$

$$\dot{v}_y = \frac{1}{2} l\omega^2 \sin \phi + \omega \cos \phi (v_x \cos \phi + v_y \sin \phi)$$

$$\dot{\omega} = -\frac{\omega}{l}(v_x \cos \phi + v_y \sin \phi)$$

(b) $t = 0.1$ $x = 0.004983$ $v_x = 0.099336$
 $y = 0.000332$ $v_y = 0.009958$
 $\phi = 0.099917$ $\omega = 0.997505$

 $t = 1.0$ $x = 0.366104$ $v_x = 0.517681$
 $y = 0.265792$ $v_y = 0.688060$
 $\phi = 0.925775$ $\omega = 0.793278$

 $t = 10.0$ $x = -4.964991$ $v_x = -0.853882$
 $y = 10.81251$ $v_y = 1.127333$
 $\phi = 2.219039$ $\omega = 0.001699$

6.8. (a) $ml^2\ddot{\theta} + ml^2\Omega^2 \sin\theta \cos\theta - mgl \cos\theta = 0$

(b)

t	θ	ω	$\dot{\omega}$
0.5	0.403383	1.264691	0.068683
1.0	0.883130	0.459049	-2.365847
1.5	0.814167	-0.727807	-2.245828

(c) $\theta_{max} = 0.927295$ rad at $t = 1.192006$ s

 $\omega_{max} = 1.264911$ rad/s at $t = 0.506431$ s

6.9. (a) $t = 0.2$ s
 $\omega_x = 14.983846$ $\epsilon_x = 0.996466$ $W = 1100.8044$ N·m
 $\omega_y = 0.880327$ $\epsilon_y = 0.010208$ $\alpha = 0.091516$
 $\omega_z = 0.499509$ $\epsilon_z = 0.044588$ $\eta = 0.070453$

 $t = 0.4$ s
 $\omega_x = 14.969975$ $\epsilon_x = 0.134382$ $W = 1500.8666$ N·m
 $\omega_y = 1.199892$ $\epsilon_y = 0.035882$ $\alpha = 0.201343$
 $\omega_z = -0.317395$ $\epsilon_z = 0.093878$ $\eta = -0.985820$

 $t = 0.6$ s
 $\omega_x = 14.999716$ $\epsilon_x = -0.977615$ $W = 145.9451$ N·m
 $\omega_y = 0.116755$ $\epsilon_y = 0.060906$ $\alpha = 0.136529$
 $\omega_z = -0.297954$ $\epsilon_z = 0.030712$ $\eta = -0.199038$

 $t = 0.8$ s
 $\omega_x = 14.995077$ $\epsilon_x = -0.266247$ $W = 607.6461$ N·m
 $\omega_y = 0.486064$ $\epsilon_y = 0.055357$ $\alpha = 0.151824$
 $\omega_z = 0.507382$ $\epsilon_z = -0.051838$ $\eta = 0.960917$

 $t = 1.0$ s
 $\omega_x = 14.962897$ $\epsilon_x = 0.942681$ $W = 1668.4876$ N·m
 $\omega_y = 1.333689$ $\epsilon_y = 0.041312$ $\alpha = 0.113436$
 $\omega_z = -0.024909$ $\epsilon_z = -0.038818$ $\eta = 0.328846$

(b) $W_{max} = 1669.4220$ at $t = 0.331667$ and $t = 0.995000$.

(c) $\alpha_{max} = 0.201447$ at $t = 0.405759$

(d) $t = 0.663333$ s

6.10. (a) $\ddot{\psi} = \dfrac{2\dot{\theta}}{\cos\theta}\left(\Omega + \dot{\psi}\sin\theta\right)$

$\ddot{\theta} = -\dfrac{1}{5}\dot{\psi}^2\sin\theta\cos\theta - \dfrac{6}{5}\dot{\psi}\Omega\cos\theta + 16\sin\theta$

$\dot{\Omega} = \dfrac{2}{3}\dot{\psi}\dot{\theta}\cos\theta$

(b) $t = 0$

$\psi = 0 \quad \theta = \dfrac{\pi}{6} = 0.523599 \quad \phi = 0 \quad X = 0$

$\dot{\psi} = 4 \quad \dot{\theta} = 0 \qquad\qquad\qquad \Omega = 2 \quad Y = 0$

$t = 1$

$\psi = 2.891062 \quad \theta = 0.096692 \quad \phi = 2.754937 \quad X = 0.156089$
$\dot{\psi} = 1.726344 \quad \dot{\theta} = -0.678622 \quad \Omega = 1.294624 \quad Y = -0.919876$

$t = 1.5$

$\psi = 3.480268 \quad \theta = -0.595660 \quad \phi = 3.225671 \quad X = -0.071231$
$\dot{\psi} = 0.534888 \quad \dot{\theta} = -2.748494 \quad \Omega = 0.831908 \quad Y = -0.910777$

$t = 2.0$

$\psi = -3.157916 \quad \theta = -0.517934 \quad \phi = 10.294732 \quad X = -0.436413$
$\dot{\psi} = 0.633105 \quad \dot{\theta} = 2.483279 \quad \Omega = 0.857619 \quad Y = -0.856059$

$t = 2.5$

$\psi = -2.533888 \quad \theta = 0.116376 \quad \phi = 10.793919 \quad X = -0.666352$
$\dot{\psi} = 1.785449 \quad \dot{\theta} = 0.645929 \quad \Omega = 1.317532 \quad Y = -0.784324$

$t = 3.0$

$\psi = -1.375078 \quad \theta = 0.380226 \quad \phi = 11.849700 \quad X = -0.844784$
$\dot{\psi} = 2.927582 \quad \dot{\theta} = 0.476119 \quad \Omega = 1.706236 \quad Y = -0.328526$

$t = 3.5$

$\psi = 0.423844 \quad \theta = 0.522848 \quad \phi = 13.623335 \quad X = -0.168626$
$\dot{\psi} = 3.993075 \quad \dot{\theta} = -0.050408 \quad \Omega = 1.998268 \quad Y = 0.013511$

$\theta_{min} = -1.542703$ rad $(-88.39°)$ at $t = 1.735130$ s

$\theta_{max} = \dfrac{\pi}{6}$ at $t = 0$ and $t = 3.470260$ s when $\dot{\psi}$, $\dot{\theta}$, and Ω also

return to their initial values.

$\theta = 0$ at $t = 1.125585$ s and 2.344675 s.

6.11. (a) $\dot{x} = v$, $\quad \dot{\theta} = \omega$, $\quad \dot{v} = \dfrac{-l\omega^2 \sin\theta + g\sin\theta\cos\theta}{2 - \cos^2\theta}$

$$\dot{\omega} = \dfrac{-l\omega^2 \sin\theta\cos\theta + 2g\sin\theta}{l(2 - \cos^2\theta)}$$

(b)

t	x	θ	v	ω
0.2	0.506533	0.113307	0.099539	0.703590
0.4	0.554881	0.314942	0.420204	1.409748
0.6	0.676823	0.712392	0.740709	2.618152

(c) $t = 0.618665$, $\quad x = 0.690667$, $\quad \theta = 0.762412$

(d) $\dot{x}_0 = 0.741316$, $\quad \theta_0 = 2.741575$

(e) $\dot{x} = 1.333450$, $\quad \dot{\theta} = 1.843889$, $\quad \hat{\lambda} = 1.833450$

(f) $\ddot{x} = g\sin\theta\cos\theta - l\dot{\theta}^2\sin\theta$, $\quad \ddot{\theta} = \dfrac{g}{l}\sin\theta$

(g) Leaves wall at $t = 0.697577$ when

$$x = 0.801676 \qquad \theta = 0.930094$$
$$\dot{x} = v = 1.446778 \qquad \dot{\theta} = \omega = 2.420338$$

Index